MW00994246

The PDMA ToolBook 2 for New Product Development

The PDMA ToolBook 2 for New Product Development

Edited by

Paul Belliveau
Paul Belliveau Associates

Abbie Griffin
University of Illinois at Urbana-Champaign

Stephen M. Somermeyer
Your Encore™

John Wiley & Sons, Inc.

This book is printed on acid-free paper. ♾

Copyright © 2004 by John Wiley & Sons, Inc. All rights reserved

Published by John Wiley & Sons, Inc., Hoboken, New Jersey
Published simultaneously in Canada

No part of this publication may be reproduced, stored in a retrieval system or transmitted in any form or by any means, electronic, mechanical, photocopying, recording, scanning, or otherwise, except as permitted under Section 107 or 108 of the 1976 United States Copyright Act, without either the prior written permission of the Publisher, or authorization through payment of the appropriate per-copy fee to the Copyright Clearance Center, Inc., 222 Rosewood Drive, Danvers, MA 01923, (978) 750-8400, fax (978) 750-4470, or on the web at www.copyright.com. Requests to the Publisher for permission should be addressed to the Permissions Department, John Wiley & Sons, Inc., 111 River Street, Hoboken, NJ 07030, (201) 748-6011, fax (201) 748-6008, e-mail: permcoordinator@wiley.com.

Limit of Liability/Disclaimer of Warranty: While the publisher and author have used their best efforts in preparing this book, they make no representations or warranties with respect to the accuracy or completeness of the contents of this book and specifically disclaim any implied warranties of merchantability or fitness for a particular purpose. No warranty may be created or extended by sales representatives or written sales materials. The advice and strategies contained herein may not be suitable for your situation. You should consult with a professional where appropriate. Neither the publisher nor author shall be liable for any loss of profit or any other commercial damages, including but not limited to special, incidental, consequential, or other damages.

For general information on our other products and sevices or for technical support, please contact our Customer Care Department within the United States at (800) 762-2974, outside the United States at (317) 572-3993 or fax (317) 572-4002.

Wiley also publishes its books in a variety of electronic formats. Some content that appears in print may not be available in electronic books. For more information about Wiley products, visit our web site at www.wiley.com.

Library of Congress Cataloging-in-Publication Data:

The PDMA toolbook 2 for new product development / edited by Paul Belliveau, Abbie Griffin, Stephen Somermeyer.
p. cm.
Includes bibliographical references and index.
ISBN 0-471-47941-1 (cloth : acid-free paper)
1. New products—Management. I. Title: PDMA toolbook two for new product development. II. Belliveau, Paul. III. Griffin, Abbie. IV. Somermeyer, Stephen.
HF5415.153.P3549 2004
658.5'75—dc22
2004019042

Printed in the United States of America

10 9 8 7 6 5 4 3 2 1

Contents

Contributors *ix*
Introduction *xxi*

Part 1

Organizational Tools 1

1

Achieving Growth Through an Innovative Culture 3
Erika B. Seamon

2

Bringing Radical and Other Major Innovations Successfully to Market: Bridging the Transition from R&D to Operations 33
Gina O'Connor, Joanne Hyland, and Mark P. Rice

3

Turning Technical Advantage into Product Advantage 71
Stephen K. Markham and Angus I. Kingon

4

Enhancing Organizational Knowledge Creation for Breakthrough Innovation: Tools and Techniques 93
Peter Koen, Richard McDermott, Robb Olsen, and Charles Prather

5

Building Creative Virtual New Product Development Teams 117
Roger Leenders, Jan Kratzer, and Jo van Engelen

6

Build Stronger Partnerships to Improve Codevelopment Performance 149

Mark J. Deck

Part 2

Tools for Improving the Fuzzy Front End 165

7

The Voice of the Customer 167

Gerald M. Katz

8

Creating the Customer Connection: Anthropological/Ethnographic Needs Discovery 201

Barbara Perry, Cara L. Woodland, and Christopher W. Miller

9

Shifting Your Customers into "Wish Mode": Tools for Generating New Product Ideas and Breakthroughs 235

Jason Magidson

10

The Birth of Novelty: Ensuring New Ideas Get a Fighting Chance 269

K. Brian Dorval and Kenneth J. Lauer

Part 3

Tools for Managing the NPD Process 295

11

Establishing Quantitative Economic Value for Product and Service Features: A Method for Customer Case Studies 297

Kevin Otto, Victor Tang, and Warren Seering

12

Integrating a Requirements Process into New Product Development 331

Christina Hepner Brodie

13

Toolkits for User Innovation 353

Eric von Hippel

14

IT-Enabling the Product Development Process 375

Henry Dittmer and Patrick Gordon

Part 4

Tools for Managing the NPD Portfolio and Pipeline 395

15

Product and Technology Mapping Tools for Planning and Portfolio Decision Making 397

Richard E. Albright and Beebe Nelson

16

Decision Support Tools for Effective Technology Commercialization 435

Kevin J. Schwartz, Ed K. Yu, and Douglas N. Modlin

17

Spiral-Up Implementation of NPD Portfolio and Pipeline Management 461

Paul O'Connor

The PDMA Glossary for New Product Development 493

About PDMA 535

Index 537

Contributors

EDITORS:

Paul Belliveau

Paul Belliveau (NPDP) is Principal of Paul Belliveau Associates. He provides entrepreneurs with strategic business and new product development counsel. Paul is a former president of the Product Development & Management Association (PDMA) and a co-founder of the PDMA Foundation. He is an adjunct professor at Rutgers Business School, Rutgers University. He was co-editor of *The PDMA ToolBook 1 for New Product Development* (John Wiley & Sons, 2002) which won a 2003 New Jersey Bright Idea Award from Seton Hall University's Stillman School of Business and the New Jersey Business and Industry Association (NJBIA). Paul has had articles published in the *Journal of Developmental Entrepreneurship* and the *Journal of Small Business and Enterprise Development.* Paul has served as a member of the Board of Advisors for the *Journal of Small Business Management* and as an ad hoc reviewer for the *Journal of Product Innovation Management.*

Abbie Griffin

Dr. Abbie Griffin (NPDP) is a Professor of Business Administration at the University of Illinois, Urbana-Champaign College of Commerce, where she teaches business-to-business marketing and first-year core MBA marketing. Professor Griffin's research investigates means for measuring and improving the process of new product development, which she has published in *Industrial Marketing Management, Journal of Product Innovation Management, Journal of Marketing Research, Sloan Management Review,* and *Marketing Science.* Abbie was co-editor of *The PDMA ToolBook 1 for New Product Development* (John Wiley & Sons, 2002). She is a member of the Board of Directors of Navistar International and was the editor of the *Journal of Product Innovation Management* from 1998–2003. Abbie holds a Ph.D. in Management from MIT.

Stephen M. Somermeyer
Steve Somermeyer (NPDP) has business development responsibilities at YourEncore and is also Principal of Somermeyer & Associates which helps firms with their innovation strategy and organizational structure. He spent over 30 years in the pharmaceutical industry at Eli Lilly and Company in manufacturing and R&D. Steve is a PDMA Director and Chair of the Nominations Committee. He was also a Core Team Member for the PDMA Body of Knowledge initiative. He was co-editor of *The PDMA ToolBook 1 for New Product Development* (John Wiley & Sons, 2002). Steve has given numerous presentations on metrics, benchmarking and outsourcing/alliances.

CHAPTER AUTHORS:

Erika B. Seamon (Chapter 1)
Erika B. Seamon is a Partner with Kuczmarski & Associates and a former Adjunct Associate Professor of Marketing at the University of Chicago Graduate School of Business. She has helped to revolutionize the thinking on how organizations can continually innovate while minimizing risk and maximizing profitability. She has worked with numerous senior management teams at a variety of Fortune 500 firms, including some of the country's top consumer products, health care, and financial services organizations. She has been quoted on multiple occasions in periodicals including the *New York Times, Crain's Chicago Business,* the *Chicago Tribune,* and *Direct Marketing News.* She has published in *Marketing Management, Electric Light & Power,* and *Food Processing Magazine.* She speaks nationally for organizations such as The Conference Board, the American Marketing Association, and the Product Development & Management Association.

Gina O'Connor (Chapter 2)
Dr. Gina O'Connor is the Academic Director for the Radical Innovation Research Project and an associate professor in the Lally School of Management and Technology at Rensselaer Polytechnic Institute. Her fields of interest include new product development, radical innovation, technology commercialization, and strategic marketing management in high technology arenas. The majority of her research efforts focus on how firms link advanced technology development to market opportunities. She has published articles in *Organization Science*, the *Journal of Product Innovation Management, California Management Review, Academy of Management Executive,* and *The Journal of Strategic Mar-*

keting, and is co-author of the book *Radical Innovation: How Mature Firms Can Outsmart Upstarts* (HBS Press, 2000). She teaches executive education and consults with firms to help them develop, embed, and sustain radical innovation management capabilities. Gina earned her Ph.D. in Marketing and Corporate Strategy at New York University.

Joanne Hyland (Chapter 2)
Joanne Hyland (NPDP) is President of Hyland Value Creation and former Vice President, New Venture Development at Nortel Networks, where she founded its internal venturing program that resulted in 12 start-ups. As a founding partner in the Radical Innovation Group, Joanne works with corporations to link innovation with strategy and to develop systems, leadership, and culture capabilities that drive corporate growth and renewal. Joanne speaks regularly at conferences on topics related to innovation and corporate venturing and has been on the faculty or a guest speaker in executive education programs. Joanne is a featured executive in *Radical Innovation: How Mature Companies Can Outsmart Upstarts* (Harvard Business School Press, 2000) and published "Using VC Experience to Create Business Value," in *From the Trenches: Strategies from Industry Leaders on the New e-Conomy* (John Wiley & Sons, 2001).

Mark P. Rice (Chapter 2)
Dr. Mark P. Rice is the Murata Dean of F. W. Olin Graduate School of Business and the Jeffrey A. Timmons Professor of Entrepreneurial Studies at Babson College, focusing on expanding Babson's strategic initiatives in innovation and entrepreneurship, linking management and technology, and promoting innovation in the curriculum. Dean Rice is co-author of *Radical Innovation: How Mature Companies Can Outsmart Upstarts* (Harvard Business School Press, 2000). He has been published in *Organization Science, R&D Management, Journal of Marketing Theory and Practice, Academy of Management Executive,* and *California Management Review*. Dean Rice previously served as Director of the RPI Incubator Program and the Severino Center for Technological Entrepreneurship. Dr. Rice holds a Ph.D. in Management from Rensselaer Polytechnic Institute.

Stephen K. Markham (Chapter 3)
Dr. Stephen Markham (NPDP) is the Director of the Center for Innovation Management Studies and Director of the Technology Entrepreneurship and Commercialization Program at North Carolina State University. He is also the President of the PDMA Foundation. Steve's research is in the areas of technology commercialization and champions of innovation.

He has been founder, director, and CFO of a number of high-technology companies. He has a Ph.D. in business from Purdue University.

Angus I. Kingon (Chapter 3)
Dr. Angus I. Kingon is Professor of Materials Science and Engineering and Business Management at the North Carolina State University. Prior to joining the faculty, he was Program Manager and Specialist Scientist at the National Institute for Materials Research, South Africa. His research specialty is electronic materials. Dr. Kingon is also Executive Director of the Technology Entrepreneurship and Commercialization Program, specializing in technology-based entrepreneurship. Dr. Kingon consults with companies, research centers, and governments around the world on technology commercialization topics. Angus has a Ph.D. in physical chemistry from the University of South Africa.

Peter Koen (Chapter 4)
Dr. Peter Koen (NPDP) is Associate Professor in the Wesley J. Howe School of Technology Management at Stevens Institute of Technology in Hoboken, New Jersey. He is currently Director of the Consortium for Corporate Entrepreneurship (CCE) at Stevens, whose mission is to stimulate profitable activities at the "fuzzy front end" of the innovation process. Peter is engaged in research directed at best practices in the front end, determining how companies organize around breakthroughs in large corporations and in knowledge creation and knowledge flow. He has 19 years of industrial experience at both large and small companies. Peter holds a Ph.D. in Biomedical Engineering from Drexel University and is a licensed professional engineer.

Richard McDermott (Chapter 4)
Richard McDermott, President of McDermott Consulting, is one of the leading thinkers, authors, and consultants on designing knowledge organizations and building communities of practice. For nearly two decades, he has worked with engineering, professional service, sales, and manufacturing firms to maximize the productivity of knowledge workers. Richard was the subject matter expert for two national studies of best practices in knowledge management—on creating a knowledge-sharing culture and on institutionalizing communities of practice—and is a frequent speaker at international conferences. His clients include numerous Fortune 500 firms.

Robb Olsen (Chapter 4)
Robb Olsen is a section manager in Corporate R&D at Procter & Gamble. Robb has 22 years of product research experience. His areas

of expertise include the innovation process, leveraging knowledge sharing for competitive advantage, upstream product design, consumer understanding, and communities of practice. Robb is currently leader of Procter & Gamble's R&D Intranet efforts, providing business leadership and strategic direction to P&G's largest intranet: InnovationNet. He has spoken at several conferences on the topic of enabling R&D and innovation via the Web, and he teaches knowledge management, consumer research techniques, and innovation methodologies within P&G.

Charles Prather (Chapter 4)
Dr. Charles (Charlie) Prather is President of Bottom Line Innovation Associates, Inc., helping organizations develop innovation as a core competency. Many Fortune 100 firms are clients, representing chemicals, paper, consumer products, high tech, government, financial services, and other segments. Charlie served DuPont some 24 years in numerous R&D and leadership positions. He is a Fellow of the Robert H. Smith School of Business at the University of Maryland, College Park, teaching in the Executive MBA program. Charlie is on the Board of Directors of DePaul University's Ryan Creativity Center, and is a frequent conference presenter. He has authored many articles, and his book, *Blueprints for Innovation* was published by The American Management Association. Charlie earned his Ph.D. in biochemistry from North Carolina State University.

Roger Th. A. J. Leenders (Chapter 5)
Dr. Roger Th. A. J. Leenders is Associate Professor of Business Development at the Faculty of Management and Organization, University of Groningen, The Netherlands. His current research focuses on the positive and negative effects of social networks on organizations in general and innovation activity in particular. Roger is mainly interested in how social networks assist or obstruct the performance of innovation teams and their members. He has authored one book, *Structure and Influence* (Thela Thesis, 1995), and edited two books (with S. M. Gabbay): *Corporate Social Capital and Liability* (Kluwer Academic Publishers, 1999) and *Social Capital of Organizations* (JAI Press, 2001). Roger's work has appeared in *Social Networks, Journal of Mathematical Sociology, Research in the Sociology of Organizations* and *Journal of Product Innovation Management*. He holds a Ph.D. in Social Sciences.

Jan Kratzer (Chapter 5)
Dr. Jan Kratzer is Assistant Professor at the Faculty for Management and Organization at the University of Groningen, The Netherlands. His

main research interests concern human factors and human networks in product development processes, particularly with regard to developments in virtual teams. He received his Ph.D. from the University of Groningen in Communication and Performance in Innovation Teams.

Ir. Jo M. L. van Engelen (Chapter 5)

Dr. Ir. Jo M. L. van Engelen is currently Professor of Business Development and Business Research Methods at the University of Groningen, The Netherlands. He received his Ph.D. from Twente University in Information Technology in Marketing Management. Dr. van Engelen also serves as a consultant and board member for several leading companies in The Netherlands.

Mark J. Deck (Chapter 6)

Mark J. Deck (NPDP) is a Director at PRTM Management Consultants, leading the Product And Cycle-time Excellence® practice there. Mark has 25 years of consulting experience focused principally on managing and developing new products and services. Mark's practice concentrates on improving the capability of companies to improve R&D innovation and development productivity by modifying front-end processes, better managing project portfolios and resources across projects, tapping the promise of co-development, planning and managing product platforms and product families and implementing enabling product development IT systems. Mark is a past President of the PDMA. He has published articles on product development in *Research-Technology Management, The PDMA ToolBook 1 for New Product Development* and PRTM's *Setting the PACE in Product Development.*

Gerald M. Katz (Chapter 7)

Gerry Katz (NPDP) is a recognized authority in the areas of new product development, design of new services and market research, with more than 30 years of consulting experience. At Applied Marketing Science, Inc., he has led more than 100 major client engagements employing the voice of the customer, Quality Function Deployment (QFD), and a large number of other marketing science applications. Gerry serves on the PDMA Board, is a contributing editor to *Visions* and has authored several award-winning papers. He has lectured frequently on the topics of new product development and market research at major business schools.

Barbara Perry (Chapter 8)

Barbara Perry is the cultural anthropologist for Barbara Perry Associates. In her 25 years of consulting to Fortune 500 companies across a broad variety of categories, the focus has been on supporting their effectiveness as a team to not only see new opportunities but to be able to realize them. Her emphasis is on facilitating the development of customer-focused, innovative organizational cultures. Her proprietary methods are being used for a large variety of purposes, including new product design and development.

Cara L. Woodland (Chapter 8)
Cara Woodland is Vice President of Customer Insights at Innovation Focus Inc. She is also President of the Philadelphia Chapter of the Market Research Association and author of numerous articles on ethnographic and observational research methods. Her emphasis is placed on teaching teams to carry out their own voice of the customer research. Cara has worked aggressively to encourage the use of both qualitative and quantitative research tools at all stages of product development and product life cycle management.

Christopher W. Miller (Chapter 8)
Christopher W. Miller (NPDP) is Founder of Innovation Focus, Inc. He is the current President of the Product Development & Management Association, and a 2003 Ernst & Young Entrepreneur of the Year Award winner. Chris has published extensively on the large variety of issues of concern within Product Development and Management. He has been recognized for his series of Growth Forum articles for the PDMA publication *Visions*. He is a psychologist, having received his Ph.D. from Case Western Reserve University by studying the lifelong learning patterns of engineers.

Jason Magidson (Chapter 9)
Dr. Jason Magidson is the Director of a global group at GlaxoSmith-Kline that works with users to develop solutions for employees in functions across the company. He is also the founder of productWish.com. For over 18 years Jason has led customer research, new product design, and product improvement projects involving internal and external users, for Fortune 1,000 companies as well as numerous government agencies, community-oriented organizations, and Web start-ups. Magidson received a Ph.D. from The Union Institute & University.

K. Brian Dorval (Chapter 10)
Brian Dorval is Vice President of Client Services at The Creative Problem Solving Group, Inc., where he is responsible for managing the development and delivery of client-focused work. He is involved in research and development on the topics of creativity, creative problem

solving, and mental imagery, and has co-authored over 50 articles, chapters, and books, including *Creative Approaches to Problem Solving* (Kendall/Hunt, 2000) and *Toolbox for Creative Problem Solving; Basic Tools and Resources* (The Creative Problem Solving Group, 1998). Brian has provided over 350 training programs, workshops, and presentations, as well as facilitated working sessions for many companies.

Kenneth J. Lauer (Chapter 10)
Ken Lauer is Vice President of Research & Development at The Creative Problem Solving Group, Inc. (CPSB), where he provides data management, statistical analysis, information retrieval, writing, and editing and manages CPSB's new product development process. He has been involved in over 25 studies published in a variety of journals, including *Psychological Reports* and *The European Journal of Work and Organizational Psychology*. Ken is certified to use a variety of psychological assessments, including the Kirton Adaption-Innovation Inventory, Myers-Briggs Type Indicator (MBTI), SOQ (Situational Outlook Questionnaire), and VIEW: A Measure of Problem Solving Style.

Kevin Otto (Chapter 11)
Kevin Otto is a Vice President at Product Genesis, Inc., a strategic innovation firm in technology consulting, marketing, and product development. He currently focuses on product platforms, modular design, and design for Six Sigma. At MIT he taught courses in product design, architecture, and robust engineering methods. Kevin is a co-author of *Product Design* (Prentice Hall, 2000), a product development textbook. He has published widely on product architectures, product families, fuzzy theory in design, and Taguchi robust design methods. Kevin has won numerous teaching awards, best paper awards in international conferences, and research rewards, including an RD100 award. Kevin has consulted widely with Fortune 100 high-technology companies. He earned his Ph.D. from CalTech.

Victor Tang (Chapter 11)
Victor Tang is a doctoral student at MIT. His research includes product development methods for complex systems, services, and strategic-decision modeling. Prior to MIT, he was Vice President of IBM China and held other executive positions in corporate strategy, product development, and business development. Vic was also IBM's IT systems manager for the 1996 Nagano Winter Olympics. He has advised the

United Nations, foreign governments, and Fortune 500 companies. Vic is also co-author of three books in product development and management. One, *The Silverlake Project* (Oxford University Press, 1992) has been translated into Russian, Chinese, and Korean.

Warren Seering (Chapter 11)
Dr. Warren Seering is a Professor of Mechanical Engineering and of Engineering Systems at MIT. His research has focused on product design and development, dynamic systems, and robotics. He has taught courses in design, product development, applied mechanics, system dynamics, and computer programming and numerical methods. He is a Fellow of the American Society of Mechanical Engineers, a member of the Board of Management of the Design Society, and holder of the Weber-Shaughness Chair in the School of Engineering at MIT. With his students, he has published over 100 papers, and with the NSF, he is the founder of MIT's Center for Innovation in Product Development. He received his Ph.D. from Stanford.

Christina Hepner Brodie (Chapter 12)
Christina Hepner Brodie (NPDP), a Principal of Pittiglio Rabin Todd & McGrath (PRTM), has over 13 years of experience working with companies to manage and develop new products and services, and improve performance through sharpened up-front definition. Co-author of *Voices into Choices: Acting on the Voice of the Customer* (Joiner/Oriel Inc., 1997), Christina is renowned for enabling business strategy and new product development teams to understand market dynamics and customer requirements firsthand. She has introduced executives, senior managers, and new product developers to voice of the customer methodologies at over 80 companies. Before joining PRTM, Christina headed CHB Consulting, working for several years in association with the Center for Quality of Management, where she evolved voice of the customer methodologies, Concept Engineering,® the FOCUS Method, and the Language Processing® Method.

Eric von Hippel (Chapter 13)
Dr. Eric von Hippel is Professor and Head of the Management of Innovation and Entrepreneurship Group at the MIT Sloan School of Management. His research examines the sources and economics of innovation, with a particular focus on the significant role played by users in the innovation development process. Eric explores how developers may best gain access to the "lead user" innovations that they need to *systematically* create "breakthrough" new products and serv-

ices. He also explores how firms can develop toolkits for user innovation to systematically share the work of new product development with customers. His methods are being used by leading-edge companies worldwide.

Henry Dittmer (Chapter 14)
Henry Dittmer is the Director of Solution Process Management and Quality Management at Avaya Communications Inc. He led the successful rollout and implementation of a product development process for managing the various division portfolios. The new product development process included portfolio planning, the product realization process, and a life cycle management methodology. A key component of this deployment was the implementation of tools to IT-enable the product development process. In addition, Henry managed Avaya's Quality Management System and all applications and implementations—including Avaya's ISO 9001-2000 registration.

Patrick Gordon (Chapter 14)
Patrick Gordon is a Director in the Worldwide Product Development Practice at global management consulting firm PRTM. He has been part of the development and improvement of PRTM's Product And Cycle-time Excellence® (PACE®) framework, particularly in the areas of product strategy and portfolio management. Patrick has led several initiatives with companies to improve the performance of the new product development process, leading to significant improvements in innovation, cycle time, cost savings, and productivity. He works with companies to IT-enable their product development management, including the emerging area of product life cycle management. Patrick has worked with companies in a range of industries, including telecommunications, data communications, and consumer electronics.

Richard E. Albright (Chapter 15)
Dr. Richard E. Albright, Principal of The Albright Strategy Group, LLC, works with organizations on roadmapping, technology futures, and integrated strategy and technology plans. He was previously Director, Technology Strategy and Assessment at Bell Laboratories, where he was responsible for development of technology strategy for Lucent Technologies. He is a Fellow of the Center for Technology and Innovation Management at Northwestern University, and he chairs the Roadmapping Task Force of MATI (Management of Accelerated Technology Innovation), an industry and academic consortium identifying and developing best practices in technology management. Dr. Albright received his Ph.D. from Polytechnic University of New York.

Beebe Nelson (Chapter 15)
Dr. Beebe Nelson (NPDP), Working Forums Founder & President, is a consultant in innovation focusing on portfolio management, life cycle management, market segmentation, technology mapping, product definition, and voice of the customer. She specializes in the design and facilitation of intercompany learning networks, including her role as Coordinator for the International Association for Product Development. She was Program Chair for the Product Development & Management Association's 2000 International Conference, chaired the People Track at the 2003 International Conference, and served as Book Review Editor for the *Journal of Product Innovation Management.* She is on the faculty of Sequent Learning Networks. Beebe holds a Ph.D. in Philosophy from Harvard University.

Ed K. Yu (Chapter 16)
Ed Yu (NPDP) is a Director in PRTM's Mountain View, California, office. He has over 20 years of experience in strategy, product development, and operations across a wide set of technology-based industries including drug development, medical device and equipment, electronics, aerospace, and energy. His work is focused on analysis of a firm's existing product development capabilities and the implementation of improved processes in project management, product strategy, portfolio and resource management, and technology management. Ed co-leads PRTM's product development practice and contributes to the continued advancement of product development management best practices through refinement of the PRTM Product And Cycle-time Excellence® (PACE®) framework. Ed has served as conference chair for the Product Development & Management Association (PDMA) and has been the keynote speaker at the ALSSA (Analytical Life Science Systems Association) conference for R&D Senior Executives.

Douglas N. Modlin (Chapter 16)
Douglas N. Modlin, PhD, has over 20 years of experience developing products for the microelectronics, pharmaceutical, and life science industries. He has held leadership roles in the development of numerous innovative microelectronics, biochip, and instrumentation products and is listed as an inventor on 31 issued U.S. patents. Dr. Modlin has held VP R&D positions at LJL BioSystems, Inc., Molecular Devices Corp., and Fluidigm Corp.

Kevin J. Schwartz (Chapter 16)
Kevin Schwartz (NPDP) is a Principal in the product development practice of Pittiglio Rabin Todd & McGrath (PRTM), a leading manage-

ment consultancy to technology-driven business. Kevin has over 10 years of experience in new product development with companies across a variety of industries. Over the last several years, he has worked with startups and Fortune 100 companies alike to develop and implement best practices for commercializing new technologies, from fuel cells to advanced materials to telecommunication devices. Kevin is a regular speaker at PDMA workshops on new product development and has been published in *Research Technology Management* and PRTM's *Insight*.

Paul O'Connor (Chapter 17)
Paul O'Connor (NPDP) is Managing Director of The Adept Group, a firm he founded in 1984. He is an expert in product development productivity. Paul has conducted assignments, implementation initiatives, and benchmarking activities with numerous firms around the world. He is also a Past-President of the Product Development & Management Association (PDMA) and teaches Portfolio and Pipeline Management for various organizations.

Introduction

Welcome to *The PDMA ToolBook 2 for New Product Development.*

As with *ToolBook 1*, this book has been written and edited by PDMA volunteers (royalties from the ToolBook series go to the PDMA) who are new product development (NPD) experts with a passion for NPD and the desire to contribute to the improvement of the NPD profession. They are NPD professionals (practitioners, service providers, and academics) who have committed their effective practice learnings to these pages. They provide you with in-depth, how-to knowledge that you can use to improve your organization's NPD operation. *ToolBook 2* is a collection of effective practice tools presented such that you can put down this book and use them immediately.

NPD has changed in the two years since *ToolBook 1* was written. Then, companies focused primarily on technology-based competition and the challenges to adapt new technologies to successful new products. The tools of ToolBook 1 thus emphasized NPD processes and their improvements.

ToolBook 2 reflects renewed interest in consumer goods and customers themselves. This book thus mirrors the maturing of the NPD process. Most organizations have installed some version of the Stage-Gate™ process or something similar, and now are addressing issues of trying to improve process effectiveness to become more competitive with new products in the marketplace. Competitive pressures have been exacerbated by a variety of factors: economic downturn, mergers, downsizing and the retirement of a generation of experienced experts, shorter product life cycles, and the increasing need to obtain expertise outside one's own organization. *ToolBook 2* thus focuses on organizations, the fuzzy front end (FFE), and learning.

As chapter drafts and revisions were edited, it became clear that much of *ToolBook 2* was going to be on the "soft" organizational issues rather than on "hard" process improvements. As NPD thought leaders, the chapter authors are much more cognizant of the importance of organizational health and culture than before. While there may appear to be overlaps among some chapters, it is suggested that readers choose specific tools within the context of their own organization's unique culture and needs.

Because of the changes in NPD's status in many firms, *ToolBook 2* begins with a part focused on organizational tools. Part 1 begins with tools to improve the NPD culture and continues with tools for converting technology into products, improving an organization's learning ability, improving its creativity, and dealing with codevelopment issues and the cultural and operational interface between NPD and operations.

Part 2 features tools for improving the FFE of the NPD process. These tools focus on the unique issues of better understanding customer needs—their *voice*—and how these shape NPD product requirements. Novelty in the FFE has its special issues, and there is a *ToolBook 2* chapter offering tailored tools to improve FFE novelty.

Part 3 includes tools to improve the NPD process and begins with quantifying the economic value of a new product. Integrating the product requirements process into the overall NPD process is difficult, but a Part 3 chapter offers effective practice tools to make it easier. This part then deals with an emerging aspect of NPD, enabling customers themselves to design their own product. Finally, this part offers a chapter dealing with the very important aspect of efficiently integrating new information technology tools into NPD.

Part 4 presents tools to improve the organization's ability to manage multiple projects simultaneously, with different scopes and at different stages of NPD. NPD program managers deal with this very complex management task every day, and this part offers a wide variety of effective practices to deal with this challenge. Using maps to better present technology platforms and tuning up decision making are valuable tools for organization leaders. Finally, the part concludes with a portfolio management maturity model chapter filled with tips on how to move your organization to the next stage.

HOW TO USE THIS BOOK

We suggest that you use this book chapter by chapter. Rather than reading the entire book, you may find it helpful to read each of the four part introductions to get a high-level perspective of each part's contents. You may want to skim several chapters that initially appeal to you. Then, as you face a weakness or shortcoming in your own NPD organization or have a specific process issue, you can go to the particular chapters that apply to your situation and try putting one or more tools to immediate use. Alternatively, you may be looking to improve some aspect of NPD on a proactive basis. In this case, we recommend that you look at the chapters that most closely fit the NPD area that you wish to improve. The chapters of *ToolBook 2* (and *ToolBook 1*) are filled with best-practice tools based upon the learnings of many organizations that have embraced NPD.

Paul Belliveau
Abbie Griffin
Steve M. Somermeyer

Part 1

Organizational Tools

This part features tools to improve the "soft" side of NPD. If there is one part of ToolBook 2 that the editors recommend NPD leaders read in its entirely, this is it. Model behavior by leadership is an important factor to any organization's performance, and this part covers a number of increasingly important NPD topics.

Part 1 starts with a chapter on innovation culture—the value and behaviors that foster creativity in developing innovative new products. This chapter leads NPD professionals through a number of tools to access the organization's culture and then offers many suggestions on how to improve your culture's creativity.

Chapter 2 tackles that frequently problematic transition of a new product from NPD to operational units. There are many war stories about the difficulties and frustrations that NPD organizations encounter when handing off their prized new product to their manufacturing and sales organizations. The authors of this chapter offer the Transition Readiness Assessment Tool to clearly delineate organizational misalignments. Highlighting the differences in new product familiarity and unit metrics, this chapter uses the transition readiness tool to lay out a detailed process on how to address many of the potential difficulties proactively.

Frequently, organizations attempting to make the leap from technology competency to product features fail. Chapter 3 addresses the difficulty that many NPD organizations encounter when attempting to convert technical competencies into successful new products. It offers a tool the authors describe as technology-to-product-to-market (TPM), which is a process to first convert new technology to new organizational capabilities and then convert these capabilities into product features. Embedded in the TPM tool are four worksheets that guide an organization through the TPM process.

The fourth chapter deals with a topic that has become a buzzword in industry: knowledge management. In a very practical approach, this chapter explains what knowledge management is and offers an approach to assess a firm's competencies in NPD. It then describes an innovation intranet and its value in R&D. Throughout the chapter there are practical suggestions on how to improve both the creation and dissemination of knowledge in an NPD organization

With the globalization of commerce and NPD, the distances between NPD team members and groups increasingly are an issue for management to deal with. Chapter 5 tackles the challenges of these increasingly virtual teams. It offers an approach to assess both the creative difficulty of an NPD project and the team interaction barriers. Then the authors offer a number of suggested actions to better align the creativity needed for the project and how the team is situated and set up.

Part 1 ends with a chapter offering practical insights on codevelopment and the management of partnerships in NPD. The chapter's first section leads the reader through a partner selection process and then offers a number of effective practices in the successful management of the resulting NPD partnerships.

1 Achieving Growth through an Innovative Culture

Erika B. Seamon

Senior management teams are continually struggling to cultivate environments that breed and nurture innovation through the good times and the bad. In the later years of the 1990s, growth through new products, services, brand strategy, and anything under the umbrella of "innovation" permeated the mind-set of business people everywhere. Big ideas, hefty venture capital investment, and the entrepreneurial spirit were the cornerstone of strategy. In the early twenty-first century, however, the new posture is protective and defense-oriented, and the focus is on cost cutting. A manager from a leading financial services firm sums it up by saying, "Cost management is more important in these times. We have been in a growth mode for five or six years, but now there is a need for continuing to focus on the bottom line."

As we've seen in years and decades past, the pendulum will continue to swing from growth mode to downsizing mode. Importantly, however, senior management teams know that a true "innovation-based culture" will outlive the whims of the economy, the turnover in management, and the changing dynamics of an industry. In times of growth, an innovation-based culture will have the people, processes, and mechanisms in place to systematically and continuously develop and launch breakthrough, category-changing new products. (*Note:* Products implies products, services, programs, offerings, etc). In a time when cost cutting is necessary for survival, effective senior management teams will use those same people, processes, and mechanisms to effectively refine and renovate existing offerings to still meet customer needs, yet enhance margins. Regardless, innovation-based cultures do not come and go; quite the contrary; they are grounded in practices that stand the test of time.

This chapter focuses on specific details of the plans and practices that senior management must ensure are in place to build and sustain a culture that is continuously innovative. While many of the tools and tasks may be coordinated and implemented by project managers and teams, it is vital that the senior management team understands, monitors, and engages in the *specific* details of what is involved. Developing breakthrough innovations that solve customers' problems and drive profitability is not a result of hands-off delegation. The implication is that in order to truly change a culture and instill a lasting mind-set across the organization, the senior management team, more than anyone, must uphold the integrity of the detailed processes and customs inherent in sustaining an innovation-based culture.

Let's first define an innovation-based culture. Can you tell if a company has an innovation-based culture by the personality of the CEO, the industry the company operates in, or the feeling you get when you walk around their offices? The answer is yes, and no. Herb Kelleher has unquestionably been a big part of Southwest Airlines' innovation-based culture, but it is also the major shifts in the airline's processes, business model, and customer services that make Southwest the fourth largest and one of the few profitable airlines in the United States. While certain industries, like high-tech, by default connote innovation, companies like Progressive Insurance that operate in less turbulent industries have been able to build innovation-based cultures by finding ways to tap into customer needs and continually create newness in their markets. And, yes, when you walk into a company like Capital One and see beanbag chairs and basketball hoops in separate innovation rooms, you get that "innovation feeling." However, their success is driven by the systematic process of continually developing and testing new product variations with their customers.

An innovation-based culture is one that systematically and continuously is able to create increased value for customers through products and services that provide new benefits. It is a culture where each manager and employee understands how to work together and what processes and procedures are required to uncover key customer insights and turn those into actionable solutions. Because an innovation-based culture has such an ingrained structure and discipline around best practices, there is the inherent comfort and understanding that new ideas and creativity have a place in the organization.

Webster's II Dictionary defines culture as "a particular form of civilization, especially the beliefs, customs, arts, and institutions of a society at a given time." Three groups of people in a company's "society" determine an innovation-based culture: the senior management team, the new product teams, and the customers (see Figure 1-1). The senior management team, for whom this chapter is intended, has a profound influence on all three groups. This chapter explores best practices or "customs" that must be employed by each group to systematically and continuously create new value for customers and establish a lasting innovation-based culture.

The following lists the groups of people and customs that, taken together, create cultural change:

Senior Management Team

- *Custom 1.* Do not delegate innovation.
- *Custom 2.* Create focused and specific innovation criteria.

New Product Teams

- *Custom 3.* Form short-term, high-powered teams.
- *Custom 4.* Recognize teams at frequent milestones.
- *Custom 5.* Take off functional hats.

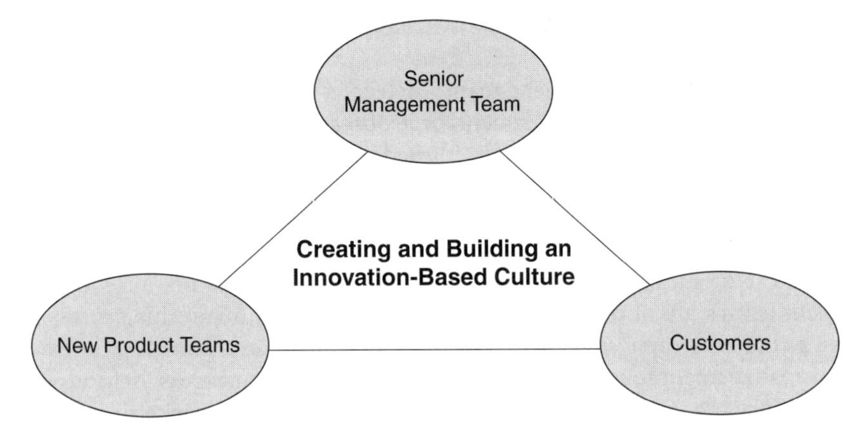

FIGURE 1-1. People transform culture: the three essential groups.

Customers

- *Custom 6.* Stretch the boundaries for problem identification.
- *Custom 7.* Have new product teams do their own research.

Importantly, the interrelationships among these three groups and the discipline each employs to effectively implement and sustain essential customs is what creates the habits and trust that transform a culture over time. First, let's examine the role and customs that a senior management team should adopt to set the tone for the organization.

SENIOR MANAGEMENT TEAM

Each member of an organization's senior management plays a crucial role in developing and supporting an innovation-based culture by linking strategy with action. The senior management team as a whole sees the big picture, understands the internal politics, can raise or lower barriers, makes final decisions, and ultimately holds the purse strings. A new product development leader at a $1.5 billion travel organization with over 3,000 employees reinforces this with her comment, "I am confident that my team and I can do anything. I just need to find a way to climb inside the minds of our top layer of management, see what they are thinking, get them aligned and get them to actually articulate what they want. Then I can start holding them accountable for helping me deliver on that." As a member of an organization's senior management team, it is essential to embrace two key customs if an innovation-based culture is going to take shape.

Custom 1. Do Not Delegate Innovation

Typically, a senior management team "delegates" new product development activities to an individual, department, or group and gets involved only after the recommendations have been already formed. Other times, there is a process in place where employees can submit ideas. These ideas are then evaluated, and those that are deemed valid will be presented to a senior management team for potential execution. In either case, shaping a portfolio of innovations is being delegated. Not surprisingly, as the investment required goes up, a senior management team's attention also goes up. Unfortunately though, this process can move a company three steps backward instead of forward. The ideal time for a senior management team's involvement is early on in the process, before significant dollars are invested, when there is more ability to influence the strategic direction of innovation efforts (see Figure 1-2).

When the early stages of innovation are delegated and the senior management team doesn't get involved until there are resource decisions to be made, it can have a severely negative impact on the culture for many reasons. First, individuals and teams become frustrated with the senior management team when they are caught off guard and not prepared to answer surprise questions in the middle or end of a new products initiative: "If they would have told me before that this was important, I would have investigated it." Second, the new product individuals are demotivated when the senior management team steps in to say that the strategy has changed and the area of focus for new innovations is no longer a priority: "Why have we spent the last three months dissecting this category if it is no longer a priority! This is ridiculous." Lastly, involving the senior management team late in the process breaks momentum and will eventually discourage people from generating new ideas: "I have all

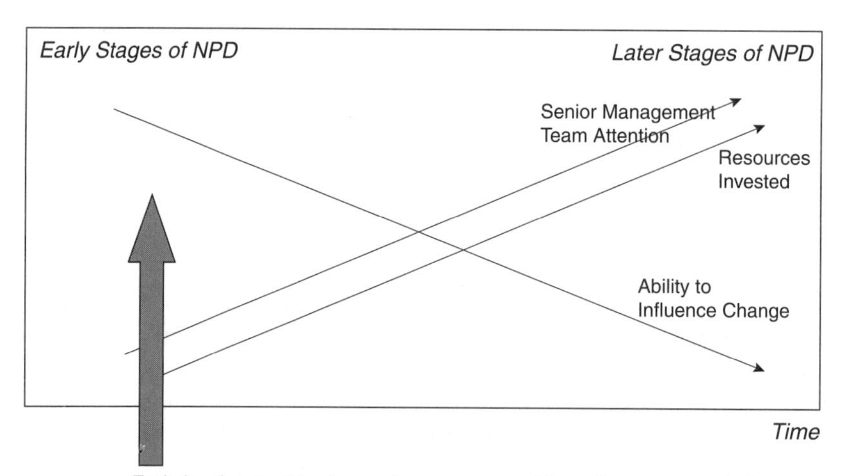

FIGURE 1-2. The ironic inverse.

these great ideas, but there is no way to get the resources to validate them or get them in front of our senior management (unless, of course, you are one of those people who know how to work the system). It's more worth my time to just do what I'm doing."

To shape a culture that believes in innovation and trusts senior management's support, the senior management team must take on an active role. The first crucial step is to form an innovation steering committee that is responsible for gaining alignment on the strategic and financial goals as well as setting expectations for any innovation/new products teams. In some instances, the innovation steering committee is the senior management team; other times, it is a subset of four to five individuals. This committee needs to stay actively involved in tracking a team's progress and meeting with that team at a minimum of one time per month. While initially the senior management team (or innovation steering committee) may feel as if they are getting too involved and perhaps stifling the freedom of new product teams, they are not; they are sending the message that innovation is important and worth their time.

In 1993, Carol Bernick, vice chairman and president of the consumer products worldwide unit for Alberto-Culver Co., realized her company needed a dramatic culture shift. As she wrote in a 2001 issue of *Harvard Business Review*, "We needed people to have a sense of ownership and urgency around the business, to welcome innovation and take risks. But in the existing culture, people dutifully waited for marching orders and thought of their bosses' needs before their customers'." In 1994, changing the company culture became the top priority for Carol Bernick and president and CEO Howard Bernick. Today, Alberto-Culver's way of business very much resembles an innovation culture. Culture is top of mind for everyone in the organization, and maintaining it is a part of everyone's responsibilities. Specific expectations for individuals and teams are defined, and people are measured and rewarded for enforcing and stimulating a positive culture within the organization. Employees recognize the impact they have on the organization. They work hard and take risks. For this, the company has been rewarded generously. Between 1994 and 2001, the company gained approximately 83 percent in sales and 336 percent in pretax profits while cutting employee turnover in half (Kuczmarski et al., 2003). Carol Bernick stays actively involved in formal and informal meetings every month with new product groups.

A senior management team, which prepares, develops, and agrees to an innovation charter that outlines the role of the innovation steering committee and sets expectations and goals for new product development across and within the organization, is taking the first step to igniting cultural change. Merle Crawford was the first to introduce the concept of an innovation charter. In its simplest form it is a document that defines what the innovation steering committee wants to accomplish company-wide relative to innovation and what role the committee promises to play. While specific innovation and new product projects will require a more refined set of financial and strategic goals, a broad-based charter relevant to the entire organization creates a "tone" from the senior management team and an overall outlook for the organization on innovation. Importantly, the innovation charter is not a document created after a senior management brainstorming session. Rather, it must be the end result and culmination of numerous senior management team discussions focused on evaluating historical company performance relative to future initiatives and plausible trade-offs. Often, the innovation steering committee will assign a task force to collect and analyze information from each steering committee member, as well as company documents, so that discussions can be data-oriented and focus on realistic plans, versus only aspirations (see Tool 1-1).

In 2003, a major food service organization aimed to develope a portfolio of new products, services, systems, and solutions to generate over $200 million in incremental revenue by 2010 and increase average margins by 3 percent. The innovation steering committee in this case was the senior management team for the organization. The team established a two-month time frame to have an innovation charter developed and agreed to each spend one full day of their time a week to help move it forward. It assigned a cross-functional team of mid-management and lower-level management "stars" to assist them in developing the charter. The following outlines what the chosen individuals did to help the innovation steering committee:

- Conduct individual interviews with members of the innovation steering committee to discuss their vision for the business; the role of new products in fulfilling that vision; the financial goals for the business and specific financial expectations for new products; key categories, segments, channels, or areas of opportunity that they would hope would be investigated; and areas that they believe are off-limits. Additionally, surface-specific criteria or hurdles for new products and any parameters for the portfolio (percent near-term vs. long-term offerings, etc.). Individual discussions uncover more in-depth and varied perspectives than group meetings.
- Analyze information from interviews to identify consistency and disconnects. Collect secondary data on market activity and company data on past product launches and financials. Evaluate goals relative to past performance and past new products developed to gain a perspective on the order of magnitude and the implications in terms of resources and changes in process. Document key findings so that the innovation steering committee can have a productive discussion.
- Provide innovation steering committee members with information before a meeting is held where they are to discuss disconnects, expectations relative to the past, and implications for pursuing these goals.

For the sake of efficiency, the steering committee was able to effectively engage assistance in the activities described in the preceding list. However, in a series of multiple meetings, it was their responsibility to develop the conclusions from the information and define their vision for innovation, objectives,

TOOL 1-1.
Innovation Charter (template)

Innovation Long-Term Aspirations:

- Innovations are defined as a new product, service, system, and/or solution that provides a new perceived benefit or value to a customer or consumer, establishes competitive insulation, and results in shareholder, employee, and customer satisfaction (*define as needed*).

Innovations at Company X will allow us to do the following:

- Enhance our market position and become a top _____ manufacturer in _______.
- Achieve financial success and beat long-term financial targets through _________.
- Gain competitive insulation and create products that are __________________.
- Better satisfy customer needs . . .
- Etc.

Innovation Strategic Vision Statement

- The New Product Portfolio Team will follow (an) ______ system that is grounded in understanding unmet and unarticulated needs of ______ in the ______ arena. The new products will align with Company X's competencies, expand our branded presence in __________ and build the xyz brand to represent a unique experience to ______ (*define as needed*).

Objectives of the New Product Portfolio Team:

- Strategic:
- Financial: Identify and shape a portfolio of concepts expected to generate $xyz million by ______ and ______.
- Market:
- Customer:
- Etc.

Steering Committee Role:

- Provide guidance, input, and motivation to help Team overcome obstacles.
- Question, challenge and reward the Team appropriately.
- Maintain big picture and ensure the recommendations of the Team are consistent with the strategy statement.
- Provide Team with the freedom and autonomy to make decisions.
- Help foster a long-term view and approach to innovation.

Steering Committee Declaration and Signatures:

______________ ______________ ______________ ______________

and their ongoing role. A senior management team, which can effectively put a realistic "stake in the ground" as to where innovation is headed in the organization, sends a strong message to the company about the commitment they have. This is an essential ingredient in shifting the culture.

One sure way for senior management to kill the spirit of innovation in an organization is to delegate it and get involved too late. Mike Gearin, who has developed over 25 innovative new products and services for Cincinnati Bell Telephone, says, "More than any single issue, teams need a sense of connectivity to corporate goals. If teams are forced to fabricate these connections, they are destined to second-guess their charter and will likely never truly focus their energies on the task at hand." The active involvement of a senior management team in driving innovation is a lynchpin to sending the right message to all employees about the importance of innovation. It is a custom that can have tremendous impact on an organization's culture.

Custom 2. Create Focused and Specific Innovation Criteria

Most of the strategic and financial goals a senior management team (or innovation steering committee) establishes are lofty, aspirational, and undefined. While at first, and only at first, these lofty goals can generate some internal excitement and chatter around the organization, they are a harbinger for failed innovation efforts. Hard questions are not answered, members of the senior management team stay on different pages, and the staff questions the credibility of achieving such goals when nothing has changed relative to the past.

For example, the divisional president of a financial services organization looking for the next five years of innovations said to his team, "I don't want to limit what you all find in terms of opportunities. Heck, if there is opportunity to meet our customers' needs, money to be made, and a fit with who we are, it's all in the cards. Sell me on the opportunity and if I'm sold, I'll bring it up (to the CEO) for his approval. You know, it will also depend on what else is being worked on in other divisions, but find me the best opportunities to bring forward." While the divisional president was enthusiastic and confident in his tone, he received multiple questions from the newly formed innovation group:

- "So, when you say if there is money to be made, what revenue and profitability levels are considered big enough to bring forward?"
- "Do you have a documented corporate strategy for where we want to take the company, so we know if an opportunity fits with who we are?"
- "Is there an expectation from the CEO on the amount he is planning to invest in growing our division versus the other divisions?"
- "Any longer-term plays will likely require us to take a loss the first few years. What's your stance on this?"

The divisional president told the team not to worry about these questions and just go out to the market and find some big opportunities. Imagine how this left the team feeling. An innovation-based culture is established when teams and individuals trust the commitment from the senior management team; they hope to begin their innovation initiatives believing that they are making a positive investment of their time.

The goal for a senior management team or innovation steering committee should be to effectively link strategy with action. The effectiveness of efforts put forth by individuals working on new products and the trust that is gained across the organization is dependent upon the senior management team's diligence in this area. Effectively doing this will require developing portfolio rollout scenarios and product development go/no-go criteria. Utilizing these tools requires hard work and active involvement from members of a senior management team. If, however, they are used in conjunction with one another and the innovation charter, they become a sound platform for innovation growth.

Portfolio Rollout Scenarios

Portfolio rollout scenarios are aimed at painting a picture for a company that illustrates the number and magnitude of products that would need to be launched over a certain time frame to reach the desired financial goal. As with the innovation charter, the senior management or innovation steering committee must stay actively involved in the building and revising process inherent in these scenarios, yet would do well to engage a team of individuals to help in the data collection and analysis. To build the scenarios, it is important for the senior management team to consider the following:

- *Typical size and types of past new products in terms of revenues, pricing, and penetration.* Establishing some benchmarks based upon the industry, category, or company helps to establish order of magnitude of the types of products that would be represented in a future portfolio.
- *Success rates.* Expecting 100 percent of products launched to be successful is unrealistic, and therefore, planning to launch one-third to one-half more products than what will become successful needs to be reflected in the scenario.
- *Market adoption.* Given history, define how long it takes for a product to be ramped up in the marketplace.
- *Seasonality.* Determine months or quarters where products can be launched.

Developing scenarios is both a science and an art. It requires looking to past benchmarks to make some high-level assumptions and then continually developing rollout configurations that could build to the desired goal. Very often, portfolio rollout scenarios create one major "aha" for a senior management team. They proactively display what often many know deep down in their psyche yet are afraid to voice. The "aha" is that there is no way to actually develop and roll out the number and type of products that would be required to reach their revenue and/or profit goal unless drastic changes are made. The portfolio rollout scenarios essentially create a platform for conversation to decide upon one of the following:

1. Revise our financial goal so that there is a scenario that is realistic; or
2. Revisit our processes, resource allocation, and overall approach to new product development; or
3. Do a combination of both.

The innovation steering committee must have an open discussion on the key insights revealed in the portfolio rollout scenarios in order to create a plan that is both realistic and credible with the organization. Having these tough discussions early on, before new product teams embark on identifying and shaping innovation opportunities, is key. The goal is for the senior management team to actively relay that they are being smart and serious about innovation. This is the type of message that can change a culture.

By way of example, a division of one of the world's largest consumer products companies was focused on growing one of their key brands from a $600 million business to a $1 billion business over a four-year time frame (see Figure 1-3a). Before any research was conducted or ideas generated, the senior management team made assumptions that there would be three broad types of products in their eventual portfolio that would differ in terms of risk/success rate, return/revenues, and time to commercialization. These assumptions were made by gathering examples of what was considered to be a smaller, closer-in new product example (in the company and/or industry) and what was considered to be a more expansive and breakthrough example. Lists of benchmarks were established, to help determine the order of magnitude around future new products that would be required. In the most simplistic form, it showed that it would take 13 of their latest new product successes (~$30 million in Year 1) to achieve the $400 million goal or three successes that were as big as the category's biggest new product in the last five years (~$130 million in Year 1). Those two statements alone began to give shape to what the senior management team was really asking the organization to achieve in terms of a financial target.

- 4.5% Base Growth from Strategic Plan. Historical CAGR (FY '93–FY '02) is 3.2%.
- **Product Type A:** Low Risk/Return Introduction ($25MM):
  - 80% of revenue taken to account for failure in the portfolio
  - 50 weeks to commercialization
- **Product Type B:** Medium Risk/Return Introduction ($60MM):
  - 70% of revenue taken to account for failure in the portfolio
  - 54 weeks to commercialization
- **Product Type C:** High Risk/Return Introduction ($90MM):
  - 60% of revenue taken to account for failure in the portfolio
  - 54 weeks to commercialization
- **Benchmarks include the following:**

	Year 1 Retail $:
Company X's Latest New Product	**$30 mil**
Competitor A's New Product A	$35 mil
Competitor A's New Product B	$43 mil
Competitor B's New Product	$98 mil
Competitor C's New Product	$110 mil
Competitor D's New Product	**$125 mil**

FIGURE 1-3a. Portfolio Rollout Scenario—Key Assumptions

This consumer products company senior management team also recognized the importance of having a balanced portfolio to house both lower-risk/-return product types (A) as well as higher-risk product types (B and C). They accounted for new product success rates by taking a percentage of revenues off each product launched. (*Note:* This is not necessarily how it happens in the market, but it is a simplistic way of accounting for success and failure without getting too complex in this initial planning stage.) After the key assumptions were made, various scenarios were developed that showed a build toward the financial target (see Figure 1-3b) and the implications for how many products, of varying types, would need to be launched (see Figure 1-3c).

These portfolio rollout scenarios raised many questions for the senior management team:

- Can we physically do this the way we are set up?
- Can our sales force aggressively roll out this many products?
- Does this timeline account for careful market testing and assessment?
- What type of resources will be required to develop, launch, and support these products each year if a scenario like this plays out?
- What types of human resources would be required to pull this off?

The senior management team was forced to address these questions before specific new product initiatives were launched. A year later, the senior management team invested in the development of a balanced portfolio of new product concepts. Throughout the process, as ideas and concepts were being shaped, the employees involved had the comfort and trust that senior manage-

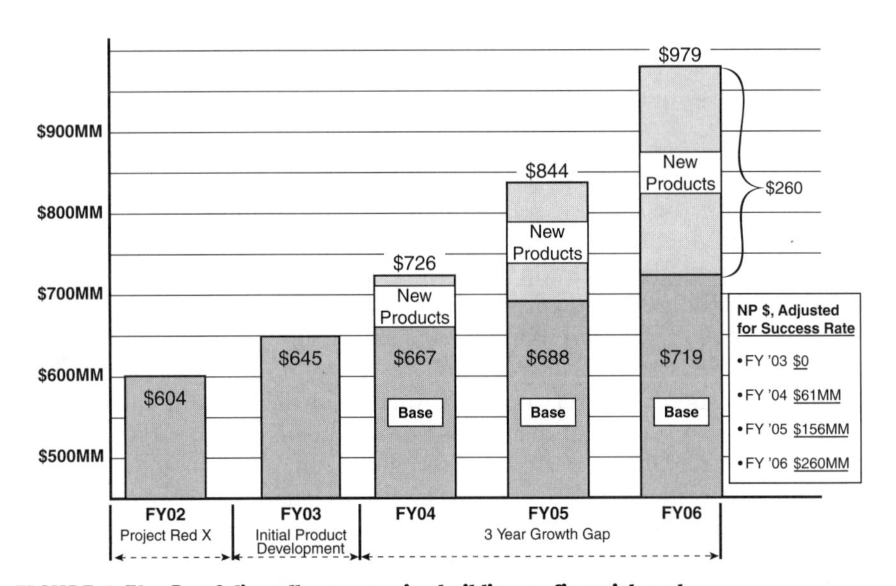

FIGURE 1-3b. Portfolio rollout scenario; building to financial goal.

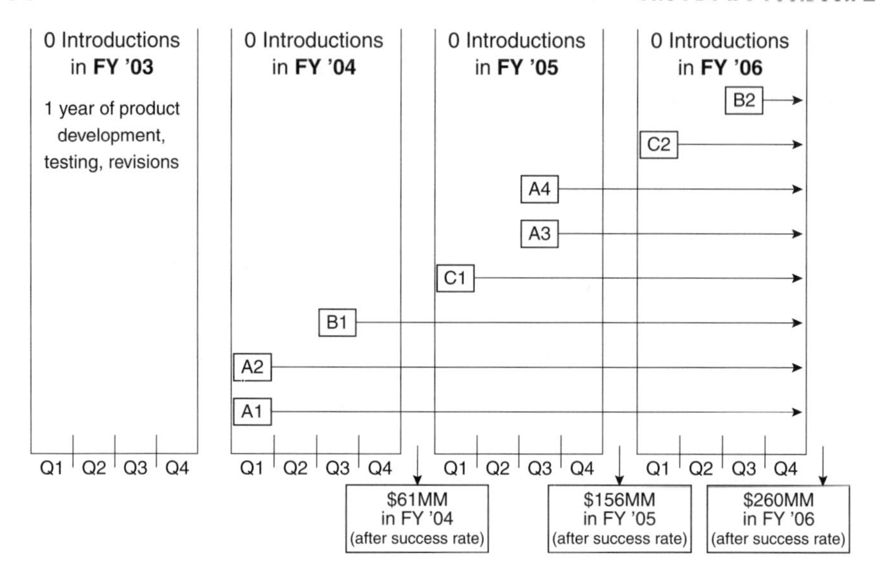

FIGURE 1-3c. Portfolio rollout scenario: launching products.

ment really understood what they had asked them to do. Portfolio rollout scenarios directly impact the culture of an organization because they, in a sense, prove that a senior management team understands the implications of their statements and goals and is prepared to put their money where their mouth is to support innovations that are developed. As a result, teams are propelled into action with a clear understanding of what they need to produce, trust and excitement are generated across the organization, and a culture begins to shift.

Product Development Go/No-Go Criteria

Another essential tool for ensuring that innovation goals are well defined is go/no-go criteria. The senior management team, or innovation steering committee, must also be the key instigator of agreeing to the criteria on which various new product ideas and concepts will be evaluated. They must do this initially along with the innovation charter and portfolio rollout scenarios and continue to revisit the criteria as new initiatives and teams are put into place. Importantly, however, criteria should be in place before ideas or concepts are generated. This is essential for creating a strong innovation-based culture because agreed-upon criteria for evaluating, prioritizing, and categorizing new product opportunities at their various stages sends a clear message to the organization that politics, personal preferences, and/or one's gut sense will *not* be the accepted method. New product individuals and employees as a whole will begin to trust that there is a method to the madness and that being a good innovator isn't just based on one's political power. Well-developed criteria ensure rigor and consistency in making decisions and help guarantee that opportunities are evaluated not only from a financial and feasibility standpoint but from a strategic, market, and customer standpoint as well. Product development go/no-go criteria are a natural outgrowth of a well-developed innovation charter, clear financial goals, and the portfolio rollout scenario discussions.

At that early stage, senior management has the information required to delineate parameters on how ideas (at a company-wide or division-wide level) will be prioritized, how concepts will be evaluated, and how prototypes will be assessed before they are launched. As ideas move through the process, criteria get more stringent and the data required to evaluate the criteria becomes more rigorous. The criteria help manage everyone's expectations so that they cannot and do not arbitrarily change over time. The criteria provide an effective and efficient method for decision making and keeping new-product people focused on what senior management deems important.

Developing go/no-go criteria entails an iterative approach of refinement by the senior management team. The senior management team and individuals focused on new product development must work on the criteria until they are comfortable that only opportunities that fulfill the defined hurdles will move forward in the process. The senior management team must agree that opportunities that fit the criteria will be supported with the promised resources for development and launch. This motivates a culture to innovate. The tool attached (see Tool 1-2) provides a basic template of criteria at the concept stage that could be used to make a decision as to whether product concepts get moved into prototype development. Detailed business cases for each product being evaluated against these criteria are necessary to effectively answer the questions. Please note the important elements of criteria:

1. There are multiple categories of criteria:
   - Strategic questions ensure that a portfolio is in line with key company, brand, and/or category goals.
   - Customer and consumer questions ensure that products are meeting broadly felt and intense problems of customers.
   - Market attractiveness questions assess the external landscape.
   - Technical questions help to gauge the feasibility of developing a product.
   - Financial questions help identify whether the product will potentially meet key hurdles.
2. Each question has a choice of answers. Open-ended questions are ineffective for the purpose of screening, so specific categories of answers must be delineated. Any necessary detail in terms of defining what a 1 versus a 5 is on a 1 to 5 scale, for example, where to get the data to answer each questions, or who would be responsible for the screening should be documented and agreed upon.
3. There are three different categories of questions:
   - Go/no-go questions have minimum hurdles (indicated by an underline) that must be obtained for the idea/concept to have any further

TOOL 1-2a.
Product Development Go/No Go Criteria—(For concepts with a corresponding business case)

	Hurdle	Type
Strategic Fit		
◆ Does the concept fit with the XXX 2010 Vision?	Yes/No	YN
◆ Does the concept fit with the XIX Vision?	Yes/No	YN
◆ Is the concept consistent with XXX's Playing Field?	Yes/No	YN
◆ How effectively can our sales forces sell this concept?	1 3 5	YN
◆ Is this concept a Close-in Line Extensions, or Line Extensions, or Innovation?*	Close-in L Ext. / L Ext. / Inn.	YN, C
Customer Fit		
◆ Does this concept help the customer better fulfill consumer needs?	1 3 5	P
◆ Is the concept flexible or adaptable for specific customer needs across channels?	1 3 5	P
◆ Within a channel, what is the breadth of appeal of this concept?	1 3 5	P
◆ Does the concept have positive customer price/value relationship?	1 3 5	P
◆ Does the concept have a strong customer purchase frequency?	1 3 5	P
Consumer Fit		
◆ Does this concept solve an unmet, latent, or unarticulated need of the consumer?	1 3 5	P
◆ Will consumers' desire for the concept likely drive higher frequency and loyalty?	1 3 5	P
Attractiveness of the Market		
◆ Relative to current competitive offerings, how unique is this idea?		
Line Extensions.	1 2 3 4 5	YN, P
Innovations.	1 2 3 4 5	YN, P
◆ What is the likelihood that XXX will be a #1 or #2 player in the market?	1 2 3 4 5	P

Key: Underline = Required result for a Go/No Go hurdle
Rating scale: *Not at All* 1 3 5 *Extremely*
YN = Yes/No; P = Prioritization; C = Category
*Close-in Line Extensions: Discontinue screening process and pass to Marketing.
Line Extensions: Continue and screen as Line Extensions.
Innovation: Continue and screen as Innovation.

TOOL 1-2b.
Product Development Go/No Go Criteria continued (For concepts with a corresponding business case)

	Hurdle	Type
Technical Feasibility		
◆ Is this concept feasible, considering all variables?	Yes / No	YN
◆ How feasible is this concept?	1 3 5	C
◆ How protectable is this concept?	1 3 5	C
◆ What is the likely timeline for development of a prototype, in months (from concept, end of step 4)?		
Line Extensions:	<6 >6	YN
Innovations:	<12 >12	YN
◆ How long will it take to develop and launch, in months (from prototype, end of step 5)?		
Line Extensions:	<12 >12	YN
Innovations:	<24 >24	YN
◆ How easily can legal/regulatory hurdles be overcome?	1 3 5	P
Financial Returns		
◆ Will the concept achieve breakeven status by target year?		
Line Extensions: By year 2	Yes / No	YN
Innovations: By year 3	Yes / No	YN
◆ In the breakeven year, what is the impact on Economic Profit?	Add Neutral Destroy	YN, P
◆ Will the concept achieve targeted Gross Proceeds of Sales (GPS) by required year?		
Line Extensions: $yM by year 2	Yes / No	YN
Innovations: $yM by year 3	Yes / No	YN
◆ Will the concept achieve targeted EBITA by required year?		
Line Extensions: y% by year 2	Yes / No	YN
Innovations: y% by year 3	Yes / No	YN
◆ If capital is required, is the payback period less than three years?	Yes / No N/A	YN

Key: Underline = Required result for a Go/No Go hurdle
Rating scale: *Not at All* 1 3 5 *Extremely*
YN = Yes/No; P = Prioritization; C = Category

consideration. These are deal breakers for the senior management team.

- Prioritization questions help weigh one idea/concept over another.
- Categorization questions help ensure a balanced portfolio in terms of feasibility and time to market. If a portfolio must have a certain percentage of the products that are near-term versus long-term, a categorization screen should account for this.

The leader of the travel organization quoted earlier also said, "Objective criteria that remain consistent though the duration of my innovation initiatives would empower my staff and create a whole different feel around here." Well-defined, documented, understood, and agreed-upon criteria do more to create an innovation-based culture than most senior management teams recognize. Creating the criteria forces tough questions to be answered by the individuals who hold the purse strings. The criteria also allow new products people to feel comforted that their new-product goals are not a moving target. And when a senior management team sticks to the criteria throughout the process and resists changing them just so an idea they like can move forward, the senior management team builds the trust and credibility that is essential to transforming a culture.

NEW PRODUCT TEAMS

It is the people within and associated with an organization that drive the culture. While the senior management team is one essential group, the second is all employees or teams responsible for new product development. If the senior management team can effectively form the right types of innovation initiatives, recognize organizational accomplishments, and consistently manage and deliver against expectations, an innovation-based culture can flourish. During the energy deregulation frenzy in the late 1990s, a Midwest utility decided that it was going to become "market-focused" and would strive to create new value-added services for its residential, commercial, and industrial customers. To kick-start this effort, the company formed a new-business development group of three people. Next, the company developed a highly sophisticated software program designed to take ideas from any employee in the organization and provide a step-by-step rigorous process for evaluating the idea and potentially carrying it through until it was approved to be developed. An employee had to fill out a form and use the drag-down menus to describe the idea and explain why it was unique, why it would help customers, and what the financial potential in terms of revenue and margin could be. The new-business development group would do research on each submission, and if the idea made it through multiple stages, it would go before the management group. If the management group approved it, the person who submitted the idea would get a $100 check and his or her name in the monthly newsletter. In two years, guess how many ideas were submitted? Only six. Obviously, this was not an effective method for creating an innovation-based culture. Four reasons explain why the Midwest utility did not succeed in developing an innovation-based culture:

1. People drive innovation, while systems and processes are enablers.
2. Innovation is a team sport, not an individual effort.
3. Innovation requires a portfolio approach, not a product-by-product approach.
4. Innovation begins with a thorough understanding of the customer, not with ideas.

The adoption of a structure that empowers, motivates, and rewards new product teams is essential for establishing an innovation-based culture. It is Pollyannaish as well as risky to expect and request that every employee be focused on uncovering customers' most intense and widely felt needs, continually generate new ideas, and take risks. While a broad mix of employees should partake in key aspects of an innovation initiative, it is ineffective and inefficient to make everyone (and thus no one) accountable. Those organizations that form cross-functional, potentially cross-divisional teams with strong team leaders and specific goals have a higher likelihood of new product success than organizations that do not. Astronaut Mary Ellen Weber has worked with teams both during space flight and on innovation projects for NASA. She recognizes the critical role that new products projects play in any organization. "The mission of a new product development team in a sense is also one of survival—their project, their jobs, and the success of the company are at stake. With so much riding on new product development, it is critical that project teams are properly supported and structured to improve project and organizational success." Effective new product teams not only launch market successes, they generate excitement across the organization, attract the best talent, and ultimately become a driving force in shaping an innovation-based culture.

Custom 3: Form Short-Term, High-Powered Teams

While many senior management teams assign a cross-functional group of individuals to the task of new product development, many times these groups are formed without an end date established. New products or innovation becomes part of their job description. When teams are formed for an undefined period of time, they by default naturally become loosely tied individuals who share information periodically and keep each other in the loop on their department's or division's efforts.

A sure way for a senior management team to make innovation hard-hitting and continuous is to create short-term, high-powered teams. The short-term nature of a team assignment is critical to creating a dynamic that is intense, yet provides a light at the end of the tunnel for team members. Every organization and industry is different, yet as a point of reference, teams operate well when they are in place between three months and ten months. A relatively short time frame with challenging yet achievable goals and a strong team leader can create a high-powered group of innovators. Essentially, a senior management team should strive to have two types of innovation teams continually in place (see Figure 1-4).

While the team members and their objectives will change with each rotation, continuity in structure and process will be a continual reminder to the organization that innovation is a priority and ingrained in the culture. The two types of innovation teams are as follows:

1. Portfolio teams are focused on shaping a portfolio of new product concepts for a market, category, brand, or business to be launched over a two- to five-year time period, depending on the pace of the industry. The portfolio rollout scenarios provide a starting point for portfolio teams. The portfolio should incorporate strategic and financial objectives, address key customer problem areas, and be supported with detailed business cases that outline specific product parameters, metrics, and the development and launch requirements and plans. A portfolio team is cross-functional, has a dedicated leader, and reports to the innovation steering committee. This team should be in place for eight to ten months in order to effectively conduct the strategic and market work necessary to define customer problems, shape ideas and concepts, and build business cases and development plans. Many members of portfolio teams become leaders of development teams.
2. Development teams are focused on developing, testing, and launching a concept or a few related concepts from the portfolio. They follow the parameters outlined in the business cases. Often, significant trade-offs and decisions need to be made by development teams to balance the feasibility and cost realities of creating prototypes while upholding the integrity of the concept in terms of meeting customer needs. This is why carryover of one or more individuals from the portfolio team, coupled with individuals with specific expertise required to develop this concept, is essential. This team should be in place for the time specified in the business case based on how long it should take to develop, test, and launch this concept. Development team leaders working on concepts that are part of the same portfolio should be in continual communication.

For example, if a business-to-business organization wanted to significantly grow its presence in the facility management market over four years, it could form a portfolio development team in May of this year (Year 1) that would be in place until December of this year (Year 1) when it recommends a portfolio of offerings and timing for development. In January of next year (Year 2), five development teams kick off, each responsible for developing, testing, and

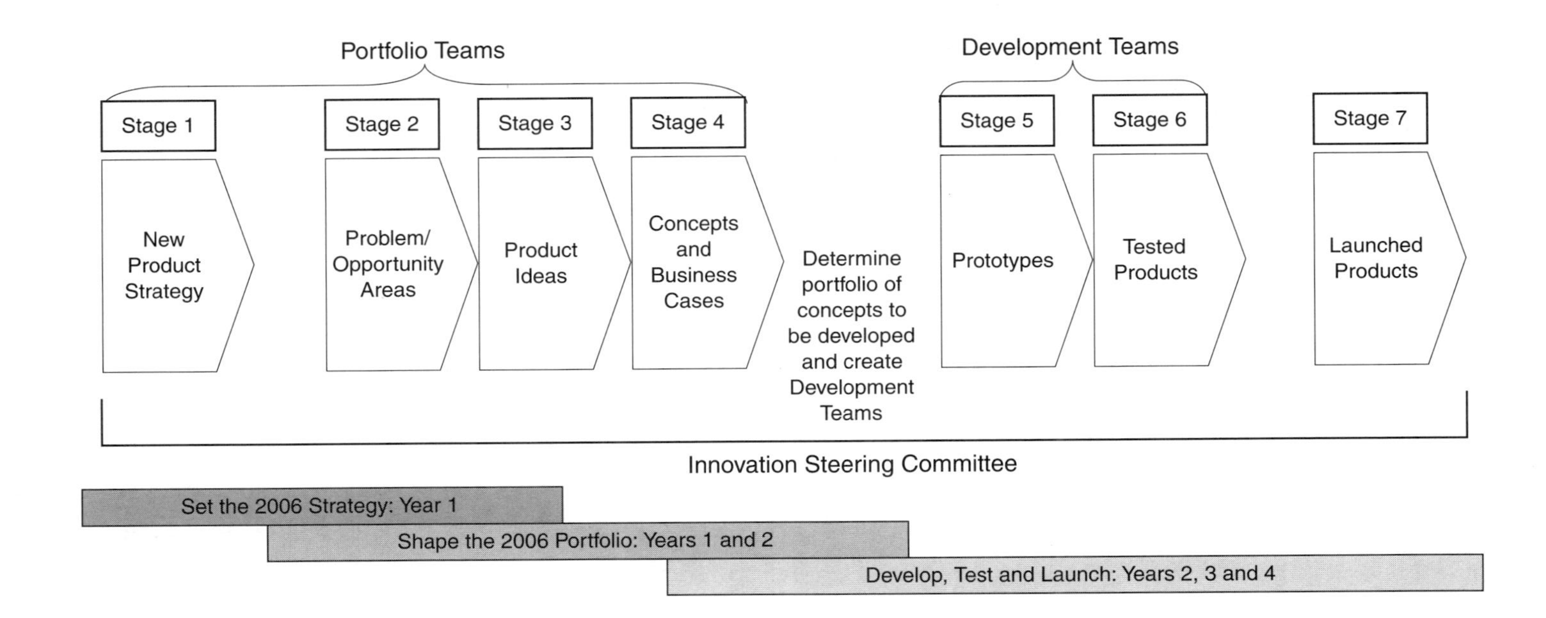

FIGURE 1-4. Portfolio teams and development teams.

launching one to two of the products in the portfolio. One team is in place for eight months, since their concept leverages existing capabilities and is set to launch during next year (Year 2). Another team may have a two-year time frame to acquire a new technology and ramp up a more expansive offering that will launch the following year (Year 3). The key here is to establish end dates.

For an organization beginning to build an innovation-based culture, it is vital that the senior management team or innovation steering committee stay actively involved with these teams. Involvement at the beginning and conclusion of each stage of the process is essential. As a rule of thumb, a minimum of one-day-a-month involvement to get updates from teams, provide perspectives, and ensure that barriers are removed and progress can be made is essential. These teams will be the seeds that are planted in the organization's culture. Therefore, the senior management team has a significant role to play in being both a champion as well as a stickler for the expectations that were set up. High expectations from the senior management team, coupled with significant support, will attract top talent to innovation initiatives and begin to change the tone of the organization.

S. C. Johnson & Son is a good example of an organization that has successfully embraced this structure. S. C. Johnson, maker of such household cleaning products as Windex, Glade, Raid, and Pledge, created a culture of innovation using "seeding." It formed portfolio teams in those categories or for those brands that were believed to have strong innovation potential for one reason or another. Members of those portfolio teams went on to lead development teams or lead new portfolio teams. This method of skills transfer and focused and high-impact short-term teams resulted in successful innovations ranging from Glade Scented Candles to Windex Outdoor Window and Surface Cleaner.

Custom 4: Recognize Teams at Frequent Milestones

Not only do new product portfolio and development teams need to have a designated end to their efforts, they must also have well-defined and frequent milestones to gain recognition, encouragement, and buy-in from senior management for the progress of their work. As a company embarks on establishing and building an innovation culture, major opportunities could fall through the cracks of the company if teams are only rewarded once they have achieved their primary goal of developing the next set of successful big hit products, or even worse, The Next Big Successful Product. Teams will not only lose motivation and perhaps become more and more discouraged with each barrier that lies in their path, but depending on when the product gets launched and the time it takes to determine its success, it could take years to see the final result. Perhaps the product fails; perhaps there has been enough turnover that none of the people who worked on the team to develop it are even around; perhaps no one in the organization ever knows of the team's achievements toward innovation. Effectively rewarding individuals and teams associated with new product development is what generates the "hallway buzz," the excitement and the enthusiasm around innovation essential for a thriving innovation-based culture.

New product teams should be viewed and rewarded as if they were climbing Mt. Everest together and hoping to reach the summit (see Figure 1-5). Getting to base camp at 11,000 feet is an accomplishment in itself. This could be equivalent to agreeing upon the key elements of innovation charter, portfolio rollout scenarios, and go/no-go criteria. Team members who helped the senior management team conduct research and reach decisions should be recognized. Reaching the next camp where the air is thinner, yet the spirit and body still strong and flexible, is similar to effectively prioritizing and defining the problem/opportunity areas to pursue. Climbing to the next camp is like developing the screened, prioritized, and categorized list of the new product ideas that best address the problem/opportunity areas. The next is having refined and shaped product concepts and business cases. With each ascent there are inevitably detours and descents, yet the cohesiveness of the team builds, and reaching each camp is an accomplishment that provides more excitement and faith that the next milestone is achievable. Importantly, however, the stakes go up too. The risks increase and the mental and physical resources invested begin to go up exponentially as the summit gets closer.

Recognition at each milestone can be provided at meetings between the innovation steering committee and the new product team. The primary motivators for new product teams are self-accomplishment, peer recognition, and top management exposure, not financial rewards. A meeting every one to two months at key milestones where new product individuals have an opportunity to share their knowledge, insights, and rationale for recommendations has a direct impact on culture. The activities at these regular meetings include reestablishing goals and key deliverables, making recommendations based upon predefined go/no-go criteria, and discussing key insights, rationale, and areas where the team needs guidance. These check-in points are imperative for many reasons:

- They rectify the ironic inverse by involving the senior management team early on in decision making.
- They send a clear message to not only new product teams but to the organization as a whole that innovation is important enough that the members of the senior management team are willing to spend time each month to drive teams forward.
- The exposure to senior people and the recognition of accomplishment at each stage continues to reenergize team members towards their goals. John Berschied, group vice president of R&D for Alberto Culver, reinforces this by saying, "The biggest reward for many people is face time with senior management. They will take on enormously risky projects just for the opportunity to have time with them." The energy, momen-

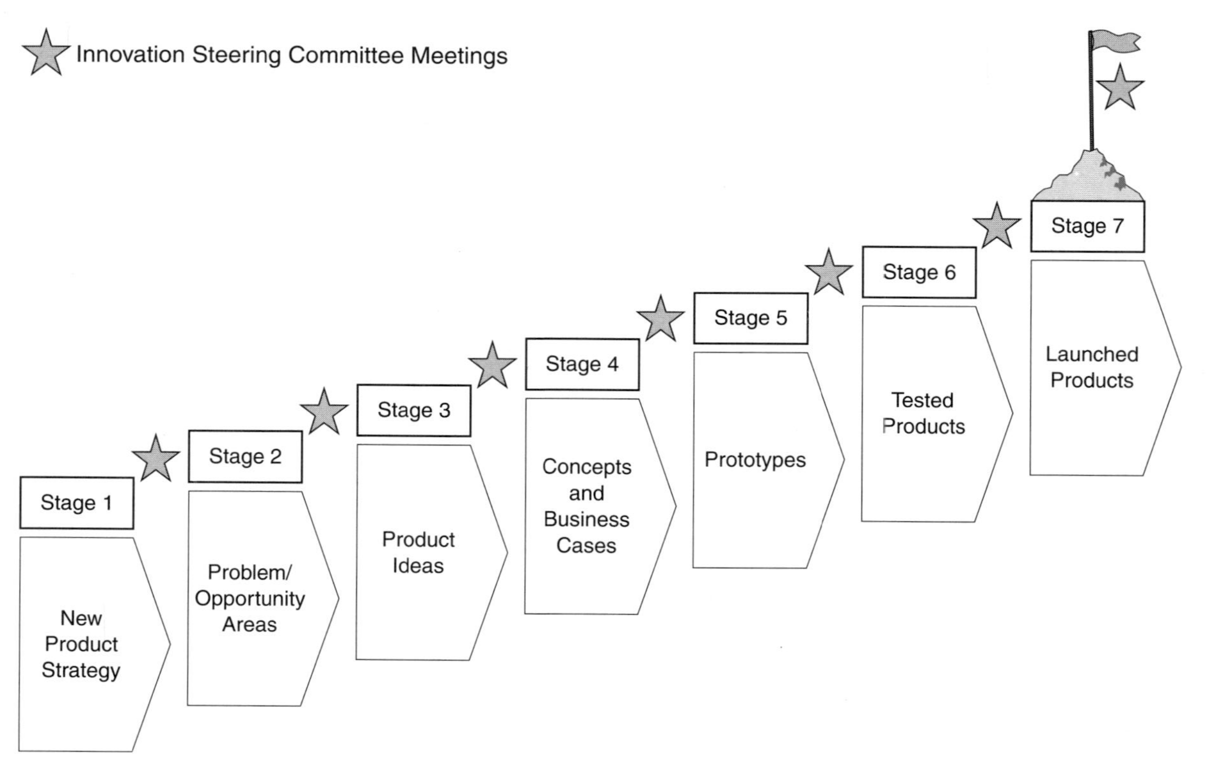

FIGURE 1-5. Recognizing teams along the way to the summit.

tum, and positive spirit of the new product teams have a direct impact on the culture of an organization.

Custom 5: Take Off Functional Hats

The senior management team not only has to form high-powered, short-term teams around innovation, it must also encourage the new product people to play a different role than they have been expected to play previously. Establishing an innovation-based culture requires getting people out of their comfort zones. If the senior management encourages individuals selected to partake in product teams to partake in roles beyond their expertise, there will be a strong impact on the culture. For example, it not only minimizes miscommunication, but more importantly, it helps a team get to a better answer. Additionally, team members learn more and are able to share new insights with others in their department. A manufacturing team member may play a role in defining the extendibility of the brand, a marketing person may be involved in understanding systems and IT implications, a finance person may need to write concepts for customer testing. Particularly in the early exploratory stages of the process, team members should be encouraged to take off their functional hats.

For example, having R&D people actively engaged in talking with customers about their needs and problems helps to bridge the gap between technology identification, acquisition and/or use, and the market for later stages of the new product development process. A senior director of R&D at a consumer packaged goods company said the following after working on a six-month portfolio team and participating in setting strategy for new products, brand extendibility research, problem and need identification and prioritization, concept shaping research, and business case development: "I have worked in R&D for fourteen years but in the last six months I was encouraged [by senior management] to be more than just an R&D guy. I now have a completely different perspective on what I do and how it fits into the rest of the organization. By being involved from the start, it has allowed me to really understand what we have to accomplish on the technology side. If someone would have just slapped on my desk the customer need research and a couple concepts to go develop, I would have no way of knowing for sure what we are aiming for and how to make the smart trade-offs. Also, if I wasn't forced to directly interact with the other functions and partake in the discussions out of my comfort zone, I would never have a true appreciation for all of the variables that must be considered."

If new product team members are encouraged to take off their functional hats and stretch their skills and perspectives, the results will be better innovations, process efficiencies, and a cultural shift. Better results come from having the team work cohesively, rather than in silos. Gaining insight from team members on areas where they have not previously been engaged inevitably leads to new perspectives. Efficiencies at the back end of the process are gained because team members, particularly in R&D and manufacturing, have a better understanding of what the team is trying to accomplish and will have already been thinking about how the product can be developed. In terms of culture, this custom can have a significant impact. Team members gain new insights and viewpoints that they bring back to their respective departments. Marketing, sales, and R&D, along with manufacturing, finance, and other departments, begin to have a new appreciation for each other and the vital role each plays in successful innovation. This is the type of relationship between employees that can shift a culture.

For a new product portfolio team, a key point early in the process where team members can take off their functional hats, is when they are doing analysis on customer problems, which is explained more in Customs 6 and 7. This type of analysis should lead to decisions about what problems are most worth focusing on for idea generation. The analysis process could play out like this (see Tool 1-3):

- Provided that qualitative exploratory customer research is conducted, split the data collected from customers into sections based on topics discussed.
- Distribute those data to each team member, including marketing, R&D, manufacturing, packaging, sales, finance, and any others.

TOOL 1-3.
Team Members Analyzing Customer Problems

Name of Problem/Opportunity Area: *XYZ*

Description of Problem:

XYZ is the most significant negative income driver, impacting many other areas of . . .

Importance of the Problem:

Breadth across segments/customer is __________ because __________.
Intensity of the problem is __________ because __________.
Frequency of the problem is __________ because __________.

Findings Tied to Key Subproblems/Issues:

1. **Lack of** __________: (define)
   Customer quote: " . . . ."
   Customer quote: " . . . ."
2. **Fear of** __________: (define)
   Customer quote: " . . . ."
   Customer quote: " . . . ."
3. ______________: (define)
   Customer quote: " . . . ."
   Customer quote: " . . . ."

TOOL 1-4.
Prioritizing Problems to Address as a Team

Problem/Opportunity Areas	Breadth of Problem	Intensity of Problem	Frequency of Problem
Top Tier			
◆ XYZ	●	●	●
◆ ABC	●	●	●
◆ DEF	○	●	●
Second Tier			
◆ ZZZ	●	○	○
◆ AAA	●	○	○
◆ Etc.	○	●	○
◆ Etc.	○	○	●
◆ Etc.	●	○	○

- Have each team member go through his or her section to uncover key findings and conclusions. Then hold a team meeting where individuals bring their analysis pieces and come prepared to discuss them and, as a group, to define and prioritize which problem areas to pursue in the later stages of the product development process (see Tool 1-4).

By having individuals actually do analysis, they will be much more engaged at the meeting and bought into why the team is pursuing certain opportunities. Additionally, they will effectively talk to about the rationale for decisions. The subtle, yet essential, reason senior management should encourage new product individuals to play broader and less comfortable roles than they normally do is that it helps the person grow exponentially as a professional. This growth fosters across a broad base of individuals the loyalty and enthusiasm that drives culture change.

CUSTOMERS

Senior management must also be concerned with the third key group of people essential to building and sustaining an innovation-based culture: current and potential customers. While many senior management teams recognize that they must be market-driven and have employees keep their fingers on the pulse of customers' needs, many do not do this. As Cooper states, "A thorough understanding of customers' needs and wants, the competitive situation, and the nature of the market is an essential component of new product success. This tenet is supported by virtually every study of product success factors. Conversely, failure to adopt a strong market orientation in product innovation, unwillingness to undertake the needed market assessments, and leaving the customer out of product development spell disaster; these are the culprits found in almost every study of why new products fail." (Cooper, 1996)

Abbie Griffin sheds some light on this disconnect: "There is great controversy in recent business publications as to whether firms should actively involve customers in the process of product development or whether customers and their input should be ignored. Those who claim that customers should be ignored argue that customers and potential product users cannot tell firms exactly what they want. Thus there is little or no use in talking to customers. Product developers on their own can best determine what products will be successful. An alternate possibility is that product developers have been asking customers the wrong questions. We have been asking them for information which they cannot provide to us. Thus, rather than totally ignoring customers, perhaps we can improve the products we develop by changing the types of information we seek from customers to that which they can provide us." (Griffin, 1997)

Breakthrough new products come from frequent use of customers in shaping a portfolio of products as well as in the actual development and testing of products. Stretching the boundaries, asking the right types of questions when uncovering and prioritizing customer problems to address, and actively involving the new product teams in this research is key. Senior management teams who are truly focused on creating an innovation-based culture must change the status quo while encouraging and enforcing this unconventional, yet highly effective, approach to customers. Employees will begin to recognize and trust that the organization actually has disciplined practices in place to ensure customer needs turn into innovations, not just big market research reports that never get used.

Custom 6: Stretch the Boundaries for Problem Identification

Many organizations that do not have an innovative culture are averse to broadening their definition of the category or industry when they search for problems and unmet customer needs. Asking customers about their likes and dislikes in their current options is a comfortable line of questioning and can lead to some strong product improvements and line extensions. Unfortunately, however, this does not lead to the breakthrough insight that can launch a new product team in the right direction or generate excitement in the rest of the organization. The senior management team plays a vital role in influencing new product teams to stretch the boundaries of thinking. Thus, it is important that the senior management team understand and appreciate the methods and objectives of market research with customers, so they can encourage teams to stretch their thinking. For the most part, team members, market researchers, and project managers will remain conservative and focus on the existing boundaries of the category unless the senior management team is actively involved in pushing them further.

In the consumer arena, doing research that will uncover lifestyle changes and high-intensity problems beyond the current category is key to breakthrough thinking. Who would have predicted the success of the cell phone, let alone our reliance upon it? Customers may not have said they needed a phone wherever they were; however, as consumers, they certainly articulated how their lifestyles were changing dramatically, with less time for more work. They said that their lives were taking them out of the home more. They complained that travel was becoming the norm and that there were tremendous problems and stresses formed by this new, on-the-go world. No wonder cell phones have become a norm in our society.

In a business-to-business arena, getting customers to talk about their problems with their businesses, their specific roles, the processes of their work, and their specific goals and barriers to achieving those goals provides the sets of problems that customers really care about addressing (see Tool 1-5). It is not our customers' job to figure out how our competencies can address those intense problems. For example, Cemex, the most profitable cement company in Mexico and number two in the world, discovered that there was significant cost and emotional frustration associated with the timing of supplier arrival at building projects, particularly with perishables. If a project was running behind, a shipment of cement could be wasted completely and would throw a team off mentally. As a result, Cemex has positioned itself as the Domino's Pizza of the cement industry by being the fastest to deliver. This delivery service ensures customers have cement when they need it, not before and not after. They can charge more for their service because they are saving builders significant costs associated with unwanted or ill-timed cement deliveries. They

TOOL 1-5.
Exploring Broad Customer Problems

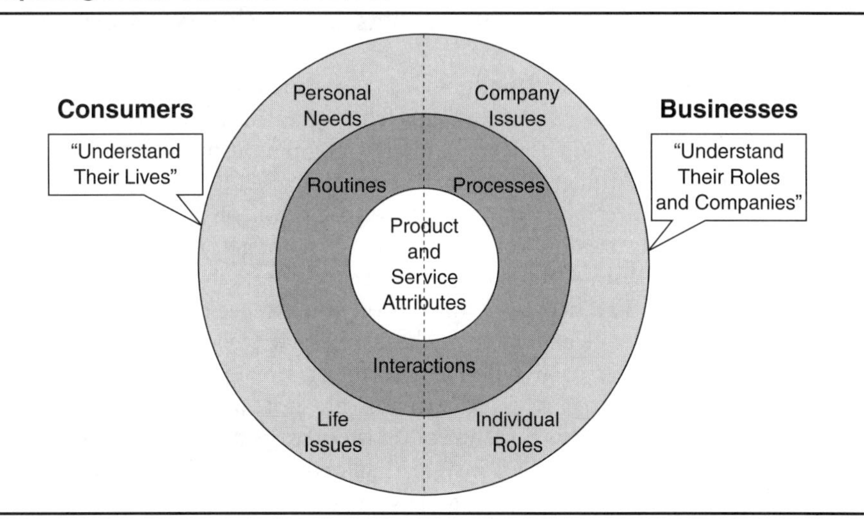

received the CIO-100 award for innovative business practices and services (Deck, 2001).

Another example is Capital One, who was able to gain deep insight into the lives of their high-debt consumers by understanding their life problems and financial woes. The insight they gained by exploring problems well beyond the boundaries of credit cards had a direct impact on their innovation efforts and the organization. The new product team, steering committee, and other key senior leaders found this insight to be so perceptive and important to the business that they shared it on a broad scale across the organization, reorganized a few of their departments to better reflect how their consumers behave, and put multiple innovation platforms into development (2000). Insights on consumer problems with credit cards alone would have merely led to gripes about late fees, customer service, and interest rates.

Therefore, the senior management team must encourage new product portfolio teams to broaden the boundaries of where they explore customer problems. The senior management team should also continually engage the team in discussions on their findings and provide insight on potential implications. If this insight is effectively gathered, documented, and analyzed in an ongoing and iterative fashion, it will lead to better platforms for innovation.

Customer insights that stretch the boundaries of the business can and should be shared across the organization. Presentations through various media like videos, newsletters, workshops, and training programs by senior management team members and new product team members not only create new thinking but instill in the culture a reminder that knowing customers, meeting their needs, and putting organizational time against innovation is top priority. Sharing these insights can generate the type of understanding and enthusiasm key to building an innovation-based culture.

Custom 7: Have New Product Teams Do Their Own Research

Many organizations spend hundreds of thousands of dollars on market research every year. This research ranges from quantitative customer satisfaction studies, to focus group concept testing, to segmentation studies, to sophisticated brand and customer behavioral research. Typically, a market research department gets a directive from a marketing, brand, or business group and then conducts these studies on their own or manages the outsourcing to research suppliers. With so much quality research done in so many organizations, why is there still a dearth of innovation and a sense across the organization that there is not a clear focus on where and how to be innovative?

Often, bridging the insights between studies is only a fraction of the issue. The bigger issue is that new product teams are not actively engaged in the research to be able to deem the "aha's" that come from reading between the lines and connecting particles of insight that market researchers do not have the liberty of doing. Senior management teams must require new product teams to be actively engaged in customer research. This requires doing in-depth qualitative research in the early stages of shaping a portfolio of new products and having the cross-functional new products portfolio team actively engaged in shaping and observing (and potentially doing) their own market research studies. The benefits of this approach are as follows:

- Research is more relevant to the specific goals of the new product team and can be tailored to the exact stage that the team is in.
- Depth of insight is significant as a result of (a) having the team of people who will use the research be involved, (b) having a cross-functional group listen and interpret the research in their own unique way, (c) having a cumulative effect of learning as research is conducted along the way to the summit.
- Impact on culture is tenfold, since team members can take the stories and insights and enthusiasm back to their respective departments.

General Mills utilized this approach and the other customs discussed, which led to the launch of many innovations including GoGurt and Milk 'n Cereal Bars, the $100 million business (Deck, 2001) that kicked-off the creation of an entirely new category called "breakfast bars." General Mills calls their portfolio teams "growth camps." It employs a cross-functional group who report directly to the top and specific goals are defined up front as to the types of portfolios they need to develop. These teams get actively involved in the research to uncover behaviors and needs of consumers in ways that challenge traditional category definitions. Senior management teams who can get new products people to actively do their own research will recognize how powerful the sharing of customer insights by a broad cross-functional group can be on shaping the culture.

CONCLUSION

While there is no easy or quick method that the senior management team can use to build an innovation-based culture overnight, the approach is clear:

1. Habits must be created that require broad and deep education of people, behavioral change, willpower, and discipline.
2. Credibility must be built, which means innovation market successes must be gained early and often to reinforce the rationale behind the habits the organization is seeking to create.
3. Trust must be gained in the senior management team to invest their time and money in innovation, to reward teams and individuals who fulfill short-term and long-term milestones, and to continue to support the innovation habits and customs even when times get tight.

As Griffin states in an article on new product best practices, "The best do not succeed by using just one NPD practice more extensively or better, but by using a number of them more effectively simultaneously." (Griffin, 1997). Changing habits takes practice, and changing mind-sets takes proven market results. Over time, when these seven customs are adopted, a culture grounded in growth, innovation, and trust in senior management is built.

BIBLIOGRAPHY

Cooper, Robert G1996. "New Products: What Separates the Winners from the Losers." In *PDMA Handbook of New Product Development* by Milton D. Rosenau et al., 12–18. New York: John Wiley & Sons.

Deck, Stewart L. 2001. "Got Soy?" *CIO Magazine.* August: 86.

Griffin, Abbie. 1996. "Obtaining Customer Needs for Product Development." In *PDMA Handbook of New Product Development* by Milton D. Rosenau et al., 153–160. New York: John Wiley & Sons.

Griffin, Abbie 1997. "PDMA Research on New Product Development Practices: Updating Trends and Benchmarking Best Practices." *Journal of Product Innovation Management* January: 131–154.

Kuczmarski & Associates. 1993. *Winning New Product and Service Practices for the 1990's.* Chicago: Kuczmarski & Associates.

Kuczmarski & Associates. 2003. *Study Examining Growth Priorities and Challenges.* Chicago: Kuczmarski & Associates.

Kuczmarski, Thomas D. 1992. *Managing New Products: The Power of Innovation,* 2nd ed. 5:163. Englewood Cliffs, NJ: Prentice-Hall.

Kuczmarski, Thomas D., Erika B. Seamon, Kathryn Spilotro, and Zachary Johnston. 2003. "The Breakthrough Mindset." *Marketing Management*, March/April: 38, 40.

Peters, Tom. 1994. *The Tom Peters Seminar: Crazy Times Call for Crazy Organizations.* New York: Vintage Books.

Webster's II Dictionary. 1984. Boston: Houghton Mifflin Company and New York: Berkley Publishing Group.

2 Bringing Radical and Other Major Innovations Successfully to Market: Bridging the Transition from R&D to Operations[1]

Gina O'Connor, Joanne Hyland, and Mark P. Rice

This chapter presents a process and tool for helping companies be more successful in transitioning radical and other major innovation projects from the R&D lab to an operating unit. The more innovative or radical an innovation, the greater the risk for companies because of the managerial complexities and higher levels of uncertainty associated with discovering new technologies and creating new markets. Seven specific radical innovation management challenges have been identified that apply to all types of innovation projects beyond incremental innovation improvements (Leifer et al., 2000). Managing the transition from R&D to operations has emerged as one of the most daunting of those seven challenges.

Companies expect that once a major innovation project is sufficiently mature, the receiving operating unit is able to employ tried-and-true project management techniques, such as the well-recognized Stage-Gate™[2] system, to bring the new products to market success. But because of the high levels of technical, market, resource, and organization uncertainty that still remain when projects are ready for transition, these well-known management practices simply do not work. Potentially game-changing innovations drop off the operating unit's radar screen, even after all of the initial investment in technical and market development has been made in R&D. A set of unique, specific activities is required to complete the resolution of these four uncertainties that neither R&D project teams nor operating unit managers are prepared to handle.

Transition managers are in a unique position to facilitate the transition management process, working simultaneously with R&D project teams, business unit interfaces, and senior management. The Transition Readiness Assessment Tool has been developed to guide this process by doing the following:

- Setting/managing expectations among the interfaces, including senior management and the project team
- Encouraging open discussions about the challenges of building and maintaining organizational legitimacy
- Establishing the strategic context for innovation projects
- Assessing the level of project transition readiness
- Identifying transition issues
- Building the transition management plan; and
- Facilitating a negotiated transition process.

This chapter briefly reviews the importance of radical innovation to business growth and renewal and the need for a systematic approach to bringing these innovations successfully to market. Second, the radical innovation management framework, including the transition management challenge, is presented. Next, the roles required for transition success, the transition management process and the Transition Readiness Assessment Tool are described. Finally, the organizational value of the transition management process for transition managers, R&D project teams, business unit interfaces, and senior management is provided based on the experiences of companies who have either followed the more comprehensive transition process or simply used the tool.

RADICAL INNOVATION, BUSINESS GROWTH, AND CORPORATE RENEWAL

Leaders of established companies have acknowledged that radical and other major innovations[3] are critical to their long-term growth and renewal. The Industrial Research Institute, a professional association of senior technology managers of large, established companies committed to R&D, conducts an annual survey of its members. In 2002 "business growth through innovation" was rated the top challenge facing technology leaders, and in 2001 the top challenge was "accelerating innovation" (IRI, 2001; 2002). Gary Hamel states "the most important business issue of our time is finding a way to build companies where innovation is both radical and systemic" (Hamel, 2002). Radical innovation transforms the relationship between customers and suppliers, restructures marketplace economics, displaces current products, and often creates entirely new product categories (Rice, Leifer, and O'Connor, 2002). Because of their ability to reshape markets and industries, companies that pursue more radical, game-changing paths have consistently outperformed competitors who are focused on incremental improvements.

Despite the importance of radical innovation for corporate growth and renewal, approaches to managing it are ad hoc and unsystematic. There is no organizational system in place to capture the learning, develop the competencies, and establish the leadership principles and managerial practices that are required for sustained success. In addition, companies have repeatedly tried to apply incremental innovation management approaches to highly uncertain ventures and have consistently failed. Companies invariably lack an infrastructure and surrounding management system to bring radical innovations as well as other types of major innovations successfully to market.

RADICAL INNOVATION MANAGEMENT FRAMEWORK

Radical innovation projects in established companies are bombarded by uncertainty, making it difficult to achieve commercial success. Earlier findings suggested that high technical and market uncertainty drove these projects (Lynn and Akgun, 1998). More recent learning suggests that there are two additional categories of uncertainty: resource and organization uncertainty. These four categories fuel seven challenges that are unique to radical innovation projects (Leifer et al., 2000) and have become the foundational elements for a new management framework, shown in Figure 2-1. This framework enables companies to move from an ad hoc approach to a system and embedded infrastructure, guided by company strategy, appropriate leadership styles, and supportive cultural characteristics. Very few companies have systems to overcome these seven management challenges and effectively commercialize radical innovations repeatedly.

This management framework applies not only to radical innovation projects but to all major innovation projects that extend beyond incremental innovation improvements—that is, platform, breakthrough, and/or radical innovations as described in Figure 2-2. This is because while technical and/or market uncertainties may be lower than for radical innovation projects, resource and organization uncertainties remain high, leading to managerial complexities that this framework is well suited to address.

TRANSITION MANAGEMENT CHALLENGE

While competencies need to be developed across all seven management challenges to build a mature innovation capability, development of a transition management competency is the focus of the rest of this chapter (see Challenge 6 in Figure 2-1). Transition management is defined as the facilitating process and enabling tools utilized for migrating projects from an R&D-based innovation project to business operating status. Securing the commitment of business units to adopt highly innovative projects from an R&D group and accelerating this transition is fundamental to project success as well as the organizational legitimacy of innovation efforts.

Transition management is a more daunting challenge than most people realize. The assumption is that an R&D project stays within R&D until tech-

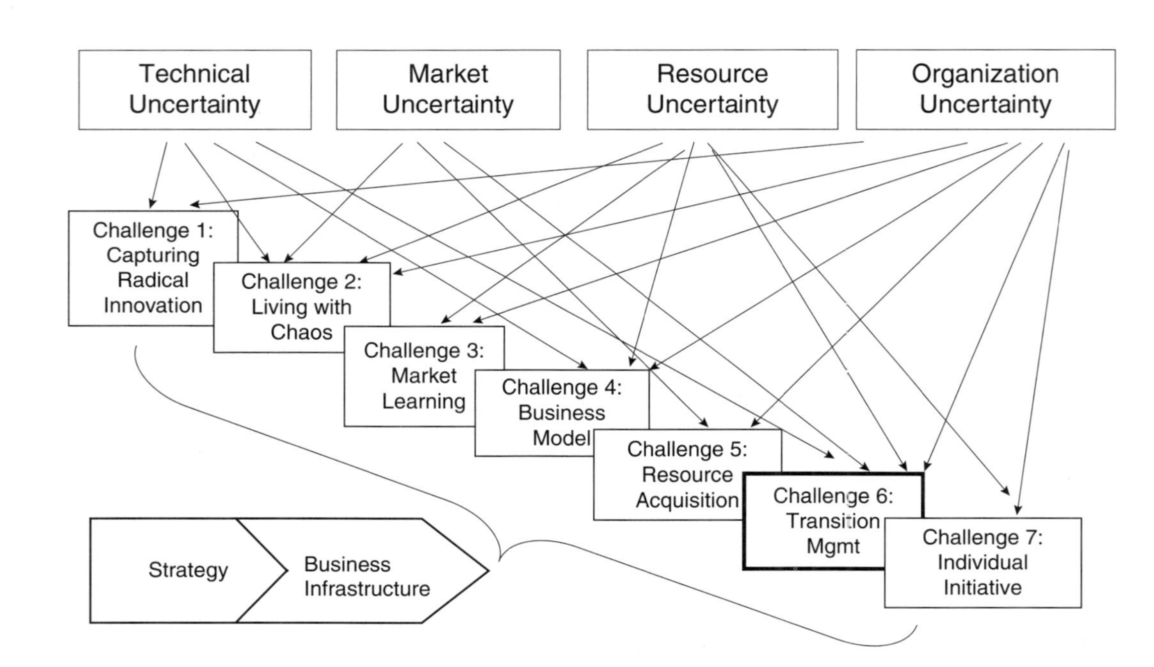

FIGURE 2-1. Radical innovation management framework.

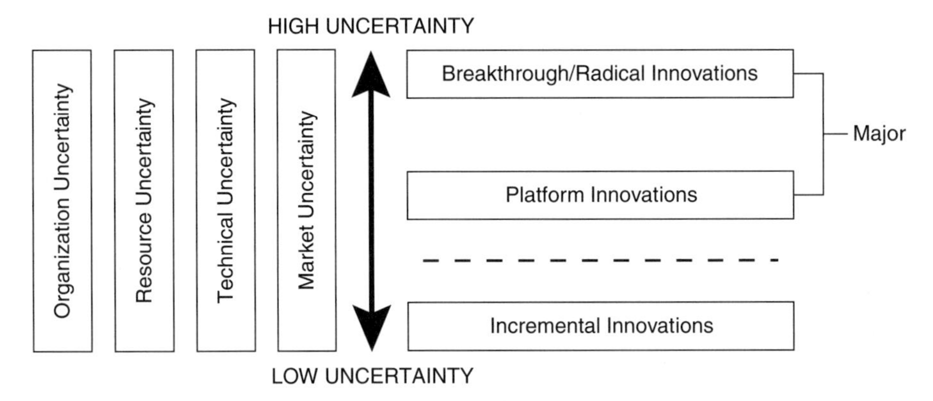

FIGURE 2-2. Uncertainty continuum by categories and innovation types.

nical and market uncertainty are low enough that the project can move into a Stage-Gate™ new product development process (Cooper, 1990) within an operating unit and be launched in a traditional manner.

Contrary to expectations, project teams in R&D and operating units, as well as managers responsible for managing innovation program initiatives, are not prepared to handle the transition activities that would reduce project uncertainties enough to increase the likelihood of business unit adoption. Markets are frequently underdeveloped, and sales forecasts cannot be determined with confidence. Operational efficiencies are not yet worked out. In addition, significant human resource issues arise, particularly if the new innovation threatens to cannibalize existing products and requires extensive new skill sets. These are not arenas that the R&D team perceives to be its responsibility. In addition, the high levels of market and operations uncertainty wreak havoc with the operating units' short-term profit pressures, so they, too, do not perceive these projects to be their responsibility. There is a middle ground of activity with radical and other major innovations that nobody "owns" in established companies. Much of this activity has to do with creating new markets, supply chains, and business models and reducing uncertainty to such a degree that a reliable sales forecast can be generated.

With the expectation that transition is the point to stop investing in development and quickly begin generating revenues from operations, it turns out that this is not only a critical but also an unexpected discontinuity in the project life cycle. Operating units are not prepared to accept these projects, as the level of risk is simply too great. While projects may have transitioned from R&D's perspective, they are generally underdeveloped for operating unit commercialization purposes and simply fall off the radar screen in the operating unit, whose primary concern is efficiency of operations. The full potential benefit of breakthrough innovations is often lost when they are integrated into current platforms or placed on a product roadmap for future consideration, never to come off the shelf because the investment required is too large. In the end, the innovation value is not realized, as these projects are "incrementalized" to fit within the operating unit paradigm.

NEW TRANSITION MANAGEMENT ROLES

It is clear through the repeated failure firms have encountered in moving radical/breakthrough innovation projects into commercial operations that several new roles should be identified to help highly innovative projects cross this bridge successfully. The first, and most critical, is the *transition manager*. Transition managers work with projects as they are maturing in the R&D organization. They report to the R&D vice president/director, who is responsible for overseeing the portfolio of innovation projects. They should have business development experience, be comfortable in an R&D environment, possess excellent interpersonal and negotiations skills, and have a strong technical competence. In working with the R&D vice president/director, transition managers should be able to leverage informal networks across the company to enroll senior management involved in growth initiatives, and/or those with an entrepreneurial bent, as transition champions. In working with senior management in the operating unit, transition managers also help identify who should take on the role of project leader in the operating unit. While some R&D project leaders transition with their project, most lack the commercialization skills required for project success. Transition managers become experts and play a critical role in organizational learning, as they build up important skills in transition management that can be disseminated across the organization over time.

In addition, new formal roles at the senior management level must be established. Each transitioning project needs a transition oversight board composed of those who have a vested interest in the project's success, as well as those that can open doors and remove organizational obstacles. The *transition oversight board* should be composed of an R&D vice president/director overseeing the innovation portfolio with cumulative experience in aiding projects through the transition process, a corporate leader responsible for strategy setting, a senior leader in the receiving business unit, a senior-level project champion supporting the project from its early beginning, and a transition manager bringing transition process expertise and project knowledge. These boards are temporary and dissolve as the project becomes increasingly accepted in the receiving unit.

The *transition team* is put in place to work with senior management and transition oversight board members to bridge the organizational gap between the R&D team and the receiving operating unit. The orientation of the former is to continue exploring, and, therefore, they have difficulty with driving to closure, whereas the latter, with its operational orientation, is looking for the transition to happen quickly and assumes that the project is ready for immediate market commercialization. The transition team should be composed of people with development and operational experience as well as transition expertise. Again, these teams are only formed for the project transition period.

In working through transition issues, strategic, portfolio, and project requirements need to be considered. With the introduction of these new roles, more strategic issues can then be addressed through the transition oversight board, the portfolio perspective is provided through the transition manager, and project transition issues are resolved with the transition team. Establishing these new transition management roles is critical to building an effective and efficient transition management process.

TRANSITION MANAGEMENT PROCESS

The transition management process provides transition managers, R&D project teams, business unit interfaces, and R&D vice presidents/directors with a systematic approach to identifying and responding to managerial challenges associated with the transition, thus improving the chances of success. Project teams are able to identify issues at the start of transition discussions and address emerging problems as they learn more about the expectations of all the stakeholders in working through the transition process. Business unit interfaces are brought into transition discussions early enough to build organizational support and their commitment to the project. With an ability to identify issues early on, track new learning over time, and secure business unit commitment, R&D vice presidents/directors are in a much improved position to manage a portfolio of high-risk innovation projects.

Table 2-1 provides an overview of the transition management process. There are seven steps to follow, a series of tasks to undertake, and four categories of uncertainty to manage for effective transition management. The uncertainties that apply to each task are identified across the four categories: Technical (T), Market (M), Resource (R), and Organization (O). The tasks are listed by step, as follows:

- *Step 1.* Identify transition senior management roles, transition champions, and oversight board members
- *Step 2.* Establish a transition oversight board
- *Step 3.* Provide transition funding, transition team members, and organizational commitment
- *Step 4.* Form a transition team
- *Step 5.* Assess a project's transition readiness utilizing the tool
- *Step 6.* Develop a detailed transition plan
- *Step 7.* Define the business model to lay the groundwork for a big market

Transition managers should facilitate this process.

TABLE 2-1.
Transition Management Process Steps, Tasks, and Categories of Uncertainty

Steps/Tasks	Uncertainties Addressed*			
	T	M	R	O
Step 1: Identify Senior Management Roles and Transition Champions				
◆ Educate senior management re. appropriate performance measures and roles.	X	X	X	X
◆ Find champions through informal networks and senior management decisions.			X	X
◆ Identify new senior level transition roles.			X	
Step 2: Establish a Transition Oversight Board				
◆ Determine right home for R&D project.				X
◆ Bridge organization structure gap between R&D team and receiving unit.				X
◆ Set appropriate expectations within the receiving unit.				X
Step 3: Provide Transition Funding, Team Members and Commitment				
◆ Provide transition funding.			X	
◆ Select transition team members.			X	X
◆ Secure organizational commitment.			X	X
Step 4: Form a Transition Team				
◆ Form the Transition Team.			X	X
Step 5: Assess a Project's Transition Readiness Utilizing the Tool				
◆ Learn when to use Transition Readiness Assessment Tool.	X	X	X	X
◆ Assess project transition readiness.	X	X	X	X
Step 6: Develop a Detailed Transition Plan				
◆ Review project strategic fit and organizational placement.		X		X
◆ Identify company value proposition, business potential, market entry/development strategy, and business model.	X	X		X
◆ Develop uncertainty reduction, resource, and operations plans.	X	X	X	X
◆ Establish critical milestones and timeline.	X	X	X	X
Step 7: Define the Business Model to Lay the Groundwork for a Big Market				
◆ Finalize the business model.		X		X
◆ Address evolution of applications and markets for market entry and development strategy.		X		
◆ Set technical specifications and resolve manufacturing issues as part of operations plan.	X			
◆ Ensure market development expectations match reality.		X	X	
◆ Understand how manufacturing challenges impact market entry objectives.		X		X

*T = Technical Uncertainty; M = Market Uncertainty; R = Resource Uncertainty; O = Organization Uncertainty

Step 1: Senior Management Roles as Transition Champions and Transition Oversight Board Members

Senior management involvement is critical to successful project transitions either as individual champions and/or members of transition oversight boards. Because of the strategic nature of innovation projects, the importance of building organizational alignment, and the need for project performance measures that are distinct from the receiving business unit's as a result of the relative immaturity and higher risk profile of these innovation projects, there are a number of activities that senior management must lead.. Finding a senior management transition champion or oversight board member with the appropriate skills, knowledge, and mind-set for guiding highly uncertain projects remains a significant organizational challenge, as many senior managers simply lack the necessary experience and are unable to see it. This is why transition managers need to be well connected within the company to leverage their own and their R&D leadership's informal networks and work with the senior leadership in R&D to establish more formal transition roles. The purpose is to find senior-level champions and board members that are motivated by the potential value of innovation projects to further a company's growth and renewal objectives, rather than those riding a political agenda because of the visibility these projects provide.

Transition managers play a pivotal role in educating senior management about the need for these different performance measures. Imposing operational metrics such as profit and loss (P&L) management and confirmed sales forecasts based on qualified leads on these immature projects will set them up for failure. Latitude is required because these projects typically are focused on new markets and new customers or, at least, at migrating current customers to very different usage situations. Extensive learning is still taking place. More appropriate metrics revolve around a milestone-based plan that incorporates measures for signing up new customers as application development partners, forming new manufacturing and distribution partnerships, filling critical team roles, converting a prototype/trial product into a commercial application, and building the operational infrastructure. While initial financial targets are set for revenue and spending, they gradually become more important over time as the project grows and matures and uncertainty is mitigated accordingly.

Step 2: Transition Oversight Board Formation

Transition oversight boards should be formed as transition discussions commence. The timing is case-specific; however, these discussions typically start three to six months prior to when a project is expected to be ready for transition from R&D to a business unit. The purpose of the transition oversight board is to guide the resolution of residual project uncertainties in order to increase the likelihood of operating unit adoption and project success. This board assumes a number of responsibilities for uncertainty reduction. Two important responsibilities are as follows.

DETERMINE THE RIGHT HOME FOR R&D PROJECT (ORGANIZATION UNCERTAINTY). Finding the appropriate home for a major innovation project is a strategic decision. There are four possible transitions paths: transfer to an existing operating unit, create a new business unit, form a joint venture, or place outside the company as a spin-off.

While transitioning a project to an existing business may appear to be the most straightforward option for developing a business internally, it is not always the case. If the project needs to be force-fitted into an operating unit's working paradigm, support will be minimal, the transition will not be smooth, and the project will likely not achieve its full potential. If the transition to an existing operating unit is the chosen course of action, then communication with that unit needs to take place early on to coordinate requirements and smooth the transition process. In most cases, the major innovation stretches the existing systems of any operating unit with new customers, new skill sets, and new business models. Further, the innovation can cannibalize current product platforms and threaten those in the receiving unit with the need to change, upgrade skills, or completely revamp their operating practices. When IBM's silicon germanium chip transitioned into its Microelectronics division, for example, the division set up an office in Boston for the primary purpose of gaining access to young MIT graduates who had been trained in chip design expertise that the division lacked. Corporate commitment is required to overcome operating unit resistance by providing resources (people and money) and adjusting performance requirements.

When a project moves beyond the existing strategic framework of the company but lies within the scope of its strategic intent and growth objectives, establishing a new business unit provides additional flexibility to stretch strategic boundaries, form nontraditional partnerships with leaders in new technologies and unfamiliar markets, and pursue new marketing and distribution channels. DuPont's bio-based materials initiative and Kodak's move to organic light-emitting diode (OLED) display technology are two examples of new organizational units that were formed to allow for growth in new domains.

Forming nontraditional partnerships also can lead to other organizational forms to drive growth objectives such as joint ventures. Corning Glass Works is a good example of using joint ventures for this purpose. Owens Corning Fiberglas was set up with Owens Glass (a flat glass producer) to commercialize Corning's fiberglass invention, and Dow Corning was set up to commercialize silicone products. There are specific transition issues unique to joint ventures that must be considered. While these ventures are effective in moving companies into new areas, there is a delicate balance of power that must be managed that comes with co-ownership. Also, the link to strategic intent, or migration away from it, depends on how the joint venture evolves. If it remains closely aligned with one of the parent company's growth objectives and serves to stretch its strategic boundaries, it may even become appropriate to establish a wholly owned subsidiary for complete strategic control by purchasing the interests of the other partner. Conversely, if the joint venture moves into an unaligned area, it would likely become a nonstrategic business investment and a divestiture candidate.

If a project moves beyond the strategic intent of the parent company, creating a spin-off business offers the parent the opportunity to cap its investment in a non-core area and mitigate its risk going forward. At the same time, it offers the potential to recoup the investment in the discovery and development phases of the project.

BRIDGE THE ORGANIZATION STRUCTURE GAP BETWEEN THE R&D TEAM AND THE RECEIVING UNIT BY SETTING APPROPRIATE EXPECTATIONS WITHIN THE RECEIVING UNIT (ORGANIZATION UNCERTAINTY). In transition, the activity is no longer a radical innovation development project, but it is not yet an up-and-running operating business (Rice, Leifer, and O'Connor, 2002). In most cases, radical innovations enter the market in niche application spaces and build momentum as the market becomes aware of the innovation's potential. These innovation projects are in the early stages of market entry ramp-up during the transition phase and, therefore, have not reached their full potential. Operating units have incentives and financial performance measures for increasing sales volumes and building market share as quickly as possible. There is little room for a major innovation project where the time to money is uncertain.

In addition, market entry strategies focus on maximizing these objectives. Dealing with niche markets with the promise of migrating to the "killer" application over time creates discomfort. The inconsistent expectations need to be discussed during the transition period, with plans put in place to provide the operating unit with growth-based incentives as well as short-term financial relief from the limitations of quarterly P&L management. Only senior management transition oversight boards can bridge this gap by being clear on the project's organizational value and addressing the expectations of stakeholders, who often have conflicting objectives.

Step 3: Senior Management Roles as Funding Providers, Transition Team Supporters and Securers of Organizational Commitment

FUNDING PROVIDERS (RESOURCE UNCERTAINTY). Funding requirements need to be addressed at the start of transition discussions. R&D project teams tap into R&D coffers, corporate innovation funds, external partners, and government sources in the early days to get their projects up and running. These sources are often insufficient or unavailable to cover the market entry ramp-up required during the transition period. New plants must be built in some cases, acquisitions of small companies are sometimes made, and entire new sales forces are hired to leverage the business opportunity appropriately. In many cases, and certainly when a proactive investment approach is not present, operating units cannot provide transition funding, as their resources are already stretched to meet existing commitments. This lack of funds slows down the project commercialization efforts and dramatically impacts the ability to effect a smooth transition. Funding can come from a combination of internal (e.g., CEO discretionary fund, corporate innovation pot, R&D, new business development organization, and business unit) and external (e.g., value chain partners and government for follow-on applications) sources. The transition oversight board also plays a critical role in securing the transition gap funding. While R&D and the operating unit are committed to the success of the project transition, each has biases and established operating modes that may compromise the effectiveness of this transition. This board can offer a more objective and strategic perspective.

TRANSITION TEAM SUPPORTERS AND SECURERS OF ORGANIZATIONAL COMMITMENT (RESOURCE AND ORGANIZATION UNCERTAINTIES). The selection of the right people to work through the transition period is critical. Only the transition oversight board has the vantage point to make the right determination as to who should be involved in the transition process. Difficulties with people and their expectations during project transition are typical (Block and MacMillan, 1993). As projects move through the transition, the skills required to take projects to operating status are very different from running a development project. They are also quite different from running an ongoing product platform, which many operating unit product managers do very well. Managing human resources during this period is particularly challenging. R&D team members leave, are reassigned, become less committed, or their interests are no longer attuned to the requirements of the project as it moves into the transition period.

Equally, there is significant uncertainty in handing off responsibility to a product manager who lacks the appropriate incentives, experience, and focus to build the market and necessary infrastructure required to ramp up the opportunity into a full-fledged business. With a product manager's short-term results orientation and multiple product lines to manage, higher-risk projects languish because they are a distraction to achieving more tangible results. Unless these issues are addressed, once again projects can end up back in R&D, where they will receive organizational attention but not necessarily the right business direction.

Step 4: Transition Team Formation

The transition team should be formed immediately after the transition oversight board is established based on the recommendations of the board members. This would be a part-time role for transition team members, requiring two to four days per month and lasting three to six months on average, depending on the complexity of the transition. The transition team should be composed of development, operational, and transition experts. Typically, this would involve at a minimum an R&D project team member, a business unit member, and a transition manager. Membership should include no less than three people to ensure there is a diversity of expertise and no more than five to ensure decisions are made in an effective and timely manner. This team facilitates the transition process by making sure both sides address the appropriate issues and reach a common understanding of the work that remains to be done to complete the transition.

Step 5: Project Transition Readiness Assessment

LEARN WHEN TO USE TRANSITION READINESS ASSESSMENT TOOL. The Transition Readiness Assessment Tool helps project teams systematically address the transition management challenges that they face. Fifteen dimensions of transition readiness need to be assessed during the transition phase. These dimensions are described in the Transition Readiness Assessment Form, included as Table 2-2. In addition, a mapping of these dimensions across the categories of uncertainty is provided in Table 2-3. This mapping provides a checklist of areas for project teams to consider when working through the sections of the tool.

Throughout the transition readiness assessment, the R&D team is referred to as the project team, and those taking responsibility for the project's market commercialization as the receiving unit team. "Receiving unit" is any division, department, or group to whom responsibility for the project's continued progress toward commercialization will be transferred. Typically, these include either a team within a current divisional operating unit, a new unit formed by the organization for the purposes of building this business, or a spin-off organization.

The objective of this tool is to help those involved in the transition management process consider the remaining work to be done and negotiate accountability with the receiving unit and senior management for completing these tasks. This requires understanding the characteristics of and expectations associated with transition readiness. The tool is designed to address the requirements for projects that operate in the regions of higher uncertainty associated with major innovations—that is, platform, breakthrough, and/or radical innovations. If applied to an incremental innovation, the tool will be perceived as overkill, since so much of what is contained is not applicable where markets are familiar, manufacturing and operations processes have already been used for similar platforms, and suppliers and appropriate partners are already established.

Assess Project Transition Readiness. An electronic version of the Transition Readiness Assessment Tool (in Excel) is available on the PDMA Web site (www.pdma.org). This tool is an enabler to the transition management pro-

TABLE 2-2.
Transition Readiness Assessment Form for Major Innovations: Preliminary Assessment—Transition Readiness Dimensions

Indicate whether or not each of the dimensions have been addressed by entering a "Y" or "N" and the extent to which each item is an important consideration for this project by entering a number 0 through 5, where:
0 = This dimension is not applicable to this project at all.
1 = This aspect of the project is applicable but is completely unimportant to its success.
2 = This aspect of the project is somewhat unimportant to its success.
3 = This aspect of the project is neither important nor unimportant to its success.
4 = This aspect of the project is somewhat important to its success.
5 = This aspect of the project is critical to its success.

Key Dimensions	Addressed Yes/No	Degree of Importance
I. Technology Readiness: The degree to which the science and technology underlying this phenomenon are well understood and incorporated into a product form in a reliable manner.	Y	5
II. Product/System Development Readiness: The degree to which the product family evolution has been planned and issues related to embedding the technical innovation into the product have been addressed.	Y	4
III. Manufacturing Readiness: The degree to which manufacturing processes have been developed and validated, and manufacturing facilities have been allocated for this project.	Y	0
IV. Software Readiness: The degree to which software and operations process not having to do with plant and heavy equipment have been developed and validated, and are ready to begin running.	Y	4
V. Partner Readiness: The degree to which the project team has clarified which aspects of the business it will outsource, and has solidified agreements with capable partners.	Y	5
VI. Clarity of Competitive Advantage: The degree to which the project offers real value in the market.	Y	5
VII. Market Entry Strategy: There is a plan for how to enter the market with respect to which application area, which value chain partners are needed and what their roles are, and how revenue will be generated across all the members of the value chain.	Y	5
VIII. Market Development Readiness: The degree to which plans and resources are in place to work with and educate the initial entry segment of the market and follow-on segments.	Y	3

TABLE 2-2. *(continued)*

Key Dimensions	Addressed Yes/No	Degree of Importance
IX. Sales Force Readiness: The degree to which the role of the sales force in this project has been clarified, planned, and communicated to them, and the sales force agrees to them.	N	3
X. R&D Readiness to Transition Project: The degree to which the R&D is ready to pass the project along because it believes it has completed its work.	Y	3
XI. Receiving Unit Commitment to Project: The degree to which the receiving unit understands the project and is committed to making it a commercial success.	N	4
XII. Human Resource Issues: R&D Project Team: The extent to which those people not continuing on with the project are well managed and recognized for their contributions thus far.	N	4
XIII. Human Resource Issues: Receiving Unit Team: The extent to which the staffing of the team for commercialization has been adequately considered and appropriate performance metrics put in place.	N	5
XIV. Informal Support Systems: The degree to which to project is championed within and beyond the receiving unit, and benefits from a strong supportive informal network.	N	3
XV. Alignment of Expectations: The extent to which the receiving unit's senior management and those individuals assigned to commercialize the project agree on goals, expectations, and appropriate measures of success for this project.	N	5

cess as part of assessing transition readiness and should be completed by a team that can, together, answer all of the questions. This requires a representative from the R&D project team or its project leader and a representative from the receiving unit team. The transition manager should provide coaching to facilitate this process, especially with the first few projects undertaken.

At a minimum, the transition team and the oversight board should be involved in a high-level review of the outcomes of the transition readiness assessment. Often, the transition team members and those who conduct this assessment will be the same. Ideally, one member of the oversight board should be involved in this detailed assessment. The transition manager should take the experiences of these project teams and convert the learning into a knowledge repository to provide a mechanism for transferring and maturing the transition management competency across the innovation project portfolio.

TABLE 2-3.
Mapping of Transition Readiness Dimensions with Categories of Uncertainty

Categories	Technical Uncertainty	Market Uncertainty	Resource Uncertainty	Organization Uncertainty
Focus:	Understanding technology drivers, value, and economic feasibility	Learning about market drivers, value creation, and business viability	Accessing money, people, and core competencies	Gaining and maintaining organizational legitimacy
Transition Readiness Dimensions:	I. Technology Readiness II. Product/System Development Readiness III. Manufacturing Readiness (for Hardware) IV. Software/Operations Readiness	VI. Clarity of Competitive Advantage VII. Market Entry Strategy VIII. Market Development Readiness IX. Sales Force Readiness	V. Partner Readiness XII. Human Resource Issues—R&D Team XIII. Human Resource Issues—Receiving Unit Team	X. R&D Readiness to Transition Project XI. Receiving Unit Commitment to the Project XIV. Informal Support Systems XV. Alignment of Expectations
Areas to Consider:	Completeness and Correctness of Underlying Scientific Knowledge Articulation of New Benefits that are Enabled Potential for Multiple Market Applications Potential Cost Saving Advantages Approaches to Solving Identified Technical Problems Manufacturing and Software Development Requirements	Clarity of Value Proposition Size of Business Potential Initial Market Entry Application and Follow-on Applications Initial Customer Partners Other Required Value Chain Agents Existence of Other Technical/ Potential Competitive Solutions Business Model Appropriateness	Availability of Internal and External Funding Project Requirements for Money, Team and Partnerships Project Lead Choice Team Competencies Aligned with Project Requirements Talent Attraction and Development Competency Acquisition In-House or External Partnerships	Strategic Context for Innovation Commitment of Senior Management Relationships with Internal Stakeholders Potential Organizational Resistors Influence with Corporate Strategy/Management Expectations of Senior Management and Transitioning Units Organizational Design

	Scalability at Acceptable Economics	Identification of Initial Sales Requirements	Partnership Identification, Formation and Management Strategies Requirements to Review Current Partnerships, Reassess Appropriateness Given Increased Maturity of the Project.	Project Home and Reporting Structure Nature of Project Guidance Process
Potential Flaws and Fatal Flaws:	Technology Proof of Concept Setback Prototype Limitations Cost Disadvantages Technology and/or Application Development Issues Development Process Major Issues	Market Attractiveness Turns Out to be False Market Test of Prototype Fails or Disappointing Inability to Secure Appropriate Customer Partner Lack of Robustness, Depth, Scope, and/or Number of New Capabilities Offered by Technology Such That Market Applications Are Limited or Constrained Inappropriate Corporate Expectations Regarding Realities of Timing of New Market Creation	Major Funding Loss because of Reversal of Overall Corporate Performance Project Team Limitations, and Inattention to Renewing Team Composition Given New Requirements Inability to Attract Required Talent Lack of Partnership Strategy Failure of Alliance Deal or Technical Partner Undefined Partnership Exit Conditions	Loss of Champion Change in Senior Management and/or Strategic Intent Change in Senior Champion/Sponsor Transfer of Responsibilities at Project Transition Lack of Strategic Marketing Communications Inappropriate Project Metrics Insufficient Runway to Demonstrate Business Results

There are four parts to the Transition Readiness Assessment Tool. Table 2-2 is the Preliminary Assessment. Column 1 of that assessment defines each dimension of transition readiness. Column 2 asks teams to indicate whether or not these dimensions have even been considered at this point in the project by entering a Y (yes) or N (no). Even though low levels of transition readiness will exist for those sections that have received no attention as of yet (i.e., N in column 2), they should be included in the assessment to commence discussions with respect to transition expectations and to determine the relative priority of requirements for development of the transition plan outlined in the next section. Finally, in column 3, teams are asked to use their best judgment to rate the degree of importance of each dimension to the project's level of success by entering a number 0 through 5, where 0 indicates the dimension is not important at all and 5 indicates a critical aspect of the project's success. This rating helps teams understand which sections of the main part of the tool they need to attend to. Those that are considered not applicable at all to the project are sections to skip. For example, project teams whose innovations are purely software-driven will find that Section III, Manufacturing Readiness, is not relevant, so they will elect to enter a 0 in column 3 for that section and then skip Section III in the main part of the tool.

Table 2-4 is the Transition Readiness Assessment, the main part of the tool. It is composed of 15 sections, each containing approximately five statements. Answers are to be filled in beside each statement based on a scale of 1, strongly disagree, to 5, strongly agree. Table 2-5 is the Transition Readiness Assessment Grand Score and Analysis. In the electronic version found on the PDMA's Web site, a profile is automatically generated, which indicates the areas where gaps exist in the project's readiness to transition. These gaps become the basis for building the work agenda by defining areas of strength and those that require additional attention. Those requiring additional attention are assigned priorities based on the level of transition readiness and the assigned degree of importance. The dimensions requiring attention are rank-ordered, moving from those with high degrees of importance to those with lower levels of importance. Within this importance ranking, the dimension with the lowest level of transition readiness is to be ranked first, the next lowest ranked second, and so forth.

Table 2-6 is the Transition Plan Requirements: Priority Areas. It lists the elements that need to be addressed in the transition plan based on strategic and operational priorities. Strategic priorities are those that require senior management attention and operational priorities fall within the purview of the project team. Transition managers and project teams are now able to see what the remaining work is that needs to be done, and they can begin negotiations about how to successfully complete the transition requirements.

A completed example is provided to offer more specific insight into the use of the tool. In this case, many of the technical and market dimensions have been addressed, whereas the resource and organization requirements have not, despite the high degree of importance assigned to many of them. This is an early indicator that the project team and receiving unit are not accustomed to having these discussions. In completing the assessment, it is quickly learned that major challenges exist in the receiving unit's transition preparedness (Sections XIII "HR Issues" at 33 percent and XI "Commitment" at 47 percent),

TABLE 2-4.
Transition Readiness Assessment

Answer each of the following questions by listing the appropriate response beside each statement from 1 to 5, where:
1= I strongly disagree with this statement.
2= I disagree somewhat with this statement.
3= I neither agree nor disagree with this statement.
4= I agree somewhat with this statement.
5= I strongly agree with this statement.

I	Technology Readiness	Response
1.	The technology embedded in this product is reliable.	5
2.	We understand how to work with the technology in this product to begin making modifications as required by the marketplace.	4
3.	The technical specifications for this product are now frozen.	4
4.	Lead user feedback has been incorporated into the design.	3
5.	We understand the science behind this innovation; we know why it works the way it does.	5
6.	Methods have been developed to assess the technology's reliability.	3
		24

II	Product/System Development Readiness	Response
1.	We are able to clearly articulate a product concept definition that uses this technology.	4
2.	System integration issues have been identified and a work plan for addressing those has been created.	1
3.	The process for product development to incorporate this technology has been clarified.	2
4.	Appropriate resources have been budgeted for accomplishing the product development plan.	1
5.	A plan has been established for evolving this technology into a platform of products.	2
		10

TABLE 2-4. *(continued)*

III	Manufacturing Readiness (for Hardware)	Response
1.	Manufacturing processes necessary for this product have been validated.	
2.	Most problems related to production scale-up have been resolved.	
3.	A pilot production demonstration has been completed.	
4.	Expenditures required for scaled-up production have been budgeted.	
5.	The facilities available for the manufacture of this product are adequate for its near-term requirements.	
		0

IV	Software Readiness	Response
1.	The required software has been developed and debugged.	5
2.	The software necessary for this product has been validated.	4
3.	Most problems related to scale-up have been resolved.	3
4.	The full-scale software production process has been adequately debugged.	2
5.	Expenditures required for scale-up have been budgeted.	1
6.	The resources required for the software production aspect of this product are adequate for its near-term requirements.	1
		16

V	Partner Readiness	Response
1.	Current partnership agreements have been reviewed for appropriateness for the operations phase.	1
2.	Partners needed for the commercialization of the first application of this technology have signed agreements to participate.	3
3.	Partners have demonstrated their capabilities to perform to the specifications.	2
4.	We have confidence in the partners with whom we have struck agreements.	3
5.	It is clear who has the responsibility for identifying and developing potential partners for this project.	1
		10

TABLE 2-4. *(continued)*

VI	Clarity of Competitive Advantage	Response
1.	Customer feedback on this product has been positive.	4
2.	Patents have been established to protect this technology.	5
3.	The value proposition of this product has been validated in the marketplace.	4
4.	There are no competing patents to this technology that have been filed by others.	4
5.	We have demonstrated that the market potential for this technology is big enough to support a business.	3
		20

VII	Market Entry Strategy	Response
1.	Potential customers for the primary application of this technology have expressed an interest in the application.	4
2.	A marketing plan for the primary application has been developed.	2
3.	Distribution channels for the primary application have been established.	2
4.	The business model that seems most logical from the market's perspective is one that the receiving unit is willing to use.	2
5.	A sales forecast has been generated that is based on an understanding of the initial entry market.	2
		12

VIII	Market Development Readiness	Response
1.	A plan is in place for exploring alternative applications for the technology.	5
2.	The first application market is aware that our product exists.	4
3.	The market has been adequately educated about the potential of this technology.	2
4.	The market has signaled that it understands what the potential benefits are that this technology might offer.	3
5.	There has been enough market development work done to safely transition project to the receiving unit.	4
		18

TABLE 2-4. *(continued)*

IX	Sales Force Readiness	Response
1.	Our sales force has exhibited enthusiasm for selling the primary application of this technology.	1
2.	We have ensured that the people hired or selected to sell this product have the appropriate technical knowledge required.	1
3.	Sales plans have been agreed upon.	1
4.	A list of potential customers for this product has been developed for our sales force.	3
5.	A sales campaign has been developed for the initial launch of this product.	1
		7

X	R&D Readiness to Transition Project	Response
1.	R&D believes it has done all it can to move this project forward.	5
2.	R&D believes this project is ready to be moved to an operating unit.	5
3.	The skills needed to take the project beyond this stage are currently not available on this R&D team.	5
		15

XI	Receiving Unit Commitment to the Project	Response
1.	There is no one in the receiving unit that we are aware of who will attempt to resist this project.	3
2.	Senior management in the receiving unit believes that this program is important for the future of the business unit.	3
3.	There is an appropriate amount of funding budgeted for the first year of this project in the receiving unit.	1
4.	Resources have been approved to cover operations until the break-even point is reached.	1
5.	This opportunity fits with the strategic intent of the receiving unit.	3
6.	Members of the team in the receiving unit perceive this as a career-making opportunity.	3
		14

TABLE 2-4. *(continued)*

XII	Human Resource Issues: R&D Project Team	Response
1.	There is general agreement on which project team members from R&D will continue on with the project to the operational stage.	1
2.	New opportunities have been discussed with R&D project team members who will no longer be involved in this project formally.	1
3.	R&D project team members who will not be continuing on with the project, but who have been involved so far, have been appropriately recognized for their contributions to the project's development.	1
4.	R&D has been recognized for moving this project so far along.	1
		4

XIII	Human Resource Issues: Receiving Unit Team	Response
1.	The receiving unit team that will be responsible for this project has been assigned.	1
2.	The receiving unit team understands the risks involved in commercializing innovations such as this one.	3
3.	Necessary skill sets that are absent from the receiving unit team have been identified.	2
4.	Efforts are being made to acquire new skill sets that are currently lacking in the receiving unit team.	2
5.	The reward structure that has been defined for receiving unit team members is satisfactory to them.	1
6.	The performance metrics against which the receiving unit team will be measured are appropriate given the level of risk associated with the project.	1
		10

TABLE 2-4. *(continued)*

XIV	Informal Support Systems	Response
1.	There is strong peer level support between the R&D project team and the receiving unit team.	4
2.	This project is supported by an informal network of people who want it to succeed.	4
3.	There is an identified champion in the receiving unit who wants to make this project happen.	4
4.	This project is viewed as fitting in to the long-run strategy of the corporation.	3
5.	This project is perceived as high-priority by the firm's senior-level management.	3
		18

XV Alignment of Expectations (Receiving Unit Team and Senior Management)

Indicate the degree to which the receiving unit team and the receiving unit's senior management are in agreement on the following issues, where:
1 = In Complete Disagreement
2 = Mostly in Disagreement
3 = Mostly in Agreement
4 = In Complete Agreement

		Response
1.	The potential of the opportunity for business growth.	3
2.	Completeness of the marketing plan.	2
3.	Expected pace of development of the overall market.	1
4.	Expected pace of development of the initial entry market.	2
5.	Appropriate applications to pursue.	3
6.	Ranking of applications in the order to be pursued.	2
7.	Level of demand/sales forecast for the first year.	1
8.	Projected revenue to be expected from the technology.	2
9.	Extent of revenue sharing with suppliers/distributors/vendors of the technology.	2
10.	Acceptability of the final business model.	2
11.	Appropriate milestones for commercialization.	1
12.	Appropriate measures of success.	1
		22

TABLE 2-5.
Transition Readiness Grand Score and Analysis

Following is the scoring profile for each of the subsections of the tool. Sections listed as Not Applicable in the Preliminary Assessment have automatically been entered and given a score of 0. The % Readiness indicates the degree to which the project is mature enough to safely transition into operations. The Degree of Importance has been automatically inserted from the Preliminary Assessment. Areas where the project is in a strong position (Strength) and that need to be addressed (Work) are identified and assigned priorities based on level of transition readiness and degree of importance. These set out the work agenda for the transition period of this project's life as per the action plan that follows.

Transition Readiness Assessment	Maximum Score	Response Score	% Possible Readiness	Degree Of Importance	Strength/ Needs Work
I. Technology	30	24	0.80	1	Strength
II. Product/System Development	25	10	0.40	0.8	Work—M (2)
III. Manufacturing (for Hardware)				0	N/A
IV. Software	30	16	0.53	0.8	Work—M (3)
V. Partner	25	10	0.40	1	Work—H (2)
VI. Clarity of Competitive Advantage	25	20	0.80	1	Strength
VII. Market Entry Strategy	25	12	0.48	1	Work—H (5)
VIII. Market Development	25	18	0.72	0.6	Strength
IX. Sales Force	25	7	0.28	0.6	Work—L
X. R&D Readiness to Transition Project	15	15	1.00	0.6	Strength
XI. Receiving Unit Commitment	30	14	0.47	1	Work—H (4)
XII. HR Issues: R&D Project Team	20	4	0.20	0.8	Work—M (1)
XIII. HR Issues: Receiving Unit Team	30	10	0.33	1	Work—H (1)
XIV. Informal Support Systems	25	18	0.72	0.6	Strength
XV. Alignment of Expectations	48	22	0.46	1	Work—H (3)
Transition Readiness Level Grand Total	378	200	0.53		

TABLE 2-6.
Transition Plan Requirements: Priority Areas

Based on the preceding analysis of areas of strength and those requiring work, as well as a mapping with the categories of uncertainty in Table 2.3, the critical elements to be addressed as part of the project's transition plan are as follows:
Strategic Priorities (system level issues requiring senior management attention)
1. HR Issues: Receiving Unit Team (Resource)
2. Alignment of Expectations (Organization)
3. Receiving Unit Commitment (Organization)
4. Market Entry Strategy (Market)
5. HR Issues: R&D Project Team (Resource)
Operational Priorities (project level issues requiring project team attention)
1. Partner (Resource)
2. Product/System Development (Technical)
3. Software (Technical)
4. Sales Force (Market)

alignment of expectations between the project team and the receiving unit (Section XV at 46 percent), and clarity of the market entry strategy (Section VII at 48 percent). In addition, it is not clear if the right partners are in place to take the project from a development to operating status (Section V "Partner" at 40 percent), some technical requirements are not as far along as originally thought (Sections II "Product/System Development" at 40 percent and IV "Software" at 53 percent), and the project team is concerned about whether HR matters will be handled appropriately (Section XII at 20 percent). While sales force requirements do need to be addressed, they are typically less of a priority at this stage in the project's life cycle. Before leaving this example, be sure not to lose sight of the project's strengths. These still need to be addressed in the transition plan but initially as a lower priority.

It is not required that each of the dimensions reach 100 percent transition readiness for the project to transition. This is a negotiated process and will depend entirely on the requirements of the receiving unit.

Steps 6 and 7: Transition Plan and Business Model

Following the methodology described previously provides the basis for building a detailed transition plan. Table 2-3 also provides a checklist of areas for teams to consider in transition plan development by categories of uncertainty. Most of the information for the plan should be available in the project team's knowledge base and from the readiness assessment discussions. This plan should define the tasks, a timetable, and roles and responsibilities of team members. In addition, the plan should indicate gaps in the current team's competency set, so that the team can be appropriately populated to effect a speedy transition. The goal is to move from a learning orientation to an operational business plan.

The transition plan should guide the efforts of the team and provide a yardstick for measuring progress. Transition managers need to be aware of their own limitations and those of others as they coach project teams and facilitate the transition process, especially in the early days of assuming this new role. The transition management requirements for major innovation projects differ markedly from incremental innovation projects, with the former requiring situational learning and the latter relying upon previous experience. The transition plan needs to be flexible regarding the timeline and fluid with respect to the resources required. Residual uncertainties will emerge during the transition, and projects will need to be redirected as a result. There also needs to be a mechanism to stop the project if progress is limited or unacceptably slow or if fatal flaws emerge.

Transition plans will be specific to the maturity level of the project and the expectations of the parties involved. Some business units will have the flexibility to accept projects with significant residual uncertainties, whereas others may not have the staffing and financial latitude to adopt these projects until these uncertainties are dramatically reduced. Therefore, it is important for project teams to focus on identifying technical, market, resource, and organization uncertainties that need to be addressed in the transition plan based on gaps uncovered in the review of the Transition Readiness Dimensions in Tables 2-2, 2-4, 2-5 and 2-6 and Areas to Consider in Table 2-3. The transition plan should include the following sections:

- Organization Plan, including Project Strategic Fit and Organizational Placement
- Market Plan, including Value Proposition for Company and Operating Unit, Business Potential, Market Entry/Development Strategy, and Business Model
- Resource Plan, including Team, Partners, and Funding Requirements
- Operations Plan, including Technology and Manufacturing Requirements
- Critical Milestones and Timeline

Five specific areas are examined in greater detail, as they are likely to become showstoppers or fatal flaws if not resolved during the transition period.

FINALIZE THE BUSINESS MODEL (MARKET AND ORGANIZATION UNCERTAINTIES). In the early phases of a major innovation project, some companies find that they need to integrate more components of the value chain into the plans than originally defined in the business model, to build the market and accelerate the adoption of the innovation. Over time, once several applications become more firmly entrenched in the market, the integrated solution can be unbundled (O'Connor and Rice, 2001).

Texas Instruments developed the Digital Micromirror Device (DMD), a chip composed of thousands of tiny mirrors, each individually directed, that can be used to radically improve resolution of display technologies. The most obvious initial application was in projection devices. Yet TI could not convince projector assemblers of the potential demand and so could not entice them to develop new projector designs that would leverage the technological capabilities of the DMD. TI therefore decided to provide the entire display engine (the DMD chip, the lens, the housing, and the power source) to accelerate market development for its innovation. It was not interested in permanent forward integration. Once several applications were well established and given that its market creation role was no longer required, TI unbundled its innovation to sell its core chip technology only to its direct customers, who, by now, saw the opportunity and began designing new projection equipment with the chip's capabilities in mind. This approach enabled TI to focus on its core technology and allow others, with superior capabilities in other parts of the value chain, to assume those positions. This is clearly a shift away from how TI had to initially approach the business model to create market demand. Project transition discussions must focus on these types of considerations to finalize the appropriate business model for moving forward, as the shift is made from innovation project to operating status.

ADDRESS EVOLUTION OF APPLICATIONS AND MARKETS FOR MARKET ENTRY AND DEVELOPMENT STRATEGY (MARKET UNCERTAINTY). While the vision of a killer application may drive project development within R&D, the realities of market creation are much more challenging. The process is one in which the firm not only learns about the market but helps the market learn about and understand the technology and its possibilities (Rice and O'Connor, 2002). Companies are on a continual quest to dominate markets through discovery of the killer application. But these kinds of applications will most likely not emerge in the early commercialization period of radical innovations (Jolly, 1997). Teams need to understand the expected level of project maturity with respect to market uncertainties in order to steer a project to the receiving operating unit at the appropriate time. Iterative learning continues through transition and even after the operating unit is up and running with new products. Most companies end up pursuing a niche entry application strategy initially and encounter time delays and a number of false starts on their way to discovering their mass market application.

Companies move from one application to another based on market learning, a concept referred to as *application migration*, migrating toward the most promising market prospects that were not initially obvious. Analog Devices' Accelerometer technology moved from an air bag sensor application for the automotive market to games on PCs, and then on to "box games" such as Nintendo and Sony PlayStation, where volumes are even bigger than in automotive. It also has received inquiries for these sensors from a variety of sporting goods (golf clubs, tennis rackets) and instrumentation products where vibration changes need to be noted.

SET TECHNICAL SPECIFICATIONS AND RESOLVE MANUFACTURING ISSUES AS PART OF OPERATIONS PLAN (TECHNICAL UNCERTAINTY). Because of the unpredictable nature of the major innovation project life cycle, technical requirements and manufacturing issues are in a constant state of metamorphosis. Continuous market learning forces project teams to redirect technical development. Early adopters drive the technical specifications and are comfortable living with a prototype in a constant state of evolution. The commitment of these prospective customers provides the impetus to commence discussions for transitioning these projects to operations because of the perceived commercial value. It quickly becomes apparent that what is acceptable to early adopters, who are experimenting and learning along with the R&D project team, is not acceptable to customers who believe they are buying a commercial product. As a result, those involved in the transition process learn that technical development is not fully completed, putting into question the readiness of the project to transition.

The complexities of moving from a "klugey" prototype designed for market learning to a product capable of scalability that meets manufacturing requirements are often unforeseen by R&D project teams. Motorola's early prototype for its paging system is an example of this type of prototype (Morone, 1993). It was of "limited practical use" and designed for a private application only. It was a low-frequency system that required extensive wiring. These wires frequently were inadvertently cut or moved. There was a "good deal of interference" and the paging receivers "were so fragile that they usually broke when dropped." Despite these limitations, in working with and learning about the market through a series of trials, Motorola eventually designed a successful commercial product that was smaller and lighter, more reliable and durable, of greater range, and with reduced infrastructure and operating costs.

By recognizing these issues up front, transition process delays can be minimized, since potential showstoppers are identified early and plans are put in place to address the technical uncertainties within the context of the transition readiness expectations of the receiving operating unit. In one company, the R&D project team relied on a manufacturing partner to develop the process innovations required to scale the operations of the business. The partner struggled for such a long time that manufacturing remained in R&D for several years to enable the process to be worked out, while sales and marketing were handled through the operating unit—a less than desirable project management scenario because of split responsibilities and inconsistent performance expectations. Involvement of the operating unit up front would have provided greater manufacturing readiness clarity and significantly reduced an extremely protracted transition timeline.

ENSURE MARKET DEVELOPMENT EXPECTATIONS MATCH REALITY (MARKET AND RESOURCE UNCERTAINTIES). Market learning for radical innovation projects is part of a cyclical "probe and learn" process (Lynn, Morone, and Paulson, 1996; O'Connor and Rice, 2001). Because this learning is more complex and time-consuming than would be typical for incremental innovation projects, the time and financial investments for market development usually are seriously underestimated. The need for application development, customer education, and user training prolongs the sales cycle and thereby increases marketing expenditures. Further, through engaging in market learning, there are many stops and starts that extend the time to discovering a potentially viable business model. This learning makes it very difficult to forecast sales for a breakthrough innovation with any degree of accuracy and can result in unrealistic sales expectations from the operating units. The pressure to conform can be extremely high. In one case, a project manager was forced to provide sales forecasts that fit the operating unit's required contribution, and was later sidelined when the numbers were not met. After investing five years in the development of the innovation, he ultimately left the company.

Often the combination of time delays and the high cost to develop new markets lead to protracted transition periods as R&D project teams work to reduce market uncertainties. Awareness of these market development realities enables R&D vice presidents/directors to direct resources to more quickly reduce market uncertainties and overcome the inertia that develops from inconsistent expectations. Even with these efficiencies, market development for major innovation projects is slower and performance metrics are different than for innovations with less uncertainty. Transition managers assume a critical role in convincing senior management of this slower process and that metrics such as new customer leads and beta test sites are appropriate interim measures on the road to actual sales and revenue.

UNDERSTAND HOW MANUFACTURING CHALLENGES IMPACT MARKET ENTRY OBJECTIVES (MARKET AND ORGANIZATION UNCERTAINTIES). Companies must make strategic market entry trade-off decisions. Pursuing high-margin, niche applications first may be the best approach to reduce manufacturing yield problems. Conversely, pursuing large mass market applications from the beginning may be necessary if creating a dominant standard is the market entry objective. Either way, there is a price to pay in short-term financial performance through reduced volume or market creation costs. In the latter case, gaining market recognition and growing sales, while establishing new standards for the industry, allows the market to become more fully developed as it learns about the technology. Transition managers and project teams need to be clear on the market entry strategy and communicate concisely the reasons for this approach to the receiving operating unit to gain organizational alignment. From the perspective of operations, limiting manufacturing yields and taking short-term financial hits are contrary to standard performance measures. If the market entry strategy is not well understood and carefully positioned to build alignment across the company, the ability to effectively transition the R&D project will become problematic, since it will not meet typical operating unit business requirements.

The transition management process involves a series of steps that senior management, transition managers, and project teams can use to guide more effective and efficient project transitions. While the Transition Readiness Assessment Tool is a key enabler, transition success depends upon the commitment of people to define transition expectations at strategic and project levels, establish new transition roles, build organizational legitimacy for innovation program activities, determine transition readiness requirements, develop a transition plan, and negotiate the parameters for a successful project transition.

ORGANIZATIONAL VALUE OF A MORE FORMAL APPROACH TO TRANSITION MANAGEMENT

IRI Member Companies' Experiences with the Transition Readiness Assessment Tool

To date, the tool has been used by more than half a dozen IRI member companies working through transition management challenges. Based on these experiences, we have managed to capture feedback on the benefits of using the tool, along with a few notes of caution. The specific insights are as follows (O'Connor, Hendricks and Rice, 2001):

- The tool is very useful for stimulating discussion between project team and receiving unit counterparts. It is substantially better than other mechanisms currently in place. "We should probably do this at least once per year to see how we have improved. Providing perspective on launching efforts very helpful." "We found this a very valuable exercise. It generates a metric, which, in itself, is quite useful."
- It can result in a defined work agenda. "Tool is especially helpful in resolving resourcing issues." "Tool is useful to show where effort needs to be focused in the future." "Tool showed our weakness in the sales and marketing area for these new to the world projects."
- It is not an easy tool to use. "Too many parts to be easy." It requires training or a process person to oversee the process until personnel are used to it. "I had to sit down with the team and walk them through it . . . couldn't just hand them a package."
- It is not worthwhile for projects that fit neatly into a business unit, for which there is a clear market and a clear technical development path. "Should only be used for breakthroughs with high uncertainty. Otherwise, overkill."

Companies' experiences indicate that projects using this tool are performing better than projects where the tool had not been utilized. The benefits certainly outweighed some of the short-term implementation issues. It is simply important to understand that the tool is designed to address higher uncertainty, major innovation projects and requires that the necessary coaching from a transition manager to help teams learn how to use the tool in an effective and efficient manner.

The most important benefit is the tool's ability to smooth the transition from R&D to operating status. Transition managers and project teams are becoming more aware of the work left to be done to effect the transition. As a result, projects are much less likely to fall off the radar screen in operating units. They move more quickly to market, are more able to attract senior management attention, and are better positioned to receive the necessary prioritization and buy-in within the receiving unit.

The tool also is helping companies identify unexpected issues and alleviate internal political problems because the issues are out on the table. It is easier to address potential staffing and capital limitations by having an early indicator of a project's expected resource requirements. This is of particular benefit to R&D vice presidents/directors and senior business unit interfaces with complex portfolio management requirements. Greater efficiency is being seen in the transition process because the organized can better time the transition, speed up the negotiation process, and thereby cut the expense of the transition. Finally, while most of these projects are still moving through the commercialization phase because of the longer life cycle of a major innovation project, initial results indicate that marketplace success of products will improve through use of this tool.

In undertaking their transition readiness assessments, companies are identifying common areas of weakness. Some or all of these weaknesses may be identified for a particular project, within a particular company, at a particular time:

- Inconsistent expectations with respect to technology and manufacturing readiness exist.
- Market uncertainties remain high in the areas of partnerships, the market entry strategy, the market preparedness, and sales force readiness.
- Human resource issues on both sides of the transition equation are not being sufficiently addressed. Weaknesses are evident in identifying project team skill set requirements, designing an agreed-upon reward structure for the new project team within the receiving unit, and recognizing the contributions of those not continuing on with the project.
- Senior management expects that these projects can be easily integrated within the existing product portfolio and therefore establishes operating unit performance metrics that are inappropriate given the high levels of uncertainty that remain in the project.

While all of these areas of weakness must be overcome to achieve transition success, the people part of the equation is often the most problematic. If building the trust and commitment of individuals working in the higher risk area of innovation is not well managed, this jeopardizes not only project but also innovation program success.

Eastman Kodak's Experience with the Transition Management Process

Over the course of the last few years, Eastman Kodak (EK) has been struggling to find its place in the digital revolution. While its R&D labs are continuing to invent new technologies, transforming these technologies into commercial value is not occurring frequently or quickly enough. In 2001, EK's System Concepts Center (SCC) identified the transition of these new business opportunities from its R&D labs to the business units as an important organizational challenge that it needed to overcome to drive the course of Kodak's long-term future.

The SCC first addressed transition requirements from an organizational capability perspective. The objective was to identify what needed to be established at strategic and portfolio levels to support a transition management capability. It then looked at project requirements based on transition readiness dimensions and use of the Transition Readiness Assessment Tool. The objective was to be able to effectively coach project teams across the 15 dimensions of transition readiness to improve their ability to transition projects more successfully.

Kodak's innovation system is complex, with multiple transition points and paths to commercialization. Therefore, the tool needed to be customized to meet EK's two most likely paths to commercialization via its recently established business accelerator for breakthrough innovation projects or its business units for fast-track projects. Significant time was spent securing senior-level business unit commitment and building SCC team alignment as to the need for a more systematic transition management capability and educating them on the value of the transition management process. The steps undertaken to build this commitment and alignment were as follows:

- A series of transition management workshops were held to provide education and build the case for why major innovations require more directed transition discussions. Topics included organizational placement options, team competency gaps, funding issues, the need for different organizational metrics, and so on.
- SCC team members became directly involved in the development of the transition management capability through a series of pilots of carefully selected projects and project leaders.

- Senior managers in R&D, corporate, and the business units were asked to become champions for the innovation program and specific projects to build organizational awareness and legitimacy.
- Senior SCC members participated in business unit strategy development meetings to better understand and influence the future direction of new business development activities.
- A strategic marketing communications program was developed to build the awareness of SCC projects and how they fit within the strategic priorities of the company.

In March 2003, EK's SCC was asked to provide feedback, for the purposes of this chapter, on its result to date in building a transition management capability to improve its transition success and the effectiveness of the Transition Readiness Assessment Tool. Over the course of the last year, several major projects, in various states of development, were involved in the following:

- Identifying senior management champions and forming guidance (oversight) boards that addressed organizational placement and funding requirements
- Establishing cross functional/cross unit transition teams to conduct transition readiness assessments to better understand and communicate the maturity of the business proposal (percent readiness to transition) and identify transition issues
- Developing formal transition plans to provide clarity and commitment as to who was doing what and address market entry and business model uncertainties

The preceding activities are resulting in successful transfer of projects at unprecedented rates and with greater revenue contribution to the business units. Projects are being so well received that they are meeting the requirements for the initial stage gate for commercialization (gate 0) on their first pass. These business opportunities include breakthrough innovation as well as fast-track incremental innovation projects. While the tool was designed for breakthrough innovation projects, Kodak and other companies are finding that a streamlined version of this tool can be applied to incremental innovation projects, especially for addressing resource and organization uncertainties. For SCC project teams, greater clarity has been achieved in the following areas:

- Understanding the business opportunity and describing its potential to key stakeholders
- Determining the maturity of the business opportunity
- Defining the work agenda for transitioning the business opportunity successfully
- Focusing attention on and more thorough assessment of the "tall tent poles"—the most important issues—first, such as team composition, organizational home, transition funding, and market entry strategy

- Assigning who is responsible for specific actions and what organization is to provide funding and resources
- Identifying what and how to communicate to gain buy-in from senior management in both the sending and receiving units

The preceding benefits are helping all stakeholders involved in the transition process (i.e., project teams, transition managers, and senior management) to be in a better position to assess the priority and assignment of resources in both the SCC sending and receiving business units. There is greater alignment of expectations, improved accountability, an increased focus on the key transition issues, and more effective resourcing of critical actions. As a result, receiving unit team members and senior management expectations are far more aligned regarding market development realities, scarce resources are being more effectively utilized, and issues are far less likely to slip through the cracks.

As stated succinctly by Greg Foust, the SCC's Innovation and Commercialization Manager, "We have clear evidence that implementing the tools and techniques associated with radical innovation does indeed increase the success rate of transitioning major business opportunities to landing zones (receiving units)." Eastman Kodak has certainly achieved, and could well be on the way to exceeding, its original objective of improving project transition success to drive EK's long-term direction.

SUMMARY OF LEARNING

Companies are establishing transition management capabilities by following a more formalized transition management process, adopting the Transition Readiness Assessment Tool, and setting up transition teams composed of the appropriate types of people. They continue to reinforce that the process and tool forces a keen awareness of how much work remains to be accomplished on many fronts before the fruits of investment in major innovations can be realized. A critical aspect of this is simply setting the appropriate expectations with the senior management of the receiving units, based on what we now know to be reality through the experiences of companies following the transition management process. Fortunately, the transition readiness dimensions force the right discussions, thereby enhancing the probability that projects in which the company has invested so heavily through R&D will not be scuttled as they move to commercial operating units.

All of the questions that the tool has been designed to address reveal that managing the transition from an innovation project to an operating business is neither simple nor easy. Resolving the remaining uncertainties during the transition, such as how to fill team competency gaps, exploit underdeveloped markets, and address inconsistent sales expectations, takes longer and requires greater investment than anticipated. However, by acknowledging that effective project transitions involve a specific set of activities and require special skills and resources, companies can accelerate the transition process, reduce the risk of failure, and improve their transition management competency.

In the end, the Transition Readiness Assessment Tool, in itself, cannot remedy innovation system challenges. It does, however, provide a conduit for addressing transition management issues as a critical enabler to a more comprehensive and systematic transition management process. In this approach, issues become more explicit and far easier to resolve. The tool provides an objective mechanism for identifying transition requirements and setting in place a plan that improves the chances of transition success. Given that most companies are struggling with transition management issues, it seems to be an area where companies can make major strides and contribute more significantly to business success.

CONCLUSIONS

Because of the complex nature of the innovation system within companies, developing a transition management capability extends well beyond the project level, where assessing transition readiness and building detailed transition plans is the primary area of focus. Transition challenges manifest themselves across strategic and portfolio dimensions, in addition to project-specific dimensions. Because of transition managerial complexities and organizational interdependencies, transition managers cannot coach project teams on how to follow the process and use the tool without first reviewing the innovation system to ensure appropriate transition management capabilities are in place for transition effectiveness.

There are a number of leadership practices that R&D vice presidents/directors and transition managers can adopt to address the broader portfolio and strategic considerations. The first is to ensure senior management transition roles are well defined on both sides of the transition equation—in other words, to push the project out of R&D and pull it into the operating unit based on effectively communicating the operating unit value. The transition process needs to be identified as a high organizational priority, backed at the top or near the top of the executive team, for these high-uncertainty innovation project transitions to be successful.

Second, there should be a separate transition oversight board that can objectively judge the performance of those involved in the transition process. This will also help to build organizational alignment and legitimacy by enrolling new stakeholders in the process and providing them with a strategic view as to the potential of the project. Performance requirements should be measured by standards that are different from those of the operating unit and different from those of the R&D project team.

Similarly, the transition budget should not be provided solely by either R&D or by the operating unit. Each of these two constituencies has a stake in the success of the transition, but each has biases and established operating modes that may compromise the effectiveness of the transition. Therefore, a more objective and strategic perspective can be offered by setting aside transition funding from other sources that are separate from the allocations to the business units, such as CEO discretionary pots, venture funds, or even external partners. Senior management's role is to determine where the transition funding is going to come from and be committed to set aside what is required for transition success. Because of a business unit's limited resources, the unwillingness on the part of the receiving business unit to commit sufficient resources needed to realize the innovation's full potential is a major threat to successful transition. There is a delicate balance between pursuing near-term profitability of the initial applications and exploring new application areas. Once again, the transition manager plays a significant role in resolving these funding issues with senior management to ensure transition success.

The fourth is the formation of a leadership-endorsed transition team populated with personnel from the R&D project team to provide the accumulated learning, the receiving operating unit to build a knowledge base to carry the project forward, and a transition manager who has expertise in facilitating transitions. If the project champion has been effective, he or she should play a key role, potentially as leader if he or she has the right skills or as advisor if not. Regardless, senior management needs to manage the leadership selection process carefully. If senior management has done this well from the very beginning, choosing carefully the project leader when the project is launched and helping him or her to grow with the project as it matures, then the company should have no problem retaining this critical individual *and* the transition team will get the leader it needs.

Another important, but difficult, role the transition manager plays with R&D project teams is to set realistic expectations among the various stakeholders about the likely evolution of the market. As the business model unfolds, there will continue to be dead ends and unexpected opportunities, as well as the applications that work out as expected. Flexibility in the ramp-up of the new business is critical. Without it, there is a risk the firm will shelve the project rather than continue to invest in the market development activity required to reap the benefits that the innovation offers. Requiring new businesses that develop from major innovations to meet high hurdle rates too soon may kill them before they have time to develop and mature. The transition manager must work with senior management and the transition team to ensure these messages are communicated and understood.

In the end, R&D leaders and transition managers must learn how to behave differently themselves so that new leadership practices for transition management are embraced and broader strategic and portfolio transition issues are considered.

NOTES

1. Special thanks to Greg Foust, System Concepts Center, Eastman Kodak.
2. Registered trademark of NPI.

3. We define radical innovations as those that have the potential for offering (a) new-to-the-world performance features, (b) significant improvement (five to ten times) in known performance features, or (c)significant reduction (30 to–50 percent) in cost of a given performance feature. Major innovations exhibit these characteristics to varying degrees along the continuum of innovation uncertainty.

REFERENCES

Block, Z., and I. C. MacMillan. 1993. *Corporate Venturing: Creating New Businesses within the Firm.* Boston: Harvard Business School Press.

Cooper, R. 1990. "The New Product Process: A Decision Guide for Management." *IEEE Engineering Management Review* June: 19–33.

Hamel, G. 2002. "Innovation Now! (It's the Only Way to Win Today)." *Fast Company* FC65 December: 114–124.

Industrial Research Institute. 2002. 2001/2002 Annual Reports. Washington, D.C.: Industrial Research Institute.

Jolly, V. 1997. *Commercializing Technologies: Getting from Mind to Market.* Boston: Harvard Business School Press.

Leifer, R., C. McDermott, G. O'Connor, L. Peters, M. Rice, and R. Veryzer. 2000. *Radical Innovation: How Mature Companies Can Outsmart Upstarts.* Boston: Harvard Business School Press.

Lynn, G., and A. Akgun. 1998. "Innovation Strategies under Uncertainty: A Contingency Approach for New Product Development." *Engineering Management Journal* September: 11–17.

Lynn, G., J. Morone, and A. Paulson. 1986. "Marketing and Discontinuous Innovation: The Probe and Learn Process." *California Management Review* 38, 3

J. Morone. 1993. *Winning in High-Tech Markets: The Role of General Management.* Boston: Harvard Business School Press.

O'Connor, G., and M. Rice. 2001. "Opportunity Recognition and Breakthrough Innovation in Large Established Firms." *California Management Review* 43, 2: 95–116.

O'Connor, G., R. Hendricks, and M. Rice. 2002. "Assessing Transition Readiness for Radical Innovations." *Research Technology Management (Nov-Dec.), 45, 6: pp 50–56.*

Rice, M., R. Leifer, and G. O'Connor. 2002. "Commercializing Discontinuous Innovations: Bridging the Gap from Discontinuous Innovation Project to Operations." *IEEE Transactions on Engineering Management,* vol 49 (Nov), no. 4, pp. 330–340.

3 Turning Technical Advantage into Product Advantage

Stephen K. Markham and Angus I. Kingon

Companies are increasingly drawing on R&D and product development professionals to increase sales through new product offerings. A major source of superior, proprietary products is technical advantage. Many companies, however, lack the personnel and structure to recognize and develop technologies and, thereafter, to develop competitive new products based upon these technologies.

The primary focus of this chapter is how to link technologies, products, and markets. It presents the technology-to-product-to-market (TPM) linkage as a critical framework to convert unique technical capabilities into product features that match enduring customer needs. The chapter emphasizes the need to examine the linkage early in the development process of innovative, new product development. Successful front-end product developers understand this TPM process and employ it regularly even if they do not have the vocabulary to talk about it. The chapter also provides concrete tools, techniques, and examples of how a company can develop technical advantages into product advantages. The goal is to give the reader tools to find, develop, and assess the correct product manifestation of the technology so it appeals to the customers the first time.

The chapter is aimed primarily at two groups of people. First, it is for multidisciplinary product development personnel within companies that are developing technology-based products and services. The tools described are suitable for individual or team use. Second, it is aimed at R&D personnel, as there is an urgent requirement for new methods for conceptualizing and testing new products that are based upon their science and technology. However, in addition to these two groups, we emphasize in the following section the fundamental nature of the TPM linkages. The implication is that the concept should be fully understood by a broad range of persons, including R&D managers, research scientists, marketing specialists, and so on.

This chapter first presents the TPM concept, emphasizing its importance. Then a discussion follows regarding definitions, implementation, personnel requirements, and policies. In addition, four worksheets are included to help convert technical advantages into product advantages.

THE TECHNOLOGY-TO-PRODUCT-TO-MARKET LOGIC

Where do new product ideas come from? It is typical to describe them as originating from a market need, or alternatively as a product concept derived from the development of new technology. These linear views both have drawbacks. It is usual for new products derived from market knowledge, and based upon existing technology, to be *incremental.* This is because it is difficult for marketing personnel to be able to conceive a new product with radically improved performance that is enabled by new technology. The marketing personnel typically do not have the necessary technical expertise or awareness of a broad range of emerging science and technology. Thus, it is generally believed that most products that are "beyond incremental" originate conceptually from the technology side. This has traditionally been termed "technology push." Of course, the drawbacks here are also well known—there are legendary cases where companies have made large technology investments, and developed new products, only to find that they completely misidentified the market opportunity and customer needs. These are costly blunders. New technology may offer great advantages, but customers and markets cannot be expected to know about them. Rather than push technologies on markets, systematic planning is required to express new technical advantages as products that meet customer needs. Thus, care must be taken to find and qualify the technologies with market potential. Companies must take the required time to understand opportunities and make well-informed and reasoned product statements.

Central to the logic and technique of turning technical advantages into product advantages is linking technical performance capabilities with enduring customer needs. This linkage requires specifying product features based on new technology capabilities and testing them for receptiveness with potential customers. Figure 3-1 depicts a model that suggests a single technology be used to create multiple product ideas and identify multiple markets. This process includes three activities: (1) finding technical advantages, (2) developing product concepts that utilize the technical capabilities to enable (unique) product features, and (3) elaborating the link from product to market. Collectively, these three steps form a concept called the "product logic."

The TPM linkage should be viewed as an *iterative* decision tool. It begins by finding technologies with unique advantages that can be recognized as new capabilities. Those capabilities in turn can be manifested as product features. The product concepts are then presented to experts in the field, and potential lead customers within specific potential market segments. If information about the product is disconfirming, that product or market segment idea is eliminated. Alternative or improved product concepts are generated and presented in an iterative manner until a product concept finds strong positive response,

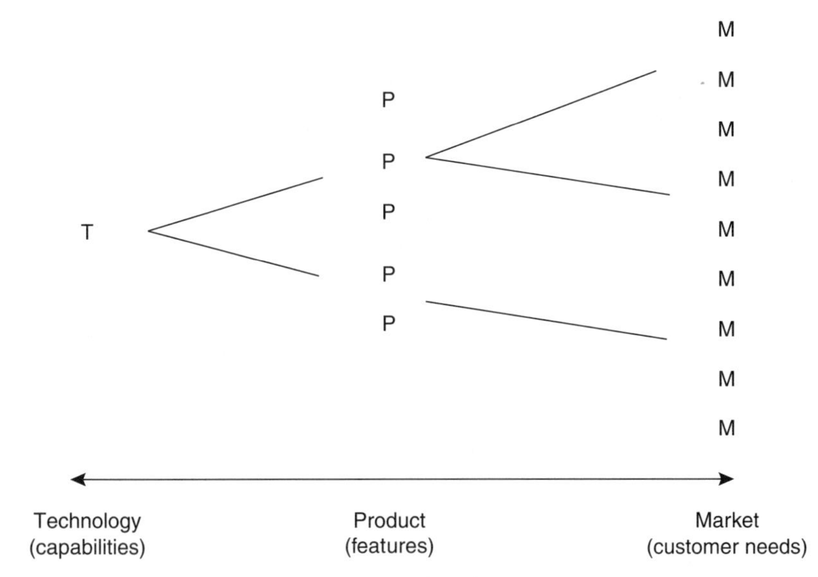

FIGURE 3-1. Technology-to-product-to-market linkage.

within one or more intended market segments. In this chapter we present the tools to carry out this operation.

The process for establishing the TPM linkages is initially time-consuming but becomes more efficient with experience. Product development teams rarely make correct TPM linkages in new areas the first few times they attempt the process. This does not mean the technology doesn't have important capabilities that some market might respond to; instead, it means that the team must persevere through iteration in order to develop viable product concepts. The information and experiences gained in the initial TPM linkage work are necessary for understanding opportunities, as well as understanding why certain product concepts will *not* be viable. Information and experience gathered in the early stages will be used repeatedly through the process of iteration.

The concepts regarding technology-product-market linkages that are presented are not entirely new. However, in this chapter they are explicitly elaborated as a systematic process for developing new product concepts and establishing the product logic. The major advantage lies with the *early* generation and testing of this product logic. This is efficient because many different technologies, products, and market combinations can be assessed before expensive technical product development work begins. It also addresses the major concern of companies regarding the very high percentage of new product failures.

The tools are intended for use by product development teams. They provide a common focus for members with necessarily different experience and roles (marketing, sales, development, operations, financial, etc.). Additionally, the tools are particularly useful for R&D staff, who may be within product development teams or independent of them within R&D organizations. This is because these personnel, who have the requisite understanding of the potential of the technology, are essential partners in the development of exciting new product concepts. However, the structure that these personnel bring to their technology development methods must also be applied to their analysis of product and market opportunities. This is particularly important as companies are downsizing their R&D organizations (Allen, 2000), necessitating greater efficiency in generating new products from corporate technology sources.

Finally, we wish to emphasize that the early establishment of product logic through the examination of T-P-M linkages is a skill that should be more widely practiced. It is an essential building block for new products and new business opportunities and should therefore be a widely understood process. The practitioners should include entrepreneurs, as well as researchers in universities and national laboratories, in addition to the company personnel mentioned earlier. Finally, the processes should be understood by R&D managers and business unit managers to provide appropriate organizational structure.

DEVELOPING AND UTILIZING THE T-P-M LINKAGES

Finding Technical Advantages

To offer a technology-based product with distinct advantages protected by patents, a company must first determine the nature of the technical advantage. Regardless of scientific discipline, only three types of technical advantage exist: (1) higher performance, (2) lower cost, and (3) new, needed capability. Technology descriptions must focus on what new capabilities are possible, since capabilities establish logical limits for the products and markets.

In determining whether a new technology has significant new capabilities, developers must determine the magnitude of its advantage. Venture capitalists often invest only in those technologies in which they can expect a ten-times performance increase over existing solutions. Radical innovation researchers suggest a three-times cost savings before an innovation is accepted (Leifer et al., 2002). While ten times faster or one-third the cost solutions are good rules of thumb, understanding the scale of advantage is not nearly so straightforward. For example, land-based turbines that generate electric power are sensitive to small increases in manifold vacuum brought on by the slow clogging of filters. A filtration system that keeps the vacuum constant can result in significant cost savings based on a small percentage of performance increase.

Our observation is that tremendous technical advantages are widely available, but many companies overlook them. This may be due to focusing on a single technical parameter and manifesting it as a single product idea to a single market segment. In many companies, technical personnel are often frustrated that commercial personnel look to other companies for a solution when the solution is available internally. At the same time, commercial personnel are frustrated that technical people keep coming up with impractical technologies. There is a gap between technical and commercial views about what is valuable. It is difficult to determine if companies lack technical advantage or lack the ability to see it. Much of the technology developed in United States' laboratories, universities, and industries is unused commercially. For example, one medical technology company has more than 350 patents, but it produces products based on only 12 of them.

Much technology is developed without a commercial application because a gap exists between technology sources and technology users. Figure 3-2 presents a development continuum between a scientific discovery and product introduction (Markham et al., 2002). For example, a scientific discovery occurred in the finding that lipoprotein metabolism and heart disease are closely correlated. A technology resulted from the ability to quickly measure lipoproteins in the blood using nuclear magnetic resonance (NMR) spectroscopic analysis. A product platform developed when the analysis of results showed levels of the various different sizes of lipoproteins is used to produce clinically valuable information for other diseases, including cancer and diabetes. Finally, the marketable product is the test results printed in a format that doctors use to treat patients with heart disease.

A problem arises when a company needs better product performance based on a new platform that requires a new technology, but the company does not have the scientific understanding to deliver (as in the case of competition). Only a handful of companies engage heavily in scientific research. A slightly larger number are engaged in developing technologies. Gaps exist when companies do not have enough technical people to interact with sources of scientific discovery or new technologies.

Companies with this technical gap may be unaware it exists. A company may have a large technical staff, but it may not have the capability to interface with scientific sources, as the technical staff members are dispersed into lines

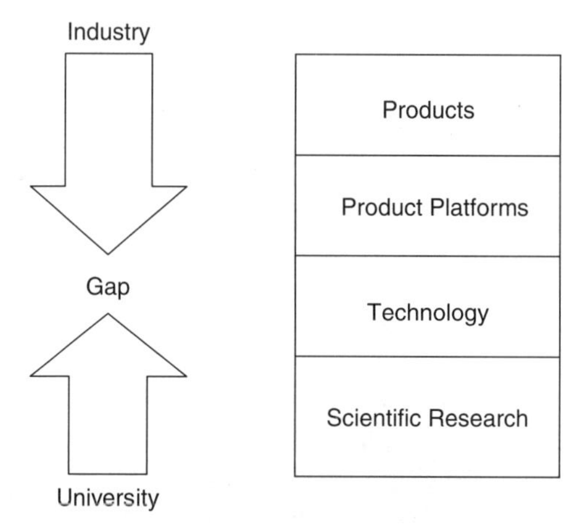

FIGURE 3-2. The gap between technology sources and technology users.

of business that concentrate only on incremental improvements. The gap between technology users and sources creates untold frustration among both the technical and business people. Technical people feel their work is valuable and addresses the very needs of the business people—namely, new high-margin products. Business people are frustrated that the technology people develop ideas that are unconnected to manufacturing, market, or financial realities.

Where do companies look for technical advantages?

1. *Internal departments.* Many companies develop more technology than they realize. Company principals must pay close attention to senior staff with a track record of innovation and young people with bright, new ideas. As one manager stated, "The next blockbuster is already in the lab; we just don't know where."
2. *Customers.* Some customers develop technology to solve a problem but subsequently decide they don't want to produce the item but would like a vendor to supply it.
3. *Suppliers.* Some suppliers develop technologies for one application that another company can use to solve the same or different problem.
4. *National laboratories.* The national laboratories have technology transfer centers that catalogue and publish summaries of their new technologies.
5. *Universities.* Nearly all research universities have technology transfer offices that catalogue available technologies.
6. *Other companies.* Many companies have technologies that might be of value to another company.

The Technology Description Worksheet (Tool 3-1) helps identify promising technologies by guiding the staff to express technology specifications (or technology descriptions) as *capabilities*. The worksheet can be used in various ways, including incorporation into a Web site and in self-report forms. Probably the most effective method is for two individuals who can express specifications as capabilities to complete the worksheet after they talk with the technologists about the nature of the research they are describing.

The Technology Description Worksheet contains three areas: (1) Description, (2)Advantage, and (3) Level of Development. For the description, the technologist may need to give a comprehensive review of the technology's underlying science. Scientists often have a difficult time talking about a technology's uses if they haven't explained its underlying science. Sometimes, talking with other technologists about the research can supply additional information. Although some of this information is not commercially useful, it will help define what the technology does and does not do, which is critical information for product development discussions.

The Technology Advantage section identifies how technical specifications translate into new capabilities. Here, the technologists simply identify the

TOOL 3-1.
Technology Description Worksheet

Technology Description Worksheet	
List technologists.	John Doe
Name technology.	TMBS Power Semiconductors
Describe technology in scientific terms.	A Trench Metal-Oxide Barrier allows greater on-state resistance for power semiconductors. Putting a trench in the surface of a semiconductor requires different process steps in wafer fabrication.
Describe what the technology does.	Allows faster switching time, allows more power to be handled by the device, decreases device size and device yield
Describe what the technology does *not* do.	It is not a logic or memory device; it is a power management device.
Technology Advantage	
In what way is the technology superior to other technologies? (Does it offer higher performance, lower cost, or new capability?)	Lower cost and higher performance.
Describe the advantages of the technology.	Lengthens the time current batteries can run a device.
List and describe possible applications.	Hand-held battery devices, motor control.
List and describe possible users.	Cell phone users, laptop computer users, PDA owners. People who need long, dependable battery life of their electronic tools. People who want more efficient and quieter electric motors.
Explain how a user would actually use the new technology.	Users would have greater use of their existing tools.
Discuss advantages to potential users of the new technology.	Longer operating time, lower operating temperature, lower operating costs, lighter weight, lower costs
Discuss platform implications for the technology. (Can the technology be a platform for multiple products?)	Can be used for multiple low-and high-power applications ranging from DC battery management to high-voltage, high-amp motor controllers.
Discuss patent efforts. (Is the technology patentable? Can the patent be policed? Can the patent be kept secret?)	Highly protected in battery applications but less protected in large motor control.

TOOL 3-1. *(continued)*

Level of Development	
Describe technology's current stage of development.	Well developed for low voltage, less than 10 amps.
Describe progress toward patenting the technology.	University is continuing to protect new applications.
Describe technology's progress toward demonstrating commercial potential.	Low-power applications still need a commercially available substrate. High-voltage applications are further away because of substrate materials.
Describe the company's current ability to commercialize the technology.	High—The company is presently well positioned and has extensive customers and experience in this market.

technology's potential uses and users. Note, however, that the technologists' answers about the technology's uses and users will likely not be the final product and market statements. Rather, these answers are illustrations of the technology's possibilities. A commercially experienced inventor with a good idea of the technology's commercial value may be able to contribute information for solid product ideas.

The Level of Development section helps assess how far the technology is from commercial use. Scientists often underestimate the work necessary to turn a new technology into a product. For example, regulatory issues, good manufacturing practices, and quality certifications all may affect product delivery, but the technologists may be unaware of those ramifications.

After using the Technology Description Worksheet to describe the technology, understand its advantages, and articulate its technical advantages as capabilities, the technologist or development team is ready to complete the technology section of the TPM Worksheet (Tool 3-2). The TPM Worksheet only identifies the issues that must be considered about a technology at this stage. We leave many other legitimate technical questions for a later time. For example, issues about cost, manufacturing, and reliability are important, but they do not always help determine the product logic. If no customers are interested in products with the new capabilities, gathering information about these issues is wasted effort and will obscure the basic product logic.

Turning Technical Specifications into Capabilities

Identifying promising technology sources is a first step in finding technical advantage. The next step is to understand a technical advantage by expressing its scientific or technical specifications as capabilities. What do we mean by the terms specifications and capabilities? We define technology *specification* as measurable performance parameters of a technology, including, for example,

TOOL 3-2.
TPM Worksheet

Technology

Technology Name: TMBS Power Semiconductors
Technical Specifications: Smaller size, high efficiency, higher on state resistance
Technical Capabilities: Longer battery life, lower operating temperature, lower production costs

Product

Product Idea: Chip set for efficient power rectification
Features: Replaceable parts for existing design
Benefits: Longer battery life

Product Idea: Chip set for down-hole-drilling applications
Features: High performance parts in hostile environments
Benefits: Able to get real-time data from drilling operations

Product Idea: Chips for electric motor control
Features: Efficient control of variable speed motors
Benefits: Operations cost savings and less motor maintenance

Market

Market Segment Description: Cell phone users
Needs: Longer talk time

Market Segment Description: Laptop computer users
Needs: Longer battery life

Market Segment Description: Companies doing oil exploration
Needs: Robust electronics for power circuits

Market Segment Description: Companies producing oil from wells
Needs: Installed devises in wells to power telemetry equipment

Market Segment Description: Companies that use electric motors for production
Needs: Motor control to reduce costs

Market Segment Description: HVAC equipment manufactures
Needs: Higher-efficiency products to be more competitive

operating temperature, speed, size, complexity, or other characteristics. A new digital signal processor might be able to compress signals and require less bandwidth to send more information. This technology could be used in many products. (The technology specifications should not be confused with the product specifications, which come much later in the new product development cycle). *Capability* is what the technology's performance parameters enable the product to do, such as allow more channels viewed on a TV. A *feature* is a concrete product attribute, such as a how many channels a new TV might be able to receive.

A capability is not necessarily a product feature or technology specification. In this case a new technology might allow a single tuner to decode more channels, and the capability is to allow more channels to be viewed. The decision about how many more channels to add to the new TV is a feature decision. The questions are, first, how well developed is the technology; second, does anyone care to have more TV channels or should the new product be in the area of wireless web applications; and third (assuming yes to the second question), how many channels would entice the customer to buy the TV? It is of fundamental importance to understand the difference between technology specifications, capabilities, and product features.

To reiterate this point, before product concepts can be developed from technologies with unique and valuable specifications, they must be identified and understood in terms of capabilities, not just scientific or technical specifications. For example, a scientific or technical specification of a new power semiconductor may be that it runs at frequencies as high as 20,000 Hz. This results in several new capabilities. One of these is that the operating frequency can be above the audible frequency range of humans. This allows the controls of electric motors to run more quietly. When the technical specification is expressed as a capability, the specification can be manifested as a product feature—a quieter refrigerator, vacuum cleaner, or washing machine. To gauge their interest, the company exposes potential customers to the concept of quieter appliances. This qualification must be completed before any more technical or product development, because concepts are considerably easier to change than technology development programs.

Technology-to-Product Elaboration (T to P)

Winning technology product ideas must be embedded with both market understanding and technical advantages. Before we can develop winning product ideas, however, we must understand what advantages a technology affords a company. It is well known that development teams must understand the needs and benefits to the customers of a new product, but the team must also understand the strategic advantage to the firm of a new product. Technologies, by their nature, define broad strategic opportunities and eliminate avenues that might be available with a different technology. For example, a hypercritical carbon dioxide process for purifying organic extracts has strategic implications that must be understood if a product idea is to be adopted by the rest of the company. If the company is looking for a new product line but the new technology only produces existing organic extracts more efficiently, the new technology will not fit with the firm's strategy even though the technology is superior. Before developing a technology into a product, developers must be certain that it meets the strategic aims of the organization. Often developers recognize the benefits of a superior technology and propose new high-margin products only to be frustrated when the company does not adopt them. This may be caused by development personnel not understanding that at this time the company does not want new products requiring whole new production facilities and marketing programs. Rather, the company wants incremental improvements and cost reductions.

Assuming the strategic intent of the organization can be understood, it is not always clear what strategic advantage a new technology offers a company. Assessing both the degree of technical advantage and ability to protect the technology, as outlined in Figure 3-3, helps teams understand if the technology they are developing meets the objectives of the firm. Understanding the strategic implications of a new technology helps platform planning and resource allocation decisions to meet objectives. We adapt Miles and Snows' (1978) strategies and David Teece's (1980) regime in Figure 3-3 to understand the interplay between technology characteristics and strategic advantages. Each cell yields directions and recommendations for strategy development. For example, companies that have highly protected technologies with moderate technical advantages cannot adopt a fast follower or innovator strategy. Similarly, a small company with a highly advantageous, proprietary technology may enjoy the benefits of being an innovator. Without this assessment, developers may wonder why their projects are not adopted even though they are technically superior. It should be noted that companies do not always have a clear strategy that can guide the selection of technologies for teams. In these cases the tool in Figure 3-3 is used to discuss what is expected from the development team.

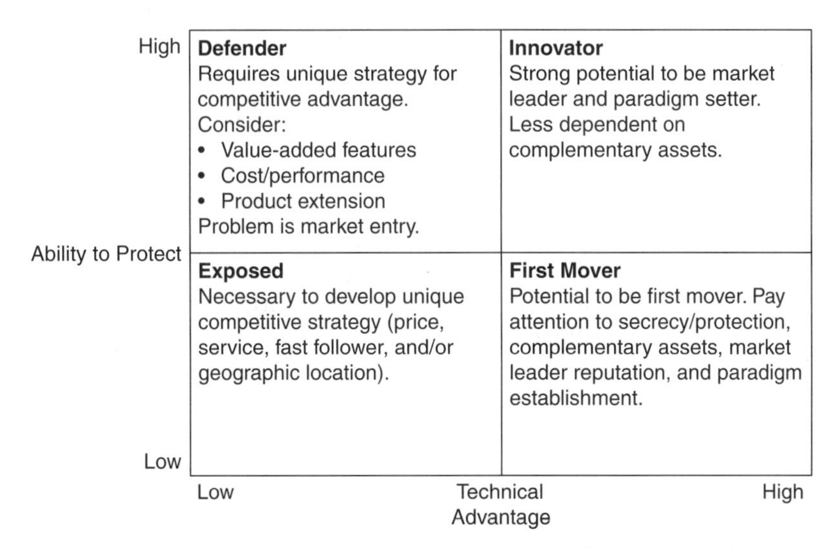

FIGURE 3-3. Technology characteristics and generic strategies.

If a team is expected to beat a competitor, they may need a technology that enables higher-performing products rather than cost reductions. Similarly, a company seeking new opportunities may want to commit to a line of development for new-to-the-world products. While it seems obvious that the product should match the intent of the company, it is not always obvious at the front end if a new technology can deliver on the objective. The tool in Figure 3-3 helps teams understand what the technology can deliver.

Once the team is certain it is working on a technology that addresses a problem or opportunity the company wants addressed, the team must then turn the technology into a winning product. There are three types of product advantage: (1) higher-performing products, (2) lower-cost products, and (3) new capability products. It is important that teams know the intended technology supports the development of the kind of product they want. This is easier for some technologies than for others.

Initial product ideas for technologies that enable higher performance or lower cost are easier and need less explanation because products and customers already exist in a company or among its competitors. A company compares the new technical capability with that of existing products and substantiates a meaningful advantage to intended customers. A financial analysis then can reveal whether the company should pursue the new product.

New capability products present a more difficult case than do higher-performing and lower-cost products. Companies cannot depend on experience and guidance from existing products and customers for technologies that enable new capabilities. Nevertheless, enduring customer needs always cause economic transactions, whether any company offers a direct product to meet those needs. For example, thousands of people calculated accounting ledgers using adding machines before the advent of desktop computing and spreadsheets.

At this point the Product section of the TPM Worksheet (Tool 3-2) is ready to be filled out. The TPM is an industry-tested process to turn a single technology into multiple products and to consider multiple market opportunities for each product idea. The challenge at this stage of idea development is to limit the task to a few variables that are the most predictable and to avoid unanswerable questions that could drive the project into premature failure. Instead, the researchers must focus on completing the product logic. The natural bias of organizational decision makers is to vote against any project at an early stage that cannot provide all the attendant information instead of focusing on the basic project logic. Therefore, the TPM Worksheet identifies only three product dimensions for consideration at this stage: (1) product idea, (2) product features, and (3)product benefit.

Companies must align new capabilities with market needs. For example, a surfactant technology has applications in paints, inks, dyes, and adhesives that enables a manufacturer to add 50 percent more solids into a liquid without changing the viscosity. All of the surfactant applications have performance drivers, but the paint market is much larger than that of the other applications.

The surfactant can be manufactured with either a styrene or an acrylic base. The acrylic base is better for paints, so a company that targets the paint market will include the development personnel on that decision. The ability to cover better with a single coat of paint reduces the volume of paint needed for a given job and saves application time. Both characteristics offer substantial advantages: The retailer can ship and stock less paint and the user has lower application costs. The manufacturer may be able to command a premium price for how this higher performance translates into tangible results for the entire value chain. Certainly, a company must address other issues about technical feasibility, production costs, and timing. However, product development personnel can focus on asking constructive questions as the R&D staff help establish the product logic.

Therefore, the TPM linkage establishes basic product logic. In other words, what capabilities need to be manifested in a product that customers want?

TURNING TECHNICAL CAPABILITIES INTO PRODUCT FEATURES. As part of the basic logic, developers must translate technical capability into product features. Technologists often present developers with reams of supporting data and impressive technical demonstrations that have little market relevance. Interesting capabilities often are buried deep in technical jargon. The challenge for both technical and business people is to focus on those few characteristics of a technology that can lead to commercial success.

Having a trusted process for the front end is critical. Decision makers often want to know information that is not available at such an early stage. Developers must establish new technology-based product ideas in a logical procedure that does not require knowing all product information at the beginning. Likewise, decision makers must delay judgment on certain dimensions if they know there is a logical process and that the rest of the product information will be gathered if the logic of the preliminary TPM linkage justifies the initial product idea. Developers will not expend additional effort to answer non-core questions or on projects that the TPM linkage does not support.

The Product Features Worksheet (Tool 3-3) helps structure the development and presentation of product ideas based on technical capabilities and customer needs (Markham, 2002). This worksheet translates technology capabilities into specific product features. In combination with other components of product specification (customers, markets, commercial path, and value chain), this worksheet helps companies develop product ideas. An idea may start in any column and be extended to the other columns. If the user cannot extend the idea to the other columns, the technology may not have a market manifestation, or the technology may not address customer requirements. This step helps the team develop more ideas and elaborate on them. As a result, the Product section of the TPM Worksheet should be revisited anytime an idea is generated or elaborated upon.

The example in Tool 3-3 is of a power semiconductor chip for power man-

TOOL 3-3.
Product Features Worksheet

Step 1: Identify the unique capabilities of the technology.
Step 2: Describe what customers need in terms of the unique capabilities.
Step 3: Specify exact product features.

Example:
Technology: *TMBS Rectifier*
Product Idea: *Power semiconductor chip for power management in cellular phones*

Technology Capabilities	Customer Requirements	Product Features
Lower on state resistance	Longer battery life	Cellular phone with 3× talk time
Lower on state resistance	Lower-temperature operations	Max. temp. 90°F
Smaller die size	Lower cost	30% device price reduction
Lower defect rate	Lower cost	30% device price reduction

agement in cellular phones. The chip allows any battery-powered device to run longer on the existing battery. Lower on-state resistance allows electricity to be converted to different voltages with less wasted energy. This has the added benefit of producing less heat. The efficient rectification of power can be manifested in many products, but we found cellular phone manufacturers would redesign a phone for a 5 percent increase in talk time. The power rectifier allows three times more talk time with the same battery. Developing the tie between technical capabilities and features cannot be technology-driven but rather must be driven by customer needs.

Product-to-Market Elaboration (P to M)

In the case of the power rectifier, the recognition that more efficient rectification allows longer battery life is good news for a variety of products, from CD players to computers to cell phones. Successfully and reliably deciding which product and market combination to develop requires a process that elaborates the opportunities and discovers flaws of the product manifestation.

When a company generates product ideas, it identifies market segments for each product. Identifying segments for products depends on the nature of the product advantage. For products with superior performance or lower costs, market segmentation is easier to determine because existing segments are a place to start. Market segments are more difficult to establish for new products because of ill-defined customers and unknown unit sales. Nevertheless, for every enduring customer need, economic transactions exist that the new product will displace.

Customer needs and technologies do not, however, automatically come together as product successes. That is, a single TPM link is not a sure thing. A company must develop the product concept so that it actually meets the need of the customer rather than merely stand as a technical capacity. For example, facsimile (fax) machines were developed long before being put into such popular use. Because they were installed only in certain locations and users paid by the page, the public rarely used them. However, when privately owned versions became available, fax machines became nearly universal. Thus, the underlying technology was introduced as a product market mix (P to M) that was unsuccessful. When the technology was used to develop a different product based on a different view of the market, it succeeded. Therefore, a company must take care to find technologies, understand their capabilities, recognize enduring customer needs, and adjust the full range of product specifications to meet market demands.

The iterative process includes visiting the market with product features based on technical capabilities, offering the marketing results to the technologist for modification and elaboration, revisiting the market with refined product features based on technical capabilities, and revising the product idea yet again. With each iteration, the process reveals whether the technical capabilities can be expressed as a product that the market values. As product/market links are strengthened or eliminated, new product concepts will generate from the insight on how technical capabilities and various intended segments respond to the new product features.

Focusing on segments and needs allows the developer to assess the fit between the product features and different market segments. The Product Attribute and Market Matrix (PAMM) Worksheet (Tool 3-4) can help a company understand which market segments are most responsive to the proposed product features (Markham, 2002). A developer can use the PAMM Worksheet to generate on one axis a list of the product attributes derived from the new technical capabilities. The markets and segments can be defined on the other axis.

As seen in Tool 3-4, the battery power market responds well to cost and efficiency, while the hostile environment market is more concerned with the ability to operate at a very high temperature. It is also obvious that the motor control market is most concerned with efficiency and amperage. What one is looking for are markets or segments that respond to product attributes. It can be seen in this example that cell phones have the most "5s." This did turn out to be the entry strategy and the first product application of the new technology. Hostile environments turned out to be a good expansion market. The technology was not yet ready for high-amperage applications, so that is a market waiting for additional technical development.

After using the worksheets to define the potential market for a product idea, the company must decide if it can address the market and if the product

TOOL 3-4.
Product Attribute and Market Matrix (PAMM) Worksheet

Step 1: On the y-axis of a matrix, list all product attributes derived from the technology.
Step 2: On the x-axis of the matrix, list all possible markets and segments for all attributes.
Step 3: Eliminate cells, rows, or columns based on obvious criteria, such as price, competitors, time to market, and low segment desirability.
Step 4: Identify cells that have high potential for a dominant product attribute in a market segment with a high need for those attributes.
Step 5: Develop a prioritized list of product attribute X market opportunities.
Step 6: Rate each product attribute from 1 = not important to 5 = very important for each segment. Ratings are derived from responses in potential customer and industry export interviews.

Product Attributes	Battery Power			Hostile Enviro		Motor Control	
	Cell Phone	Lap-Tops	CD Players	Oil Wells	Monitors	HVAC	Motor Control
More efficient	5	5	4	1	1	4	4
Lower temperature	5	3	2	1	1	1	1
Operate at 20,000 Hz	1	1	1	1	1	3	5
Operate at 600C	1	1	1	5	5	3	3
Lower cost	5	5	5	2	2	3	3
Less than 10 amps	5	4	4	3	3	1	1
Greater than 10 amps	1	1	1	2	1	5	5

Markets and Segments

idea represents a large enough market opportunity. For example, one microchip manufacturer found that a new microchip could be used for image recognition of fresh grocery store produce. The manufacturer identified the customer population and found that the market need was clearly present to the customers who were willing to buy the final product. Nevertheless, the microchip manufacturer decided against production because the volume of chips would fall under a company threshold. The team then reassessed the market and found that the same image recognition capability could be used for quality control inspection in a variety of production settings. This reassessment resulted in a decision that favored production because of an increase in the manufacturing volume.

PERSONNEL AND IMPLEMENTATION

The best processes in the world will not help a company without capable, motivated people. This section examines the people that innovate and how they fit into the organization. We identify personality characteristics, experience and background, intuition, formal position, informal position, source of influence, and organizational support as important personnel characteristics for people involved in the front-end.

PERSONALITY CHARACTERISTICS. Much has been written about creativity and personality characteristics in the innovation literature. Success in the front end does require certain personality characteristics, but success depends on a lot more than personality. In fact, some characteristics that are correlated with innovativeness, such as being late to meetings and not following up with people, work against you when seeking support from other people in the organization for the product concept. People engaged in front-end activities do not differ from the norm in Meyers-Briggs-type indicators or in the Fundamental Interpersonal Relationship Orientation–Behavior (FIRO-B). People engaged in front-end innovation do have a higher propensity for taking risks; they also have a higher need for a wide variety of life experiences, as measured by the Jackson Personality Inventory. Front-end people are not oblivious to risk—if anything they are more sensitive to it. They just choose to take the risk anyway (Markham, 1998).

EXPERIENCE AND BACKGROUND. The person must have a breadth of experience to make connections between technologies, products, and markets over a period of years. Some people never make connections even with extensive work experience. Often it is unique people that engage in the front-end of product development. A technical background is not absolutely essential, but technical people can understand marketing more easily than marketing people can understand science. A variety of different life experiences is likely to increase one's ability to make connections that other people can't. Varity in life gives one different perspectives because there is a broader range of experiences to draw on. Thus, simply being in product development or R&D for many years is not the best indicator that one will be successful turning technical advantages into product advantages. The broad experiences individuals have amassed are critical when developing diverse linkages between technology, products, and markets. Most front-end developers relate stories about how they thought of an idea drawing on an unrelated activity they are engaged in outside of work. For example, one highly successful early stage pharmaceutical developer is a wine expert who scuba dives in Africa to collect exotic fish. He sometimes gets ideas about potentially new biochemical pathways to investigate from his observations of wine fermentation and fish schooling patterns.

INTUITION. Interviews (Markham, 1991; 1998; 2000) suggest that successful front-end product developers understand this TPM process and employ it regularly even if they do not have the vocabulary to talk about it. The process is done intuitively; thus, it is hard to explain to others. Front-end developers often describe the process as tumultuous and disordered, but in fact the same steps are taken by different people in the front-end on a regular basis. Since the average person does not engage in this process on a regular basis, it looks like a bewildering maze of sometimes contradictory activities. Since there are not a lot of people engaged in the front-end in a given company, the developers usually feel isolated and that no one understands the nature of their work—which is true for them. When we look across multiple people and projects, a clear pattern emerges from their collective intuitive behavior.

FORMAL POSITION. Generally, companies have relatively few formal positions for people engaged in this activity. If formal positions do exist, they are usually for a select few individuals with such a strong track record that the company finally recognizes these unique people should focus on new opportunities. Although many companies have innovation or internal venture programs, the link between the daily activities of the majority of people and front-end innovation is still quite distant. People in high, formal positions may take the role of sponsor. For example, one top executive in a major automotive company said innovation was his responsibility and that the way he accomplished it was by reaching down in the organization to find innovative people, then charging them with innovating, and protecting and rewarding them.

INFORMAL POSITION. In many organizations engaged in developing new ideas, many of the people engaged in front-end work are self-appointed. They see an opportunity and they begin to promote the idea, often at substantial risk to themselves. They promote the idea well beyond their current job requirements. It is ironic that even in companies that espouse being innovative, the roles of people that actually are engaged in creating enormous value in the front-end for the company are so little known and their activities so little understood. Notwithstanding the formal organization being somewhat oblivious to these people, the informal organization is well aware of who the innovators are. Innovation often happens in spite of the organization.

SOURCES OF INFLUENCE. Whether these idea developers are in formal or informal positions, they all use their personal contacts and friendships to get what they need to develop their ideas. They rarely engage in overt or covert influence tactics; rather, they simply ask for a favor from someone they know. They also leverage "higher-order" statements from the company mission statement to extract comments from senior management. They then use these to justify or legitimize their activities and rationalize asking for support from other people. For example, they may ask a friend to help do something because the CEO said the company was dedicated to being innovative. In reality, this may be at odds with their managers' production requirements.

Regardless of how these people seek support, they bring with them some form of credibility. In the case of highly experienced people, they bring a track record of success that inspires others to trust them. Less experienced people usually have to develop a compelling case for others to understand and trust what they are doing to continue support beyond a simple favor.

ORGANIZATIONAL SUPPORT. Organizations provide little support for front-end innovation, and for good reason. The company must maintain an efficient production system to maintain profitability. Innovation is inherently inefficient because so much learning is required before the new idea is properly developed for its intended market. Therefore, too much support of innovation may lead to a loss of focus on profitable operations. Companies often make grand statements about being innovative and introducing leading products, yet they go to great lengths to avoid disrupting their existing production, marketing, distribution, and financial plans. Hence, front-end innovators should not expect much organizational support.

Implementation Recommendations

Product innovation is accelerating, and integrated technical and commercial functions are more crucial than ever. Therefore, an effective organizational response would be to initiate a TPM Function. This must be interdisciplinary and multilevel. A TPM manager might liaise with existing technical managers, project managers, and product managers. The TPM Manager focuses on what projects to start, whereas project managers focus on completing projects. A liaison role helps coordinate efforts, but the TPM linkage does not fall under one person's responsibility. Rather, parts of it reside in different areas of the organization. Additionally, few, if any, people are trained or experienced in this activity. Therefore, a TPM management function would include an experienced staff that would coordinate with the technical and commercial personnel to develop TPM links incorporating people across the company. In addition, this group would be responsible for training people in all departments about the linkage. The TPM function would also include less experienced people as trainees.

In addition to TPM managers and trainees, a TPM Eexecutive as a peer with R&D, marketing, and production executives is necessary to establish directions and priorities for product innovation. Senior TPM Mmanagers would report to the TPM Eexecutive and liaise with senior technical staff, business managers, and project managers for a portfolio of projects. TPM managers would be responsible for individual TPM projects as assigned.

Finally, fully implementing the TPM function requires people who would actually do the work to turn technical advantages into business advantages—the people who would actually do the work in the worksheets. Such people might be called TPM Analysts. These people would actually find technologies, turn technical specifications into capabilities, and elaborate on the T-to-P and the P-to-M linkages by using the worksheets contained in this chapter. Overall, if everyone involved in product development participated in the TPM work, it would take about 1 percent of the development personnel to manage and facilitate the TPM function. In units where the work of the TPM link is done by the TPM function, the percent should be in the 2 to 3 percent range. In large companies there may be both divisional TPM groups and corporate-level TPM personnel. This number would more or less depend on the amount of new technology in the industry and how aggressive the company is at finding new opportunities.

SUMMARY AND CONCLUSION

New technology can be a source for new products. Yet the technology is of limited value without the ability to manifest that technology as a product needed by a set of customers. The technology-to-product-to-market (TPM) linkage represents a core skill for a company with a strategy to be a product leader, or that relies on new technologies for new products.

The TPM model provides a process with tools to identify promising technologies, express the advanced technical specifications as capabilities, express those capabilities as product features and benefits, and, finally, identify market segments most receptive to the benefits of those features. Companies waste effort in developing new products when developers misplace effort, fail to complete the product logic, or don't clearly identify the market opportunity for a particular technology. Similarly, developers often expend too much development effort on products before they recognize the product logic and make appropriate continuation decisions. The challenge is to focus on a minimum set of information necessary to make high-quality decisions. By focusing on the product logic rather than on all the other information associated with new products, a developer can assess many more combinations of technical capabilities, product features, and market segments. Thus, the probability of success increases.

The product logic is developed before additional technical work is undertaken. Rather, a clear understanding of the products and markets should guide technical development. Nevertheless, unique technological capabilities can also compel whole new product lines if the capabilities offer higher performance, lower cost, or new features that address enduring customer needs.

REFERENCES

Allen, K. R. 2003. *Bringing New Technologies to Market,* p. 8. Upper Saddle River, NJ: Prentice Hall.

Howells, J. 1997. "Rethinking the Market: Technology Relationship for Innovation. *Research Policy.*

Jolly, V. 1997. *Commercializing New Technologies: Getting from Mind to Market.* Boston: Harvard Business School Press.

Leifer, R., C. M. McDermott, G. C. O'Connor, L. S. Peters, M. Rice, and R. W. Veyzer. 2002. *Radical Innovation: How Mature Companies Can Outsmart Upstarts.* Boston: Harvard Business School Press.

Markham, S. K. 1998. "A Longitudinal Examination of How Champions Influence Others to Support Their Projects. *Journal of Product Innovation Management* 15, 6: 490–505

Markham, S. K. 2000. "Championing and Antagonism as Forms of Political Behavior." *Organization Science* 11, 4: 429–447.

Markham, S. K. 2002. Moving Technology from Lab to Market." *Research Technology Management* 45, 6, November–December: 31–42.

Markham, S. K., and A. Griffin. 1998. The Breakfast of Champions: Associations between Champions and Product Development Environments, Practices, and Performance. *Journal of Product Innovation Management* 15, 5 : 436–455.

Markham, S. K., A. I. Kingon, R. J. Lewis, and M. Zapata III. 2002. "The University's Role in Creating Radically New Products." *International Journal of Technology Transfer and Commercialisation* 1:163–172.

Miles, R. E., and C. C. Snow. 1978. *Organizational Strategy, Structure and Process.* New York: McGraw-Hill.

Stalk Jr., G., and A. M. Webber. 1993. "Japan's Dark Side of Time." *Harvard Business Review* 71, 4: 93–103.

Teece, D. 1980. "The Diffusion of an Administrative Innovation." *Management Science* 26: 464–470.

4 Enhancing Organizational Knowledge Creation for Breakthrough Innovation: Tools and Techniques

Peter Koen, Richard McDermott, Robb Olsen, and Charles Prather

Breakthrough innovations are typically unique, original, and unexpected, as opposed to those that are predictable or incremental and require the sharing of tacit knowledge (Polanyi, 1966). Tacit knowledge is beneath the surface of conscious thought and is derived from a lifetime of experience, practice, perception, and learning. This is in contrast to explicit knowledge, which can be transferred through language (speech, writing, and graphics.) Language is typically the primary mechanism for sharing explicit knowledge. In contrast ". . . language is not the primary mechanism" (von Krough et al., 2000, p. 83) for sharing tacit knowledge. Tacit knowledge ". . . is bound to the senses, personal experience and bodily movement . . ." (Burt, 1992; p. 83) and ". . . requires close physical proximity . . ." (Burt, 1992; p. 83) to the people doing the work. But how do product development managers and process owners optimize tacit knowledge exchange in an organization in order to enable breakthrough innovation? What methods and tools are available to do this?

To answer this question we provide in the first section of this chapter an explanation of how knowledge creation actually occurs and the differences between explicit and tacit knowledge. The necessary organizational conditions for creating new knowledge: innovation vision, and a collaborative environment are discussed in the next two sections. An innovation vision is an essential part of knowledge creation, since breakthrough innovation is often unpredictable and frequently described as accidental. Without a common vision, discovery may be of little value to the company. A collaborative environment that encourages creativity, collaboration, and irreverence and enhances people-to-people contact is also an essential organizational characteristic. The remaining three sections present organizational tools and techniques for enhancing knowledge creation: competence-based communities of practice, competence and capability mapping, and innovation intranets.

Competence-based communities of practice are groups of people informally bound together by shared expertise and passion directed at a specific competence and/or capability of the company. Competence and capability mapping is needed to select the areas the company is willing to build communities of practice around. Innovation intranets capture the current and past learning of the organization. They provide a forum for people to establish new contacts within a company when the information is captured in easily searchable databases that link the capabilities and competencies to the people working on the projects. The concluding section ties the methods together.

KNOWLEDGE CREATION

What is knowledge? Knowledge is defined as a justified belief that increases a person's capacity for action and is different from data and information (Nonaka, 1994). Data are raw facts collected, for example, from experiments. Other examples are chemical specifications and material limits. Information is a collection of organized data that may inspire action. Data processed and viewed through a filter becomes information—such as a graph showing the increased rate of copper removal as pad pressure is increased. But information alone cannot make the action a reality. Knowledge is information combined with experience, context, and reflections that may be used to make decisions and take actions. It is a human act that is created in the present and belongs to and circulates with people. Nevertheless, knowledge is the interpretation of information and is based on the mental model or view of the world. Knowledge will typically vary between individuals, since we each have different views of the world.

The difference between data, information, and knowledge may be illustrated using the invention of the "not too tacky" adhesive used in Post-It Notes,® which was invented by Dr. Spencer Silver, a scientist at 3M in 1964 (Nayak and Ketteringham, 1994). Data would represent the adhesive force, or what scientists call "tack," created between surfaces. Silver determined the adhesive force of his new molecule following a polymerization catalysis experiment where he used an unusually large amount monomer. When this adhesive force was compared to that of other adhesives, it became information. He also found that it tended to stick to only one surface—more information about how the monomer behaved. Knowledge was created when Silver realized that he had created a new type of monomer that was more cohesive than adhesive. Based upon this knowledge, Silver spent five years trying to find an application for this new adhesive.

Knowledge may be classified as tacit or explicit (Polanyi, 1966; Nonaka and Takeuchi, 1995; and Spender, 1996). Tacit knowledge is context-specific, personal, hard to formalize, and difficult to communicate, and it is stored semiconsciously and consciously in peoples' heads. In contrast, explicit knowledge is codified, structured, and accessible to people other than those originating them. In actuality, most knowledge lies between explicit and implicit (Leonard and Sensiper, 1998).

Nonaka and Takeuchi (1995) provide an example of how tacit knowledge may be converted to explicit through observation, imitation, and practice. Matsushita Electric was in the process of developing an automatic bread-making machine. Initial bread produced by the mechanized process failed to taste the same despite efforts to replicate the dough-kneading process. No meaningful differences were found based on the analysis of the dough prepared by a master baker when compared to that made by the machine. The team then went to the master baker and volunteered to apprentice themselves and tried to make the same bread. At first they were unsuccessful, despite numerous attempts to replicate the process. After several days of trying to create the tasty bread of the master baker, one of the team members noticed that the baker was not only stretching the bread but simultaneously twisting it. This turned out to be the secret. The tacit knowledge of twisting the bread now became explicit. The master baker never realized he was doing it, and his hand movement was so subtle that the product development team never realized it until they tried to replicate the process.

Polanyi (1966) indicated the personal nature of tacit knowledge and how it belongs to the individual and is derived from the unique experiences of the individual. The tacit knowledge of inventors plays a key role in invention. A study of several biotechnology firms showed that the development of the industry is closely correlated to the presence of "outstanding scientists having the tacit knowledge to practice recombinant DNA" (Zuker et al., 1988). Histories of important discoverers often refer to ". . . inexplicable mental processes . . ." (Leonard, 1998, p. 115) or "hunches" that occur—as when Watson and Crick dreamed of the formulation of the double helix. Development of tacit knowledge often follows an implicit type of learning. Studies in psychology indicate that people often learn complex information informally and then use it to make decisions—which become interpreted as intuition.

Knowledge creation for incremental innovations that incorporate product improvements using existing technologies in market channels well understood by the organization built on evolutionary product improvements are typically based on the exchange of explicit knowledge that is within and external to the company. For example, the need of a customer for a longer-lasting battery for her cell phone is explicitly defined, as well as the technical pathways to making moderate battery capacity technical improvements. In contrast, breakthrough innovations are derived from individuals' tacit knowledge (Mascitelli, 2000; Leonard and Barton, 1995; Leonard and Sensiper, 1998; and Zein and Buckler, 1997). Just as there is a continuum from incremental to breakthrough innovations, there is a likewise a continuum of knowledge derived from explicit and tacit knowledge for these innovations.

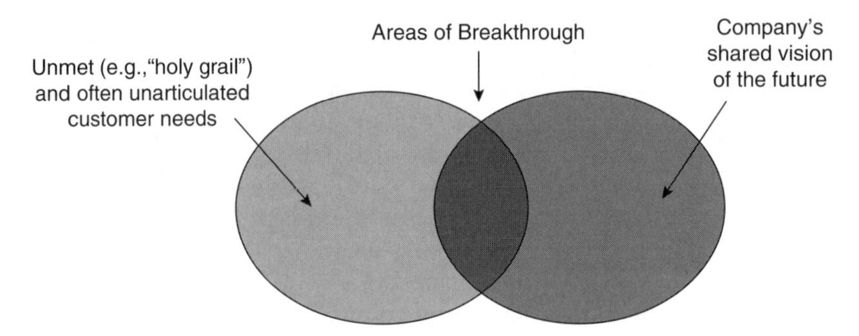

FIGURE 4-1. The insight for a breakthrough innovation comes at the intersection between the unmet customer needs and the company's shared vision of the future.

INNOVATION VISION

Imagination for breakthrough innovation comes from the tacit knowledge of the individual guided by the compelling benefits of the technology to potential marketplaces combined with an understanding of the future vision of the company. This is schematized in Figure 4-1. On the left are unmet and often "unarticulated" customer needs. Breakthrough innovations occur when well-known, significant (i.e., "holy grail") unmet customer needs are satisfied (e.g., cure for prostate cancer) or an unarticulated customer need is satisfied (e.g., Sony Walkman or 3M Post-It Notes®.) On the right is the shared vision of the company. It is impossible to anticipate all of the needs and interactions needed for a breakthrough innovation. Lacking a vision, individuals will rely on their own mental models of the future and pursue many disparate directions that have little value to the company. Deep research for breakthrough innovation relies on guiding individuals with a vision that extends beyond explicitly stated goals.

"A group of people who are committed to a common vision is truly an awesome force" (Senge, 1990; p. 221). Vision allows the committed person to bring ". . . energy, passion and excitement . . ." to a project (Senge, 1990; p. 221). Peter Senge (1990; p. 149) further asserts ". . . that nothing happens until there is a vision." The vision should be a specific destination and a picture of a desired future. Corning senior management shared a vision of the future to develop ceramic substrates for catalytic converters that enabled them to develop a cellular substrate that became the standard of the industry. The process of instilling a vision will ultimately unleash the tacit knowledge of the organization to create new knowledge (von Krough et al., 2000).

von Krough, Icihjo, and Nonaka (2000) indicate that a knowledge vision should provide the following seven attributes:

- *Commitment.* Senior management needs to be committed to the vision.
- *Generativity.* The knowledge vision should provide new thinking and generate new organizational imagination.

- *Specific style.* Each organization needs to determine its own style that is consistent with its culture for presenting their vision. Some organizations will use a simple bold phrase, while others may use a series of workshops.
- *Restructuring the current knowledge system.* The knowledge vision should indicate how the current company's knowledge may be used to develop products in new fields.
- *Restructuring of the current task system.* The knowledge vision should indicate how the company may need to change.
- *External communication of values.* The vision should communicate what new knowledge the company is seeking.
- *Commitment to gaining an advantage.* The vision should enable the new knowledge to create a competitive advantage.

McGrath (2000), in his popular book on product strategy, describes the core strategic vision as the "keystone" of successful high-technology companies. The author also indicates that visions may become obsolete and need to change. Ken Olsen, who founded Digitial Equipment, original and very successful vision of a minicomputer market became obsolete over time. Digital Equipment's failure to see the evolution of the PC market was one of the principal causes of its failure.

Lynn and Akgun (2001) indicate that an effective project vision should have three components:

MOST EFFECTIVE METHODS, TOOLS AND TECHNIQUES FOR CREATING AN INNOVATION VISION

- A future project vision is essential so that the tacit knowledge of the individuals may be harnessed in outcomes beneficial to the corporation.
- A future project vision should provide clarity of direction and remain stable.

HOW TO

An innovation vision is typically created in a workshop consisting of senior executives. However, prior to the workshop, the boundary conditions, which shape the vision, need to be understood. These boundary conditions consist of the following:

- *Core competencies and capabilities of the corporation.*
- *Financial.* What are the goals for expected revenue and growth?
- *Corporate.* The vision needs to be in strategic alignment with the business.
- *Technology trends.* New emerging technology may enable a company to expand or redirect its vision.
- *Market trends.* New market trends may allow the company to expand its current product line or take advantage of unmet customer needs.
- *Product strategy.* A company needs to modify its product strategy so that it is aligned with the vision.

The innovation vision over time needs to change and be grounded in the boundary conditions. Refer to McGrath (2000) for a more extensive discussion of how to create an innovation vision.

- *Vision clarity.* Refers to having a well articulated and easy to understand target.
- *Vision support.* Implies commitment from people throughout the organization to support the vision.
- *Vision stability.* Refers to having the vision remain stable over time.

Lynn and Reilly (2002), in one of the few studies that related project success to the vision process, found that both vision clarity and stability are needed for breakthrough innovations.

While the importance of establishing a vision at the organizational level is well accepted, there is much less work on the role of visioning at the project level. However, without a product vision, individuals will rely on their own vision and the company will not be able to effectively mobilize their tacit knowledge nor instill a commitment to hurdle the high barriers associated with breakthrough innovations.

COLLABORATIVE ENVIRONMENT

Sharing of tacit knowledge and creating new concepts hinges on individuals willing to share their true beliefs. "Justification becomes public" (von Krogh, 1998; p. 135). This process of justifying one's own true beliefs—which may or may not be factual and is often based on hunches—". . . makes the knowledge creation process a highly fragile process" (von Krogh, 1998; p. 135). Most people quickly relate to care—as it might describe the way that a parent relates to a child. von Krough et al. (2000) indicated that care in an organization is based on five factors:

- *Mutual trust.* One needs to trust the individual you are sharing tacit knowledge with not to embarrass you when you have shared your hunches.
- *Active empathy.* Empathy may be described as an attempt to "walk in the other person's shoes" and share his or her pain and frailties.
- *Access to help.* von Krough (2000; p. 52) indicates that a ". . . knowledge creating company thrives on the pedagogical skills of its caring experts . . ." In other words, the experts in the organization are willing to provide help, as in the relationship between a master carpenter and his or her apprentice.
- *Lenience in judgment.* Harsh judgment, laughter, and criticism will convey an environment that far-fetched concepts are unwelcome and prevent the sharing of one's own true beliefs.
- *Courage.* Individuals should not be afraid to experiment and to allow their concepts to be exposed to fierce judgment. Conversely, individuals should be able to provide feedback—even if it is disruptive to the colleagues to which it is directed.

Leonard (Leonard and Sensiper, 1998) suggested that another barrier to sharing of tacit knowledge is inequality among the individuals. She provides the example of how nurses are hesitant to suggest treatments to physicians ". . . not only because the doctors have a higher status but because the nurses base their diagnosis on a different knowledge base." The nurse's perceptions are based on observations over time by spending time at the bedside. The physician's judgments are based on cross-sectional data such as blood tests, X rays, and so on.

Prather (2000) and Prather and Gundry (1995) emphasize the importance of trust and openness. They found that it is critical to a collaborative culture. Without trust and openness, none of the other dimensions matter very much. Collaboration will not occur when fear reigns because of past experience with retaliation or punishment when things didn't go as expected. When trust is low or nonexistent, people will not want to collaborate, nor will they want to offer up their best ideas to anyone. Trust is inextricably linked with openness, and vice versa. Low trust results when leaders play their cards close to their chest, revealing as little as possible about organization issues and plans, and revealing almost nothing about their own personal lives. Leaders must take the initiative to break this vicious cycle where low trust leads to low openness, and low openness leads to low trust. They can do this best by leading by example.

MOST EFFECTIVE METHODS, TOOLS, AND TECHNIQUES FOR CREATING A COLLABORATIVE ENVIRONMENT

A collaborative environment for encouraging tacit knowledge exchange is built upon one that cares for the individual. Care may be enhanced by five factors:

- Mutual trust and openness.
- Active empathy.
- Access to help from caring experts.
- Lenience in judgment (i.e., the holding back of harsh judgment and criticism).
- Courage to experiment and face fair, but possibly harsh judgment.

A collaborative culture may also be enhanced by architectural and furniture designs that encourage socialization.

HOW TO

Weak job security, unjust behavior, highly individualistic incentive systems combined with intense competition between employees will destroy a caring and collaborative organization. A collaborative culture may be created (von Krough, 1998) by doing the following:

- Creating an incentive system that rewards behavior that builds organizational relationships
- Establishing mentoring programs that give senior members of the organization responsibility for mentoring junior members
- Ensuring that trust, openness, and courage are explicitly stated, valued and practiced by senior management
- Holding project debriefings that allow individuals to share their project experiences and learn what parts of the project went well and those that did not
- Scheduling social events where people can explore the interests of their colleagues.

Trust results when people know and respect each other as people and believe that leaders have their best interest at heart. Trust also results when leaders believe that their people are being honest and direct about how things really are. To build trust, the people need to start by building genuine openness. Prather further emphasizes that providing an environment for collaboration *always* begins with senior management.

COMPETENCE-BASED COMMUNITIES OF PRACTICE

Communities of practice (COPs) are groups of people informally bound together by shared expertise and passion (Wenger, 1998; Wenger and Snyder, 2000; Wenger, McDermott, and Snyder, 2002). Communities provide the rich socialization experience needed for sharing tact knowledge. Competence-based COPs have three characteristics: They are organized around a particular technical domain, the members know and relate to each other, and over time they develop a sense of "good practice."

The competency area could be geology, biochemistry, or civil engineering. Or it could be a topic that crosses several disciplines, such as transportation in Third World countries, a new technology platform, or a biological cell. Communities often focus on topics people have spent years studying and developing and can tap a genuinely passionate interest of members. Product Development Management Association is an example of a cross-firm COP focused on new product development.

What ties a COP together are the relationships people form within it. COP's spring from people's natural need to learn from and help each other. As members help each other by sharing insights and information or thinking through a problem together, community members get to know each other. As one community member said, "The real value of the community is that I now know who in the world knows what, so I know who to call when I need help." Over time, community members can learn a great deal about each other's areas of expertise, thinking style, strengths, and weaknesses. One may be so meticulous in his analysis that you can recognize it at a glance. Another may be less technically rigorous but more deeply insightful. This intimate knowledge of each other's work makes community discussions richer. They become more than exchanges of information. They become a dance of the interaction of styles. And through that dance community members, come to appreciate others' contributions, energy, interest, perspective, and humor and develop a kind of "craft intimacy" with each other.

Over time, community members also typically develop a shared expectation about what constitutes "good practice" in their field. Sometimes they do this formally by creating a set of guidelines or procedures. More often they develop this sense of common practice informally, by learning from each other's insights. Even when communities develop official best practices, they usually find that most of the knowledge they share is tacit, previously unarticulated ideas and insights they discover in the course of helping each other solve problems and think together. Engineers at Daimler Chrysler who developed an Engineering Book of Knowledge said that the most useful part was not the best practice entries they created but the thinking and learning they experienced while developing the best practice. Relationships, not documents, form the rails along which knowledge passes in communities. As a result, the community members are capable of transferring tacit knowledge more easily than other organizational mechanisms.

The next three sections on COPs discuss the structure of communities, how they work, and critical success factors.

Community Structure

Voluntary communities do arise naturally in most organizations. However, they often have a difficult time surviving the urgency of project work. During the crunch time on projects, it is difficult for a community to hold its member's attention, particularly when they are being drawn to contribute to other projects. Project managers often offer only mild support for team members to participate in communities, particularly when they feel that the member's time will be spent helping other teams. The community typically requires official recognition or assignment of a community leader, with time allocated to community leadership, assignment of core member's roles and time, regular meetings, availability of electronic tools, and clear support from senior management.

Because communities focus on sharing practitioner knowledge as it is needed by members, COPs can become a powerful way to link professionals within a discipline who are distributed among different development teams, and even in different locations. When Shell Exploration & Production Company organized into permanent development teams, the scientists and engineers on those teams felt a need to talk regularly with other members of their discipline so they could help each other solve everyday technical problems and maintain a high level of technical excellence. This avoided many of the problems of matrix organizations while linking people together who serve on separate teams. While matrix organizations use the same kind of structure—a reporting relationship—on both axes of the organization, COPs weave the organization together using different kinds of structures: tightly knit teams with reporting relationships on one side; loosely-knit knowledge-sharing communities on the other. As in many product development companies, communities are a way to preserve a technical focus even while the organization is structured around development teams.

Discipline-focused communities provide one dimension of innovation. By combining insights learned on different development teams, they are able to bring new perspectives to individual member's work. A community of petrophysicsts (i.e., physicists with domain knowledge in petroleum or natural gas), for example, identified upcoming issues in their discipline and invited suppliers and others to address them. To create credibility, the issues that the team should focus on should meet three criteria:

- *Important to the business.* The stakeholders are more likely to maintain support for the community.
- *Passion.* The community members should feel passionate about the issues. This helps ensure that the community will maintain its current members and be attractive to new ones.
- *Breadth and focus.* The issues should be wide enough to bring new people in but narrow enough so that interesting topics will be discussed. Over time the community can choose to go into new areas; but at the early stages, open channels of communication have a better chance of occurring in a focused area of interest.

Innovations often come from the borders, not the center, of a discipline, technology, or industry. To maximize innovation, some companies have formed communities around topics that intentionally cross boundaries. One of Shell's most effective communities focused on a kind of geological structure: turbidites. Geoscientists from different disciplines thought together about geological structures and reservoir characteristics in order to identify the sweet spots for development. As one member commented, nowhere else could he hear so many different perspectives on a single issue. Daimler Chrysler intentionally mixed people from marketing and engineering into a single community to share ideas about new products. Members sat in one another's work area and participated in product idea discussions. The company even limited membership to one year in order to regularly bring in fresh ideas.

Because communities are grounded in relationships, they can develop the trust and mutual understanding needed for a clash of perspectives to create deep, rich, and useful insights. They create the container in which mixing perspectives can create new knowledge.

How Communities Work

Most communities, even those that are geographically dispersed, emphasize person-to-person contact over documents and databases. Typically, COPs have seven ways of connecting: e-mail, electronic bulletin boards or threaded discussions, one-on-one teleconferences or meetings, small group teleconferences, face-to-face meetings, a leader who networks among members and connects those with problems to others with insights or solutions, and a document library. In the most common form of community interaction, a member posts a question on a bulletin board and several people respond. Sometimes they attach data or a report. But frequently the responses are followed by one-on-one e-mail or teleconferences. One community at Shell found that questions received an average of 4.5 responses within the first 48 hours, but much of the subsequent knowledge sharing happened offline. Community leaders often play the role of cajoling members to post and answer questions, connecting members, finding new members, finding opportunities to connect the community with outside resources, and managing the interface between the community and the organization. Because community participation is usually voluntary, the active engagement of the community leader has been found to be critical to overall community success.

Critical Success Factors

There are ten critical success factors for building a community (McDermott, 2000).

Management Challenges

- *Focus on topics important to the business and community members.* A community of practice without value to the company will be unable to get the support it needs to survive.
- *Find a well-respected community member to act as a co-coordinator.* Communities are held together by people who care about the community. For a large community, the role needs to be full-time. As McDermott (2000) indicates, when it is less than 25 percent, the leader often abandons his leadership role. It was also found that the best leader was a senior practitioner—not usually the company's expert. The primary role of the leader was to connect people, not to provide answers. Often the expert ends up doing the latter rather then the former.
- *Make sure people have the time and encouragement to participate.* People need to be given time and encouraged to participate. At the very least they should not be penalized for their participation.
- *Build on the core value of the organization.* Make sure that the structure of the community is consistent with the culture of the organization. Trying to create a microculture in the community that is different than the organizational culture will cause the community of practice to fail.

Community Challenges

- *Involve thought leaders.* Getting respected thought leaders involved will enable it to have the required energy to survive. Once the community is running, participation of the thought leaders is not as critical.
- *Create forums for thinking.* Documented reports, mini-symposia, bimonthly phone conferences all create opportunities for sharing in the community.
- *Maintain personal contact among community members.* Maintaining social contact is key to the continued success of the community. "People don't contribute to the community because it is good for the company.

They do it because the . . . [coordinator] . . . asks them to" (McDermott, 2000).

- *Develop an active passionate core group.* To survive, a community needs to have a core group of passionate members, though a large group of "lurkers" will often be part of the community as well.

Technical Challenges

- *Make it easy to connect, contribute to, and access the community.* Ease of connecting with other members of the community is a critical criteria for success. Ease of use has little to do with software functionality. For example, one of Shell's COPs chose software that was less than ideal because most of the community was already using it. Ultimately, communication platforms should be chosen that are seamless to the community members.

Personal Challenge

- *Create real dialogue about cutting-edge issues.* Relationships happen through dialogue and socialization—not reports about best practices.

An Example—Technical Community at Schlumberger

Schlumberger, a large international company, created a Production and Reservoir Engineering Community (Edmundson, 2001). The community consists of 536 members globally dispersed around the world whose scope is centered on Schlumberger's core competence in production and reservoir engineering. The goal of the community is to better optimize the value of each well. In particular, they have begun to develop case histories, which they put on a common Web site, that explain how the output of different wells were optimized. In addition, the members have begun cataloguing the industry best practices into three areas: Good Idea, Local Best Practice, and Schlumberger Best Practice. Overall, the community creates a network of support for Schlumberger's technical experts in this area and maximizes the company's knowledge sharing and creation potential.

COMPETENCE AND CAPABILITY MAPPING

Competitive advantage is often achieved when a company has unique core competencies and capabilities that are valuable, rare, immutable, and nonsubstitutable—so-called VRIN attributes (Barney, 1991; Conner and Prahalad, 1996; Wernerfelt, 1995). Development of technology-based COPs are inherently expensive, since they must be managed and consume critical human resources. Core competence maps define the core competencies, capabilities, and gaps of an organization and permit the organization to determine the areas

MOST EFFECTIVE METHODS, TOOLS AND TECHNIQUES FOR CREATING A COMPETENCE BASED COMMUNITY OF PRACTICE

A competence-based community of practice allows technical discovery to occur across and within boundaries of the company by providing a rich environment for sharing tacit knowledge. No similar structure exists in companies for sharing across disparate functions of the company. Further, the mode of communication in a well-functioning COP is based mostly on socialization—which is the primary method for sharing tacit knowledge. Critical success factors for a COP are as follows:

- Focus on topics that are important to the business.
- Ensure the COP is led by a full-time, well-respected member of the community—not necessarily the expert.
- Make sure community members are encouraged, and not penalized, for participating.
- Ensure that it is built on the culture of the organization.
- Ensure, at least initially, that the thought leaders are part of the community.
- Create forums for thinking. Be careful that the COP does not become another project team.
- Maintain personal contact among the community members.
- Develop an active passionate group. Remember that a COP is a voluntary community. So the COP itself needs to create the excitement.
- Make it easy for the community members to connect.
- Create real dialogue about cutting-edge issues.

HOW TO

Refer to Wenger et al. (2002) for a more detailed discussion. The basic steps, once a competence area has been selected, are as follows:

- Plan the community by defining its scope, building a case for action, and identifying potential coordinators, thought leaders, and members.
- Create a preliminary design for the community that would include its structure, hot topics, knowledge sharing practices, and key members.
- Appoint a community co-coordinator with a dedicated budget and time commitment of at least 25 percent.
- Deliver early value, once the community is formed, since the communities are fragile in the beginning.
- Initiate regular community events to help "anchor" the community.
- Engage managers and stakeholders, as the community begins to coalesce, in order to legitimize participating and assure its long term importance to business.

for developing a COP. As new capabilities and competencies are needed, new COPs should be established. Sticking to a competence or capability that is no longer valued by the consumer (i.e., core rigidity) may be the reason that once-successful companies fail. Polaroid, for example, stuck to its competence in the chemistry of developing films in real time and did not develop new competence and capabilities in digital photography—which provided the customer similar real-time experiences but did not utilize Polaroid's chemical competencies.

Following is a three-step methodology for obtaining a consensus view of the firm's core competencies and capabilities, with the outcome being the development of a competence and capability map for the company:

1. *Identification of "enabling" competencies and capabilities.* A cross-functional team—of R&D, operations, and marketing—identifies the enabling competencies and capabilities of the firm. These are the enabling competencies and capabilities that are needed for current and future products. The term enabling is used since they "enable" the development. *They may or may not be core.* A $3 billion food company started with a list of over 45 enabling competencies and capabilities. This list was then narrowed down to 16 by a consensus view of which were critical and after eliminating ones that were defined differently but were essentially the same. The key challenge for the team was to develop clear and concise descriptors and boundary conditions for each of the enabling competencies and capabilities. In addition, the team also identified a single competitor within its industry who had the "best" competencies or capabilities with respect to the one being evaluated. Typically, each competency or capability will have a different competitor. In this way, the rarity and degree to which the competence or capability is immutable, which are two of the VRIN criteria, may be evaluated.
2. *Creation of a competence/capabilities and needs map.* A broad cross-functional team composed typically of 20 to 30 people representing all divisions of the company (i.e., R&D, marketing, operations, finance) are assembled at an off-site meeting. During the first day of the meeting, the group creates a competence/capabilities needs map for all of the enabling competencies and capabilities determined from step 1. An example is given in Figure 4-2 comparing competencies of the Product Development & Management Association to that of the Project Management Institute. A consensus view of the skill level of the company

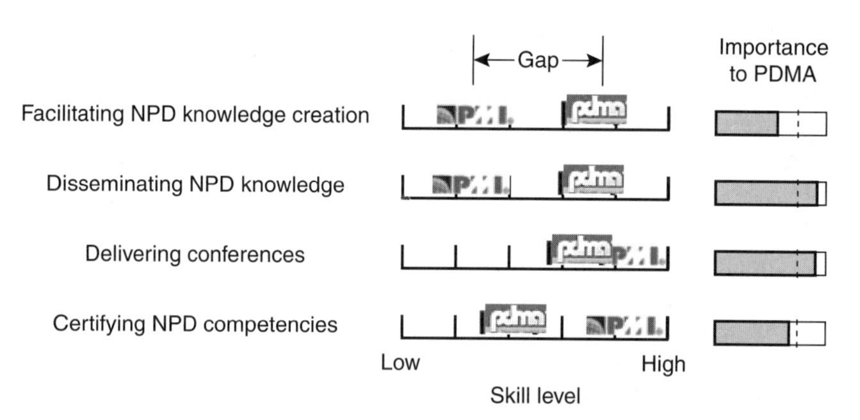

FIGURE 4-2. Hypothetical capabilities/competencies needs map created in step 2 comparing competencies in New Product Development (NPD) of the Product Development & Management Association (PDMA) to a potential competitor-the Project Management Institute (PMI).

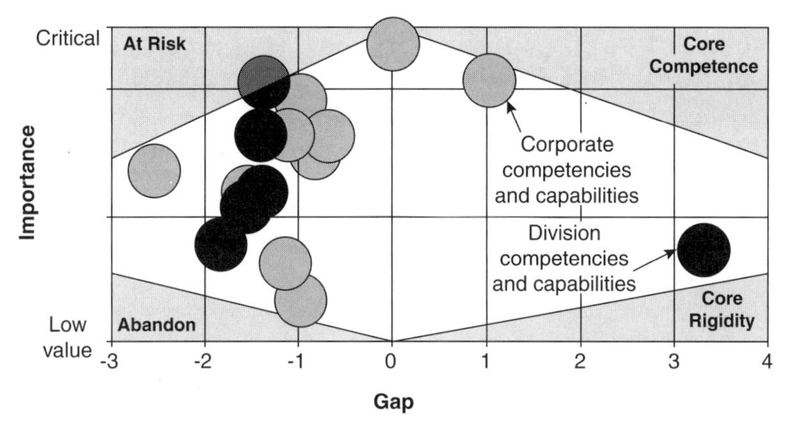

FIGURE 4-3. Competence/capability map created at a $3 billion food company.

relative to the targeted competitor is determined for each competency/capability. In addition, the importance (i.e., value) of the skill level to the company is also determined. The relative skill levels are determined by anonymous "voting" of all the participants. Typically, there is consensus of the gap between the company and the competitor, but there is often a discrepancy in importance. These differences represent a point of discussion, which is often evaluated by additional research after the meeting.

3. *Creation of competence/capabilitiesmaps.* From the data in step 2, a competence and advantage map is determined by reconfiguring the data. The importance level becomes the ordinate, and the gap becomes the abscissa. An example competence map from the $3 billion food company is shown in Figure 4-3, with all of the 16 competencies and capabilities. This company has *no* competencies or capabilities that are core (see upper right-hand corner). In addition, the company has a competency/capability (see lower right-hand corner) that is superior to the competition but not valued by the company. This represents a competency that should be leveraged. Alternatively, it could represent a core rigidity that should be abandoned. Further, there were a number of competencies and capabilities (see upper left-hand corner) where the company was at considerable risk from competition.

The critical success factors for determining core competencies and capabilities are as follows:

1. *Developing clear descriptors for the enabling competencies and capabilities.* Ultimately, people need to "vote" on the competence and determine its gap relative to a competitor and relative importance. The discussion around the competence becomes nonproductive if the team cannot agree on what it is.

MOST EFFECTIVE METHODS, TOOLS, AND TECHNIQUES FOR DETERMINING CORE COMPETENCIES AND CAPABILITIES

Core competencies and capabilities need to be determined to determine where Communities of Practice need to be established.

HOW TO

A three-step methodology for developing a consensus view involves the following:

1. Identifying enabling competencies and capabilities using a cross-functional team of the firm. Enabling competencies enable development. Whether they are core is determined in step 3.
2. Creation of a competence map. A broad cross-functional team determines the relative gap between a targeted competitor and the relative importance of each enabling competence and capability.
3. Determination of core competence. The core competencies and capabilities are then determined by plotting the responses from step 2. The enabling competencies are core if they have a large positive gap and are rated as important.

2. *Identifying a single competitor to measure the gap against.* Competencies become core when they meet the VRIN criteria. There are often multiple competitors, and measuring against a single competitor may create an overly high hurdle. The single competitor is needed to gain consensus. Later analysis can add additional competitors to the evaluation.
3. *Verifying the results.* The initial series of gaps between the company and the competitor are determined based on assumptions and consensus. In many cases, they need to be verified by actual tests, market research, and reverse engineering.
4. *Repeating the analysis at regular intervals.* Competencies and capabilities do change over time as new skills are brought into the company and competitor's strengths wax and wane and the relative importance changes. The competence map should be re-created at least every two years.

Perhaps the biggest pitfall is in not creating the map and presuming that "everyone" knows what the core competencies and capabilities are. Performing one of these analyses often tests the organization, since long-standing assumptions about core competencies are challenged. The company often realizes that many of the competencies are not core—since a competitor has rapidly gained skills.

INNOVATION INTRANET

Knowing who knows what, who has worked on what, what was done, and what was learned can stimulate innovation, help solve tough technical problems, and avoid duplication of effort. But sharing this kind of information can be very difficult to achieve in a large organization. Individuals throughout the company generate valuable information and expertise but cannot know through normal social interaction all the members of the organization, their expertise, and the work they do in enough detail to leverage it in their own work. Further, geographic dispersion and organizational silos are significant barriers to information sharing.

Across a large organization, similarities in technical disciplines, problems faced and solved, and lines of business mean that members of the organization will generate information and expertise of value to other organization members who are unknown to them. Effectively leveraging collective information and expertise is difficult, but far from impossible. Outlined in this section are some proven principles, models, and examples that have worked at Procter & Gamble (P&G), as well as other large organizations, for enhancing connections from people to people and people to information. InnovationNet is a collection of information and applications designed to accelerate the rate of successful technical innovation at P&G.

Applications Contained in P&G's InnovationNet

TEAM SPACES. By this we mean electronic collaboration tools that provide project teams with the ability to share documents, discuss issues, and schedule activities. Team spaces help project teams be more effective by helping team members work issues, share learning, and reach decisions when they can't all be together at the same time and place. Examples of commercial team spaces include those from Intraspect, Lotus, and EDS.

Some of the requirements for success include the ability to do the following:

- Limit access to team members only, if desired by the team. This is an important enabler of the trust and social contract that is required for teams to successfully share information electronically in a central location.
- Publish documents and data in their native formats.
- Push new information to team members—for instance, via e-mail.
- Integrate with personal and group scheduling tools.
- Make the publishing process extremely simple, requiring as few clicks and as little manual data entry as possible.
- Leverage ubiquity of Web browsers for delivery.

Team spaces are heavily used throughout industry. For example, P&G has over 3,500 active team spaces at any given time, and team members place high value on the ability to move projects ahead more quickly and effectively with this tool.

PERIODIC REPORTS. If they are generated and shared routinely, periodic reports (status reports, trip reports, experimental summaries, technical reports) can be powerful information and expertise sharing tools. Identifying what is already being written and then capturing and sharing it effectively in a central repository can be a very quick way to start or expand information sharing by leveraging existing work processes and organizational culture. This kind of repository serves as a people-to-information tool because of the information shared in the reports. It also serves as a people-to-people tool, where all the reports a person writes can be seen as one large curriculum vitae, thereby helping others identify whom to contact on specific subjects.

WEB SITES FOR COP. Every COP at P&G has a Web site. Information is also aggregated across COPs into "knowledge centers." Applications to help COPs conduct symposia and training and to capture the information shared at these events are also included. There are also COP areas in the online Q&A tools (i.e., AskMe.)

AD HOC QUESTION AND ANSWER TOOLS. These tools provide the ability for people to find out who knows what and to quickly ask these resources a question or seek a solution to a problem. By providing a central repository and associating organization members with specific skill or experience areas, personal networks can be quickly and effectively expanded. The full collective information of the organization can be accessed. AskMe is one example of this type of application.

Principles for Success

BE CONSUMER NEED-DRIVEN (I.E., IF YOU BUILD IT, THEY WILL NOT NECESSARILY COME). Carefully studying the needs of your target audience, and continually testing to be sure your information sharing efforts are delivering against these needs, is the most important factor for success. This type of consumer needs assessment and testing needs to span the range from idea generation through development and ultimate delivery. The same consumer understanding and assessment tools that are applied to the development of new products should be applied to the development of internal information sharing programs. Some of the key areas for understanding include current and desired experiences (i.e., need gaps), assessment of ideas and concepts against the needs, usage and attitudes, and usability testing of specific tools.

PARTNERSHIPS BETWEEN ALL KEY STAKEHOLDERS. Effective partnerships between the key stakeholders involved in the information sharing program are essential to success. Shared vision, goals, and success criteria can help ensure that information and expertise sharing efforts will ultimately succeed.

GETTING INFORMATION TO SHARE

- *Extend, enhance, or simplify existing work processes to capture information.* Everyone is too busy, and tasks that are not seen as critical to getting daily work done are at risk, including incremental work to capture and share information. One very effective way to address this is to make information capture and sharing a natural part of the everyday or existing work process in a way that simplifies or enhances the work process for the individual. For example, InnovationNet has been very successful at capturing and sharing periodic reports from R&D members by providing an electronic means for publishing these reports that is easier for the individual to use than the old paper-based process.
- *Build on the positive knowledge and expertise sharing elements of the organizational culture.* All organizations have some existing information sharing practices that have become part of the organizational culture. Studying these, identifying the sponsors and thought leaders, and building on these positive cultural elements can be a very effective strategy for expanding an information capture and sharing program. Conversely, failure to identify the cultural norms can lead to programs that are perceived as off target, intrusive, and poorly thought out, and that fail.

SHARE EFFECTIVELY AND HELP INDIVIDUALS MANAGE THE FLOOD OF INFORMATION

- *Be as open as possible.* The value of collective information is multiplied by sharing the information as broadly as possible. In large organizations that have valid information security goals, the two objectives can conflict. Information sharing and security will have to be balanced through discussion and compromise, but the general trend for information sharing is to expand access to multiply value. Periodically revisiting past agreements as members of the organization become acclimated to broader sharing can lead to steady increases in breadth of access and therefore value.
- *Aggregate effectively; avoid tool proliferation.* In many larger organizations, people feel flooded with electronic tools and Web sites. Aggregating collections of information and expertise in ways that make sense to the target audience can help reduce the number of places people have to "go look" and will be appreciated by end users of the tools.
- *Provide multiple ways to find what you need.* People process information in different ways, and a collection of tools to help find information is required. The following range of tools are available:
  - *Search.* Our experience is that full text search capability across all collective repositories is essential. This is in part driven by people's Internet experiences, (i.e., the common experience is that all sites have search, and people are used to comprehensive search tools like

Google.) In other words, the standard to which internal searches for information and expertise will be held is the best of the search tools from the public Internet.

  - *Browse.* Browse complements search by providing a set of information classification terms that can be organized to help people click through easily recognized terms until they find what they are looking for. The downside to browsing large collections of information is that the classification terms will not make perfect sense to everyone in the organization.
  - *Expert-guided view.*This is a view or collection of information that has been pulled together by an expert or team of experts in an area. This kind of collection or view can be extremely helpful to those seeking information, but absolutely requires direct human intervention to set up and maintain.
  - *Subscription.*Simply remembering to go look for information is not nearly as effective as having a machine bring you the new information in your area of interest. By entering subscription terms related to your interests or needs, you can keep current on areas of interest automatically.

- *Who and what are equally important.* Consistent with the "one big curriculum vitae" statement made earlier, it is important to associate each information object or document with one or more human contacts or authors. All information shared across the organization should have contact names associated with it to allow direct contact with the authors/experts and encourage people-to-people connections.

SUMMARY AND CONCLUSIONS

Breakthrough innovations involve the sharing of tacit knowledge among individuals. However, this type of knowledge cannot be shared through language and is inherently fragile. Organizations can facilitate the sharing of breakthrough tacit knowledge by setting the right organizational conditions and implementing tools to enable the exchange.

The organizational conditions require both an innovation vision and a collaborative culture. The innovation vision provides the overall direction to all of the activities. Since it is impossible to predict the outcome of a breakthrough, an innovation vision is needed to guide, empower, and excite the inventors to areas that are strategically important to the company. The speed and vitality in which knowledge is created is dependent on the collaborative culture of the organization. A collaborative environment that fosters care for the individual enhances the sharing of tacit knowledge—since it is an extremely fragile process. Conceptually, the speed of tacit knowledge exchange depends on the degree that the environment is collaborative. In a noncaring environment, tacit knowledge will remain tacit and not be shared.

MOST EFFECTIVE METHODS, TOOLS AND TECHNIQUES FOR CREATING AN INNOVATION INTRANET

An effective intranet to maximize people-to-people and people-to-information connections needs to include the following:

- Team spaces.
- Places for periodic reports.
- Web sites for communities of practice.
- Ad hoc question-and-answer tools.

To achieve success the innovation intranet must be

- User-driven;
- developed and guided by all of the key stakeholders;
- aligned with the business strategy of the functions served;
- seamless in information sharing (i.e., users do not have to go to any extra steps to share information); and
- Able to help individuals manage the flood of information.

HOW TO CREATE AN INNOVATION INTRANET

Creating an innovation intranet as discussed in this section requires the following:

- Focused effort to meet the consumers need (i.e., R&D people who will be using the intranet) of getting and sharing the knowledge and information that will help them be more effective in their innovation work.
- Carefully selected set of tools and content designed to increase the number of people-to-people and people-to-information connections that people find useful in their innovation work.
- Adequate staffing. This will require a substantial number of full time staff from IT (i.e., those building the tools) and R&D (i.e., the organization who will be providing the content and served by the tools), and other technical functions that work closely with R&D.
- Staying close to all of the stakeholders, which include the end users, content owners, and management.
- Developing the innovation intranet based on substantive consumer research. This would include understanding the ways in which people find and share information and make connections today, identifying unmet needs and regularly tracking progress against meeting the needs through qualitative and quantitative consumer research techniques.
- Ensuring that people can easily find who and what they need by using full-text searches across all file types and site locations, as well as tools to connect with in specific areas of expertise.
- Personalizing the intranet through individual tools. For example, the intranet uses a push technology that highlights recommended reading for the user when a new publication is posted in that user's area of interest.
- Utilizing external IT platforms that best meet the needs of the users of the intranet, are cost effective, and that provide flexibility to meet future needs.
- Constantly driving awareness and usage by working with senior managers, thought leaders and early adopters, using it to help enable key events, and providing training for new hires and training on new tools and content.
- Balancing the needs for information security with the benefits derived from sharing information broadly.

Competence-based Communities of Practice and innovation intranets provide the mechanisms for sharing tacit knowledge when the innovation vision is well articulated and a collaborative culture exists. Competence-based communities enhance the company's core competencies and capabilities by leveraging the collective knowledge of the company. Competency and capability mapping is required to determine where the Communities of Practice should be focused. Socialization, which involves the sharing of tacit knowledge, is enhanced through Communities of Practice. This process may be facilitated by an innovation intranet—which is an organizational tool that captures the current and past learning of the organization and provides a forum for individuals to exchange explicit knowledge and develop and enhance their existing relationships.

Those companies that have superior competence-based Communities of Practice and innovation intranets would be expected to repeatedly reap the rewards of breakthrough innovations with rapid cycle time. These organizational tools require a significant investment and commitment. Implementation and sustainability of these tools can only occur when there is a clear, supported, and stable innovation vision combined with a collaborative culture.

REFERENCES

Barney, J. B. 1991. "Firm Resources and Sustained Competitive Advantage." *Academy of Management Review* 17, 1: 99–120.

Burt, R. 1992. *Structural Holes*. Cambridge, MA: Harvard University Press.

Conner, K. R., and C. K. Prahalad. 1996. "A Resource-Based Theory of the Firm: Knowledge vs. Opportunism." *Organization Science* 7, 5: 477–501.

Edmundson, H. 2001. "Technical Communities of Practice at Schlumberger." *Knowledge Management Review* 4, 2: 20–23.

Leonard—Barton, D. 1995. *Wellsprings of Knowledge*. Boston: Harvard Business School Press.

Leonard, D., and S. Sensiper. 1998. "The Role of Tacit Knowledge in Group Innovation." *California Management Review* 40, 3, Spring: 112–132.

Lynn, G. S., and A. E. Akgun. 2001. "Project Visioning: Its Components and Impact on New Product Success." *Journal of Product Innovation Management* 18: 374–387.

Lynn, G. S., and R. Reilly. 2002. *Blockbusters*. New York: HarperCollins.

Masitelli, R. 2000. "From Experience: Harnessing Tacit Knowledge to Achieve Breakthrough Innovation." *Journal of Product Innovation Management* 17: 179–193.

McDermott, R. 2000. "Knowing in Community: Ten Critical Factors for Community Success." *IHRIM Journal* March: 4(1) 19–25.

McGrath, M. E. 2000. "*Product Strategy for High-Technology Companies*, 2nd ed. New York: McGraw-Hill.

Nayak, P. R., and J. M. Ketteringham. 1995. *Breakthroughs*. San Diego: Pfeiffer and Co.

Nonaka, I. 1994. "A Dynamic Theory of Organizational Knowledge Creation." *Organization Science* 5, 1, February: 14–37.

Nonaka, I., and H. Takeuchie. 1995. *The Knowledge Creating Company*. Oxford: Oxford University Press.

Polanyi, M. 1996. *The Tacit Dimension*. Gloucester: Peter Smith.

Prather, C. W. 2000. "Keeping Innovation Alive after the Consultants Leave." *Research Technology Management* 43, 5, September–October: 17–22.

Prather, C. W., and L. K. Gundry. 1995. *Blueprints for Innovation*. New York: American Management Association.

Senge, P. M. 1990. *The Fifth Discipline*. New York: Doubleday.

Snow, J. 1855. *On the Mode of Communication of* Cholera. London.

Spender, J. C. 1996. "Competitive Advantage from Tacit Knowledge? Unpacking the Concept and its Strategic Implications." In *Organizational Learning and Competitive Advantage*, ed. B. Moingeon and A. Edmondson, 56–73. Thousand Oaks, CA: Sage Publications.

von Krough, G., K. Icihjo, and I. Nonaka. 2000. *Enabling Knowledge Creation*. Oxford: Oxford University Press.

Wenger, E. 1998. *Communities of Practice*. Cambridge: Cambridge University Press.

Wenger, E., and W. Snyder. 2000. "Communities of Practice: The Organizational Frontier." *Harvard Business Review* January–February: 139–145.

Wenger, E., R. McDermott, and W. M. Snyder. 2002. *Cultivating Communities of Practice*. Boston: Harvard Business School Press.

Wernerfelt, B. 1995, "The Resource-Based View of the Firm: Ten Years After." *Strategic Management Journal* 16, 3: 171–174.

Zien, K. A., and S. A. Buckler. 1997. From Experience Dreams to Market: Crafting a Culture of Innovation. *Journal of Product Innovation Management* 14: 274–287.

5 Building Creative Virtual New Product Development Teams

Roger Leenders, Jan Kratzer, and Jo van Engelen

This chapter provides new product development (NPD) project managers with a simple tool that will help manage NPD teams toward greater creativity. In NPD, creativity is of preeminent importance. Many NPD projects start with only a vague idea of what the final product will be like; the more innovative the project, the more vague the starting point tends to be. Creativity is indispensable to fill in the blanks. In addition, the market success of a company's NPD effort is strongly and positively related to the uniqueness of the product, both in terms of product functions and technical aspects. This increases the level of creativity that is expected of the company's new product professionals. Moreover, the more innovative the new product, the less it is possible to rely on set procedures and routines. NPD is often argued to be the epitome of a lack of routine, as evidenced by competing goals, unstable environments, long time horizons, incompleteness of operational specifications, and unclear applicability of past experience. But where does creativity come from? How can creativity be managed? More specifically, how can a manager control and direct the level of creativity that is reached by the NPD team? Many NPD managers have no idea how to do this. They sometimes try to enhance a team's creativity by adding "creative individuals" to the team, but this seldom helps to reach a consistently high level of creativity in the teams. To make matters worse, managing creativity seems to become even more difficult when it needs to be generated by a "virtual" NPD team. The fact that its members reside at different locations around the globe makes managing the team toward outstanding creative effort a daunting task.

This chapter assists NPD project managers in solving this task as follows. First, we show that there is no such thing as a "virtual" NPD team. It simply doesn't exist. Virtuality is a continuum; every team is more or less virtual. The trick is to understand how virtuality shapes your team. We call this the team's *Virtuality Fingerprint*. Then we describe how this fingerprint supports or impedes NPD team-level creativity. No matter how many creative individuals are assigned to a team, some fingerprints make consistently high levels of creativity impossible. Conversely, by working with the fingerprint that is ideal for your needs, the team's creative performance can be enhanced without having to add creative, but often difficult to manage, individuals to the team.

The tools in this chapter are most directly useful for those managing NPD projects. The tools can be used to set up new product teams (in terms of infrastructure, location, and team tasks) in a manner that supports the level of creativity that needs to be generated by the team. More importantly, the tools point to what to change when higher or lower levels of creativity are needed. With this knowledge, the project manager can use these tools to anticipate future team creativity requirements and take measures accordingly. Moreover, the creativity requirements of projects are rarely—if ever—stable over time, as the nature of projects and the demands imposed on them vary over time. Generally, the early stages of new product development projects, in which the design of the new product and its functions is central, require much higher levels of creativity and flexibility of new product teams than do projects that are in the final stages of the new product process. The tools presented in this chapter provide insight into how to manage the creativity of NPD teams over the course of the project. As such, the tools are useful for both NPD team managers and project managers. The tools and recommendations we present here are based on extensive scientific research and on a total of over 75 years of hands-on experience in NPD in various industries.

Let's begin by discussing NPD *teams*.

THE TRUE TEAM

Sit back for a second, relax, and think of what defines a true team. Chances are, you'll think of a true team as a set of people who work closely together, appreciate each other's presence, and create solutions to problems out of their mutuality. When managers are asked to give us examples of true teams, they often point to examples from sports (e.g., ice hockey or basketball teams), where team members fight together to win and are more concerned about the team winning than whether they themselves score points. Often, examples are provided from our favorite TV series in which teams encounter the most extreme situations and master them by holding on to each other. The members of the 4077 M.A.S.H. unit and the bridge crews of the Starship Enterprise in any generation typify the true and successful team. Sometimes, "true teams" are described in accordance with the words of Benjamin Franklin: "We must all hang together, or assuredly we shall all hang separately." A recent example is the war against Iraq in which small groups of highly specialized and highly trained soldiers have no other option than working together in order to save their lives and do their job.

But what about teams in NPD? The view of the "true" team in NPD is one in which NPD team members work together, are located close to one another, communicate face-to-face frequently, and together solve the design tasks at hand, coordinating through input from all team members. However, NPD managers are usually quick to point out that such teams have become rare because of dramatic changes in the business environment. These changes have had their effect on the way in which NPD projects are executed. The knowledge required for the development of most new products has become increasingly specialized and detailed. Given the pace at which knowledge becomes outdated (and new knowledge is developed around the globe), NPD increasingly requires in-depth mastery of specialized knowledge areas. Given the rate at which technology develops, remaining a specialist in one's field now often demands even more effort than it used to take to become one. Compared to only a few decades ago, even a moderate variation to an existing product requires a vast amount of in-depth knowledge and expertise. The specialized skills and talents required for the development of new products often reside (and develop) locally in pockets of excellence around the company or even around the world.

Firms therefore have no choice but to disperse their new product units to access such dispersed knowledge and skill. Regardless of whether these skills and talents are acquired from within the company, or through codevelopment contracts or alliances, the result is that many NPD projects consist of individuals residing in different places.

In addition, trends such as increasing product complexity, shortening product life cycles, and increasingly rapid competitive response have made design tasks more complex. Adding further functionality to a product frequently calls for additional expertise in new areas. For example, making a shaver suitable for use in the shower requires additional expertise in, among other things, the use of new materials and mechanics and the use of shaving emulsion—knowledge traditionally absent in dry shaving teams—and it affects the design of the battery and other components. Add to this the increased tendency toward modular product design—which results in the need to take into account a variety of potential product functions, each supported by different kinds of knowledge—and it becomes clear that the NPD environment has changed tremendously. As a result, managing NPD projects has become increasingly difficult and challenging. In fact, it would often not be possible at all to manage such complex tasks if communication technology had not matured so rapidly. With the help of computer-driven communication technologies, globally scattered knowledge can be transferred easily and the communication of NPD team members grouped in subunits across the city, country, or globe becomes smoother.

The ideal of the "true" team appears to have become just that: an ideal. Most NPD teams no longer consist of members located all together; they are now often dispersed across the globe. They no longer rely mainly on personalized face-to-face interaction; instead, they largely communicate through electronic means. Finally, rather than solving all problems together, members of modern NPD teams work on strongly decomposed design tasks—the team structure deriving from the structure of the overall design task.

Many companies have realized that the design and management of NPD teams is vital to NPD success. They realize that teams that are left to their own devices are rarely truly successful and are rarely able to manage their creativity over the course of their existence. As a result, several companies have installed managers whose task is to design teams and assist the persons managing them. For example, Royal Philips Electronics works with dedicated team start-up managers and dedicated project-continuation managers. The question of how to make NPD teams function better is a burning issue in most companies, regardless of size, age, profit motive, or industry.

THE VIRTUAL TEAM

When it comes to NPD teams, the new magic word is "virtual." Before continuing to read this chapter, close your eyes one more time, relax, and note what impressions first come to mind when thinking about a virtual team. What does it look like? How does it function? Odds are, you will think of a team with members dispersed around the world, living and working in different time zones, and its members rarely or never get to see each other in person. You may think of a team with marketeers located in the United States (dispersed across several locations, of course), software engineers in India, mechanical engineers in Europe, and production liaisons in Taiwan. In your mind's eye, the members of your virtual team probably communicate mainly through e-mail. As a consequence, the design task is highly structured, because it would be a hopeless task to have everyone coordinate with everyone individually through electronic means.

"Virtual teams" are a relatively recent phenomenon (Andres, 2002). Virtual teams are groups of individuals collaborating in the execution of a specific project while geographically and often even temporally distributed, possibly anywhere within (and beyond) their parent organization. Virtual teams work across boundaries of time and space by utilizing mainly communication modalities that are asynchronous (e.g., an e-mail message is sent today but may be read and answered tomorrow) and apersonal (e.g., the team member sending the e-mail message cannot use visual nor vocal cues to evaluate how the receiver interprets the message). Although virtual teams have become increasingly common in most business areas, the use of virtual teams is outright inevitable in NPD.

Compared to traditional teams, virtual teams are significantly different. In the proverbial traditional team, the members work next to each other; in the proverbial virtual team, members work in different locations. In traditional teams the coordination of tasks is simple and performed by the members of the team together; in virtual teams tasks are much more highly structured. And virtual teams rely on electronic communication, as opposed to the face-to-face communication used in traditional teams. Table 5-1 summarizes the distinc-

TABLE 5-1.
Fully Traditional Teams versus Fully Virtual Teams

	Fully Traditional Team	Fully Virtual Team
Proximity	Team members all co-located.	Team members all in different locations.
Communication mode	Team members only communicate face-to-face (i.e., synchronous and personal).	Team members only communicate through asynchronous and personal means.
Team task structure	Team members coordinate the team task together, in mutual adjustment.	The team task is so highly structured that coordination by team members is rarely necessary.

tions between the two extreme forms. As you will undoubtedly notice, these extreme forms are rare and most teams have characteristics that are indicative of both traditional and virtual types of teams.

THE VIRTUALITY FINGERPRINT

It has become common for people to talk about virtual teams as though teams are either virtual or not virtual. However, although virtuality has become important in NPD, purely virtual forms are extremely rare. In the purely virtual NPD team, its members are fully geographically and temporally dispersed and communication is maintained solely through apersonal, asynchronous means. But this structure will rarely arise in NPD environments: Even in large-scale projects, such as in the development of a new airplane or a satellite, with specialists scattered and consulted across the world, chunks of the work are done by individuals located in the same building or on the same complex. Representatives (or, sometimes, all members) of dispersed teams often travel around the world and meet face-to-face. While virtuality in NPD has strongly increased and will likely continue to increase, the entirely virtual NPD team still remains scarce. Even in the development of Linux—the most common example given by people who herald the power of a virtual team—many personal face-to-face contacts were necessary over the course of the project. The other extreme, the conventional, fully co-located, face-to-face new product team in which all specialists work under the same physical roof and all communication is personal and synchronous is also increasingly unlikely. In reality, most NPD teams employ specialists (and customers) from various parts of the world or from other buildings and communicate through electronic means—at least to some extent.

Since aspects of virtuality occur in any modern NPD team and virtuality is a continuum, it makes no sense to either manage the team as if it is "virtual" or manage it as if it is a "nonvirtual" team. Rather than treating a team as either virtual or nonvirtual, one gains much better insight in how to manage a NPD team by considering how virtual it really is. For this purpose, this chapter introduces an instrument called the team's Virtuality Fingerprint. The team's Virtuality Fingerprint covers the three factors summarized in Table 5-1: team member proximity, communication modality, and team task structure. These factors are exactly equal to the model by Urban et al. (1995) that has recently become influential in the academic literature on virtual team design. Finding a team's Virtuality Fingerprint is easy. Figure 5-1 contains characteristic descriptions, or vignettes, for the NPD team. For each box, tick the vignette that is most appropriate for your team. To each box "virtuality points" have been assigned, running from 1 to 5 (for now, discard the "flexibility points"). These scores are used to fill out Table 5-2, for later reference.

The higher the scores, the more virtual the team. In our experience, after determining their team's Virtuality Fingerprint, most managers realize their virtual teams really aren't that virtual after all. More importantly, they realize that their teams may score high on one or two virtuality dimensions but low in another. Conversely, the Virtuality Fingerprint often makes managers of nonvirtual teams realize that their teams are not really that nonvirtual. For example, team members may be co-located in the same building but still communicate with each other largely through electronic means. Such a "nonvirtual" team then has some fairly virtual characteristics. Determining a team's Virtuality Fingerprint is easy and takes only a few minutes. But it is often extremely powerful, because it helps explain why certain "virtual" teams do not function well. In the remainder of this chapter, we focus on how a team's Virtuality Fingerprint tends to affect NPD team *creative* performance (see the sidebar titled "What Is Team Creativity?" for a brief conceptual discussion).

TEAM VIRTUALITY AND CREATIVITY

In this section we discuss how the three dimensions of virtuality each affect NPD team creativity. This section provides a brief overview of the latest findings in this area (see the sidebar titled "Individual and Team Creativity" for some references on creativity research). In particular, it will show how one can determine how much creativity a team can generate and how much virtuality is required for the team to generate the level of creativity that is needed.

First, we discuss how the three virtuality dimensions affect NPD team creativity. The dimensions discussed have been shown repeatedly in scientific research to have strong effects on team-level creativity. There are other factors that affect team-level creativity as well; however, the three discussed are among the most important and are related to the virtuality of NPD teams.

In the following three boxes characteristic descriptions (vignettes) for your product development project are presented. For each box please tick the vignette that is *most appropriate* for your project. At the end, plot the indicated scores along the corresponding sides of the virtuality pyramid in order to get your real-time project diagnosis.

Box 1: Team Member Proximity

□ All members of the NPD team are located in the same geographical location. Furthermore, they can and will meet each other in person almost every working day. They even might reside in the same room or laboratory (Virtually: 1 point; Flexibility: 1 point).

□ For the greater part the team members are located in the same geographical location, where they can and do meet in person. The distributed members of the team, however, experience difficulties in changing their location and therefore cannot attend most of the plenary meetings in person or join laboratory research at the main office of the team (Virtuality: 2 points; Flexibility: 2 points).

□ All members of the NPD team have a suitable home base, but they can and do change their geographic location whenever required. The flexible team members can meet in all offices and all laboratories when required; anytime, anywhere (Virtuality: 3 points; Flexibility: 3 points).

□ The members of the NPD team are partly located in the same geographical location. Other members of the team are on different sites. It is difficult—however, not impossible—for the team members to travel and to meet in person in the office or the laboratory (Virtuality: 4 points; Flexibility: 2 points).

□ All members of the NPD team are all sited on different geographical locations. They can or will never change their location and will therefore never be located closer to one another. For this reason, the members of the team will never meet in each other in person (Virtuality: 5 points; Flexibility: 1 point).

Box 2: Communication Modality

□ Face-to-face contact is the main communication modality used by the members of the NPD team. Up to 80 percent of the communication is generated in personal contact. This type of contact is characterized by visual and synchronous interaction (Virtuality: 1 point; Flexibility: 1 point).

□ The emphasis is on face-to-face and telephone contact. From time to time, e-mail contact is used. The NPD team focuses on synchronous and personal visual and vocal contacts (Virtuality: 2 points; Flexibility: 2 points).

□ All team members use all communication modalities, e.g, face-to-face contact, e-mail, postal mail, groupware systems, telephone, and the like. The modalities are used in a flexible way; they make use of every modality when required. This flexible communication infrastructure balances synchronous and asynchronous contacts, as well as visual and nonvisual contacts (Virtuality: 3 points; Flexibility: 3 points).

FIGURE 5-1. Virtuality Vignettes

□ The emphasis is on postal mail, fax, e-bulletin boards, and e-mail contact. The NPD team focuses on asynchronous and nonpersonal (nonvisual, nonvocal) contacts (Virtuality: 4 points; Flexibility: 2 points).

□ The members of the NPD team use postal mail services as their most important communication modality. This type of contact is characterized by (almost) exclusive use of nonpersonal and asynchronous interaction (Virtuality: 5 points; Flexibility: 1 point).

Box 3: Team Task Structure

□ The team assignment is well established—that is, the new product is well defined and all action relevant aspects are covered. Relatively simple tasks result from the assignment and are executed under strict coordination of the team leader (Virtuality: 1 point; Flexibility: 1 point).

□ For the most part the team assignment is well established. However, some aspects of the new product are not specified in full detail. This results in relatively simple tasks that are executed under relatively strict supervision of the team leader. Team members demonstrate some mutual interaction to cope with the few emergent aspects of the NPD project (Virtuality: 2 points; Flexibility: 2 points).

□ The team assignment partially has an emergent character and is partially well established. This results in a mix of relatively simple tasks and more complex tasks. These interrelated tasks are coordinated in a flexible way, with more strict coordination of some team members and more autonomy for other team members. The team leader changes his coordination plan whenever necessary according to the requirements of the NPD project and facilitates and supports input by team members on project structure and definition (Virtuality: 3 points; Flexibility; 3 points).

□ For the most part the team assignment has an emergent character—that is, the new product is not fully specified. Relatively complex tasks are assigned to the team members. The team members are relatively autonomous and demonstrate mutual interaction to cope with all the emergent aspects of the NPD project (Virtuality: 4 points; Flexibility: 2 points).

□ The team assignment has an emergent character—that is, the new product is ill defined, since it is impossible to specify all action-relevant aspects at this time. Very complex tasks result from the assignment and are executed by very autonomous team members. Team members coordinate tasks largely through lateral communication (Virtuality: 5 points; Flexibility: 1 point).

FIGURE 5-1. *(continued)*

Team Member Proximity and NPD Team Creativity

Most managers think that NPD team members need to be located closely to each other if they are to generate high levels of creativity together. This is correct—at least, to some extent. The basic argument is that the more proximate team members are, the better the opportunity to make contact with each other. This provides them with opportunities to become acquainted and find out common interests. Moreover, proximity supports trust. NPD professionals are

TABLE 5-2.
Team Virtuality Fingerprint

DATE: TEAM NAME:					
Proximity Communication modality Team task structure					

often more prone to ask questions of others they trust and feel comfortable with. In fact, proximity increases overall communication. Communication has been described in fostering the creativity of teams, because the dissemination of ideas and information is facilitated when team members communicate more. In other words, the closer the members of the multidisciplinary team are located to one another, the more they can and will exchange ideas and knowledge, and thus the more novel ideas and knowledge that can be generated.

While this argument makes much sense, it is also easy to argue against the positive effects of team member proximity on team-level creativity. First of all, the presence of others can serve as a distraction. It has been suggested that scientists need to work alone and that their social needs distract them from their work (Lovelace, 1986; Baron, 1986; Guerin, 1986). As a consequence, high levels of interaction should decrease creativity. This effect has often been observed and described by many psychologists in their studies of creative groups (both in NPD settings and in other settings). Moreover, distracted team

WHAT IS TEAM CREATIVITY?

The most common definition of creativity in organizational settings is that of Amabile (1988, 1996), who argues that creativity is exhibited when a product or service is generated that is both novel and useful with respect to the firm. With respect to teams, creativity refers to the team's ability to come up with new ideas, methods, approaches, inventions, or applications. Whereas Amabile's definition has become generally accepted and used by researchers in the field, the way in which the novelty and usefulness can be measured empirically remains open to a lively and unrelenting debate. In our own studies of how virtuality affects NPD team creativity, we have used the measures suggested by Leenders, van Engelen, and Kratzer (2003), who use the subjective measures of team creativity provided by outside team leaders and inside team members and measure NPD team creativity as the extent to which the team is better at developing new ideas, methods, approaches, inventions, or applications than other teams in the same industry or company.

INDIVIDUAL AND TEAM CREATIVITY

In this chapter, we show how the virtuality of the team affects team-level creativity. Of course, other factors affect team creativity as well. Most studies of creativity focus on the creativity of individuals, and thousands of books and academic articles have been written about this topic. The study of team-level creativity is fairly new, because of its empirical complexity. Still, the current findings and theories are remarkably congruent and differ fairly little from the studies of individual creativity. The main recent studies of team-level creativity (and some other group-level creativity studies) show that the main determinant of team creativity is in the interaction among team members, rather than, for example, in the levels of individual creativity. NPD team creativity requires the combination and integration of input from multiple NPD team members. Through effective communication, building on the knowledge of other team members, team members exchange information and create new knowledge and insights. To achieve innovation, there must be ideas, and these initially appear from among individuals in the team. A new idea dies unless it finds a breeding place. Developing, refining, testing, and in the end implementing these ideas further rests on interaction among the team members. Creativity does not happen inside people's heads, but in interaction (Csikszentmihalyi, 1996).

You might refer to, for example, Leenders, van Engelen, and Kratzer (2003); Albrecht and Ropp (1984); Nemiro (2002); Kratzer (2001); Amabile (1996); Woodman, Sawyer, and Griffin (1993); King and Anderson (1990); Glynn (1996); Drazin, Glynn, and Kazanjian (1999); Lovelace, Shapiro, and Weingart (2001), Agrell and Gustafson (1996); Csikszentmihalyi (1996), Payne (1990); Visart (1979); and Brauner and Scholl (2000).

members are likely to start distracting other team members and, therefore, decrease their creativity as well. As a result, when team members start distracting others, the team as a whole spirals into low levels of creativity, as the distraction effect will accelerate among interacting members (Shalley, 1995). This effect occurs in many "open" offices and often before management senses it happening.

In addition, the more NPD team members run into each other and discuss new ideas and potential solutions, the more teams may be carried along by the momentum of their enthusiasm for an innovative idea, rather than by a clear understanding of its real value. This can decrease critical thinking and can push team members toward mutual beliefs and thereby reduce the number and quality of problem solutions generated by the team as a whole. In other words, proximity may introduce "group-think" into the NPD team, and this effectively kills the creative achievement of the team. The time it will take before groupthink becomes an issue often depends on the extent to which the team is cross-functional. Teams that are not (or hardly) cross-functional are likely to fall victim to groupthink much earlier than highly cross-functional teams. But even highly cross-functional teams tend to groupthink after having worked together on a new product for a sufficiently long time. It is no coincidence that Irving Janis, the psychologist who introduced the world to the term "groupthink," has based most of his analysis on the study of cross-functional teams. In his opinion, groupthink lurks whenever a group of qualified experts convenes to make a decision, group cohesiveness is high, and a feeling of mutual cooperation is present. The groupthink effect arises much quicker than teams and team managers expect. In fact, by the time the team itself realizes (or is willing to admit) the presence of groupthink, its effects have often been present for weeks or even months.

These kinds of arguments, in conjunction with practical experience, show that holding on to either extreme (team members being either always highly co-located or constantly highly dispersed) impedes NPD team creativity. However, when a team can flexibly move team members between co-location and dispersion and is able to extract the benefits of both extremes, it maximizes its creative potential. In such teams, team members interact frequently enough to discuss ideas and disseminate knowledge and information in a timely manner, are not distracted from their individual tasks, and strongly reduce the risk of groupthink. Such a position, in which the use of co-location and dispersion are balanced in a flexible manner, is highly conducive to creativity (see A in Figure 5-2).

Communication Modality and NPD Team Creativity

Face-to-face communication is the oldest form of communication. This type of communication has many advantages. For one thing, the message is received at

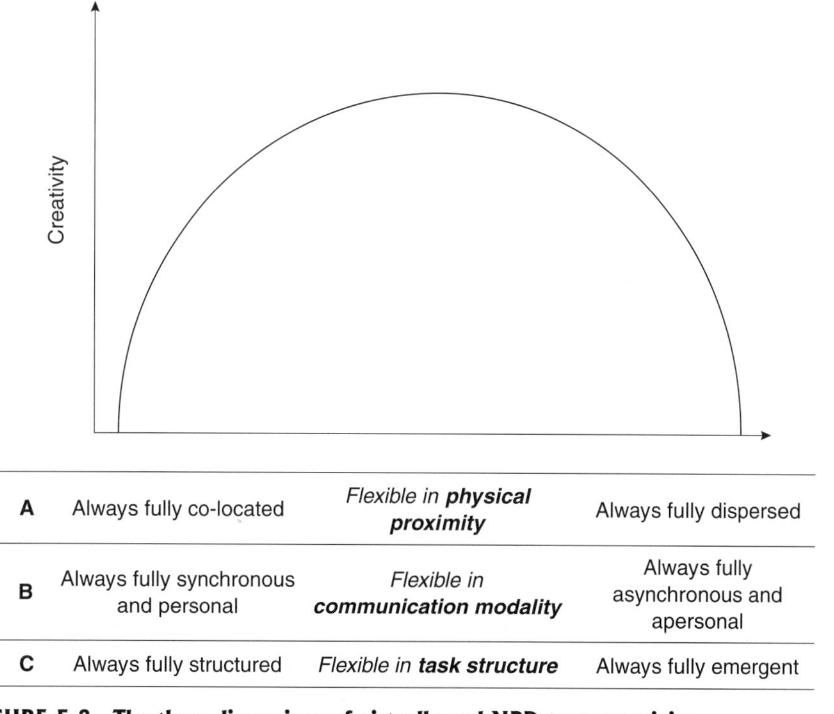

Creativity
A
Always fully co-located
Flexible in physical proximity
Always fully dispersed
B
Always fully synchronous and personal
Flexible in communication modality
Always fully asynchronous and apersonal
C
Always fully structured
Flexible in task structure
Always fully emergent

FIGURE 5-2. The three dimensions of virtually and NPD team creativity.

the same time as it told, and response also follows immediately. Therefore, the medium is fully synchronous. Also, both parties can use vocal and visual cues to check whether the other person understands the message, agrees with it, or has some other noteworthy response to it. Therefore, the medium is fully personal. This also helps in creating trust among team members, which is known to be advantageous to joint creativity. Moreover, since team members actively and directly interact with each other, face-to-face communication allows team members to discuss topics of a highly complex nature. This is advantageous if creative solutions need to be found for complex problems.

Unfortunately, face-to-face communication can also take much time, especially when the number of team members involved is large. Organizing meetings or coordinating large groups through bilateral face-to-face communication can be extremely time-consuming and, as studies have shown, generally decreases the creativity of the NPD team as a whole.

As teams become increasingly virtual, they tend to rely more on electronic means of communication. These can reduce or alter some of the temporal, physical, and social constraints on communication. To some extent, this is beneficial to NPD team creativity, as a lot of information can be easily disseminated in a relatively fast and reliable fashion. It makes sure that everyone receives the same information. It does, however, not ensure that everyone interprets the information in the way it was meant. Also, it does not ensure that everyone receives and reads the message at the same time; the time lag can sometimes be quite large. This can stall the creative process, as it makes people wait for responses. In addition, researchers have found that computer-mediated teams tend to take longer to complete a complex creative task. Discussing complex problems via e-mail or other asynchronous and apersonal modalities simply often takes longer. In general, teams that exclusively employ asynchronous and apersonal modalities in their effort to generate creative solutions to team-level tasks do as poorly as teams that attempt to solve these problems through constant and extensive joint face-to-face meetings. Research shows that both extreme forms take longer to find solutions and solutions are just not as creative.

The highest levels of creativity tend to be found in teams that move flexibly between modes of communication. Such teams use periodic face-to-face encounters or information technologies that simulate face-to-face contact, or both, in addition to the use by these teams of highly asynchronous and apersonal communication—all whenever each mode is most needed. In other words, the most creative teams tend to be those that flexibly employ and have at their disposal an array of communication modalities. In such teams, members get together and discuss matters face-to-face when complex multidisciplinary issues are to be resolved and decisions have to be taken, but use electronic and even postal means when they need to resolve more individualized matters or when issues are relatively simple (see B in Figure 5-2).

Team Task Structure and NPD Team Creativity

The structure of the team task defines the nature of the interactions that ensue in order for the NPD team to complete its team task. The creation of work-related interdependencies within NPD teams has an immediate effect on team member communication. If tasks in NPD teams are not interdependent, there is no need or pressing reason for them to work together and to communicate (DeSanctis et al., 1999; Cohen and Mankin, 1999). This is often detrimental to team-level creativity. At the same time, if everybody depends on everybody else, it is a miracle if a solution is reached at all (Kratzer, 2001)—not in the least because this would make highly complex overall tasks practically unmanageable. It would be more of a miracle if this solution were truly creative as well.

By actively shaping and reshaping the interdependencies, the team manager can alter who communicates with whom and, thus, who shares information, ideas, and knowledge with whom. This has an immediate and strong effect on the level of creativity the team can sustain. In general, strongly decomposing the team task into largely independent subtasks increases time to market but tends to reduce the innovative character of the overall solution. On the other hand, no decomposition at all makes everyone dependent on everyone and is equally detrimental to team creativity. Also, the more clearly and deeply defined the product specifications are that the team needs to adhere to, the stronger tasks tend to be decomposed and the less challenging the resulting tasks tend to become. This tends to lower the team's creativity. On the other hand, leaving out any definition at all may allow the team with the most creative freedom, but the lack of any guidance and boundaries tends to make the overall task so complex and overwhelming that creativity can no longer be expected.

Team managers, in their attempt to resolve these problems, often choose either of two solutions: taking a fully central position in the team or placing themselves in the periphery. Neither is advisable when it comes to creativity. Managers that place themselves at the heart of the team, are central to all the action, and determine every task and assignment by themselves effectively demolish the creative process in NPD teams to such an extent that most team-level creativity is destroyed. Similarly, when the team manager always stays at a distance and does not interfere with the team process at all, teams operate without any boundaries to guide them. Such teams are likely to come up with solutions that do not fit the company's strategy or are not feasible for other reasons. These teams rarely produce truly creative ideas.

As research and practice have shown, teams are most creative when their tasks are somewhat defined and structured in advance but leave them with enough creative space to work with. They are most creative when the team leader is clearly present but strongly facilitates team initiative. There are boundaries, and these are clear to all, but they are wide and flexible enough to allow for and support the creative process in the team. Creative NPD teams have considerable leeway in defining their project, both in terms of the product to be developed and the way in which they structure their approach. In other words, NPD team creativity is supported when it navigates along the "team task structure" dimension and moves between the extremes in a flexible fashion (see C in Figure 5-2).

Taking the Arguments Together

As we have argued, strict extreme forms tend to be detrimental to creativity. The flexible situation in which the advantages of both extremes are harvested (by using elements of both) and the disadvantages are prevented (by not moving too far to either extreme) is by and large the most conducive to NPD team creativity. Figure 5-2 shows this graphically. Flexible teams actively, dynamically, and purposefully employ the benefits of the entire virtuality spectrum. For example, a new product team that operates in a truly flexible manner with respect to communication modality will make use of fully synchronous and personal communication when the most complex, multidisciplinary issues need to be discussed and final decisions have to be taken and it will move away from that extreme as soon as possible. Similarly, the team will make use of fully asynchronous and apersonal means of communication when members are working on relatively simple or specialized issues. Teams that skillfully navigate between the extremes of each dimension boldly move to either extreme when (and only when) necessary and move away from there whenever possible have the highest creative potential. Teams that are less flexible and are stuck close to either extreme are seldom able to reach a truly and sustainable creative performance.

Providing teams with high flexibility when high creativity is demanded and lower flexibility when lower creativity is demanded requires skillful leadership. This is no easy task. The fact that only few leaders are consistently successful in managing team creativity seems to be largely due to two factors. First, most team leaders do not know what makes a team more or less creative. As a result, they have no systematic means at their disposal that they can use if a team needs to be more or less creative. The tools of this chapter should help to solve this factor. Second, it requires leadership to apply the tools and implement its recommendations. To do this, a leader needs to both know *what* to do (the recommendations in this chapter provide options as to what can be done) and *how* to make it happen in his or her company. The ability to do all this is an important determinant of how skillful and successful one truly is as a NPD team or project leader.

In the next section we discuss how you can determine the creative potential of NPD teams.

THE VIRTUALITY PYRAMID

Having established how each of the Virtuality Fingerprint's dimensions determines the team's level of creativity, in this section we show how much creativity a given team may potentially support. Some Virtuality Fingerprints support much creativity, making it possible for a team to be profoundly creative. Other Virtuality Fingerprints, however, make it nearly impossible to attain genuinely creative solutions, let alone sustain such creative performance over extended periods of time. Given the level of creativity that is required in the NPD project, it is now possible to analyze whether a team will be able to be creative enough or is likely to be overly creative and therefore inefficient and ineffective. The Virtuality Pyramid has been designed to enable team leaders to find how much creativity their NPD teams can generate and sustain; it takes a team's Virtuality Fingerprint as a starting point.

The first step that needs to be taken is to translate the team's Virtuality Fingerprint into "flexibility scores." These scores can be found in the vignettes of Figure 5-1. The relation between the team's virtuality scores and its flexibility scores are straightforward: 1 point is awarded to either extreme and 3 points to the middle position. As you may recall from the previous section, flexibility relates to the extent to which a team actively, dynamically, and purposefully moves along each axis. These flexibility scores can be used to fill out the first three rows of Table 5-3, for later reference.

Now imagine a triangle with each side representing the flexibility score on either dimension (see Figure 5-3). If a 3 is scored on each dimension, the triangle will be large and have equal sides (see A in Figure 5-3). If a 2 is scored on each dimension, the triangle again has equal sides, but its surface is smaller (B

TABLE 5-3.
Team Flexibility Score

DATE: TEAM NAME:					
Proximity Communication modality Team task structure Required creativity					

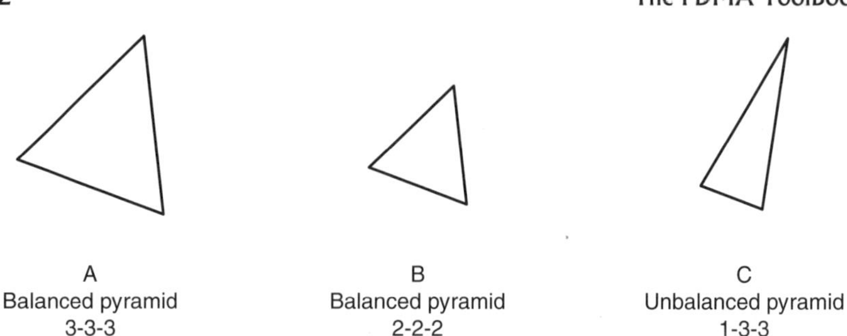

FIGURE 5-3. Several flexibility configurations.

in Figure 5-3). Since the higher the flexibility on a given dimension, the better it is for team creativity; teams with a larger triangle can be more creative. The 3-3-3 team can therefore generate and sustain more creativity than the 2-2-2 team.

If a team scores a 1 on one dimension and a 3 on two others, the triangle will lopsided (see C in Figure 5-3). This team could have supported much creativity if it wasn't held back by one dimension. For example, if a team is partly co-located and partly dispersed, with excellent traveling facilities, its flexibility score for proximity is 3. Suppose that the team also has a wide array of communication modalities at its disposal and uses all of them flexibly. The flexibility score on communication modality then is 3 as well. However, if the leader of the team strictly coordinates all team processes, insists that communication occurs vertically, strongly predefines the team's outcome, and leaves the team without room to develop its own processes and define part of its own task, the flexibility score on team task structure is 1. It is clear immediately that such a team cannot perform highly creatively. Its lack of flexibility in task structure eliminates most of its creative potential. In such a team, no travel budget will make the team generate more novel solutions.

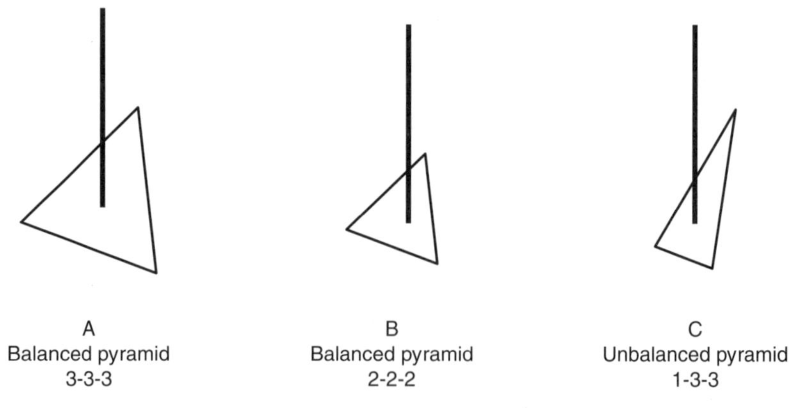

FIGURE 5-4. Several flexibility configurations and creativity.

The relation between creativity and flexibility can be visualized by considering these triangles as the base and the level of creativity as a mast that is to be supported by this base. In Figure 5-4, we have added a mast to each of the three triangles, the height of which demonstrates how much creativity the configuration can support. The 3-3-3 triangle can easily support a high mast; it can support high levels of creativity. This level of creativity is too much for the 2-2-2 triangle. A slightly shorter mast, however, is again would be supported in a stable fashion. The 1-3-3 triangle would tip over if it had to support such a high mast; it is only stable enough to support a low level of creativity.

A tool called the Virtuality Pyramid brings all of this together. The Virtuality Pyramid shows whether the extent to which an NPD team acts flexibly along the three virtuality dimensions allows the team to generate the creativity that is required of it. The Virtuality Pyramid is shown in Figure 5-5. The pyra-

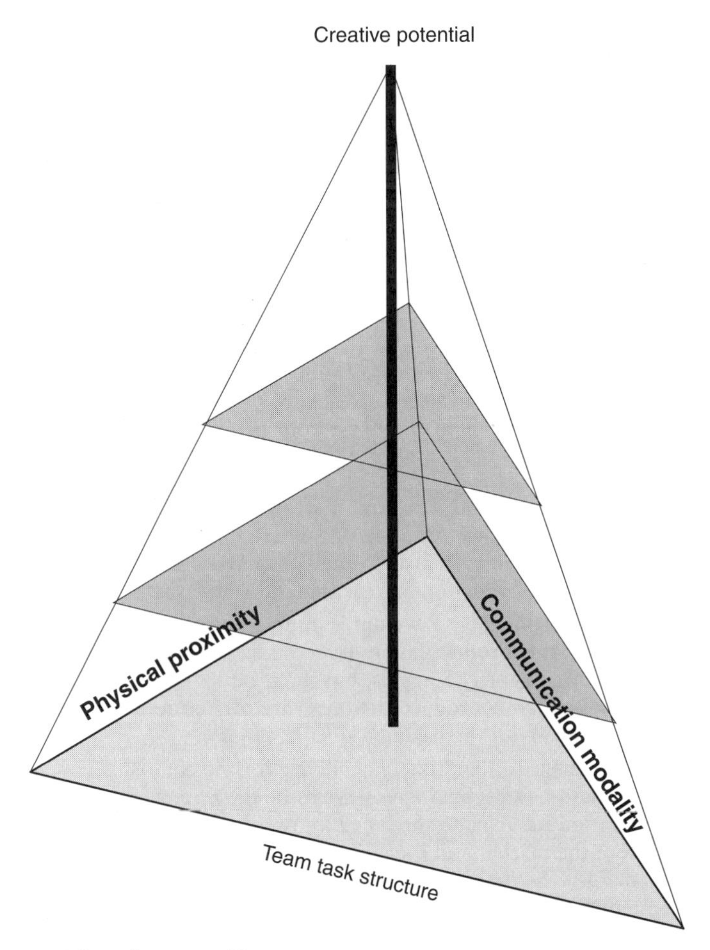

FIGURE 5-5. Virtuality pyramid.

mid has an equilateral triangle as its base. The height of the pyramid represents the amount of creativity that can be supported by the team. It runs through the center of the base (the mast). In addition to the triangle that forms the base of the pyramid, you also see two smaller triangles. The top triangle represents the 1-1-1 flexibility configuration, the middle triangle represents the 2-2-2 team, and the base of the pyramid represents the 3-3-3 team. The Virtuality Pyramid is applied as follows. First, refer to Figure 5-6 and consider the vignettes that describe the level of creativity that is required of the team. Tick the vignette that fits the team's situation best. Each vignette has a number of points attached to it, running from 1 to 3. The scores can be recorded in the bottom row of Table 5-3, for later reference.

If a required creativity score of 1 is found, this means that team flexibility 1-1-1 best fits its creativity needs. It is stable, cost-efficient, and keeps the creativity of your team in check. After all, only a very limited amount of creativity is desired of the team. If the team has a score of 2 for creativity, the best-fitting flexibility configuration is 2-2-2. This kind of team has the most optimal setup for generating a reasonable amount of creativity. If it were less flexible, it would not be able to generate enough creativity; if it were more flexible, the team may be tempted to entirely new, highly creative technologies that do not fit with the firm's strategy. If score of 3 is most appropriate, full flexibility on all dimensions (3-3-3) is necessary to generate enough creative achievement. If any dimension were less flexible, the team would not be able to consistently achieve solutions of high enough creativity.

In the following boxes characteristic descriptions (vignettes) regarding the creativity required by your product development project are presented. For each box please tick the vignette that is *most appropriate* for your project.

Box: Required Creativity

☐ The design assignment of the NPD team is straightforward and based on existing and well-known technologies. All design goals are fully established and task uncertainty is extremely low. In fact, the new product is a redesign (new version of release) of an existing product on an existing and well-known market (1 point).

☐ The design assessment of the NPD team is quite ambitious but is limited to be an extension of an existing product family (or product line). Subsystems of the new product may require new technologies, but the product is mainly based on existing technologies. New product attributes are offered to users that are generally well known with the basic product (2 points).

☐ The design assignment is the ultimate challenge for the company and the NPD team. Several brand-new technologies have to be developed and engineered to create the new product. A wide variety of knowledge and capabilities are to be originated. Both the market and the customer needs for the new product are starting to emerge. The design task is ambitious but in very general and therefore nonspecific terms (3 points).

FIGURE 5-6. Creativity Vignettes

Unfortunately, teams rarely are fully balanced. More likely, your team's score is too high or too low on one or two dimensions and only just right on the other one or two. We advise to try to fit your triangle into the pyramid. It may have to go diagonally. With the creativity axes running through its center of gravity, it then immediately become clear how easy it will be for it to tip over.

After you use Figures 5-1 and 5-6 to determine the virtuality of a team and the level of creativity that is to be generated, you need to ask yourself two questions:

- Is the flexibility triangle stable, in other words, does the team have the same score for each dimension?
- Is the flexibility triangle large/small enough to support the level of creativity required?

Most likely, the answer will be no to at least one of these questions. Most team leaders find themselves answering no twice. In the next section we briefly discuss what can be done about that.

MANAGING TEAMS TOWARD CREATIVITY

If a team does not reach a perfect 3-3-3 for the creativity need of 3, or a perfect 1-1-1 for the creativity requirement of 1, the solution is straightforward: Shrink or stretch the failing dimensions. In other words, given the level of creativity that is required of the team, decrease or increase the flexibility of the delinquent dimensions such that it corresponds to the team's creativity requirement. How to do this is often highly specific to the situation. Your company may have several options in place, or you may only have a very limited set of alternatives at your disposal. This section provides some suggestions on what to do. Of course, each manager can adjust this list with the options provided by his or her own company/situation.

In the following we briefly describe some options. We do this separately for each dimension. The arguments are summarized in Figures 5-7A, 5-7B, and 5-7C. In these figures, the five scores of the Virtuality Fingerprint (you can replace them with the flexibility scores if this is easier) are printed on the inverse U's. Between the numbers are some highlights of what one can do to move from one score to the next. Above the inverse U, the figure moves from left to right, increasing the level of virtuality. Below the inverse U, it moves from right to left, decreasing the level of virtuality.

The options we provide are fairly general and are meant to inspire the reader to consider alternatives that fit his or her own unique situation. Before we present them, we make two remarks. First, we do not discuss any options that entail changing team members. This is, of course, an option that may be valid in some situations, if you have the authority to do so. However, the options that follow also work without changing the composition of the team. Second, the implementation of our recommendations can be done in a drastic

PROXIMITY MANAGEMENT (EXAMPLE)

A team that was set up to develop a new product line in the household electronics industry went about proximity management in a very natural manner. Many of them were physically located in the same complex (although not in the same building) at the beginning of the project. In the early stages of the project, they spent much time together, in hospitals, research labs, congresses, and local watering holes. This allowed them to get to know each other well (many of them did not know each other at the outset) and discuss and come up with ideas in a less formal setting. This allowed them to develop some genuinely creative broad ideas on the products that were to be developed. Within the strict guidelines for the product the company had set, the team used these face-to-face sessions to design rough outlines of the "solution space" and potentially lucrative directions. After that, tasks are decomposed and divided (not solely at the discretion of the team members) and the team members start the work on the functional design of several of the product's components. From here on, much of the work occurs in different offices, some involved scientists work at home, and others are even moved to different parts of the company in other parts of the world. The performance of the team remains high, but the proximity is reduced considerably. Of course, at various times during the project, several team members travel to meet with each other face-to-face, sometimes for weeks, and return home when face-to-face meetings are no longer needed. By starting off with high levels of proximity, reducing proximity when possible, but increasing it again when required, the team was able to remain highly creative over its lifetime. Interestingly, another team in the same company was abandoned after having worked on a related project. The reason: its work was not innovative enough, according to the company's management. Management did not understand why, because they had put their best manager on the project and had put most involved researchers on the same floor in one building.

or subtle manner. For example, while a team leader may not have the authority or budget to change team member proximity by physically moving members to different buildings, team member proximity can also be effectively altered by moving some team members to different areas of the same building, to a different floor, or by reassigning desk configuration (e.g., Allen, 1977; Keller, 1986; Van den Bulte and Moenaert, 1998). Such actions are often possible within team managers' authority and budget. We therefore recommend that you consider how these options can be applied in your own specific case; all this requires is a little creativity! We provide some examples in sidebars titled, "Proximity Management (Example)," "Communication Modalities (Example)," and "Managing Task Structure (Example)," and Managing NPD Team Virtuality (Example)."

Team Member Proximity

Referring to Figure 5-7A, note the following:

1 → **2.** The team members are fully co-located. To move up one notch on the proximity-flexibility dimension, locate parts of the team in other loca-

COMMUNICATION MODALITIES (EXAMPLE)

A team at a large European developer of copier machines was to develop a new printer engine—the heart of the printer. To manage the team's creativity, the team's use of communication modalities was managed, largely, as follows. During the conceptualization phase, in which the demands for creativity are perhaps highest, specialists from various disciplines (R&D, marketing, production, services, sales, control, etc.) work on the conceptual design of the printer engine. While this could be done through electronic devices, it is more fruitful to exchange sketches, graphs, simulations, and so forth through face-to-face contact. This also allowed for the use of intensive visualization techniques with all team members. This created "short and fast lines" and allowed for the quick and reliable transfer of "thick" knowledge and information. In this manner, the team was able to reach a good and shared overview of the new product. On the other hand, in the phase of detail design, face-to-face contact would be too distracting and unproductive. Moreover, it would not help to increase creativity and, as history had shown, was even likely to decrease creativity. Here, electronic modes of communication were recommended, along with the use of couriers (within the company, within the same city) for intra-team communication. When decisions had to be made—especially design decisions that related to the interfaces between the decomposed tasks—more personal and synchronous modes of communication were used, such as face-to-face contact or video conferencing.

tions. As some level of dispersion is introduced to the team, the team leader also needs to create some, albeit minor, traveling facilities.

2 → 3. Disperse team members further. The team now consists of chunks that are dispersed; members within a chunk are primarily co-located. Co-locating is primarily based on competencies and subtasks. Offer excellent traveling facilities (both in terms of funding and logistic support), and support and stimulate any traveling of team members between locations.

MANAGING TASK STRUCTURE (EXAMPLE)

A pharmaceutical company was determined to find to find a new treatment for a specific coronary disease. This form of stepwise innovation requires the team to combine highly specialized fields of knowledge in biochemistry, medical science, and production (e.g., for producing the active ingredients). In this case, the design tasks are highly interdependent. The team task can barely be structured and is highly complex. After synthesizing the active ingredients, several subprojects are set up to run in parallel, including the upscaling of production, the testing of the medication's effect on animals, and starting the procedures for FDA approval. These tasks are hefty, but of comparatively low complexity. Many of these tasks can be performed based on existing procedures, the manner in which (sub)tasks are decomposed are rather well known in advance. The tasks can be divided over small groups, and each group can perform its work largely independent of the other groups. Team management tracks the progress of the subtasks and brings their results together. The high level of structure keeps the team going. Note that during this part of the project the task structure remains largely unchanged. This is done because a high level of creativity would not be productive at this stage.

MANAGING NPD TEAM VIRTUALITY (EXAMPLE)

At NASA and ESA (European Space Agency) the project structure has changed tremendously in the last decades. To be able to afford the instruments, which are developed with state-of-the-art performance, large consortia of institutes were formed. But also a wider variety of technical expertise needed for the instrumentation required a larger number of participating institutes in a consortium. Presently the largest instrument consortium at ESA comprises 25 institutes in 15 countries. This project, however, did not start out like this. First, a small group of highly experienced scientists and engineers was located at only one institute to set up the product concept and the general design of the required consortium. These early project members got to know each other and learned to understand each other. After the start-out phase of the project, the large consortium was formed according to the technical expertise demanded. A consortium of 25 institutes emerged after the product structure of the instrument was decomposed. So, the task structure was established, and because the institutes are geographically distant, most communication took place using electronic modes of communication. However, soon the intended project progress was not met, mainly because creativity was lacking, as management later diagnosed. Since most space developments are at the forefront of what is technically possible, they require high levels of creativity. To increase the creativity of the whole consortium, the project management decided to change the following things. First of all, consortium meetings were established allowing all members from all institutes face-to-face contact. Moreover, videoconferencing was used much more often to communicate between the institutes. Second, scientists and engineers from some institutes were temporarily located at other institutes, especially if there were communication problems diagnosed between them. And third, the structure of tasks was (partly) adapted to temporary demands. Sometimes one institute needed to work together with some other institute only temporarily, because of an agreement about a certain material or a certain test design. This changing task structure was now very explicitly managed. All in all, the investments of making all ingredients of virtuality more flexible were rewarded. As of this writing, the project is back on track and is likely to finish on time.

Strive for an optimal mix of team members being co-located and dispersed, of them traveling, and of them staying at a location.

3 → **4.** Largely disperse the team. At the same time, reduce traveling facilities (both in terms of budget and logistic arrangements). Traveling between locations should be restricted to only essential visits.

4 → **5.** Fully disperse the team. Only co-locate team members if there is no other way. Tell the team members to remain in their locations and retract the traveling facilities. Team members are expected to stay put.

5 → **4.** The members of the team are fully dispersed. The team leader can now co-locate small parts of the team but should only do this with members with highly interdependent tasks. In addition, some, albeit minor, traveling facilities are necessary.

4 → **3.** Co-locate team members further. The team now consists of chunks that are dispersed; members within a chunk are primarily co-located. Co-locating is primarily based on competencies and subtasks. Offer excellent traveling facilities (both in terms of funding and logistic support), and

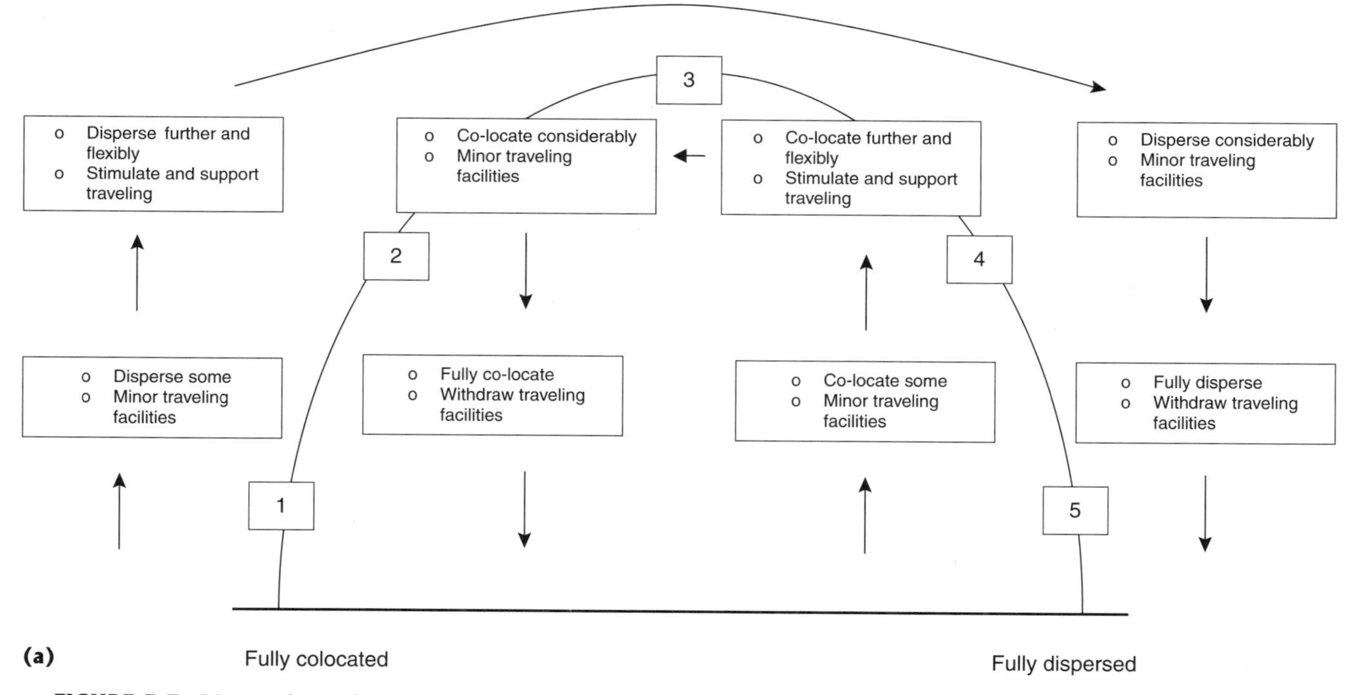

FIGURE 5-7. Managerial options for managing the NPD team's Virtuality (a) Physical proximity, (b) communication modality, and (c) task structure.

support and stimulate any traveling of team members between locations. Strive for an optimal mix of team members being co-located and dispersed, of them traveling, and of them staying at a location.

3 → **2.** Largely co-locate the team. At the same time, reduce traveling facilities (both in terms of budget and logistic arrangements). Traveling between locations should be restricted to only essential visits.

2 → **1.** Fully co-locate the team. Only allow dispersed members if there is absolutely no other way. Since team members are all in the same location, all traveling facilities can be canceled.

Communication Modality

Referring to Figure 5-7B, note the following:

1 → **2.** To move from 1 to 2, the team leader can introduce and support the use of the telephone and e-mail. Still, it is stipulated that most communication remain face-to-face.

2 → **3.** Introduce new communication modalities—in particular, electronic communication (including all kinds of groupware). After these additional modalities have been introduced, recommend the use of all and many different modalities of communication. Make sure the use of the modalities is flexible and balanced; electronic communication cannot take over face-to-face or other personalized forms of interaction. Insist on a mix of visual and nonvisual communication and on synchronous and asynchronous interaction.

3 → **4.** Emphasize the use of asynchronous, apersonal communication modes. In particular, urge team members to send e-mail, use other electronic communication modes, send faxes, and insist team members replace most face-to-face and telephone contact with these modalities.

4 → **5.** Disable most synchronous modes of communication. Insist that postal mail and other fully asynchronous and apersonal modes of communication be used—that synchronous and direct personal communication is unnecessary and a waste of resources.

5 → **4.** The team leader can introduce and support the use of e-mail and bulletin boards. This slightly decreases the asynchronous character of the team's communication. Still, stipulate that most communication remains through postal, apersonal, and asynchronous means.

4 → **3.** Stipulate face-to-face and telephone contact, to provide a flexible mix of multiple communication modalities. Insist that personal and synchronous modalities of communication are just as important as apersonal and asynchronous modalities.

3 → **2.** Emphasize the use of synchronous, personal communication modes that allow for visual and vocal cues. In particular, make sure to urge team

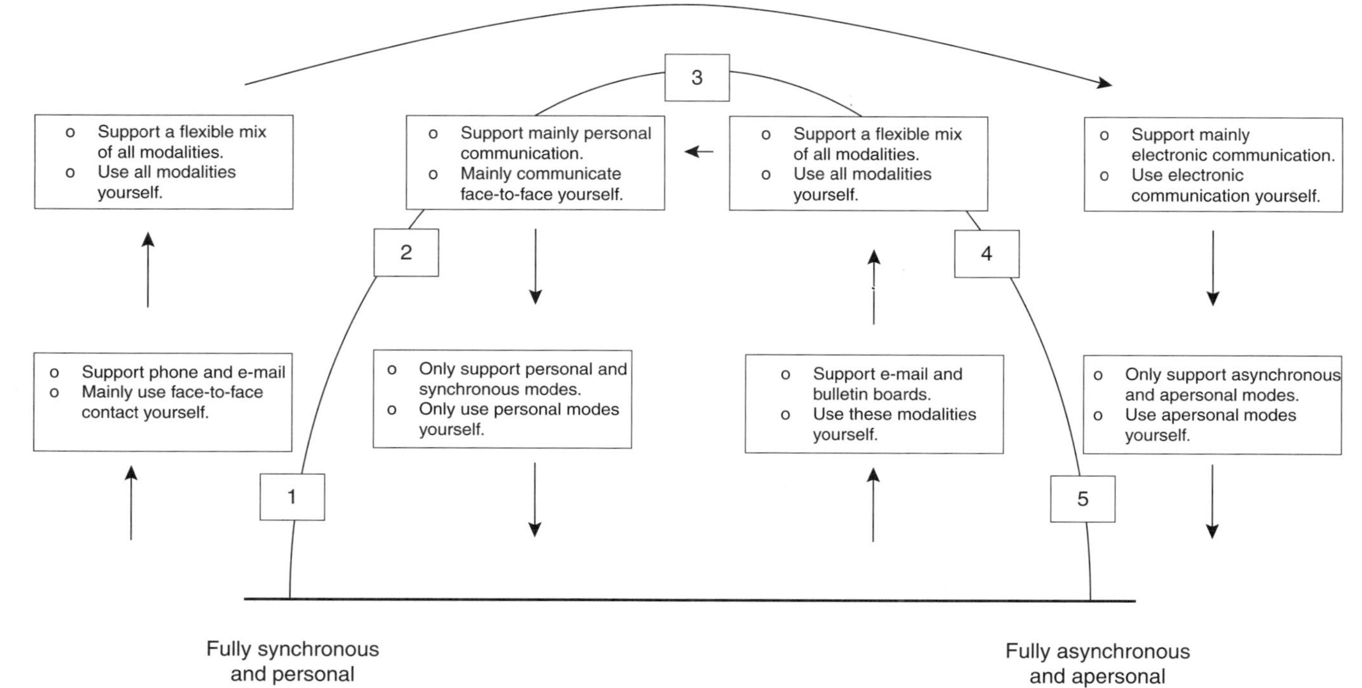

(b)

FIGURE 5-7. *(continued)*

members to mainly communicate face-to-face or by telephone. Insist they replace most electronic communication modes.

2 → **1.** Disable most asynchronous and apersonal modes of communication. Insist that only face-to-face contact be used.

Team Task Structure

Referring to Figure 5-7C, note the following:

1 → **2.** Introduce several compound subtasks to the team. Provide the team members with some freedom to redefine small parts of the team task. Loosen the definition on some aspects of the team task.

2 → **3.** Allow the team autonomy, wherever and whenever you deem possible. While the team leader maintains control over the team, he or she should gladly accept input by team members. Foster and nurture a flexible mix of vertical and lateral communication and coordination within the team. The definition of the team task structure is loosened to a considerable extent. Still, keep a keen focus on whether the team remains on track. Make sure the team task is structured and decomposed into coherent but not fully independent subtasks.

3 → **4.** Team member input on tasks and processes are very important. The team leader needs to facilitate and support this. Make sure that only vital aspects of the task are prespecified. Make sure to not interfere with the team's process, until things clearly run out of hand. Increasingly complex tasks can be assigned to the team.

4 → **5.** The team leader does not prespecify the product and task. This is left entirely to the team. The team members themselves handle coordination; the team is empowered to take all relevant decisions. Coordination is entirely lateral, and the leader does not interfere.

5 → **4.** Although much of it is left to the team, the team leader does prespecify parts of the manner in which the team structures its work. Also, provide the team with some early definitions of its task. This allows you to reduce the number of compound subtasks to be carried out by the team. Reduce the team's freedom to redefine its task.

4 → **3.** Increasingly (pre)determine the team's tasks. Also start to actively take part in the team's process. Still, team member input on tasks and processes are important. The team leader facilitates and supports this. Make sure to foster and nurture a flexible mix of vertical and lateral communication and coordination within the team. The definition of the team task structure is tightened to a considerable extent and increased in depth.

3 → **2.** The team leader increases his or her control over the team but still accepts some input by team members. Vertical coordination becomes

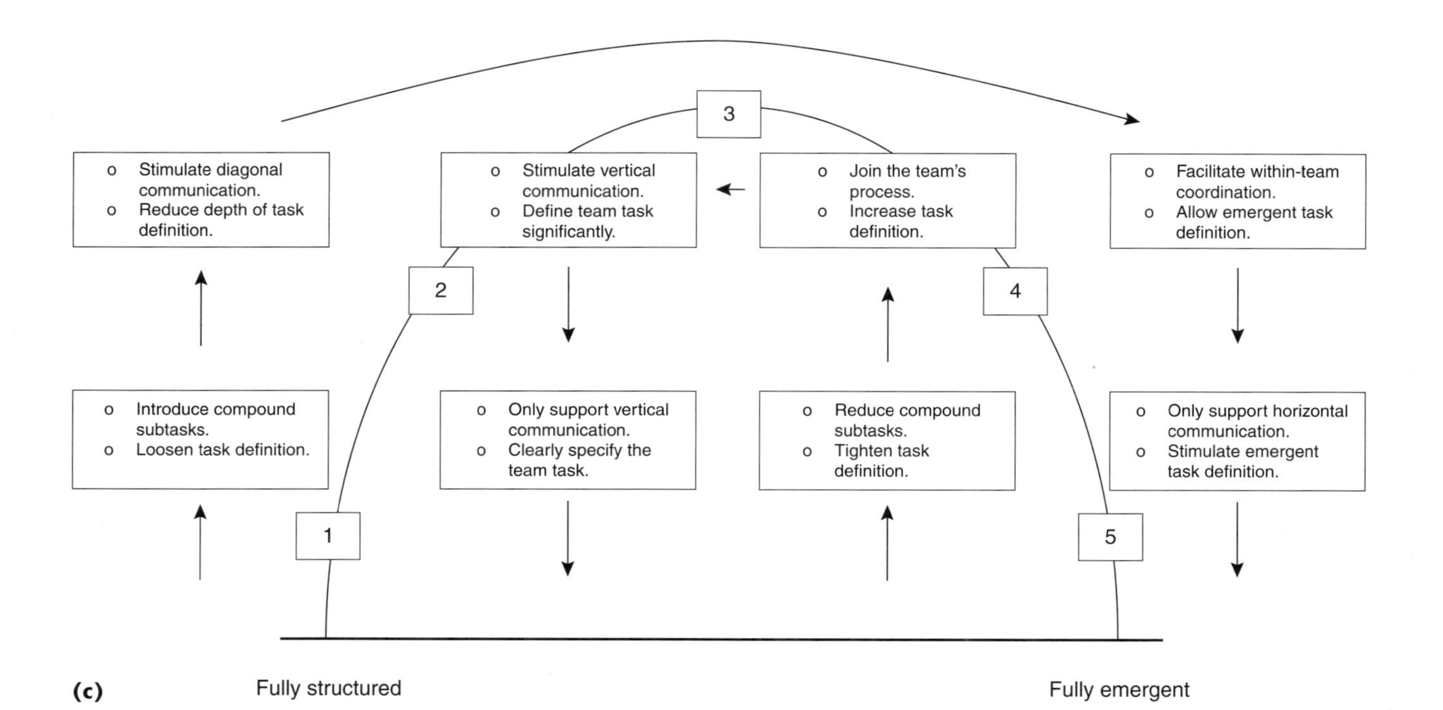

FIGURE 5-7. *(continued)*

increasingly important. Strongly define the team's task, structure the team's processes, and make sure that the tasks assigned to team members are increasingly simple and clear.

2 → **1.** Make sure the team's tasks are well specified. Tasks are clear and constant over the team's life span. The team task that remains is relatively simple and transparent. The team leader handles any necessary coordination. Task-related communication is therefore almost entirely vertical.

Some Considerations

The Virtuality Pyramid candidly points to the reason why a "virtual" team isn't as creative as management wants it to be. Frequent application of the Virtuality Pyramid allows team leaders to diagnose their team's creative potential and to keep track of its creative developments. If the team's creative achievement needs to be enhanced, even if only temporarily, you need to optimize the flexibility of the team's virtual design, at least temporarily. As the project progresses, the creative potential of NPD teams can be managed by increasingly virtualizing or devirtualizing the team along the three virtuality dimensions.

In any case, one cannot assume things will go right by themselves. Teams do not reach high creativity out of the blue; they require a creativity-nurturing environment. Such an environment is created by the 3-3-3 flexibility configuration. Applying the pyramid is fast and easy. Since even the best teams can stumble along the way, it is your task to spot these hurdles and act on them in time.

By the same token, poor teams can be put back on track by effective management action. Therefore, we urge team and project leaders to apply the Virtuality Pyramid several times over the lifetime of the project. Doing this will reveal trends. This enables you to quickly measure the effect of any management action taken. To truly learn the effect of each intervention option, we strongly recommend that only one intervention option is applied at a time. Guided by the options mentioned in this section and by today's time-compressed environment, one may be tempted to apply several interventions at the same time. This may well work. However, by applying multiple interventions simultaneously, a host of changes are created. This may make it impossible to gauge the effect of an intervention option separately, because its effect will be moderated by the effects of the other interventions. Even though the results may be beneficial, you still really haven't learned what to do next time. Therefore, we urge the reader to be patient: Apply only one intervention at the time, document it well, discuss its effect with other NPD managers, and allow some time for the effects to show. Then feel free to add a second intervention later if necessary. Of course, if time constraints do not permit such a learning cycle, multiple actions can be applied simultaneously. But if at all possible, it is highly advisable to perform interventions sequentially. This will make the team more effective and make you a better-skilled manager.

CONCLUSIONS

Over recent decades, managers have been forced to adapt the way in which they organize their new product development efforts. Several market developments have propelled NPD to be executed in increasingly virtual settings. This has created a new challenge: how to manage the creative achievement of teams in which members meet each other less frequently in person, where discussions occur in asynchronous modes, and where task structures are largely emergent. Many managers have told us they initially thought it impossible to actively manage and stimulate creativity well in virtual settings. This chapter has shown that this thought is partly correct: Fully virtual teams are rarely highly creative. Constant high levels of creativity tend to be designed out when NPD is organized in a fully virtual fashion.

However, one of the important arguments in this chapter is that virtuality is a matter of degree. It makes no sense to refer to a NPD team as "virtual." It also does not make sense to talk about a nonvirtual NPD team. All NPD teams are virtual to some extent; the key to managing the NPD team is to know *how* virtual a given team is. The Virtuality Fingerprint has been designed for this exact purpose: It reveals how virtual a given team really is and what the virtuality-determining dimensions are. The team may be more virtual in one respect but less virtual in another.

The second way this chapter helps you manage NPD teams is by showing how the three virtuality dimensions affect the creativity of NPD teams. Armed with this knowledge, leaders can diagnose the team's creative potential. In addition, it provides team and project leaders with actionable cues as to how to increase (or decrease) the team's creativity. Since innovativeness is perhaps the most important factor to NPD success, knowing how the creativity of NPD teams can be managed is vital to the success of NPD teams and to that of the team or project leader.

Virtuality is not bad for creativity per se. In fact, adding a serious dose of virtuality to an otherwise essentially nonvirtual team generally significantly increases the team's creative potential. But there can be too much of a good thing. If the NPD team becomes highly virtual, it is at risk of losing much of its creative potential. Flexibility of virtuality is critical to creative achievement.

The tools we presented in this chapter are based on extensive research and practical experiences. The tools are simple to use but extremely powerful. They show that achieving creativity in virtual settings is certainly achievable. More importantly, NPD team creativity can be managed. This chapter has shown how.

REFERENCES

Agrell, A., and R. Gustafson. 1996. Innovation and Creativity in Work Groups. In *Handbook of Work Group Psychology,* ed. M. A. West, 317–343. Chichester, UK: John Wiley & Sons.

Albrecht, T. L., and V. A. Ropp. 1984 "Communicating about Innovation in Networks of Three U.S. Organizations." *Journal of Communication* 34: 78–91.

Allen, T. J. 1977. *Managing the Flow of Technology.* Boston: MIT Press.

Amabile, T. M. 1988. "A Model of Creativity and Innovation in Organizations." In ***Research in Organizational Behavior,*** ed. B. M. Staw and L. L. Cummings, 123–167. Greenwich, CT: JAI Press.

Amabile, T. M. 1996. *Creativity in Context.* New York: Westview Press.

Baron, R. S. 1986. "Distraction-Conflict Theory: Progress and Problems." *Advances in Experimental Social Psychology,* Vol. 19, ed. L. Berkowitz, *1–40.* New York: Academic Press

Brauner, E., and W. Scholl. 2000. "The Information Processing Approach as a Perspective for Group Research." ***Group Processes and Intergroup Relations*** 3: 115–122.

Cohen, S., and D. Mankin. 1999. "Collaboration in the Virtual Organization." In *Trends in Organizational Behavior,* ed. C. L. Cooper and D. M. Rousseau, 105–120. Chichester, UK: John Wiley & Sons.

Csikszentmihalyi, P. 1996. *Creativity: Flow and the Psychology of Discovery and Invention.* New York: HarperCollins.

DeSanctis, G., N. Staudenmayer, and Sze Sze Wong. 1999. "Interdependence in Virtual Organizations." In *Trends in Organizational Behavior,* ed. C. L. Cooper and D. M. Rousseau, 81–103. Chichester, UK: John Wiley & Sons.

Drazin, R., M. A. Glynn, and R. K. Kazanjian. 1999. "Multilevel Theorizing about Creativity in Organizations: A Sensemaking Perspective." *Academy of Management Review* 24: 286–307.

Glynn, M. A. 1996. "Innovative Genius: A Framework for Relating Individual and Organizational Intelligences to Innovation." *Academy of Management Review* 21: 1081–1111.

Guerin, B. 1986. "Mere Presence Effects in Humans: A Review." *Journal of Experimental Social Psychology* 53: 1015–1023.

Keller, R. T. 1986. "Predictor of the Performance of Project Groups in R&D Organizations." *Academy of Management Journal* 29: 715–726.

King, N., and N. Anderson. 1990. Innovation in Working Groups. In *Innovation and Creativity at Work,* ed. M. A. West and J. L. Farr, 81–100. Chichester, UK: John Wiley & Sons.

Kratzer, J. 2001. *Communication and Performance: An Empirical Study in Development Teams.* Amsterdam: Tesla Thesis Publishers.

Leenders, R.Th.A.J., J. M. L. van Engelen, and J. Kratzer. 2003. "Virtuality, Communication, and New Product Team Creativity: A Social Network Perspective." *Journal of Engineering and Technology Management* 20: 69–92.

Lovelace, K., D. L. Shapiro, and L. R. Weingart. 2001. "Maximizing Cross-Functional New Product Teams' Innovativeness and Constraint Adherence: A Conflict Communications Perspective." *Academy of Management Journal* 44: 779–793.

Lovelace, R. F., 1986. "Stimulating Creativity through Managerial Interventions." *R&D Management* 16: 161–174.

Payne, R. 1990. The Effectiveness of Research Teams: A Review. In *Innovation and Creativity at Work,* ed. M. A. West and J. L. Farr, 101–122. Chichester, UK: John Wiley & Sons.

Shalley, C. E. 1995. "Effects of Coaction, Expected Evaluation, and Goal Setting on Creativity and Productivity." *Academy of Management Journal* 38: 483–503.

Urban, J. M., C. A. Bowers, J. A. Cannon-Bowers, and E. Salas. 1995. "The Importance of Team Architecture in Understanding Team Processes." *Advances in Interdisciplinary Studies of Work Teams* 2: 205–228.

Van den Bulte, C., and R. K. Moenaert. 1998. "The Effects of R&D Team Co-location on Communication Patterns among R&D." *Management Science* 44: S1–S18.

Visart, N. 1979. Communication between and within Research Team Units. In *Scientific Productivity,* ed. F. M. Andrews, 223–251. Cambridge: Cambridge University Press.

Woodman, R. W., J. E. Sawyer, and R. W. Griffin. 1993. "Towards a Theory of Organizational Creativity." *Academy of Management Review* 18: 293–321.

6 Build Stronger Partnerships to Improve Codevelopment Performance

Mark J. Deck

Working across corporate boundaries to codevelop products and customer solutions is becoming a critical aspect of product development in a wide range of industries. By leveraging the expertise of their value chain partners, companies can focus on their own core competencies, make better use of their R&D resources, develop more innovative and complete offerings, and achieve faster time to market. Moreover, advances in technology have finally made effective codevelopment—and the integration it demands—truly feasible. But too many companies jump into codevelopment efforts without thinking strategically about who to partner with, how to leverage key capabilities, and what outcome to aim for. In this chapter I highlight the pitfalls of this approach and provide an overview of a structured partner selection and management process that can prevent them.

Those at the forefront of codevelopment see close collaboration with other companies in the development chain as fundamental to their business model. Such companies develop a strategy and relationships to support codevelopment. They seek to leverage their own internal strengths as well as the core competencies of their development partners. They analyze the value chains they operate in, target the areas where they can add the most value, seek partners for help in developing complete solutions, and actively manage their relationships with these partners.

Millennium Pharmaceuticals, founded in 1993, typifies this approach. Through its proprietary technology and R&D platform, the company identified dozens of potential new therapeutic and predictive drugs. Its early-stage development pipeline has been full of promising projects that target these new drugs. Taking a pharmaceutical project from concept to approval is a long and costly proposition, however. From its inception, Millennium has focused on drug discovery and partnered with established pharmaceutical companies—with the right assets in place—to help with the development process. If Millennium had tried to perform all of the development chain activities by itself, only a small fraction of its projects would have proceeded to the later stages of development. "Collaboration is essential for us to be able to unleash our early-stage pipeline," says Keith Dionne, vice president/general manager of Technology Business at Millennium.

Cisco Systems is another example of a company that emphasizes collaboration in nearly every element of its operations. The company has an explicit strategy of leveraging the advantages its partners provide, thus gaining a larger market share for its products in a shorter time frame. "The ultimate benefit of collaboration is that we make the pie bigger, faster," notes Michael Frendo, vice president, Technology Center. "By working with other folks, we can accelerate the development of a new technology or a new service, and if we get to market sooner, we always get more of the market."

A codevelopment practices survey conducted by PRTM in conjunction with the Product Development & Management Association (PDMA) and The Management Roundtable (MRT), at PDMA's Co-Development Conference Series (PDMA/MRT, 2003), further reflects this emerging codevelopment trend. More than half of the respondents said that at least 20 percent of their projects involved a codevelopment partner. More importantly, almost half said that this percentage would likely double over the next two years. Across all industry segments in the study, the top reason given for codevelopment partnerships among smaller companies (less than $500 million in revenues) was "faster time to market"; for large companies, the number one motivator was "innovation." But almost 70 percent of the respondents said they were unsatisfied with their codevelopment efforts. The top two reasons given were "poor foundation for collaboration" and "inadequate executive leadership," two key ingredients for building strong partnerships.

Effective codevelopment is clearly a major challenge. Many companies struggle even to define objectives and guidelines for their partnerships. The following definition and characteristics provide a good starting point.

In a collaborative development partnership, two or more independent enterprises work together to design and release a new product, service, or technology for mutual benefit. Such partnerships share the following characteristics:

- *The parties will interact closely over a period of time.*
- *All parties are willing to invest—time, energy, and money—in each other for their mutual goals.*
- *All parties will benefit from the success of the partnership.*

Obviously, not all relationships are strategic or require such an investment. Doing business with a supplier of commodities, for example, is unlikely to warrant the same level of partnership as a relationship with a company intimately involved in the conceptual design of a new product or solution.

Optimal codevelopment performance results from process excellence in three dimensions: strategy formulation (where to partner and why), project execution (how to execute projects with partners), and partner selection and management (how to initiate and cultivate productive partnerships). I've chosen to focus this chapter on this last dimension because of its relative newness as a corporate capability and the challenge that most companies face in dealing with it effectively. You'll see how to create a process for selecting and managing partners and the benefits this process provides, and you'll receive helpful advice for making it work. You'll also explore the topics of partner selection criteria, joint development agreements, partner relationship management, partner performance metrics, and partner portfolio management.

A MODEL FOR CODEVELOPMENT

Based on recent research (Deck and Strom, 2002), we've PRTM identified three key sets of practices required for codevelopment success: *strategy,* which includes development chain design and partner selection and management; *execution,* which includes governance, metrics, teams, and processes; and *infrastructure,* which includes the IT systems that enable effective codevelopment. These practices form the basis of the three-level model for managing codevelopment, shown in Figure 6-1.

The research at the Co-Development Conference Series validates this model. Of the top seven practices followed by participating companies, as

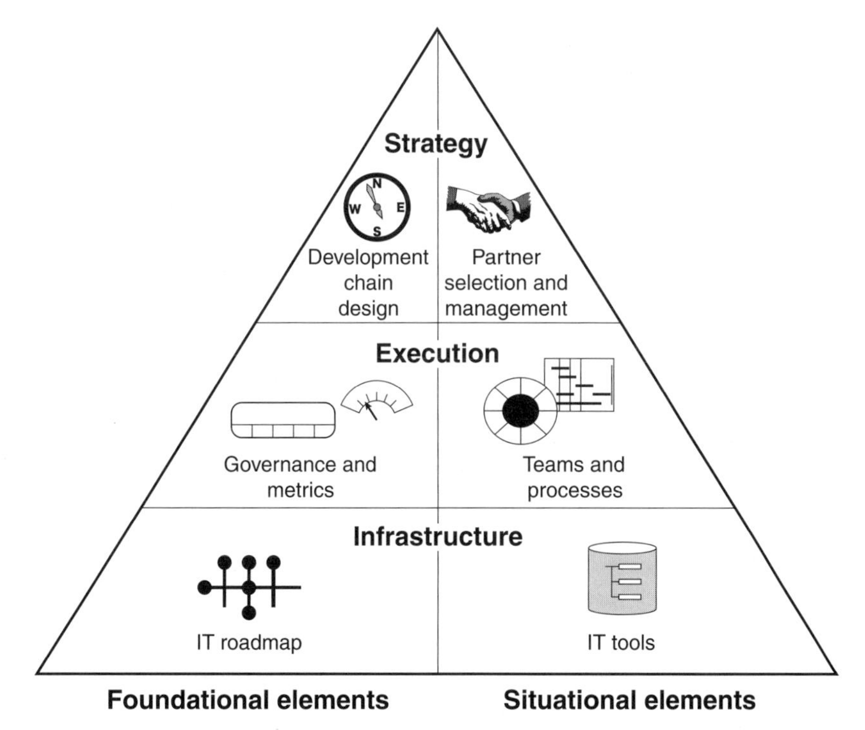

FIGURE 6-1. Codevelopment model.

Codevelopment Practice	Importance Percent Top Score	Practice Percent Top Score
Integrated process extends to partners, with common deliverables and definitions.	85%	22%
Key partners represented on project teams and given well-defined roles.	82%	18%
Designated executives manage partner relationships and resolve problems.	70%	16%
Partner management is a defined process with standard agreements and templates.	65%	18%
Process and criteria for partner selection and evaluation is defined.	65%	15%
Metrics are defined to evaluate projects and partner relationships.	62%	8%
Product and technology roadmaps are jointly defined with partners.	58%	11%

FIGURE 6-2. Top 7 practices for successful codevelopment.

shown in Figure 6-2, two focused on project execution, three had to do with partner selection and management, and two focused on metrics and strategy. While all of these practices were viewed as important, they were not yet widely used by most participants.

Other research shows that partner management leads to better codevelopment performance. Dyer, Kale, and Singh (2001) researched more than 1,500 alliances in 200 corporations and found that companies with a dedicated function and process for partner management had both more alliances (by about 50 percent) and more successful alliances (39 percent success rate improvement).

A STRUCTURED PROCESS

For many companies, the need to collaborate is sporadic and usually driven by a particular project or specific market circumstances. As a result, many companies are building codevelopment skills at the project level. While leaving partner selection and management to codevelopment teams on individual projects can work when only a few such projects are in the pipeline, this approach prevents realizing the maximum value from codevelopment when many products with multiple partners are involved. Following are some common problems:

- While project-level issues may be highly visible, performance problems related to relationships may go unnoticed.
- Companies waste time and effort on building new partnerships without having fully leveraged their existing relationships.

- While navigating the minefield of contracts, companies keep "reinventing the wheel," effectively starting each new codevelopment deal from scratch.
- Opportunities to leverage synergies from other initiatives are missed.
- Changing relationships strain once-strategic partnerships.
- In the flurry of activity, companies fail both to identify the best partners and to weed out the worst.

The key to avoiding these problems is to establish a repeatable process for partner selection and management, a process with a clear definition of strategy inputs, process steps, inputs and outputs for each process step, decisions, governance and process ownership, organizational approach, and metrics. I recommend a process with three phases, outlined briefly in the following and subsequently detailed with examples:

Assessment and Selection

- Identify codevelopment opportunities based on product and codevelopment strategies.
- Define selection criteria, their relative importance, and where appropriate, a scoring methodology.
- Evaluate partners against the selection criteria using site visits and interviews.
- Begin preliminary selection based on quantitative and qualitative assessments against selection criteria and prepare for contract negotiation to look for deal breakers in advance.
- Identify needed infrastructure.

Relationship Initiation

- Validate partner capabilities.
- Negotiate formal agreements (e.g., contractual terms, treatment of Intellectual Property, etc.).
- Define how the partnership will work (Joint Development Agreement, or JDA).
- Develop supporting infrastructure.

Relationship Management

- Manage the partnership within all programs.
- Manage the portfolio of partnerships.
- Resolve ongoing issues.
- Install methods for measurement, feedback, and continuous improvement.

These three phases ensure that the partner selection and management process leads to optimal codevelopment project results.

The sections that follow describe the critical aspects of each of these phases, along with examples from companies involved in codevelopment. The intention is not to provide a comprehensive treatment of the subject here, but to explain the important issues you'll need to address when creating a partner selection and management process.

PARTNER ASSESSMENT AND SELECTION

High on the list of needs for an effective partner assessment and selection process are a clear codevelopment strategy, a clear definition of partner selection criteria that supports this strategy, and a decision on who chooses and negotiates with the partners. A codevelopment strategy answers several critical questions:

- Is this opportunity appropriate for a codevelopment partnership?
- What kind of partner relationship would be appropriate for this situation?
- Are there other related opportunities or partners to consider?
- What do we hope to achieve with such a partnership, and what benefits would each party receive?

Without a guiding codevelopment strategy, partner assessment and selection becomes an exercise in deal making rather than establishing a foundation for lasting value. Selection criteria should be applied equally to all partners to keep them aligned with strategy. For example, whether considering a major channel partnership, an integrated relationship with a contract manufacturer, or an alliance with a small software developer, Cisco uses the following four criteria: short-term returns for both companies; clearly defined, long-term potential for both companies; shared vision of technology and market developments; and shared destiny of cooperation, not competition.

Cisco discusses these criteria with each potential partner and will not launch a codevelopment project unless both parties share these views. As Cisco's Frendo explains: "We try to win something together and get more tightly bound as companies within the first six to nine months—even sooner, if possible. That early win is essential. If we start out trying to 'boil the ocean,' to create a solution that's many, many years out, somewhere along the line both companies will stray from the original goal and lose interest."

If a short-term win seems possible, then the potential for a longer-term relationship is evaluated. Cisco will not pursue a significant codevelopment effort solely on the basis of a short-term win. Frendo thinks partners need to share a view of where their market is headed to collaborate successfully. As he says, "Putting all this together, getting our cultures aligned, and doing the ramp-up of getting two teams to work together is just too expensive a process if you don't also have a longer-term relationship in mind. So we're looking for both the short- and longer-term win-win."

	Short-term	Long-term
Tactical	Are current projects meeting the expectations established when the projects were initiated?	Is the relationship meeting the expectations set when it was initiated?
Strategic	Are the current intentions and goals of the relationship clear to all parties involved?	How does each partner's strategic roadmap impact the future direction of the relationship?

FIGURE 6-3. Dimensions of partnership management.

How do you know a shared vision when you see it? "Our definition," says Frendo, "is a compatible outlook on issues of competing technologies, operational paradigms, etc." The notion of a shared destiny is vital because it allows for the sharing of the intellectual property that is required for a truly collaborative effort to succeed. "We want a destiny that is shared, not a destiny that is competitive," Frendo says. "If companies start by trying to learn from each other, but a year down the road they're really planning to compete, it dooms the relationship."

As shown in Figure 6-3, I advise looking at partnership attractiveness along two dimensions: the tactical and the strategic. Moreover, based on PRTM research and client experience, PRTM recommends these additional practices for the assessment and selection phase:

- Ensure that the executive team agrees on the strategy and is fully behind the initial assessment of partnership opportunities.
- Create a clearly defined set of core selection criteria. Additional situation-specific criteria can always be added as needed. Some examples might include the following:
  - Caliber of management staff
  - Existence of required skill sets, both technical and managerial
  - Ability to meet financial obligations of the relationship (including capital and expense requirements)
  - Ability to support IT infrastructure requirements
  - Extent of the partner's collaboration with competitors
- Craft a mutually rewarding value proposition before approaching potential partners.
- Assign competent relationship builders to conduct candidate meetings and initial contract discussions; these responsibilities call for experienced people who carry some authority in the organization.
- Use publicly available information such as financial statements, analyst reports, and published articles to screen candidates, and be sure to eval-

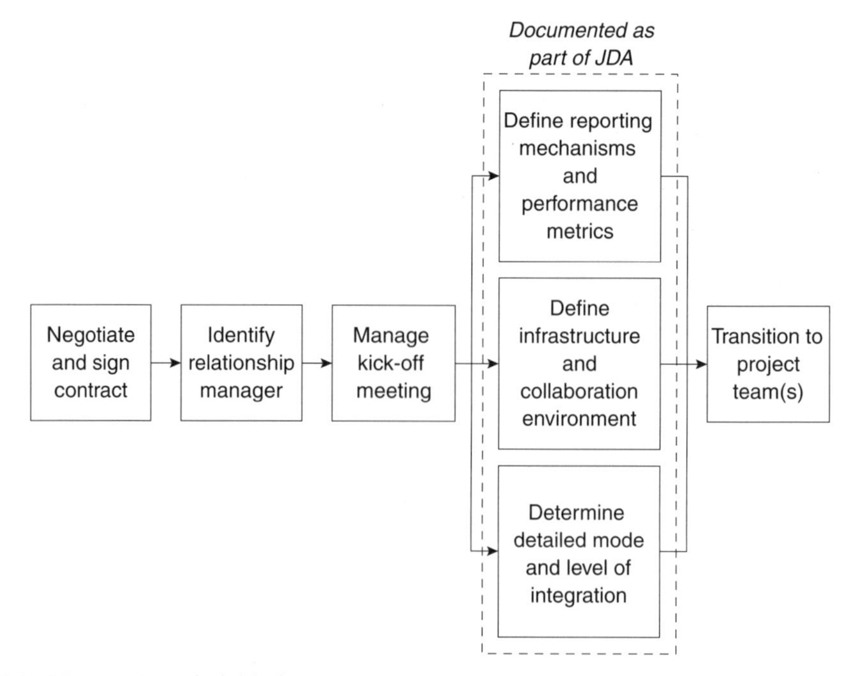

FIGURE 6-4. Steps in initiating the relationship.

uate their codevelopment track record. You can learn a lot about their "partnership culture" from how they've treated other partners in the past and how well those partnerships have succeeded. When feasible, interviewing them and their past partners is an extremely valuable evaluation mechanism.

- Get broad, cross-functional input on partnership recommendations. In addition, create a cross-functional decision team to approve the final choice and make go/no-go decisions for the projects.

Once you've found a strategically sound codevelopment opportunity and an appropriate partner, it's time to consider the governance and contractual questions that will determine the success or failure of the initiative: Who pays for what (people, tooling, launch costs, etc.)? Who owns what? Is the relationship exclusive? Who can make commitments about resources, timing, and so on? How will decisions be made and conflicts resolved? How will the two entities work together on projects? How will performance be measured and reviewed? (See Figure 6-4.)

RELATIONSHIP INITIATION

The objective of the relationship initiation phase is to set up a governance structure that balances financial, legal, and risk requirements without making either partner feel constrained by the terms. An effective way to set up this environment of collaboration is to divide the contract into two parts: the formal business contract and the less formal and non-legally binding joint development agreement (the JDA).

The business contract is legally binding and usually addresses financial, ownership, and critical performance parameters (see Figure 6-5A). Its purpose is to manage business risk and protect the interests of each party. Because the business contract and amendments to it often require lengthy negotiation, they should be limited to the "critical few" legally binding issues. Candidate topics for inclusion in the business contract include intellectual property (IP), ownership, and confidentiality; funding, payment terms, and purchasing commitments; incentives and penalties; exclusivity and subcontracting; and end-of-life options.

Unlike the business contract, the JDA provides guidelines for the day-to-day interaction between the partners and is generally focused on project logistics instead of relationship structure (see Figure 6-5B). It describes the rules of engagement, including guidelines for organization, communications, issue resolution, and evaluation. Typical JDA topics include program objectives and scope; program timeline and key milestones/deliverables; post-launch support requirements; process management, including project management, issue resolution, change control, JDA maintenance, and process integration; organization of teams and decision making; measures and targets; communication and use of information technology; and terminology and standards.

Besides these two documents, a number of other factors should be considered when initiating a partnership:

- Validate your initial assessment of partner capabilities during the relationship initiation phase.
- Make contracts and JDAs fair, but not onerous. If they're too burdensome in trying to protect against all contingencies, they can jeopardize the relationship.
- Make sure the JDA explains how the project will be managed, evaluated, and terminated.
- The organizational and IT infrastructure needed to support the partnership should be put in place early.

PARTNER RELATIONSHIP MANAGEMENT

Without effective partner management, any codevelopment effort will suffer in terms of time, leverage, and productivity. In fact, you would do well to revisit your existing partnerships before pursuing new ones. Companies must ask themselves these key questions: Have we fully leveraged the capabilities of our current partner? Have we looked for new ways to extend our partnership for mutual benefit? Have we fully integrated our businesses to the extent that we can for mutual benefit? Is the partnership effective, or should it be terminated?

As shown in Figure 6-6, partnerships require two layers of management:

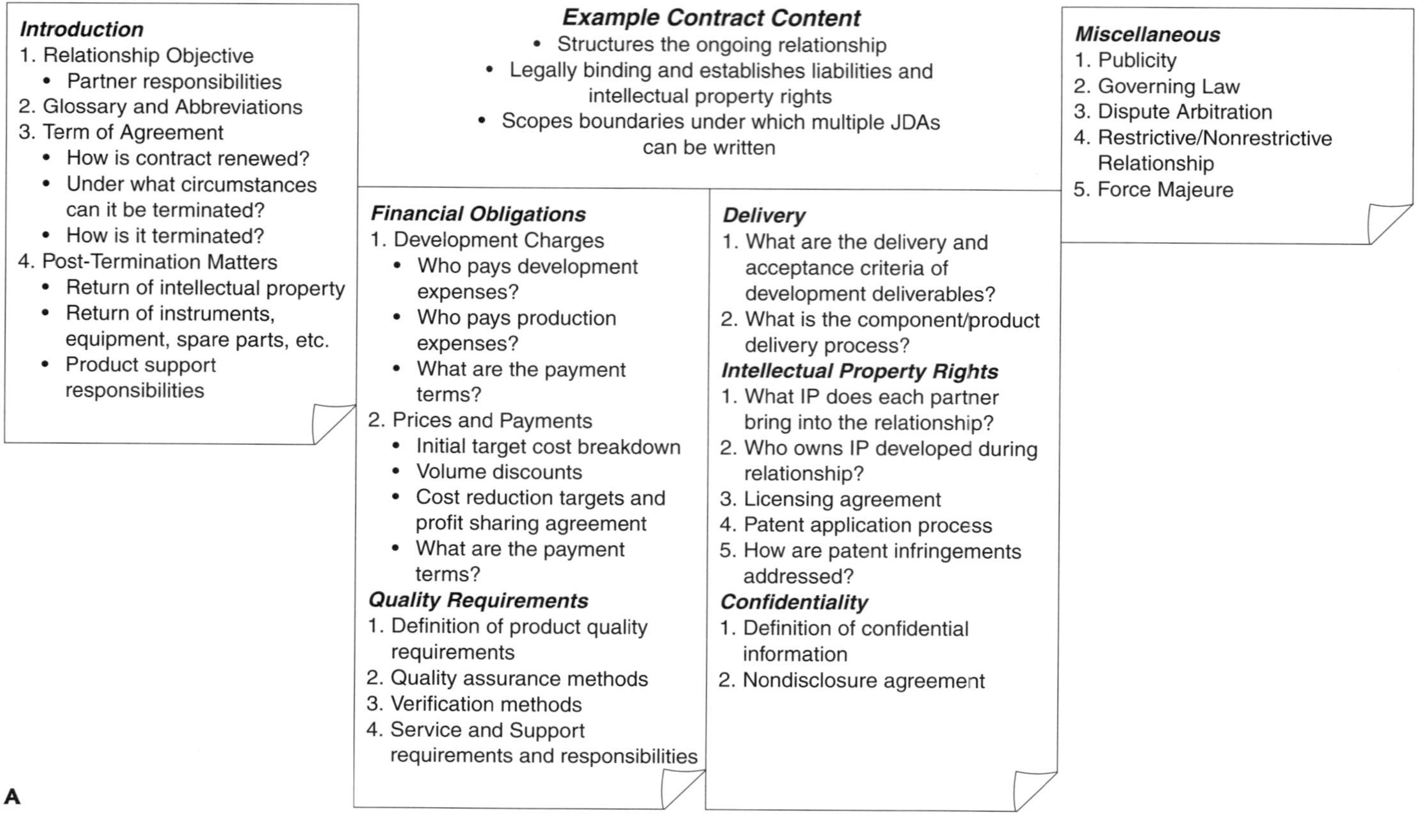

Example Contract Content

- Structures the ongoing relationship
- Legally binding and establishes liabilities and intellectual property rights
- Scopes boundaries under which multiple JDAs can be written

Introduction

1. Relationship Objective
   - Partner responsibilities
2. Glossary and Abbreviations
3. Term of Agreement
   - How is contract renewed?
   - Under what circumstances can it be terminated?
   - How is it terminated?
4. Post-Termination Matters
   - Return of intellectual property
   - Return of instruments, equipment, spare parts, etc.
   - Product support responsibilities

Financial Obligations

1. Development Charges
   - Who pays development expenses?
   - Who pays production expenses?
   - What are the payment terms?
2. Prices and Payments
   - Initial target cost breakdown
   - Volume discounts
   - Cost reduction targets and profit sharing agreement
   - What are the payment terms?

Quality Requirements

1. Definition of product quality requirements
2. Quality assurance methods
3. Verification methods
4. Service and Support requirements and responsibilities

Delivery

1. What are the delivery and acceptance criteria of development deliverables?
2. What is the component/product delivery process?

Intellectual Property Rights

1. What IP does each partner bring into the relationship?
2. Who owns IP developed during relationship?
3. Licensing agreement
4. Patent application process
5. How are patent infringements addressed?

Confidentiality

1. Definition of confidential information
2. Nondisclosure agreement

Miscellaneous

1. Publicity
2. Governing Law
3. Dispute Arbitration
4. Restrictive/Nonrestrictive Relationship
5. Force Majeure

A

FIGURE 6-5a. Example co-development contract.

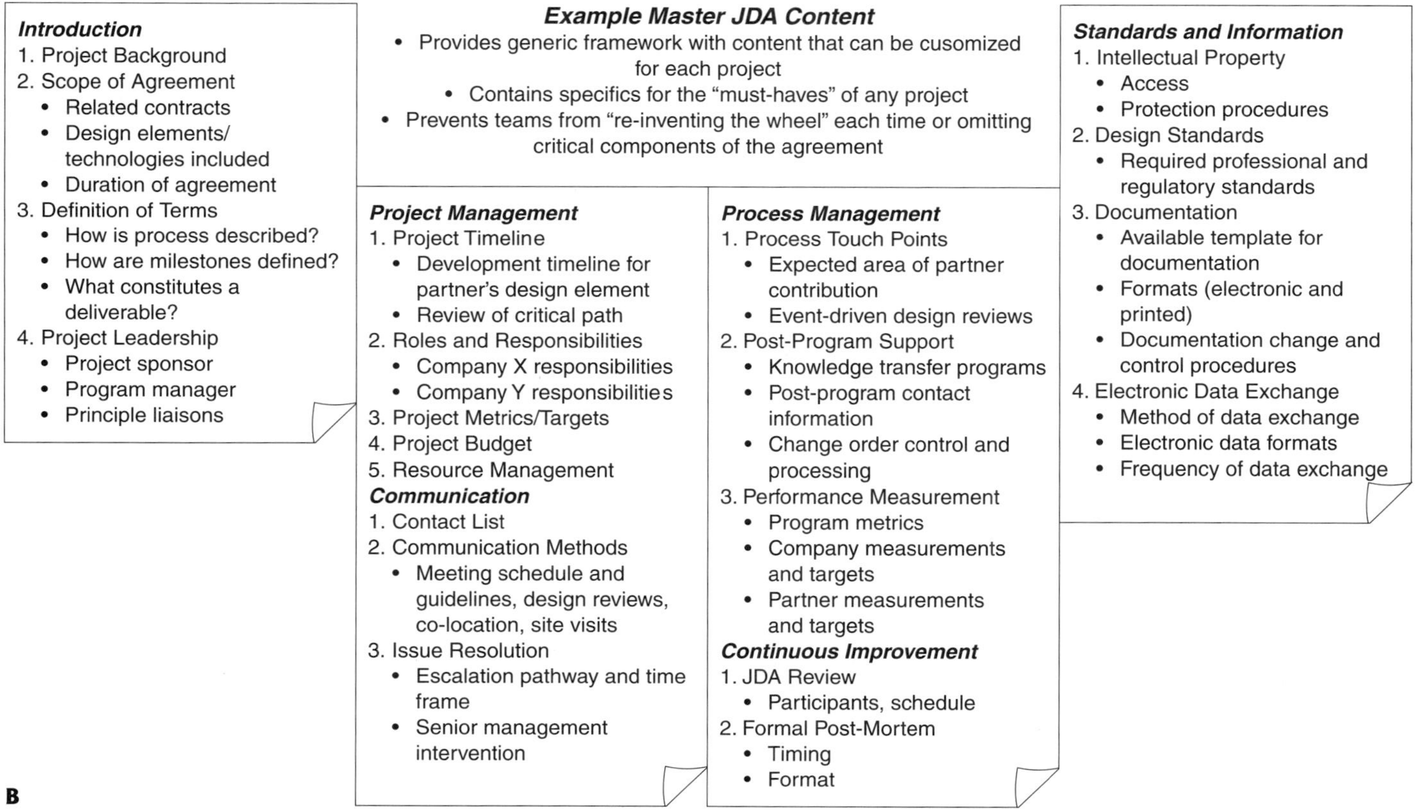

Example Master JDA Content

- Provides generic framework with content that can be cusomized for each project
- Contains specifics for the "must-haves" of any project
- Prevents teams from "re-inventing the wheel" each time or omitting critical components of the agreement

Introduction

1. Project Background
2. Scope of Agreement
   - Related contracts
   - Design elements/technologies included
   - Duration of agreement
3. Definition of Terms
   - How is process described?
   - How are milestones defined?
   - What constitutes a deliverable?
4. Project Leadership
   - Project sponsor
   - Program manager
   - Principle liaisons

Project Management

1. Project Timeline
   - Development timeline for partner's design element
   - Review of critical path
2. Roles and Responsibilities
   - Company X responsibilities
   - Company Y responsibilities
3. Project Metrics/Targets
4. Project Budget
5. Resource Management

Communication

1. Contact List
2. Communication Methods
   - Meeting schedule and guidelines, design reviews, co-location, site visits
3. Issue Resolution
   - Escalation pathway and time frame
   - Senior management intervention

Process Management

1. Process Touch Points
   - Expected area of partner contribution
   - Event-driven design reviews
2. Post-Program Support
   - Knowledge transfer programs
   - Post-program contact information
   - Change order control and processing
3. Performance Measurement
   - Program metrics
   - Company measurements and targets
   - Partner measurements and targets

Continuous Improvement

1. JDA Review
   - Participants, schedule
2. Formal Post-Mortem
   - Timing
   - Format

Standards and Information

1. Intellectual Property
   - Access
   - Protection procedures
2. Design Standards
   - Required professional and regulatory standards
3. Documentation
   - Available template for documentation
   - Formats (electronic and printed)
   - Documentation change and control procedures
4. Electronic Data Exchange
   - Method of data exchange
   - Electronic data formats
   - Frequency of data exchange

B

FIGURE 6-5b. Example co-development master JDA.

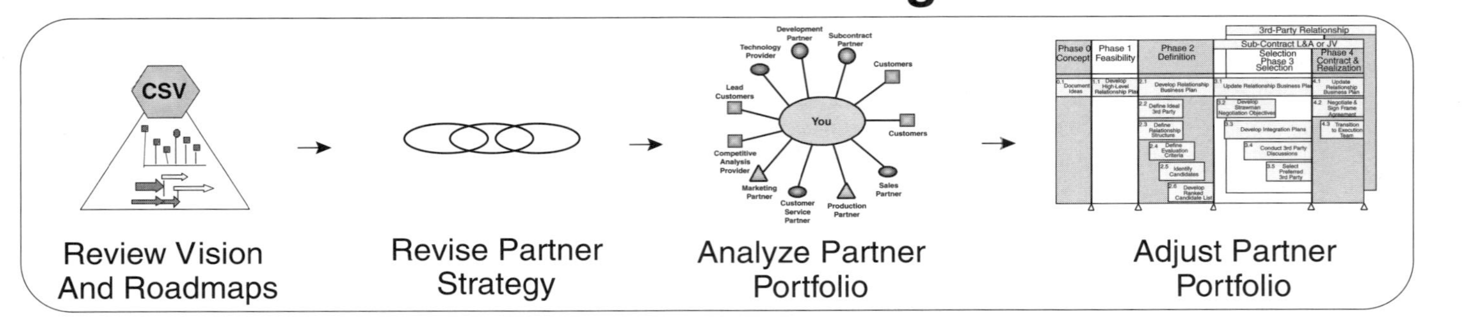

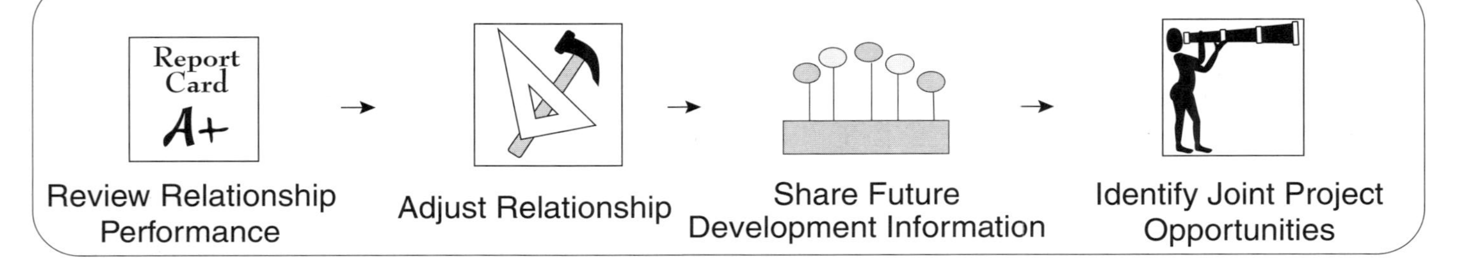

FIGURE 6-6. Effective partner management at the project and portfolio levels.

at the individual partner level and at the partner portfolio level. Can your current set of partners meet your company's strategic partnering goals? Beyond strategic fit, the portfolio of partners should be examined for balance and mix. Are there opportunities to configure partnerships differently to take better advantage of unique capabilities and scale economies? Finally, relative performance must be compared against goals to determine whether partnerships should be changed or ended.

Eli Lilly created an Office of Alliance Management to achieve these goals (Futrell, Slugay, and Stephens, 2002). One particularly innovative practice it launched was a periodic survey of partnership "fit" along three critical parameters: culture, operations, and strategy (see Figure 6-7).

It's important to measure partnership performance in a well-defined, rigorous way, as Lilly does. Other practices to consider when building your own partner management process are as follows:

- Start simple, then deepen partner relationships as performance warrants.
- Conduct regular executive sponsor meetings to monitor the relationship and head off issues before they become critical.
- Build organizational trust and understanding with regular, broad-based joint communications.
- Prepare and manage professional development plans to grow the organization's relationship management skills.
- Foster relationships at multiple levels—executive-to-executive, director-to-director, and team-to-team.

GETTING STARTED

Putting a partner selection and management process in place requires making a broad commitment to codevelopment as a practice. If your company is working with codevelopment partners on more than a handful of projects, the potential for improving performance is great. This chapter outlined a process structure, key concepts, and proven practices.

Success also requires that some entity within the organization be accountable for managing the process. For some companies, this might be the business development function. Others create an office of partner management that establishes common contract templates and selection criteria, administers the partner relationship and portfolio management processes, assists project teams as they ramp up new projects with new and existing partners, gets involved with partner negotiation, and becomes a source of relationship-building expertise.

In the end, it's performance you're looking for in a partnership. But remember, performance is a two-way street. A partnership is only as strong as the capabilities of each partner and the effective management of the relationship. Codevelopment is tough to do well, but the benefits make it well worth

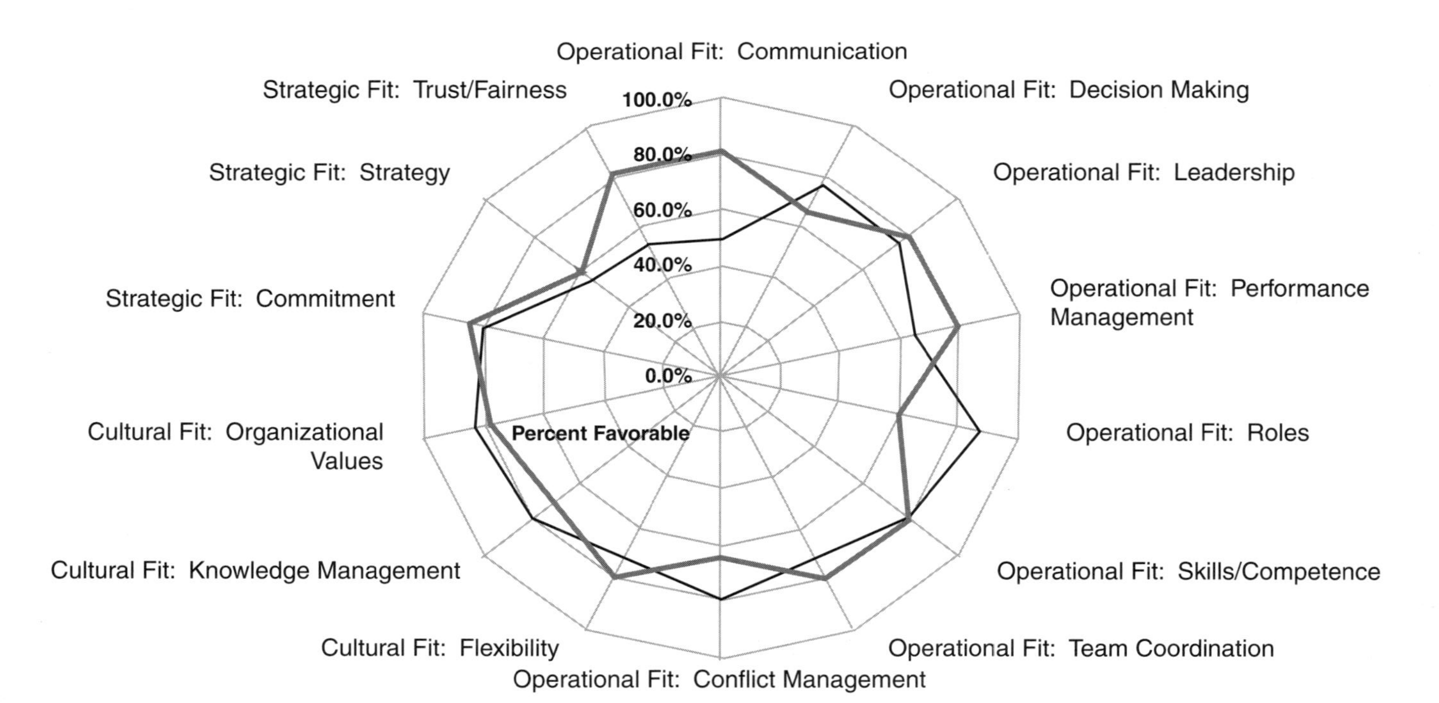

FIGURE 6-7. Partner management at Eli Lilly.

the effort. By following the guidelines outlined in this chapter, your company can turn the promise of collaboration into real, sustainable results.

REFERENCES

Deck, M., and M. Strom. 2002. "Model of Co-Development Emerges." *Research-Technology Management*, Vol. 45, No. 3, May–June, 2002, pp. 47–53.

Dyer, J.H, P. Kayle, and H. Singh. 2001. "How to Make Strategic Alliances Work." *MIT Sloan Management Review*. 42, 4, Summer: 37–43.

Futrell, D., M. Slugay, and C.H. Stephens 2002. "Becoming a Premier Partner: Measuring, Managing, and Changing Partnering Capabilities at Eli Lilly and Company." *Journal of Commercial Biotechnology* June.

PDMA/MRT. 2003. Co-Development Survey Results. *Conference Proceedings of the Second International PDMA/MRT Congress on Co-Developing Products with Partners, Suppliers, and Customers: "Achieving the Promise."* Scottsdale, AZ. January 27–29.

Part 2

Tools for Improving the Fuzzy Front End

Part 2 provides tools to be applied in the fuzzy front end (FFE). Many NPD organizations have come to realize painfully that just creating innovative technology does not ensure successful new products. Somehow the opportunities of the marketplace and the needs of customers must be ingrained in FFE activities to increase the probability of commercial success. The NPD process is probably mistakenly visualized as a linear process beginning with the FFE and ending with commercialization. Perhaps a more useful picture is a loop showing a linkage between the marketplace and the FFE. The chapters in this part largely deal with building or strengthening this link

Chapter 7 deals with another frequently used phrase: the voice of the customer. It is a how-to guide for determining which customers to interview and how to translate their articulated needs to technical product attributes. It walks the NPD professional through the creation of an interview guide and then the processing of the customer data generated. The data processing includes affinity diagramming and prioritization.

Chapter 8 continues with best practices for determining customer needs. Increased competition has driven the need for better understanding of customer needs. The use of anthropology tools in NPD has begun to emerge. This chapter provides a clear process to apply these anthropological/ethnographic tools. For each of the process steps, suggested examples and illustrations are provided.

Chapter 9 continues the customer needs theme and provides tools to go beyond understanding customer needs to uncovering customer wishes. Customer wishes are a means to determine the ideal product features from the customer's perspective. A customer idealized design (CID) process is described, with emphasis on both conducting the CID data collecting session and the iterative processing of the data to result in a new product design. Two case studies are included to foster understanding and to assist in implementation of the tools described in the chapter.

The last chapter in the FFE part, Chapter 10, addresses a common weakness in the widely used tool of brainstorming. Brainstorming is a popular NPD tool and lots of novel insights and ideas result from brainstorming sessions. Yet the eventual NPD outcome is frequently disappointing. The brainstorming weakness is how its results are converged or focused to result in a manageable number of ideas to deal with. The chapter's authors describe an enhanced brainstorming technique called SWIFT, which stands for Strengths, Weaknesses, Individuality, Fixes, and Transformation. SWIFT is an effective practice to help ensure that novel solutions are not lost in the focusing activities. Although specifically aimed at FFE activities, this tool could be widely applied across the NPD process.

7 The Voice of the Customer

Gerald M. Katz

This chapter describes what have been identified as "best practices" for gathering the *Voice of the Customer*, or VOC. It is essentially a how-to guide geared toward new product development professionals—the typical cross-functional core team made up of scientists, engineers, marketing managers, salespeople, customer service managers, financial managers, IT professionals, or anyone else within the organization responsible for customer needs assessment as the basis for new product and service design. While VOC is hardly new, this chapter tries to provide, perhaps for the first time, a condensed, yet cogent, practical methodology for just how to go about it.

BACKGROUND

Until about 1980, new product development was generally viewed as the exclusive domain of scientists and engineers—the technical R&D staff within the organization. Their job was to come up with interesting new technical innovations that would excite customers. They would then "throw it over the wall" to other functions within the organization—engineering, manufacturing, marketing, sales, and so on—whose job it was to carry this technical innovation through to the marketplace.

But then something dramatic happened: Japan. By the early 1980s, Japan had emerged as a major force in new product development, capturing large market shares in industries as diverse as automotive, consumer electronics, and heavy manufacturing. Much has been written about how this was accomplished, but most of it led back to an emerging emphasis on *quality*. However, following the principals they had learned from W. Edwards Deming, the Japanese expanded their definition of quality to go well beyond simple manufacturing defect reduction. The Japanese extended their quality processes all the way back to the initial product design phase. And at the top of the list was an almost fanatical emphasis on studying the *customer* as the first step in any new product development initiative, and then tying all of their subsequent design decisions back to a clear understanding of customers' wants and needs.

To do this, the Japanese developed a technique that came to be known as *Quality Function Deployment* (QFD), or the *House of Quality*. QFD was developed at a Japanese shipbuilding firm in the early 1970s as a way to present and organize customer needs and to weigh the trade-offs among all of the potential design parameters (Cohen, 1995). QFD starts with a detailed list of customer wants and needs, and over time, this list of needs, and the process by which they were obtained, came to be known as the Voice of the Customer.

In the early days of QFD, product development teams simply created a list of what *they* thought was important to their customers. However, it was soon recognized that this was highly biased and that a far more thorough job could be done by actually observing and listening to customers. To a novice in new product development, this might simply mean going out and visiting a few key customers. However, an entire science has grown up around this discipline—that is, how best to go about listening to and observing customers, and processing what is learned into a usable Voice of the Customer for product and service design. The first, and still one of the most important, studies of this area was reported by Abbie Griffin and John Hauser in their 1993 paper, "The Voice of the Customer" (Griffin and Hauser, 1993). Since then, many others have weighed in with important and useful works (Burchill and Brodie, 1997; McQuarrie, 1998; and Ulwick, 2002).

It is also important to note that most of the VOC techniques that have been developed over the years were originally intended for manufactured goods. Today, however, as the service economy has emerged, they are now being applied just as frequently to services. Keep in mind that almost all physical products have a strong service component, and it is difficult to separate the two when talking with customers. As will be discussed later, services present a few additional challenges. However, the basic techniques have been shown to work just as well for services as for manufactured goods.

The Development Team's Dilemma

If one accepts the premise that it is important to begin the new product development process by asking the customer about his or her needs, what's so difficult about it? Unfortunately, what a development team really needs is fairly different from what conventional market research usually provides. Conventional market research practices tend to focus more on the evaluation of existing ideas and solutions rather than providing information that can be used as a foundation for further innovation. In addition to all of the logistical problems of how to gather needs, there is an even subtler issue for development teams who are faced with the task of turning those needs into creative new features and solutions. What they need is to be able to come up with innovative new solutions and to make trade-offs at the detailed specifications level, since few companies can afford to do everything for everybody.

The problem is that customers speak in fuzzy consumer language, while product developers must translate this into explicit technical and measurable design parameters. For instance, if a customer asks for a more powerful computer, the developer must first decide which technical specification to emphasize. Should it be MIPS, RAM, storage capacity, or downloading speed? If the customer wants good on-time performance from an airline, how should that be measured? If the customer wants a good ergonomic chair, what should be measured here? In each of these cases, the design specification chosen will have a dramatic impact on the ultimate design. And in order to pick the right design specification, it is necessary to have a really deep understanding of how the customer defines a powerful computer, good on-time performance, or a good ergonomic chair.

The Voice of the Customer is very different from the words that engineers or product planners use to define a new product or specify service improvements. For example, automotive customers may say they want a "roomy front seat," but engineers must deal with inches of legroom, shoulder room, hip room, and headroom, as well as many other dimensions—and none are labeled "roomy." What does "healthy-looking hair" mean in a shampoo? What does "hassle-free service" mean for an insurance company? What does "convenience" mean for a bank? And how should these things be measured?

Making assumptions about what the customer means is fraught with danger. For instance, if the company comes up with three alternative chair designs intended to address the need for a good ergonomic chair, how will it decide which one is superior in the eyes of the customer? The company simply can't afford to build prototypes and do product testing with customers for every single issue. Thus, product designers and engineers have a deep need to really understand how the customer defines these things.

Many new product developers are inherently skeptical about this area, and many engineers insist that customers cannot tell you what they want, or that they won't know what they want until they see it. For instance, no customer would likely have told us that he or she wanted a microwave oven. But these statements confuse *needs* with *solutions* to those needs. In the parlance of QFD and new product development, in general, a microwave oven is not a customer *need*; it is a technical *solution* to a set of needs. For instance, customers have long expressed the need for an oven that (1) preheats more quickly, (2) won't dry out leftovers when reheating them, (3) doesn't heat up the kitchen on a hot summer day, and (4) heats food quickly throughout without burning it on the surface. These detailed needs concerning ovens have been around forever, and the invention of the microwave oven was simply a new and better technical solution to these needs. Most customers are not very good at describing the exact features and solutions that they want, but they are quite good at articulating their underlying wants and needs and the problems that they are currently encountering with today's products and services. As will be seen, a good Voice of the Customer makes a clear distinction between *needs* and *solutions* to those needs. And in doing so, product developers are more likely to think deeply about new ways of addressing customer needs.

Nine Myths about the Voice of the Customer

A number of practices and beliefs about the Voice of the Customer are addressed explicitly in this chapter:

1. The Voice of the Customer is primarily a *qualitative* market research process, not a *quantitative* one.
2. Voice of the Customer interviews should be conducted primarily with your most important current customers—that is, your key accounts.
3. Noncustomers will be reluctant to talk with you.
4. The best way to elicit wants and needs is to ask the customer what he or she wants and needs.
5. Customers can't really tell you what they want; they won't know it until they see it.
6. A separate note-taker is the best way to record needs.
7. Customers will not allow you to tape-record their conversations.
8. The process of organizing or "affinitizing" the needs is best done by the product development team itself.
9. Prioritization of the needs is also best done by the product development team.

As you will see in reading this chapter, these beliefs are erroneous. If followed, they will lead to a less than complete understanding of the Voice of the Customer. This chapter provides information necessary for teams to capture the VOC in a way that more closely reflects customers' true wants and needs. These myths will be returned to at the end of the chapter.

DEFINING VOICE OF THE CUSTOMER

We begin with a formal definition. The Voice of the Customer is:

- a complete set of customer wants and needs;
- expressed in the customer's own language;
- organized the way the customer thinks about, uses, and interacts with the product or service; and
- prioritized by the customer in terms of both importance and performance—in other words, current satisfaction with existing alternatives.

A Complete Set of Customer Needs

A *customer need* is defined as "a description of the benefit to be fulfilled by the product or service" (Griffin and Hauser, 1993). It is generally expressed in the form of a phrase that describes anything the customer wants, needs, demands, requires, or wishes for (whether consciously or unconsciously) in a particular product, service, or process in order to better deliver that benefit. Doing so will make things better, faster, cheaper, or easier for the customer to achieve his or her desired outcome. In most product or service categories that have been studied thoroughly, there are at least 75 to 150 unique, detailed customer needs. Good product development requires that they be extracted as thoroughly and completely as possible.

In the Customer's Own Words

There is a strong tendency on the part of product developers to translate the customer's words into their own company jargon, losing much of the richness and subtlety of meaning. For instance, airlines use words like "emplaning" and "deplaning" to describe the process of getting on and off the plane. Interestingly, customers almost never use these words and may not even know what they mean! To best understand the customer's needs, you must try to maintain their vernacular as much as possible.

Organized the Way the Customer Thinks about the Product

Since 100 need statements is too high a level of detail for a product development team to work with, they will need to organize the needs into some higher level of aggregation. Research has shown that customers generally organize the needs differently from the way people within a company might. Therefore, it is important to get the customers to perform this step.

Prioritized by the Customer

Likewise, dozens of cases done in parallel have shown that customers tend to prioritize the needs quite differently from people within a company, and so, again, it is important to get customers to perform this step as well. There are two major types of market research: qualitative and quantitative. *Qualitative* market research is expressed in words, reasons why, feelings, benefits, and motivations. *Quantitative* market research—on the other hand, involves the counting of something, and thus, is expressed in numbers—who, how many, or how much? One of the greatest misconceptions about the Voice of the Customer is that it is only a qualitative process. In fact, the Voice of the Customer involves both qualitative and quantitative market research. Too often, the emphasis is placed on the former, with short shrift given to the latter. As you go through the process in this chapter, both types of research needs are addressed.

PLANNING FOR THE VOICE OF THE CUSTOMER

Determining the Scope of the Project

The first decision in gathering a Voice of the Customer is to decide whom to interview. Unfortunately, this is one of the areas that can create the most difficulty later on if not done carefully. There is often a knee-jerk reaction among the inexperienced to simply begin by going out and talking to some key, well-known customers without giving it much thought. Another sometimes problem-provoking method is to ask the sales force for the names of some good customers to interview. In general, companies tend to focus their interviews on their largest, happiest, most loyal customers. However, this could be a potential trap, as these people are likely to be biased because they are among the most satisfied with your current products (McQuarrie, 1998). In general, you will learn more from talking to your competitors' customers, your non-customers, and your ex-customers, as they will likely be able to tell you about needs that you haven't addressed very well or that your competitors have been addressing better.

To choose the "right" set of customers to interview, it is important to back up a little and begin by establishing some clear definitional project boundaries. For instance, what will this project encompass? Is the goal of your new product development project to understand all office seating products, or just some subset, such as high-end ergonomic desk chairs or conference room chairs? And if so, are you focusing only on the needs of the end user who will use the chair, or others who also play a key role in the decision process, such as facilities managers, financial decision makers, architects and designers, and office furniture distributors.

Consideration also needs to be given to the question of how hard it will be to find these types of people and to recruit them to participate. In general, consumer products are easier to recruit for than B2B products, because the number of people who might qualify is so much greater. For instance, almost everyone would qualify as a customer for a residential long-distance telephone service VOC study or an electric utility VOC study. But only a few hundred people in the entire world would qualify for a VOC study among "chiefs of cardiac surgery at major medical centers that perform at least 600 open-heart procedures per year" (an actual project done a few years ago). Of course, there are lots of exceptions. Some consumer products have very low incidence, for example, orchid growing hobbyists or chamber music concertgoers. And some B2B products have very high incidence, e.g. fax or copying machine users. But overall, these are the exception, and you will need to consider this question as you plan for your VOC.

Which Customers to Interview

Once you have put some clear boundaries around the project definition, you can then ask yourself who all of the relevant players might be in gathering the Voice. Consideration needs to be given to geographical differences, different industries or types of users, different roles among the various users and decision makers, different demographics, and so on. This is always a judgment call and usually requires lots of compromise in that you will never be able to interview enough people to cover every subgroup as completely as you might wish. The goal is to try to cover all of the bases a little bit, or at least well enough that you feel you have heard from that particular subgroup. Note that, in all of these questions, the key decision is not *who* to interview, but *what types* of people to interview. Once interviewing is underway, these decisions can always be altered based on what you've heard to date. If one type of respondent is not proving to be very productive, some of the remaining interviews with that group can be eliminated and new ones scheduled with the types of people who are proving to have more valuable things to say. For instance, in a study of blood analyzers used in hospitals, purchasing managers had little understanding of the actual functionality of the device, leaving those decisions to the physicians, nurses, and laboratory technicians who have hands-on experience with them. The purchasing manager's role simply was to negotiate the lowest price and best terms possible. So partway through the study, it was decided to eliminate any further interviews with purchasing managers, substituting a greater number of physicians, nurses, and laboratory technicians, who were proving to be far more productive interview subjects in terms of understanding performance needs.

A good rule of thumb to follow in selecting the type of customers to interview is to ask yourself who actually comes into functional contact with the product or service and who has any actual decision-making authority or influence over what product gets purchased. In addition, consideration should also be given to the entire supply chain for the product—in other words, those who distribute or sell the product.

These sample design decisions are usually a judgment call but are critical to the success of a Voice of the Customer process. They are nontrivial and almost always deserve a lengthy discussion among people from many different functions within the organization before even a single interview is scheduled.

How to Interview Customers

An initial question in gathering the Voice of the Customer is whether to interview people individually or in groups. Market researchers refer to these as *one-on-one* interviews versus *focus group* interviews.

While there are clearly some key pluses and minuses involved in each method (see Figure 7-1), Griffin and Hauser looked at this question empiri-

Focus Groups

- Advantages
  - + Group dynamic—One member's comments trigger comments from other members.
  - + Takes less time—Several customers are interviewed in the time it would have taken to do a single one-on-one interview.
  - + More entertaining for the people who are observing.
- Disadvantages
  - – One or two people may dominate the discussion.
  - – Limited air time for each participant.
  - – "Group think" phenomenon.
  - – More difficult to get multiple high-level people and/or competitors to show up at the same time.

One-on-One Interviews

- Advantages
  - + Each respondent gets 30 to 60 minutes of airtime.
  - + More data is generated per respondent than in a focus group.
  - + Easier to coordinate scheduling.
- Disadvantages
  - – Not very entertaining for observers.
  - – More time-consuming.
  - – More work for the interviewer/moderator.
  - – More analysis time required.

FIGURE 7-1. Focus groups vs. one-on-one interviewing

cally and clearly came down in favor of individual interviews. Their argument was partly financial; focus groups simply cost a great deal more, particularly for categories in which the customers are higher level and, therefore, harder to recruit, such as physicians, executives, or high net worth individuals.

But there are also a number of practical issues. Most important among these was their empirical finding that an average hour of "airtime" produces about the same number of needs regardless of which method you use; that is, focus groups do not produce appreciably more needs statements per hour than do individuals—even though many more respondents are present. This is in no way a criticism of focus groups per se. Focus groups are generally the preferred medium when the goal is to search for a broad *consensus*. This is why they are so often used in concept testing and political research. But the goal in Voice of the Customer research is to generate as broad a list of detailed needs as possible, and this requires the ability to go off on tangents. Many people assume that frequency of mention of a need implies higher importance. However, research has shown that frequency is not necessarily reflective of importance. Whether a need is expressed many times or just once, it will simply end up as one need statement in a large database of potential need statements. (Prioritization, measuring the relative importance of the needs, will be addressed later on.)

So, if your goal is to elicit as broad and detailed a list of needs as possible, this requires that the interviewers have the ability to go off on tangents. And it is far easier to do this in an individual interview than it would be in a focus group interview in which all but one of the respondents might be left out while the interviewer goes off on the tangent with the one individual who has expressed that need.

Thus, it is generally recommended that Voice of the Customer interviews be conducted individually. A common question is whether it is okay to interview in small teams, such as two interviewers with one respondent. This is generally acceptable so long as only one of the individuals is leading the interview with the other person simply tossing in an occasional follow-up question. In fact, this approach can actually be beneficial in ensuring continuity should the primary interviewer need to pause to regroup his or her thoughts. However, anything more than two can begin to feel like an inquisition to the respondent and, thus, should be avoided.

In general, it works well to schedule respondents at one-hour intervals, with the average interview lasting about 45 minutes. This gives the interviewer some rest in between interviews, along with the flexibility to go a little longer if a respondent is unusually interesting and productive.

Where to Interview Customers

Another area of broad debate is whether the interviews are best conducted *on-site*, that is, at the customer's actual product usage location, or in a *central location*, such as a market research facility, a hotel meeting room, or even at a corporate facility. Again, there are some clear pluses and minuses to each (see Figure 7-2). The key advantage to on-site interviewing is that you get to observe the customer in his or her natural environment actually using the product or service. If usability issues are key, this is extremely important. The disadvantage, however, is that on-site interviews are much more expensive and time-consuming to conduct, because of all the travel expenses and scheduling difficulties that on-site interviewing usually entails.

A third alternative that sometimes works well for B2B markets is to conduct the interviews during a major conference or convention at which a broad cross-section of the target audience is present. This can often help to decrease travel costs and time significantly.

Most teams conducting a Voice of the Customer simply assume that they will have to travel to the respondent. In fact, Edward McQuarrie's book *Customer Visits* even assumes it in the title. However, it turns out that for many products and services, the concept of on-site interviewing is either unimportant or completely impractical. For instance, it would be quite difficult to conduct on-site interviewing for a medical device that is only used in rare emergency situations (you would have to wait around for a long time!). Likewise, there is rarely an "on-site" for most insurance products. For these types

On-Site Interviews

- Advantages
  - + Can observe the customer actually using the product or service.
    - "Contextual inquiry"—especially good for "usability" issues.
  - + Helps interviewer to appreciate customer needs and environment more deeply.
  - + PR benefits from showing customers how important they are.
- Disadvantages
  - – Interviews are more time-intensive because of travel and scheduling requirements.
  - – Finding customers willing to have you come and observe them in their workplace—Confidentiality and privacy issues.

Central Location Interviews

- Advantages
  - + Very efficient use of time
  - + Can accommodate many observers—Via one-way mirror or closed-circuit TV
  - + Easy to record interviews—audio and/or video
- Disadvantages
  - – Not "contextual"

FIGURE 7-2. On-site vs. central location interviewing

of products, and even for some in which usability *is* part of the issue, central location interviewing often works quite well. Most people do remember the big themes and can talk about them quite clearly, even though they are not physically next to the product. Finally, it should be noted that sometimes a combination of both on-site and central location interviewing can be a good compromise. However, if you choose to go this route, it is recommended that you do the on-site interviewing first to get a better understanding of the context and the physical environment in which the product is used.

Another common question is whether telephone interviewing can be substituted for either on-site or central location interviewing. In general, most VOC practitioners feel that face-to-face techniques work better in that they afford the interviewer the ability to read facial expressions, body language, and functional usability issues more clearly. However, conducting some of the interviews by telephone is a perfectly reasonable way to include a key individual who is either too far away to justify the travel expense or unavailable at the time you are nearby, or to fill in a few last minute holes in your sampling plan.

Finally, a warning: If you do plan to conduct the interviews at the customer's location, it is suggested that you notify them beforehand that you would like to spend at least some of the time at the physical site of the product. If all you are going to do is sit in a conference room, that conference room could just as easily be located anywhere. We once traveled a great distance to observe a world-famous medical clinic, only to be told upon arrival that patient confidentiality and privacy prevented them from allowing us to even enter the main care facility!

How Many Customers to Interview

The number of customers to interview is an area in which there is actually a good deal of consensus among both academics and practitioners alike. Again, Griffin and Hauser examined this question empirically. At the outset, remember that the initial one-on-one interviewing is a *qualitative* research step in which the concept of significance testing is meaningless. Thus, you can set aside, for the moment, all of your instinctive notions regarding sufficient sample size in order to obtain tolerable error ranges around what you learn. Griffin and Hauser's approach was to ask the question: Of the total set of needs generated from a sufficiently large number of interviews, what percent of those needs would have been generated from *n* randomly chosen interviews? Their conclusion: 30 one-on-one interviews, each lasting about 45 minutes, produce nearly 100 percent of all of the needs, and 20 interviews produce nearly 90 percent of all of the needs. Again, experience seems to confirm this. In most categories, the interviews generally feel like they are starting to become redundant somewhere in the range of 15 to 20. And even in more complicated categories, with lots of different segments and user types, it is rarely necessary to go beyond 30 to 40 interviews.

CONDUCTING INTERVIEWS

Developing an Interview Guide

While a good interview should feel like a natural, extemporaneous conversation, it should not by any means be random. It is critical that you go in with a detailed plan as to the content of that conversation. A good *interview guide* is the answer.

An interview guide should be thought of as an outline, not a script. It is, in a sense, a crib sheet for the interviewer. It should begin with more general topics and then gradually shift to more specific ones. A good interview guide sets the stage and helps put the respondent at ease. It is useful to explain at the outset: (1) the purpose of the study, (2) how the respondent was selected, (3) what other types of people are being interviewed, and (4) that the respondent's identity will remain private to the greatest extent possible (see Figure 7-3).

Many customers initially may misunderstand the purpose of the interview and assume that it is either a veiled sales call or an opportunity for them to pressure you to add some unique feature they specifically need or to fix certain problems that they have been struggling with for a long time. It is critical that you put these misunderstandings to rest right at the beginning with an appro-

My name is ____________________ and I'm with ____________________ located in ____________________. First, I want to thank you for agreeing to meet with me today.

Let me tell you a little about what we're trying to accomplish here. I'm part of a team that is working on the development of a next-generation office seating product for regular use at a desk. Now, the process we follow starts with a series of one-on-one interviews, and that's what we're here for today. We'll probably talk for about 45 minutes or so, and the discussion will be very open-ended. What we really want to hear about are your experiences with the chairs you've had in the past or seen elsewhere—your likes, dislikes, wants, and needs—anything that could be done to make your chair better and your workday a little easier.

We've been traveling around the country interviewing people like you—that is, people from all types of companies who spend a good part of their day sitting at a desk or a computer.

I have a rough outline of things we might talk about, but we should feel free to talk about whatever you think is important. There are no right or wrong answers here. Our primary objective is simply to hear the actual words and phrases that people use when they talk about their desk chair. And so, at times, I'll probably ask you some seemingly obvious follow-on questions, and I hope you'll bear with me when this happens.

Before we begin, I'd like to ask your permission to tape-record our conversation. The reason for this is that it's really hard to take good notes or try to remember what was said after the fact. And so we've found that it's just a lot easier to tape-record everything and then analyze it later. I'm not going to be asking you about anything very sensitive, but if at any time you want to say something to me "off the record," just let me know and I'll be glad to stop the tape.

I also want to assure you that we are not here to sell you anything today, and nothing you say will be used for any direct sales or marketing purposes—that is you won't be getting a call from our sales person next week! Also, I just want to be clear that I'm not in a position to commit to any timetables or fixes on any of our current products. My focus is entirely on our next-generation products.

Any questions before we begin?

FIGURE 7-3. Sample interview introduction

priate introduction that explains the purpose of the interview. You must make clear to them that you are not there to sell them anything and that you do not have the ability to commit to any types of timetables or fixes. Your focus is purely on your "next-generation" product.

What Questions to Ask the Customer

While it might seem a little ironic, the worst way to extract customer wants and needs is to ask the customer directly, "What are your wants and needs?" Even worse is to ask the customer, "What are your *requirements*?"—a word that is often used as a synonym for "wants and needs." Why is this so?

If you simply ask the customer, "What are your wants and needs?", what inevitably happens is that the customer immediately goes into solution mode, spitting back what he or she believes to be the best current solutions available, such as "it's gotta have cupholders" (for cars), "a 10-minute snooze alarm" (for a clock radio), or "10-second incremental billing" (for a long-distance plan). Furthermore, the word *requirements* connotes "must-have" needs, and what you really want to hear are their "wished for" needs that no one has yet addressed very well. So, for instance, if you asked customers about their requirements in a copying machine, they would probably go on and on about two-sided copying, a good collator, enlargement and reduction, and so on. And focusing on these items alone is likely to produce nothing more than a good "me-too" product.

Instead, the interview should focus on experience and desired outcomes. It is critical to understand how the respondent uses the product or service, what it is he or she is trying to accomplish, and what things get in the way now. What are some of the most difficult tasks you're trying to accomplish with this product? What are the areas that are most problematic? Asking about likes and dislikes, about bests and worsts, and about other extremes usually yield rich results. If asked what they look for in a business hotel, most people will talk about a comfortable bed, a good TV, a clean bathroom, and so on. But if you ask them about the best and worst experiences they've ever had in a hotel, they might talk about the noise coming from the room next door, having to get down on their hands and knees to find a phone jack for their computer, the broken showerhead, or the noisy heater/air conditioner. These are areas that might provide the best focus for product and service improvement (see Figure 7-4).

Market researchers often speak of *aided* versus *unaided* questions. An unaided question would be "Describe your use of this product." This might provoke any of a number of responses, both expected and unexpected, from the respondent. An aided question, on the other hand, suggests a specific topic, such as "Have you ever had any *maintenance problems* with this product?" You should begin your interview on a more unaided basis and then gradually probe on a more aided basis into some of the areas that you are particularly interested in, assuming that the respondent has not already brought them up on an unaided basis.

The goal is to get beyond generalities and drill down to the detail level for each distinct topic of the interview. So, good interviewing technique requires lots of probing and good follow-on questions. For instance, anytime the respondent uses a nonspecific adjective, you should ask a follow-on question to uncover the specific definition of the word. So, when customers say that something is "good quality," you need to understand how *they* define "good quality." Likewise for words such as "flexible," "consistent," "convenient," "appealing". What is it that makes a product or service "user-friendly," "reliable," "complicated," or "attractive"? If you simply assume that you know what these words mean, it's likely that you'll leave a lot of important information behind.

1. First, tell me a little bit about yourself—what do you do here, how long have you been with this company, and so forth?
2. Now, tell me about your office and, in particular, your desk chair. What does it look like/feel like, what features does it have?
3. What do you like about it? And what do you dislike about it? Why is that?
4. How does it compare to some of the other chairs you've ever had or seen in other people's offices? What's better or worse about it?
5. Thinking about other types of chairs or places that you sit that are not at your desk or not even in your office, how does your office chair compare to those? Are there things about these other types of seating products that you wish your office chair had?
6. Tell me about the *best* experience you ever had with an office chair, one where it made your day easier or more productive. Now tell me about the *worst* experience you ever had with an office chair, one where it got in the way and made your day harder or less productive.
7. What are some of the tasks or things you need to do in your job that your chair makes easier or more difficult to accomplish? Why is that?
8. What are some of the things that make one chair more *comfortable* than another? More *ergonomic*? More *functional*? More *aesthetic*? Anything else that would make one chair more desirable than another?
9. If you could change one or two things about your chair, what would they be? Why would that be better?
10. Is there anything else you would like to tell me about office seating products, or is there anything you thought I was going to ask you about that I didn't?

THANKS!

FIGURE 7-4. Sample interview guide

In general, open-ended, indirect questions work best. Direct questions too often produce one-word answers: yes, no, good, bad, okay. These are not very useful for product developers. Probably the single best probing technique is simply to ask "why?" For instance:

- Why do you say that?
- Why do you feel that way?
- Why do you prefer that one?
- Why is that important to you?
- Why would that be better?

You should not expect many customers to be able to discuss—let alone guide your actual decision-making—on the latest technologies. They may not even be aware of them. But after all, that's not their responsibility—it's yours! Their job is simply to articulate their wants, needs, likes, dislikes, and the problems they are encountering today. These "why" questions help to elicit these things, leading to a more creative melding of current customer needs and new technical solutions.

CREATING THE LIST OF NEEDS

How to Capture Customer Needs

Historically, most Voice of the Customer interviewing makes use of the concept of a "scribe"—that is, a sophisticated note-taker. However, I strongly recommend that you audiotape the interviews instead. For consumer products and services, this rarely raises eyebrows. But for commercial or industrial products and services, many people worry that their customers will not allow them to do this. Experience shows otherwise. If explained honestly and carefully, most customers have few problems with the idea, and almost all forget that the tape recorder is there after the first few minutes. If the respondent does express any doubt, a good technique is simply to allow the person to ask you to stop the tape if there's anything he or she would like to say off the record. A good Voice of the Customer interview is not an exercise in corporate espionage. Any questions that begin to sound like you are pumping the respondent for competitive intelligence will result in the respondent quickly shutting down or possibly reducing his or her willingness to speak candidly. So, it is critical to never ask any questions that might arouse suspicion as to your objectives. Remember, your goal is simply to understand the customer's wants and needs in support of your next-generation product. By allowing the respondent to control the on/off switch of the tape recorder, you'll almost always gain his or her trust right from the outset.

So why is tape recording suggested? There are several problems with note-taking. First, respondents are speaking quickly, and it is almost impossible to capture everything they say, even with an extremely fast note-taker. Second, there is a tendency to translate the customer's words into "company-speak," losing much of the richness of meaning in the process. And third, if the interviewer is also the note-taker, it adds one more nearly impossible task to what is already a multitasking type of process. Audiotaping allows for a full transcript of the interview (usually costing about $50 to $70 each) and the ability to more carefully analyze the interview afterward—a far more thorough and clearer way to extract the needs. In addition, audiotaping allows others within the organization to easily and quickly access the full content of the interview at any time in the future.

Highlighting transcripts using a colored marker is a laborious but enlightening process (see Figure 7-5). Anything that suggests some kind of a need should be highlighted and later can be "wordsmithed" into the form of a need statement, preserving the customer's language as much as possible (see Figure 7-6). You should also have two or more readers for each transcript. Research at MIT showed that a single analyst tended to identify only about 40 percent of the attributes in a given set of interviews; thus, multiple analysts were needed to identify a sufficiently high percentage of attributes. And, finally, because people from different functional areas commonly see the needs from

TELL ME A LITTLE ABOUT YOUR CURRENT OFFICE CHAIR. WHAT DOES IT LOOK LIKE? WHAT FEATURES DOES IT HAVE?
Well, it's pretty old, but I'm kinda used to it. It's beige with a low back. It has these stupid little tufts that sometimes get caught on my slacks or my skirt. But it's pretty comfortable most of the time.

WHAT MAKES IT COMFORTABLE?
Well, there's lots of room to move around in it, you know, to change position when you've been in one position for too long, especially when I'm working on my computer for a long time. I like to lean forward for while, then back, then to the sides, and so on. It seems to accommodate that and sort of moves with me.

IS IT FABRIC OR LEATHER?
It's fabric, and unfortunately, it's starting to wear out on the arms. That's probably partly from my own use, but also I think 'cause the cleaning people always shove them under the desks and tables after they vacuum, and so the arms get banged up every day. We've asked them not to do that a thousand times, but you know how it is . . .

DOES THE SAME THING HAPPEN WITH LEATHER?
Well, I used to have a leather chair at a previous company I worked for, and I don't remember the arms getting beat up on that one. But I hated that chair.

WHY IS THAT?
Because on a warm day, if I perspired, I sometimes stuck to the chair and it really hurt to get up!

SO SHOULD I ASSUME THAT YOU PREFER FABRIC?
Oh, yeah, definitely.

AND WHY IS THAT?
Because it breathes more. I don't get as sweaty and I don't stick to the chair when I want to get up.

WHAT ELSE DO YOU LIKE ABOUT YOUR CURRENT CHAIR?
Well, it's pretty easy to adjust, at least for the up-and-down part. Although I had already been using this chair for over a year when someone showed me how to adjust the stiffness of the tilt when I lean back. I didn't even know it had that.

SO IT WASN'T THAT OBVIOUS?
No, you had to read the manual or have someone show it to you. And I'm not a manual reader. In fact, I have no idea where the manual even is! I don't think I've ever seen it.

YOU SAID IT WAS EASY TO ADJUST FOR THE UP AND DOWN PART. WHAT IS IT THAT MAKES IT EASY TO ADJUST?
It's just this little paddle underneath on the side that you pull on and the height rises or goes down if you sit on it. I used to have this one chair where the adjustment knob was underneath in the front, and I always felt a little unladylike when I reached down there to adjust it.

FIGURE 7-5. Office seating products transcript and highlighting example
Melissa Edwards, Director of Product Marketing, XYZ Corporation

IS THERE ANYTHING THAT YOU PARTICULARLY DISLIKE ABOUT YOUR CHAIR?
Oh, yeah. It's a pain when I want to move around, like from my desk over to my computer table.

WHAT DO YOU MEAN WHEN YOU SAY "A PAIN"?
You know, it's only about four feet, but the wheels are so stiff, that I have to do one of these—you know, clomp, clomp, clomp—to get over there. I wish it were easier to scoot around in my workspace.

FIGURE 7-5. *(continued)*

different perspectives, it is advisable for people from many different functions to participate in the reading of the transcripts.

What You End Up With

One of the thorniest problems in the process comes from the fact that most customers don't speak in nice, clean need statements. A skilled interviewer is often able to move the respondent in that direction, but many of the needs remain buried in words and phrases that fall slightly short of a clean need statement.

Many of them fall into one of several categories that are close to a need but not quite truly a need:

- *Engineering Characteristics or Solutions.* Many customers will offer a solution, an engineering characteristic, or a laboratory measurable technical specification that they think does a good job of addressing their need. For instance, "The exterior wall should be an alloy of aluminum and titanium." Whenever this happens, a good interviewing technique is to ask them *why* they think that would be a good solution. This often

- Nothing on the surfaces that can get caught on my clothing.
- Lots of room to move around within the chair.
- Easy to change positions within the chair—leaning forward, backward, or to either side.
- Arms that don't wear out, even when getting banged around or shoved under a desk or table.
- A chair that "breathes"—I don't stick to the surface even if I'm feeling sweaty.
- Easy and intuitive to adjust the chair into whatever position I want.
- Instructions on how to adjust the chair are always accessible if needed.
- I never have to get into an embarrassing position to adjust the chair.
- Easy to scoot around within my workspace.

FIGURE 7-6. Extracted needs statements

provokes them to state the real underlying need: "It needs to be really lightweight and strong at the same time." Likewise, if a customer were to say, "My computer should come with an uninterruptible power supply" or "The software should contain automatic file saving," it is important not to mistake these for needs. They are just that customer's suggestion of what would be a good current solution. If the interviewer were to delve a bit deeper by asking the question, "Why would that be a good solution?" the customer would likely say something like, "Well, then I would never lose any of my files." This is the real underlying need, and hopefully, there will be even better technical solutions in the future.

- *Target Values or Specifications.* Many customers will not only talk about engineering characteristics but will actually state what they believe to be appropriate numerical target values for some of the technical specifications. For instance, "The sheet metal at this point should be no more than *n* microns thick." Or, "We should never have to wait in line for more than ten minutes to buy our tickets at the movie theater." While this is certainly useful information as to what they believe is a reasonable expectation, five minutes would clearly be better, and two minutes would be better yet. The interviewer should probe as to *why* they believe ten minutes is appropriate. Then the customer might say something about "having enough time to buy some popcorn and get seated before the movie starts." That's the *real* need. Suppose someone could take their refreshment orders while they waited for their tickets, and they paid for and received both simultaneously! Or suppose they could order their refreshments over the Internet at the same time that they bought their tickets, paying with their credit card all in one transaction.
- *Opinions.* Many customers will state their opinion as to how something ought to be. Again, this is not the same as a need, and the interviewer needs to simply ask how they think things ought to go. So, if the customer says, "I hate doing maintenance on this device," the appropriate follow-on question is to ask "why," and how they wish the maintenance process would go. This will usually elicit the appropriate underlying need statements.

This distinction is subtle but critical. If you accept every opinion from respondents as to what are the best solutions, engineering characteristics, or target values, you are probably heading for a "me-too" product. The goal of Voice of the Customer is to understand the most minute details of an underlying need, and then for the team to try to find new and creative ways to address that need that others have not yet thought of. This takes a fair amount of practice, but if understood, it will make your interviews infinitely richer.

Although a skilled interviewer often can turn these engineering characteristics, solutions, target values, and opinions into cleaner need statements, the truth is that many times, team members will forget to try; and even if they do try, the customer will fail to produce the desired type of statement. In these cases, the reader who is highlighting the transcript is going to have to take some liberties. And while this is somewhat less than desirable, one shouldn't fear doing so. The rules are simply that it is important to try to maintain the customer's language as much as possible and be careful not to read one's own interpretation into the phrase. Very often, pieces of sentences and sentence fragments in a larger paragraph can be "crafted" into a cleaner needs statement. Sometimes, the interviewer's paraphrasing is a clearer statement of the need than anything the customer has said, and it's okay to highlight that. Likewise, it is perfectly okay to take a phrase that was stated in the negative and turn it into the positive. For instance, "We need a machine that isn't so heavy and difficult to move around" might be turned into "A machine that is easy to move around." Finally, it is important that the need statements be clear and understandable outside the context of the interview itself. The acid test is to be able to show it to someone who has not read the entire transcript and have them know exactly what you mean.

Once all of the transcripts have been highlighted and those phrases that fall slightly short are crafted into a good form, the phrases are entered into a simple database of phrases. This can be done in almost any type of file format—text, word processor, or spreadsheet—thus creating a database of captured phrases in the form of true customer need statements.

How to Edit the Customer Needs

The database of phrases that results from, say, 20 to 30 interviews with two or more readers per transcript is likely to be quite lengthy. It is not uncommon to end up with anywhere between 750 and 1,500 such phrases if done thoroughly. Needless to say, there is much duplication in such a database. Many needs will have been articulated by multiple respondents, and it is likely that several readers on each transcript may have highlighted the same phrase. So, the next step in the process is to edit these statements down to a final set of *unique* phrases. This is largely a judgmental process. It is best carried out initially by one individual who has a great memory and the ability to hold a large amount of qualitative information in his or her head at one time. If a need statement is duplicated five to ten times, the criteria used as to which phrase to keep is simply to pick the one that is worded best—the clearest, most understandable description of the need. This process, referred to as "winnowing" the phrases, may take this person one to three days until it gets down to about 150 to 200 remaining unique phrases.

At this point, it is often useful to print the phrases out on little slips of paper or Post-it® Notes and then bring the rest of the team back into the process. The team can now assist in removing any final duplicates, cleaning up the wording for clarity, and searching for any remaining missing needs. The final resulting list of unique need statements is usually somewhere in the range of 70 to 140 statements, with a median right around 100.

HOW TO ORGANIZE CUSTOMER NEEDS

Affinity Diagrams

While most product and service categories do indeed have in the neighborhood of 100 such need statements, this is still too high a level of detail for a product development team to work with efficiently. So, an additional level of aggregation is needed. A common tool to do this is the so-called *affinity diagram*, a fancy name for "bucketing" of the needs (see Figure 7-7). In this process, the need statements are simply organized into groups that go together in some logical means of organization. Traditionally, the product development team itself carries out this process of affinitizing the needs.

To do this, all of the need statements are copied onto cards or Post-it® Notes, which are then placed on a large table or wall. The team members then begin to organize the needs into groups by moving them around according to whatever logical constructs they are thinking about. Initially, this is done in complete silence, and after a while, it becomes apparent that different team members are thinking about different constructs or ways of organizing the needs. It is not at all uncommon for certain needs to be moved back and forth between various groups by the participants, and this tends to highlight different insights about how people think about the product or service. Once all of the needs have been placed into groups, the silence is broken and the team then

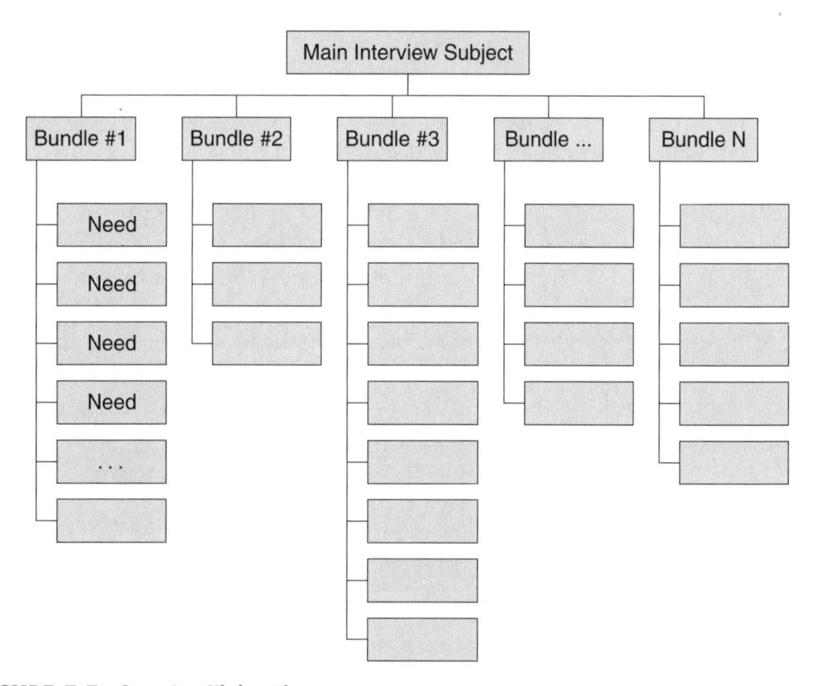

FIGURE 7-7. Sample affinity diagram.

begins to discuss their resulting affinity diagram, placing titles on each of the groups and trying to resolve any remaining ambiguities or disagreements.

However, an interesting wrinkle put forth by Griffin and Hauser, and confirmed in many real-world cases, is that customers are likely to organize the ideas slightly differently from the way an internal product development team would. Their explanation is that people from inside the company are likely to organize the needs consistent with their organization chart, which reflects the way in which the product is made and delivered. Customers, on the other hand, have no knowledge of this organization chart and are more likely to sort the ideas according to the way that they acquire, use, and experience the product. And this contrast can have a major effect on some of the downstream product development decisions. For instance, across a large number of cases involving computer systems, the internal sorters almost always made a clear distinction between hardware-related needs and software-related needs, while external sorters almost never made this distinction.

So, the safest way to accomplish this is to engage customers in the affinity diagramming process. This can be done in several ways. The easiest is simply to organize a group of three to six customers and let them do it as a group. This can be followed by a focus-group-like discussion in which they can be quizzed as to how they classified and labeled each of the buckets, and which needs were the subject of ambiguity and might have been put elsewhere. If any of the groupings are still quite large, that is, a group with many need statements, the customers should be pressed to subdivide them into several subcategories. Ideally, they should finish with somewhere between 15 and 25 of these affinitized, bucketed groupings. Most product development teams report that this is about the right level of "granularity" for ongoing product development.

Statistical Methods for Organizing

An even more elegant way of carrying out this process makes use of multivariate statistical methods. Griffin and Hauser prescribe a process whereby a large group of customers (usually about 50 to 100) carry out the exercise individually at home or in their office. In this method, the participants are prerecruited by telephone or e-mail and asked if they would be willing to participate in an exercise that will be mailed to them. Again, they are offered either cash or a substantive gift to thank them and to encourage their participation. If they agree, they are sent a packet with materials, instructions, and a prepaid return envelope. The materials include a "card deck" with one need statement per card, which they are asked to sort into their own individual affinity diagram. They are then asked to select the one card in each pile that best summarizes the ideas in that pile (referred to as the "exemplar") and to place it on top of the pile. The selection of these exemplars is useful in the naming of the various groups that result when all of these individual affinity diagrams are statistically "merged" with one another. (They are also asked to prioritize—or indicate the relative importance of—each of their piles, as discussed in the next section.)

The proportion of times that two attributes are placed in the same pile of cards can be thought of as a measure of their similarity of meaning. A matrix is then created in which each row and column intersection contains the frequency with which each pair of cards is sorted together. For example, if 57 of the respondents put card 3 and card 4 in the same pile (along with other cards), then the matrix would contain "57" in row 3, column 4 (and row 4, column 3). Cluster analysis is then used to group the phrases that are most similar into a hierarchy of attributes, expressed in the form of a tree diagram (technically, a *dendrogram*) that shows the various levels of aggregation. Needs that are frequently sorted together will appear close to one another on this tree diagram, that is, on the same or a nearby "branch." And those that were not very frequently sorted together will appear far apart on the tree diagram, that is, on an entirely different "limb." This cluster analysis provides a more representative affinity diagram of customer needs, independent of any imposed structure or bias from the internal core team sorters.

While somewhat more complicated and expensive, this method solves one important problem. In any given product category, there are an almost infinite number of ways to sort the need statements, all of them looking fairly logical on the surface. So, when carried out by the core team itself or just a few small customer groups, one can never be sure whether their affinity diagram is truly reflective of the population as a whole. The statistical method solves this problem.

The final step in organizing the needs is simply to label each group. While it is fairly easy to create simple one- or two-word labels for each of the groups, it is even more useful to create a complete need statement in itself, which summarizes the ideas in that pile. For instance, there's likely to be a pile having to do with "customer service" or "ease of use." It is preferable to borrow some of the phraseology from the individual need statements in that group and turn the label into something like, "Provides good, accessible customer service whenever I need it," or "A product that is easy and intuitive to learn and use."

HOW TO PRIORITIZE THE CUSTOMER NEEDS

We know intuitively that all needs are not created equal. So, the final step in the Voice of the Customer process is to prioritize the need statements. This prioritization is ideally conducted on two different dimensions: *importance*—how important is each need cluster, relative to the others?—and *performance* or *satisfaction*—how satisfied are people with the products, services, or alternatives they use now?

Getting the Numbers

Historically, this exercise has also usually been carried out by the product development team, judgmentally assigning the prioritization themselves. However, this is potentially an enormous mistake, as it has been shown in dozens of cases that customers usually will prioritize the needs quite differently from people within the organization. People within a company often assume that the most important attributes are those at which their company already excels. Similarly, there is a strong tendency to rate the competition's performance on many of the attributes much higher than customers actually would, probably because the competition always looks so much bigger, tougher, and better "buttoned up" than they are in reality.

Fortunately, this problem is easy to overcome with a simple survey that can be conducted either via mail, telephone, fax, or the Web (see Figure 7-8). The questionnaire simply lists the 15 to 25 bundled needs statements, along with a rating scale—typically a five-, seven-, ten- or hundred-point scale. Customers are then asked to rate each of the clustered needs for how important it is in their purchase decision, and second, how satisfied they are with whatever product or service they currently use. Finally, some classification data—such as demographics about the individual or the company—can be used to segment the prioritization data during analysis. Such a survey should take no more than ten minutes to administer on any individual customer, and an overall sample size of 50 to 100 customers is usually sufficient, unless there is a need to perform a more sophisticated segmentation. If the latter is the case, and it is not prohibitively expensive to do so, you may want to consider a much larger sample size—perhaps 30 to 50 per segment and anywhere from several hundred to as many as a thousand overall.

Understanding the Numbers

A useful way to analyze the data is to calculate averages for each of the individual needs clusters and then graph them on a two-dimensional scatter plot with one dimension for importance and the other for current performance (see Figure 7-9). Bundles that fall into the upper right-hand quadrant are those that customers tell us are high in importance and high in current performance. These are attributes that must be maintained in the new product but generally are not the ones deserving of the highest level of focus in ongoing product development. Those in the upper left-hand quadrant, on the other hand (high importance and low performance), are clearly those upon which the greatest degree of effort must be placed. These are the needs that customers tell you are above average in importance, but where current performance is lagging. The ability to innovate on these needs and come up with better solutions will almost surely be highly attractive to customers and lead to a winning new product. Those bundles that fall into the lower right-hand quadrant are needs that customers are telling you are not very important, and upon which they are already reasonably satisfied with what they have. These attributes generally do not deserve any additional investment or product development effort, and may be areas in which some savings can be harvested in the creation of the new product. Finally, those in the lower left-hand quadrant—that is, those that the customers are telling us are low in importance and low in current satisfaction, should not be dismissed too quickly. Sometimes these attributes are rated low in importance only because the customer assumes that no one could possibly do better; in a sense, they are giving you a "pass" on these attributes. However, these sometimes present inter-

This survey should take you no longer than 10 minutes to complete

IMPORTANCE

The following list contains some items that people have told us are important to them in an office chair. Please rate each of the following items on a scale of 1 to 7, where **1** is **Not at all important** and 7 is **Extremely important.** If a phrase does not apply to you, circle "NA."

	Not At All Important		Average Importance			Extremely Important		
Is comfortable to work in	1	2	3	4	5	6	7	NA
Easy to scoot around in my workspace	1	2	3	4	5	6	7	NA
Easy and intuitive to adjust	1	2	3	4	5	6	7	NA
Is a nice chair to look at; blends well with my office decor	1	2	3	4	5	6	7	NA
Strong, durable covering (fabric, leather, etc.)	1	2	3	4	5	6	7	NA
Gives me plenty of room to move around in	1	2	3	4	5	6	7	NA
Pleasant for leaning back and putting my feet up	1	2	3	4	5	6	7	NA
Arms are at the right height and width for my elbows	1	2	3	4	5	6	7	NA
Doesn't show dirt and dust	1	2	3	4	5	6	7	NA
Makes me feel important within the organization	1	2	3	4	5	6	7	NA
Base is strong and able to withstand a "beating"	1	2	3	4	5	6	7	NA
Works well for unusually tall, short, heavy, or light people	1	2	3	4	5	6	7	NA
Easy and quick to repair	1	2	3	4	5	6	7	NA

FIGURE 7-8. Sample office seating questionnaire

PERFORMANCE

Thinking specifically about your current office chair, please rate ***how good a job*** that chair does on each of the following items. Rate each of the following items on a scale of 1 to 7 where **1** is **Poor** and **7** is **Excellent**. If the phrase does not apply to you, circle "NA."

	Poor						**Excellent**	
Is comfortable to work in	1	2	3	4	5	6	7	NA
Easy to scoot around in my workspace	1	2	3	4	5	6	7	NA
Easy and intuitive to adjust	1	2	3	4	5	6	7	NA
Is a nice chair to look at—blends well with my office decor	1	2	3	4	5	6	7	NA
Strong, durable covering (fabric, leather, etc.)	1	2	3	4	5	6	7	NA
Gives me plenty of room to move around in	1	2	3	4	5	6	7	NA
Pleasant for leaning back and putting my feet up	1	2	3	4	5	6	7	NA
Arms are at the right height and width for my elbows	1	2	3	4	5	6	7	NA
Doesn't show dirt and dust	1	2	3	4	5	6	7	NA
Makes me feel important within the organization	1	2	3	4	5	6	7	NA
Base is strong and able to withstand a "beating"	1	2	3	4	5	6	7	NA
Works well for unusually tall, short, heavy, or light people	1	2	3	4	5	6	7	NA
Easy and quick to repair	1	2	3	4	5	6	7	NA

How satisfied are you *overall* with your current office chair?
(Please circle one number below.)

Very Dissatisfied						**Very Satisfied**
1	2	3	4	5	6	7

Please take a few minutes to answer the following classification questions. They will help us to better interpret your answers. Be assured that all of the information you provide is kept strictly confidential.

FIGURE 7-8. *(continued)*

How old are you? *(check the appropriate box)*
☐ Under 18
☐ 18 to 24
☐ 25 to 34
☐ 35 to 44
☐ 45 to 54
☐ 55 to 64
☐ 65 to 74
☐ 75 or over

Gender *(check the appropriate box)*
☐ Female
☐ Male

How would you best describe the work that your company does? *(check the appropriate box)*
☐ Communications
☐ Education
☐ Finance/Insurance
☐ Grocery
☐ Health Care
☐ Lodging
☐ Process Industries
☐ Real Estate
☐ Retail
☐ Restaurants
☐ Other (specify)

____

How many people are employed at this business location? For this question, we mean full time employees *working 30 hours per week* or more. *(Check the appropriate box)*
☐ Less than 100 *full-time* employees
☐ 100 to 250 *full-time* employees
☐ 251 to 500 *full-time* employees
☐ 501 TO 1,000 *full-time* employees
☐ More than 1,000 *full-time* employees

Which of the following categories best describes the approximate annual gross revenue of your company? *(Check the appropriate box)*
☐ Under $500,000
☐ More than $500,000 but less than $1 million
☐ $1 million or more but less than $5 million
☐ $5 million or more but less than $10 million
☐ $10 million or more, but less than $25 million
☐ $25 million or more but less than $50 million
☐ $50 million or more

Thank you for participating in this survey.

FIGURE 7-8. *(continued)*

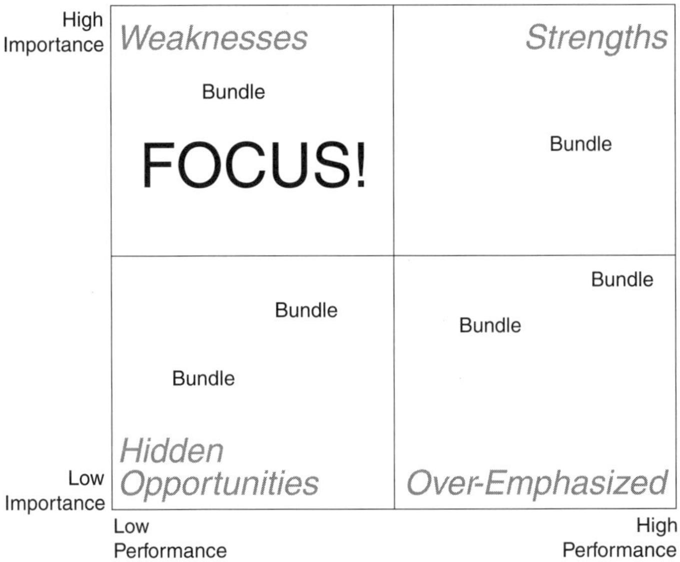

FIGURE 7-9. Importance vs. performance.

esting hidden opportunities. Consider the famous Kano model (Cohen, 1995); the so-called "delighters" often come from this quadrant in that someone comes up with a clever way to innovate, and customers respond with excitement—a sort of "wow" reaction. And over time, these needs take on greater importance and move into one of the upper quadrants.

TIME AND RESOURCES NEEDED

In general, a thorough VOC project takes about two to three months to carry out, although some report doing it a little faster, and some take much, much longer (six months or even more). A typical timeline might break out as follows:

Activity	# of weeks
Project and sample design	1
Recruiting of interview respondents	2
Interviews	1–4
Transcription and highlighting	2
Editing	1
Creation of hierarchy	1
Prioritization (survey)	2
Total	10–13

During the process, the team members' time commitment will vary from week to week and from person to person. In most teams, one or two people "lead the charge," spending as much as 30 to 50 percent of their time on VOC, while others spend about 10 to 25 percent of their time on it—not trivial, but not overwhelming.

Out-of-pocket expenses can range from about $10,000 to $20,000 if you do it yourself and don't have to travel too much, to as much as $60,000 to $75,000 for a big international case. (Also, some training and coaching are essential at the beginning.) A number of presentations at various conferences have reported a median somewhere around $40,000. If you choose to outsource some of your VOC activities to professional market researchers, the out-of-pocket costs range from about $70,000 to as much as $200,000 for a complicated international case, although this is partially offset by a reduction in your staff's own time commitment. The median cost for outsourcing is usually around $100,000.

ADDITIONAL CONSIDERATIONS

What about "New-New" Products?

Most of the work that has been done in the area of Voice of the Customer market research involved relatively well-defined existing product categories. A common question is this: what to do when such a category does not exist, that is, for what are often referred to as "new-new" products. Clearly this is a more complicated case; but it *can* be done. First, even if the category does not yet exist, there are usually alternatives that people use to accomplish the same type of task now. A good example concerns a company that was trying to develop a technology using extremely bright light to fight the effects of jet lag. The company was trying to commercialize products that delivered carefully measured doses of bright light, which research had shown could help individuals adjust their internal time clocks more quickly when traveling across multiple time zones. While no such product or technology existed up to that point, frequent international travelers were easily able to talk about the various tools and techniques they used to fight jet lag now—diet, medication, toughing it out on the first day, and so on.

However, the company needed to go one step further. While people were easily able to describe their current methods of fighting jet lag, in order to understand their wants and needs around the new technology, the company needed to present some initial conceptual designs (concepts) in order to "open up the space" to discussion. While presentation of concepts is typically not a part of Voice of the Customer research, in new-new products, it may be necessary to use them. The nature of the interview, other than that, is fairly typical. The first half covers their current ways of dealing with the problem. Then the concepts are presented, and the remaining time is used for people to express their likes, dislikes, questions, and doubts. Clearly, this is a more complicated task than for an existing product category. But it is usually the best one can do, and there is still tremendous value in the resulting Voice of the Customer data. So long as the emphasis is not so much on *whether* they like or dislike the concepts, but more about *what* they like and dislike in the concepts, the same level of need statements usually results.

What about Manufactured Goods vs. Services?

Throughout this discussion, I have been talking about gathering the Voice of the Customer for both manufactured products and services. Yet because these technique's origins were more with physical products, and because most of what has been written involves products, there remains a bit of doubt as to their applicability for services.

This controversy is largely unwarranted. First, keep in mind that, even for physical manufactured goods, there is always a service component that customers will usually want to talk about, even if your primary focus is on the physical product—that is, it is almost impossible to separate the two. Second, many "products" are, in fact, services, for instance, telecommunications products, energy products, and financial products such as banking, insurance, loans, and credit cards. The techniques described here work just as well for services as they do for manufactured goods. The only noteworthy difference is that the resulting needs and performance measures are usually a bit "softer" in nature. Whereas with products, you will be translating customer needs into (mostly) laboratory measurable specifications such as weight, thickness, or force needed, with services the measures are usually things like courtesy ratings, percent taking a certain action, or difficulty/time/number of attempts it takes to accomplish something. These things are usually a little more difficult to measure, but that does not relieve you of the obligation to do so if these turn out to be the critical metrics to satisfy the customer's needs.

You should not be shy about using the Voice of the Customer for services. Nor should you avoid including discussion of the service components when using it for manufactured goods.

What about Global Products and Services?

Many products and services are intended for a worldwide marketplace today, and this creates an entirely new challenge in that the Voice of the Customer must be gathered globally. This almost always results in a much higher degree of complication and cost, but it may be entirely necessary. To ignore the inevitable differences that exist across borders and cultures is to invite the kinds of disasters we have all heard about, such as steering wheels on the wrong side! And it is simply foolish to assume that anything that is well received and popular in the home market will automatically be loved elsewhere.

Sometimes, the underlying customer needs themselves are different in each market. Other times, corporate "politics" require that the market research be conducted in a number of foreign markets to get buy-in from overseas management. But the most common (and best) reasons to do VOC internationally is to better understand the following:

- How attitudes about the product may differ depending on geography
- How customers in different geographic markets weight various product attributes
- How product usage differs among customers in various countries
- How climate, market conditions, regulatory issues, and other environmental factors influence product preferences

The goal is to determine when to create an entirely separate product or marketing strategy for an international market, and what product changes are necessary.

Most of the research methodologies described earlier in this chapter translate well to international environments, with only minor procedural differences. However, there are a few idiosyncrasies that may require special consideration:

- Research etiquette may differ from country to country. For instance, research materials may need to be hand-delivered rather than mailed in certain countries.
- While gifts to thank people for their participation may be appropriate or even expected in some countries, they may be out of the question (or even illegal) in others.
- Results can be significantly different when the interview is not conducted in the respondent's native language. You should always try to conduct them in the native language of the respondent to ensure that the words and semantics are properly understood and captured.
- To ensure that the meaning of the customers' words and phrases are preserved exactly, all materials—the interview guide, the needs statements themselves, and the prioritization survey—should be "double-translated" from language A into B and then, independently, from B back into A.

It may not be practical to go to this level of detail in every single foreign market. But many companies include an additional three to ten countries (besides the home market) in their regular VOC activities.

Revisiting the Myths

We summarize the chapter by returning to the nine myths outlined at the beginning of the chapter and recap what you have learned.

1. *The Voice of the Customer is primarily a qualitative market research process, not a quantitative one.* Clearly, this is incorrect. While the interviews and needs extraction is a qualitative process, the organization of the needs into an affinity diagram is highly quantitative if done statistically, and the process of prioritizing the needs is entirely a quantitative one.
2. *Voice of the Customer interviews should be conducted primarily with your most important current customers—that is, your key accounts.* This is potentially a source of enormous error. It is important to include lots of average customers, both your own and your competitors', in the process. You will usually learn more from your noncustomers than your best customers. It is not meant by this that you should exclude your key accounts, only that they should simply be one part of the sample.
3. *Customers can't really tell you what they want; they won't know it until they see it.* This confuses solutions with needs. Customers are, in fact, very good at articulating their needs; they simply can't foresee all of the new technical solutions that will address those needs. But, remember, coming up with good technical solutions is your job, not theirs.
4. *The best way to elicit wants and needs is to ask customers what they want and need.* Unfortunately, this form of direct questioning usually forces customers to spit back what they believe to be the best current solutions to their needs. You should avoid it. Instead, ask them to talk about the outcomes they desire, the tasks they are trying to accomplish, and what they've liked and disliked about current products and solutions that they use. This will take you where you want to go.
5. *Noncustomers will be reluctant to talk with you.* While noncustomers may initially be a little suspicious about your motives, so long as you explain to them honestly what you are trying to do and stick to that agenda, most noncustomers will be glad to sit down with you, particularly if you give them some incentive or gift for doing so.
6. *A separate note-taker is the best way to record needs.* Audio recording and verbatim transcription creates a far more thorough, detailed set of needs that are less subject to a note-taker's biases and interpretations of those needs.
7. *Customers will not allow you to tape-record their conversations.* Again, this is simply not true. If explained clearly and carefully, and by

giving customers control of the recorder when they want to say something off the record, most customers will, in fact, allow you to record the conversation. This is particularly true if the interviews are being conducted at a neutral location but is even the case when the interview is conducted at the customer's site.

8. *The process of organizing or "affinitizing" the needs is best done by the product development team itself.* Research has shown that customers will affinitize differently, and so it is important to include them in the process.
9. *Prioritization of the needs is also best done by the product development team.* In dozens of cases in which a side-by-side comparison was conducted between customers' ratings and internal company ratings, the results were almost always quite different. Thus, it is imperative that this prioritization be carried out by customers using a simple survey instrument. Otherwise, the team is in danger of focusing on the wrong needs.

SUMMARY

At the end of the Voice of the Customer process, you should have the following:

- A detailed list of about 70 to 140 detailed, unique customer need statements (that came from the 20 to 40 individual interviews).
- Phrases organized/affinitized (preferably by the customers) into 15 to 25 "buckets" or affinity groupings
- Needs clusters that are prioritized in terms of their relative importance and performance.

Clearly, this is not an easy process. It requires a great deal more than just a few visits with a few key customers. But if done properly, the richness of the data and the amount of learning will amaze you. Almost everyone who has ever gone through this process reports that it was eye-opening, even elating, and clearly worth the effort, resulting in better new products and services, more satisfied customers, and higher long-term profitability.

REFERENCES

Burchill, Gary, and Christina Hepner Brodie. 1997. *Voices into Choices*. Madison, WI: Joiner Publications (Oriel Inc.).

Cohen, Lou. 1995. *Quality Function Deployment: How to Make QFD Work for You*. Reading, MA: Addison-Wesley Publishing.

Griffin, Abbie, and John Hauser. 1993. "The Voice of the Customer." *Marketing Science* 12, 1, Winter: 1–27.

Katz, Gerald. 2001. "The 'One Right Way' to Gather the Voice of the Customer." *PDMA Visions* 25, 2, October Accessible at:http://www.pdma.org/visions/oct01/voc.html.

McQuarrie, Edward F. 1998. *Customer Visits*. Thousand Oaks, CA: Sage Publications.

Ulwick, Anthony. 2002. "Turn Customer Input into Innovation." *Harvard Business Review* 80, 1, January: 5–11.

8 Creating the Customer Connection: Anthropological/ Ethnographic Needs Discovery

Barbara Perry, Cara L. Woodland, and
Christopher W. Miller

The real voyage of discovery consists not in seeking new landscapes, but in having new eyes.

—Marcel Proust

In recent years, increased competition has driven the need for deeper customer understanding. There has been an increasing awareness that traditional means of knowing the customer fall short in delivering the deep insight necessary for truly innovative product development. In response to this reality, the trend has been to augment traditional market research with more direct, empathic ways of understanding the customer, most of which have their roots in anthropology. These methods go by many names: ethnography, fieldwork, immersion, and observational research.

There are many excellent models and variations on ethnography that are built on a shared set of principles. While the use of ethnography is growing, stereotypes, assumptions, and misunderstanding exist about its use in business applications. Some examples include "it costs too much;" "it takes too much time;" "it's difficult to extract actionable findings"; "it's too anecdotal for upper management to 'hear;' " "the sample size is too small to be valid."

This chapter's purpose is to demystify the practice of ethnography and to spell out in a useful way what it is, how to perform it, and how to apply it. Our hope is to advance current practices and provide processes that are state-of-the-art to make this research approach more user-friendly, streamlined, practical, easier to apply, more "do-it-yourself," and producing better results. The chapter begins with a definition of ethnography and its applications in business and new product development, followed by the five steps to conduct ethnographic research. These five steps are as follows:

Step 1. Plan the project
Step 2. Get started
Step 3. Conduct the fieldwork
Step 4. Analyze the data
Step 5. Use the outcomes

WHAT IS ETHNOGRAPHY?

Ethnography is the research method used by anthropologists to understand behavior *in context.* Today, it is being applied in a wide variety of nonacademic settings, including corporate marketing, product development, and governmental and nonprofit organizations. The goal of ethnography is to develop and convey a well-rounded "insider" perspective of people's values, customs, beliefs, behaviors, and motivations. The ethnography process is a catalyst for innovative thinking. The outcomes are actionable insights, those "aha's" that come from having the ability to "see with new eyes," as well as a renewed sense of energy and commitment to the audience.

Organizations have come to the realization that while they often have reams of data about their customers, what really matters (and remains elusive) is the ability to understand their customers. What are the complex processes involved in the formation of your audience's perceptions and choices?

Ethnography differs from other qualitative research methods in several fundamental ways:

- First, it is an *inductive* approach to understanding. The research ends with hypotheses, rather than beginning with them. Through the use of an open-ended, discovery-oriented approach to the research topic, ethnography allows knowledge to emerge rather than have it forced into predetermined categories. Ethnography is most likely to result in completely new and unexpected learnings because it is open to what is really *there*, instead of being constrained by what is already known or presumed to be relevant. As the famous anthropologist Clifford Geertz said, "The trick is to figure out what the devil *they* think they're up to."
- Second, it *focuses on context.* The learning is a result of the time spent in the environment in which the behavior takes place, not in a lab or behind a glass wall. Careful on-site observation leads to informed questions, which provide a deeper level of understanding and new directions for action. *The meaning is embedded in the context in which it occurs.*

For example, a conglomerate of tableware manufacturers wanted to learn what tableware truly meant to people. They focused not on the artifact (the products) but on the past and present experience of eating at home. An example of this came to be called the "mother role void (MRV)." As the traditional role of mother is changing, how do people learn about things such as how to choose tableware, set the table, display proper table manners, and cook recipes? The team found men and women avidly searching for ideas in a knowledge void filled by Martha Stewart and others. They began to theorize that their industry could fill this need. Their findings revealed a large variety of symbolic connections that would never have surfaced otherwise (Perry, 1998).

- Third, it is *holistic*. In contrast to other approaches, inquiry starts very broadly and then narrows to the specific focus. From the time the site visit begins, analysis and interpretation inform the questions and lead to testable interpretations. When the data are analyzed, themes and relationships emerge and an actionable framework can be constructed. This synthesis, or way of "seeing the old landscape with new eyes," merges the insider's perspective with the researcher's insights.

  While ethnography is a qualitative method, the definition sidebar defines the difference between qualitative and quantitative research.

DEFINITION

Quantitative research: *Answers the questions that begin with "what," "how often," and "how many," and uses tools such as surveys, polls, and statistical analysis. While it can cover many people quickly, the researcher is often left with questions about what it all means.*

Qualitative research: *Answers questions that begin with "why," "what influences," and "how," and uses fewer, open-ended and in-depth interviews to give findings an added dimension to uncover the meaning and motives behind peoples' actions.*

As a qualitative tool, ethnography allows us to understand what an issue is like from an insider's point of view. The ethnographer tries to get the most complete and accurate perspective of the customer's experience and how it affects and is affected by people's beliefs and attitudes. The goal is to better understand how the customer feels, why they do a certain behavior, how the customers make choices, and what do their choices mean in the context of their broader goals and needs. (See Figure 8-1).

As anthropologist Mike Agar said, "formulating the nature of the problem is where ethnography shines. Such creativity comes out of numerous cycles through a little bit of data, massive amounts of thinking, slippery think-like intuition and serendipity. Ethnography emphasizes the interrelated detail in a

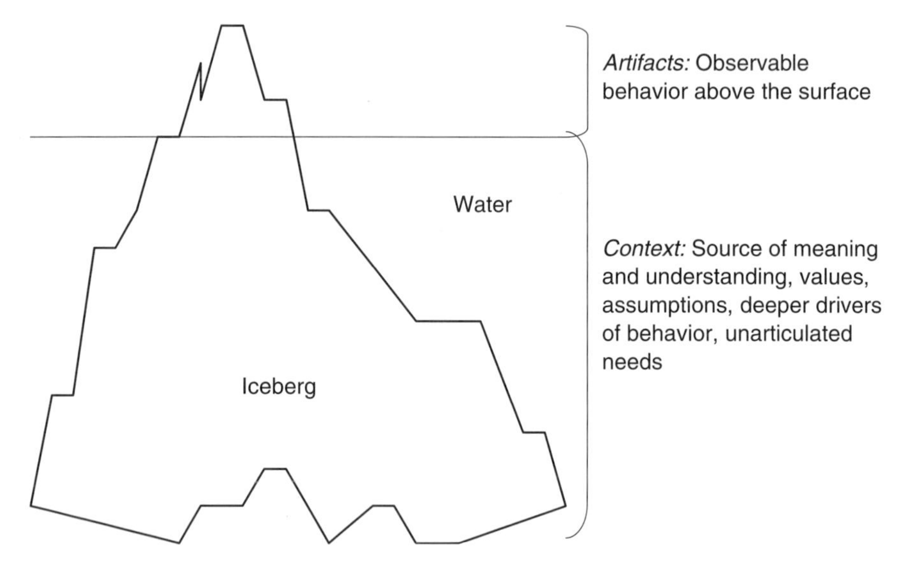

FIGURE 8-1. Behavior is the visible tip: 90 percent of insight is below the surface.

small number of cases, rather than common propositions across a large number." (See Figure 8-2).

APPLICATIONS FROM THE BUSINESS TO PRODUCT DEVELOPMENT ENVIRONMENT

Ethnographic methods are applied proactively to see beyond immediate product development concerns to broader issues. Innovation begets innovation. As each new product is introduced, it raises the bar for all the other products that relate to its category. Team-based ethnography in the product development process, from the fuzzy front end to post-launch, is a proven means to develop insight into the customer to do the following:

- Discover actionable unarticulated needs.
- Understand the emotive function of a product feature.
- Develop new products, brand extensions, and improvements on current products.
- Build hypothetical customer requirements.
- Determine the "real" problems, when the problems are unclear or embedded in a complex or multisystem structure.
- Build a cross-functional alignment around the customer need.
- Spur insight, innovation, creativity, and confidence across the team.

A note of caution: Ethnography does not replace other parts of the new product development process or the research continuum. It is best used to build beginning hypotheses, which are themes that are repeated throughout the

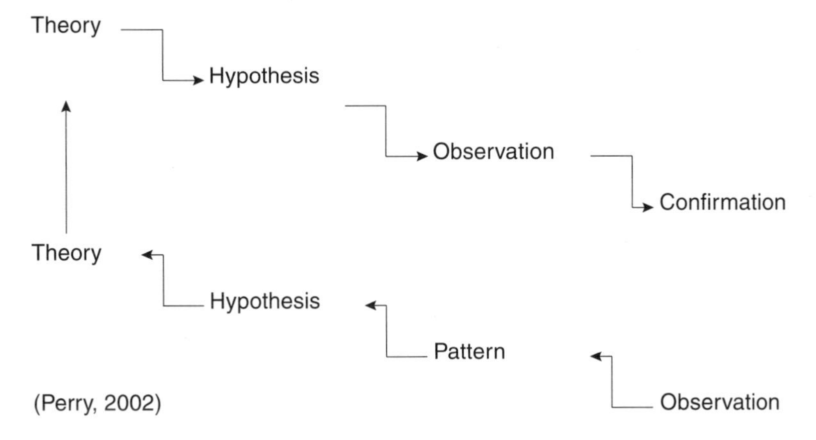

FIGURE 8-2. Deductive vs. inductive market research.

research, about the marketplace. Ethnography is not applicable to the larger population without further quantitative testing (see Figures 8-3 and 8-4).

STEP 1: PLAN THE PROJECT

As with any research project, it is important to set expectations and determine what you hope to gain from the project. There are three parts to plan for any

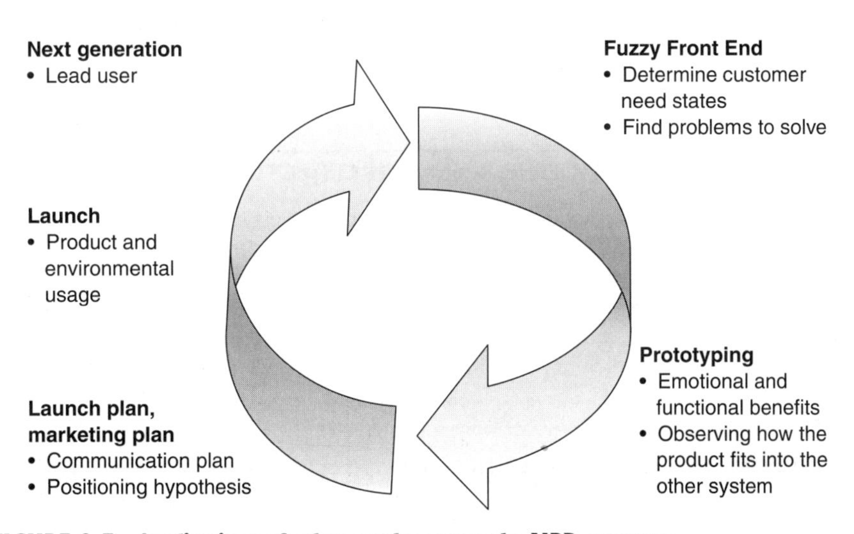

FIGURE 8-3. Applications of ethnography across the NPD spectrum.

Objective	One-on-ones	Focus Groups	Ethnography
Idea generation	Black	White	White
Concept development and screening	Grey	Grey	White
Evaluation of clarity or comprehensibility	White	Black	White
Opinion lead/lead user panels	White	Black	Black
General market education and immersion	Black	Black	Black
Hypothesis generation	Black	Black	Black
Research design	Grey	Grey	White
Questionnaire construction	White	Black	White
Clarification of responses from quantitative tests	Grey	Grey	Grey
Reconciliation of differences in findings between studies	Grey	Grey	Grey

Black=Key strength Grey=Strength White=Secondary

FIGURE 8-4. When to use ethnography.

ethnographic research study, which make up the components of a research charter. They are as outlined in the following sections.

Part 1. Determine the Research Objectives

The objectives need to be *specific enough to give direction but broad enough to enhance the discovery process.* Objectives define the parameters of where a research team will and will not explore and the products' context of use. For example, to study cosmetics, the research team should first understand the meaning of beauty; to understand toys, it's important to first understand the meaning of play; to understand a mortgage, it's important to first understand the meaning of the home. After the research objective is defined, then the research method should be selected.

CASE STUDY: LIPSTICK, THE RESEARCH OBJECTIVES

One lipstick company, who needed to maintain and enhance their lipstick position and product portfolio, wanted to explore the world of lipstick to uncover insights to develop new lipstick products and brand extensions. Ethnography was chosen because it would allow them to better understand the context around the lipstick product and also understand lipstick usage. When determining their research objectives, they decided to focus solely on women in the context of their lipstick usage. Their learning objectives were two-fold:

- To understand how, when, and where women used lipstick (specific lipstick usage)
- To understand the needs, drivers for women who used lipstick (contextual behavior)

These objectives helped shape the research design, types of observations, and questions asked during the site visits (Wellner, 2002).

Part 2. Determine the Research Design

Remember, the goal is to *gain insight to build hypotheses*, not create statistically significant numbers. After the research objectives have been determined, the second most important part of research is to determine whom to interview and observe, how to recruit them, and which research and data collection methods to select.

1. *Determine the type of customer.* In most businesses, 80 percent of sales come from 20 percent of the customers. In ethnographic research, the research objectives are used to determine the type of customer needed for the study. If the research objectives are to better understand the current customer, it is appropriate to pick the target market. However, if the objective is to better understand the needs of emerging or new customers, the type of customer could be considerably different. Since ethnography is to better understand the broader context, it is important to understand all those who impact the use of the product, whether they use it or not. Use the exercise that follows to help determine the type of person or site you would like to visit.

EXERCISE: WHO IS YOUR CUSTOMER?

List all those who touch or interact with your product from its conception to its disposal.

List where you could go to interact with those people in their "real" world.

CASE STUDY: LIPSTICK, THE TYPE OF CUSTOMER

To sustain current share, to defend against new competitive entries, and to further grow and broaden appeal, Procter & Gamble's brand Cover Girl decided to use team-based ethnography. They selected segmented targets of women who were the following:

- *Loyal lipstick users*
- *Occasional lipstick users*
- *Those who rejected the usage of lipstick*
- *Aware of lipstick but did not use lipstick*

2. *Determine the recruiting method:* Expect that up to 20 percent of the site visits will be unsuccessful for reasons such as a misrecruit, inarticulate customer, inappropriate timing, and the inability to find the site visit location. Great care should be taken to determine the appropriate

target to recruit and then to validate that the recruited customer is the appropriate person for the research study. Particular attention should be paid to the recruiting process because of the small sample size.

As with most recruiting methods, a screener is a beginning tool used to determine if the person is a qualified customer for the research study. A screener includes questions to screen the appropriate demographic and psychographic profile of the customer needed for the research. Following are some hints to create a screener:

- Start with broad questions and move to more specific questions.
- Ask nonleading, multiple-choice questions.
- To save recruiting time, ask disqualifying questions early within the screener.
- Screen the person for past participation in market research and competitive employment.
- Unless it is a qualifying question, save personal questions such as income and number of children until the end of the screener (Groves et al., 2001).

Recruiting methods are as follows:

- *Experienced recruiters.* These are recruiters who have had past experience in recruiting for ethnographic research. Recruiters need to be carefully screened and instructed on the extra care to be taken with recruiting customers, administration, and logistics. After the initial recruit is completed, the research team should conduct an open-ended rescreening of the recruited customers.
- *Known customers/friends of friends.* Depending on the nature of the study, this type of person can work well. They have an inherited rapport and trust with the researchers, but thought needs to be given on how to reduce the amount of bias from the customer.
- *Customers from past focus groups.* Interesting and articulate focus group participants may be good candidates for doing an ethnography research project. They have already been screened, are known to be articulate, and are more likely to be responsive participants.
- *Ads in newspapers, newsletters, postings, and online bulletin boards.* This method is one of the most risky; therefore, careful screening of potential customers is essential. Some people look for these types of ads as a moneymaker and may not be the type of person to include in a study.
- *Associations, clubs, and support groups.* Go where the target customer would be. Many organizations will offer to participate in a study as a fundraiser. Again, careful screening of potential customers is important.

3. *Determine the research method.* The basic approach for ethnography is participant observation (see Figure 8-5). This consists of a combination of interviews and observation to understand customers' behavior.

CUSTOMER CONCERN WITH PRIVACY AND SAFETY

When recruited, the customer is not always aware of the organization sponsoring the research. Customers may have concerns about participating in the research for fear of safety and invasion of privacy. Following are ways to alleviate this concern:

- For any data collection method beyond field journal (e.g., audio, video), it is necessary to get permission from the customers who are recruited.
- Reassure customers that the information they provide to the research team will remain confidential and will not be used publicly.
- Tell customers in advance the names of the research team members who will visit them, and have the research team show identification at the door.
- A research team member should call just prior to the meeting to reconfirm, check directions and begin the personal contact.
- Encourage customers to have a friend, spouse, or family member present during the time of the site visit.
- Set expectations with customers. Tell them what will happen during the visit and why the research team is there. Also, formally introduce the research team and describe their roles (e.g., observer, note-taker) and the length of the visit.

Participant observation is the process of immersion into an environment in which the product behavior occurs. To gain understanding, both observation and the experience of the whole environment are needed. Detailed observations guide the research team to new connections, patterns and insights.

Interviewing is the other part of the equation. The research team is able to have real-time exploration and probing of the customers' actions and beliefs. This allows the research team to check their assumptions, which lead to new and robust interpretations. Interviews are generally one-on-one and in-depth, but sometimes the interviews are done in groups, if appropriate; for example, research that is conducted with a teen best friend pair, family unit or an adult group of friends who share a passion or influence each other. The interviews are open-ended, which allow customers to cover all the relevant topics based on their own way of seeing the world. Subsequent interviews are

CASE STUDY: LIPSTICK, THE RECRUITING METHOD

For the lipstick company, who sought to better understand the lipstick usage of women, a multi-tiered recruiting method was used. The steps used were as follows:

Step 1. An experienced recruiter was brought on board to screen potential customers.

Step 2: Friends, family members, associations, clubs, and support groups were enlisted to locate potential customers for site visits.

Step 3. Focus groups were used to screen for those who met the customer specifications.

Step 4. Permission was gained from the customer to use the data collection methods.

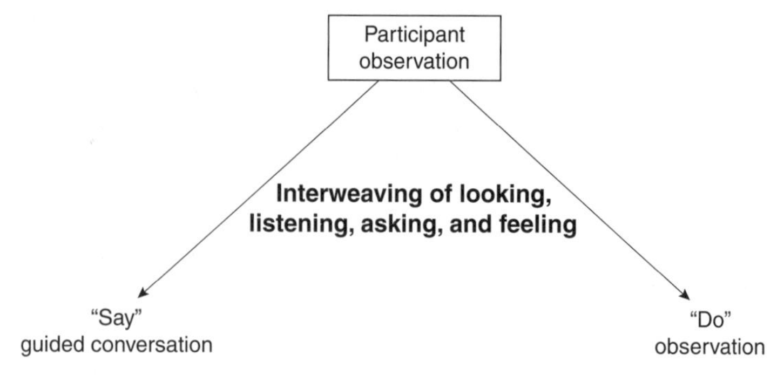

FIGURE 8-5. Anthropology toolkit.

typically more structured, as knowledge and relevant questions are developed that need testing.

The amount of time spent with the customer and the number of visits depend on the research objectives, but some general rules are as follows:

A. *Longitudinal versus short study: A* longitudinal study is defined as a study with more than one cycle of fieldwork, and a short study is one cycle of fieldwork. If a subject topic is intimate or of a private nature, such as feminine hygiene or finances, a longitudinal study is appropriate. In this case, it is important to build rapport with the customer over time before delving into more personal topics; thus multiple visits to a site would be required. A short ethnographic study is appropriate for everyday products that are not private and do not require a lot of cognitive thought.

B. *Number of site visits or interviews.* Ethnographic sampling is very different from sampling methods used in quantitative work. An ethnographic sample is quite small and broadly representative of a constituency. The goal is to understand and gain insight, not the kind of generalization that only comes from more quantitative and larger samples. Like a detective, ethnography looks for patterns and clues that get closer to the reality of others. As more time is spent with the customer, more insight is gained. From experience, it takes

CASE STUDY: LIPSTICK, LONGITUDINAL VS. SHORT STUDY

Lipstick itself is a subject that most women feel can be discussed quite openly, which lends itself to a short ethnographic research study. Although a longitudinal study was not needed, the field team interacted with most respondents in multiple locations, such as their homes, a retail setting, and a social situation. This allowed the team to ask any unanswered questions, observe new behavior, or probe on emerging themes discovered from other site visits.

CASE STUDY: KIMBERLY-CLARK DEPENDS "AND POISE" LONGITUDINAL STUDY

The Kimberly-Clark Adult Care Sector found difficulties in locating and communicating with the customers of the Depends "and Poise" adult incontinence product line. An ethnographic study was conducted to more fully understand their customers, language, product usage, and unmet needs in varied situations.

Adult Care researchers repeatedly entered the home of target customers and conducted loosely guided conversations over an eight-month period. The repeated visits allowed the research teams to build rapport, reduce the embarrassment of talking about private issues such as adult incontinence, follow up and clarify issues not covered in earlier visits, use more obtrusive methods of data collection such as video and photography, and be involved in out-of-the-house activities such as store and doctor visits. The longitudinal study allowed the research teams to gain skills as they went and to change the process as necessary, while also learning a tremendous amount about their customers. The findings had direct applications to product development, testing, packaging, and marketing (Sherrod, 2001).

about 9 to 20 in-depth site visits to gain the most insight from the customer. Variance in the number of site visits depends on the amount of time spent with the customer (Griffin and Hauser, 1992). Frequently, a site visit can involve several parts of the value chain or family unit (e.g., child, brother, and caregiver).

4. *Determine the data collection methods.* There is a balance between the intrusiveness and the creation of an archival record of the visit when the appropriate data collection method is determined. The data collection method chosen should be as unobtrusive as possible but also provide a sense for the customer and site for those not present for the site visit. Some customers will be comfortable with any data collection method, but others will modify their behavior and give an unnatural response. Either way, the research team needs permission from the respondents for each method used.

   The data collection method chosen should be adaptable to many different types of environments that could be encountered during the site visits. For example, does the method require a plug, batteries, or a flat surface to rest upon? A struggle with the equipment can stress the customer and research team and increase the perception of intrusiveness. For this reason, the research team should experiment with the equipment prior to the site visit.

   Do not let the process of documentation inhibit your ability to understand. Anthropology is first and foremost experiential. If you find yourself hiding behind the camera, consider a less obtrusive technique. It is always important to know why a data collection method is chosen, and the value it adds to the project. The five primary data collection methods are as follows:

- Field journal. A written record of the site visit. The research team takes verbatim shorthand notes during the site visits and then revisits those notes to fill in any gaps and highlight the most important points. Notes are critical to the team's ability to debrief after their field experience and create an archival record. This method should be used in most ethnography studies.
- *Audio recording.* An audio recording of the conversation of the site visit. Choose high-quality equipment that is as unobtrusive as possible. Although this method does not allow others to experience the physical environment, it does allow for an in-depth objective review of the content.
- *Photography.* A visual picture of the environment of the site visit. This method combined with other methods provides a visual of the environment and allows others to get a feel for the context of the entire site. Equipment varies for this method, but again, unobtrusiveness is important. One benefit is that this method can be relatively inexpensive, depending on the type of camera, yet still provide a visual record of the site visit. The common use of digital photographs provides an easy way to communicate the research to others.
- *Video.* A visual record of conversation, behavior, and environment of the site visit. Video provides the visual and verbal record of the conversation and gives the most realistic perspective for those not involved in the research. This method can be more obtrusive and expensive and needs to be weighed with the goals of the research and comfort of the consumer. Coding, editing, and transcription of the tapes can be time-consuming, but if done correctly, they can provide a wonderful overview of the research.

CASE STUDY: LIPSTICK, THE DATA COLLECTION METHODS

The application of lipstick can take place in very close quarters such as a bathroom or car, which limits the suitable data collection methods. The lipstick company chose to use homework, field notes, and photography, as their data collection methods for the following reasons:

- *Homework.* A homework packet contained a lipstick log and was sent to the respondents prior to the site visits because the team spent a limited amount of time with the customer and lipstick usage is spread throughout the day. The lipstick log asked for customers to capture what, when, where, and why lipstick was applied.
- *Field notes.* A standard and necessary data collection method for any ethnographic research study.
- *Photography.* The application of lipstick is a very visual topic that needed visual images to show the steps in the process, which helped control the bias in recording the process.

- *Respondent homework prior to the site visit.* This can take the form of a picture diary, journal entry, or collage. Prework often helps prep the customer to start thinking about the topic before the site visit. In some situations, homework is given between two site visits. Biasing the customer should be weighed against the overall research design.

In all cases, it is important to have the research design reviewed by an experienced ethnographer to ensure that methods are legal, ethical, and valid.

Part 3. Determine the Level of Sponsor Involvement

Firsthand experience always makes a stronger case than secondhand information. Sponsors are those who provide the funding for the project, and their involvement early on can be the key to the success of the project. The role of the sponsor is to provide the validity and credibility to link the findings to the larger goals of innovation and the business strategy. Sponsors can also rally the team during the low points of the process. Typical low points are when there is an overwhelming amount of data, a change in the strategy for the product line, or non-project-related activities such as a company reorganization. Sponsors are also the ones who position this work as strategic and approve the time spent. Key sponsors and research teams involved in the project gain not only process ownership but ownership in the outcome. They help to define objectives, spend time in the field with the customer, and interpret and apply the findings. In this way, the customer's voice becomes the sponsor's and team's voice, making the sponsor and team better aligned in their ability to represent the customer.

Forms of Involvement

- *Abstract participation.* A professional conducts the visit and sponsor representatives observe. The sponsor representative's responsibility is to observe the research, collect data, and possibly ask a question or two of the customer.
- *Evolving participation.* Sponsor representatives share equal responsibility with a professional to conduct the site visit. The professional party acts as a coach to work with the team to conduct the research.
- *Full participation.* Sponsor representatives conduct the research. A professional may only be involved to train and coach the team through the process. If this method is used, it is important that teams are adequately prepared to enter into their customer's world and get the most from their experience (Woodland, 2002).

CASE STUDY: LIPSTICK, THE LEVEL OF SPONSOR INVOLVEMENT

To get company sponsored buy-in to the results, a cross-functional team of product development, brand and customer insight professionals, and a cultural anthropologist conducted the lipstick research. This allowed the sponsoring company to be directly involved in the analysis and interpretation of the data after the site visits.

Think carefully about the situation of your team. Who should do the field research?

Situation	Ideal	Second Option
You have the budget but not the time.	Use a mixed team of internal and external experts to do the research and independent professionals to handle the management.	Use an independent professional.
You have the time but not the budget.	Use a professional to coach your team to do the work.	Use a mixed team of internal experts and independent professionals.

COACHING TIPS TO PLAN THE PROJECT

- The research will only be as good as the research objectives.
- Address the privacy and safety issue with customers in a straightforward manner.
- Do not expect customers to fit exactly into segmentation silos. Most customers have traits of multiple market segments.
- Be flexible in scheduling the site visits. Balance convenience with the research objectives of the project.
- Expect that for *every hour* spent in the field, *ten hours* of preparation, administration, and follow-up work will need to be done.

STEP 2: GET STARTED

Part 1. Develop the Field Research Team

The authors' bias is to have professional ethnographers act as research designers, coaches, trainers, and facilitators. Experience tells us that the most productive use of ethnography involves direct participation by those who must act on the result. Within many organizations, the gatekeeper of customer knowledge is the market research and customer insights department. Product developers make a mistake when they allow themselves to be left out of the market research process because then they are left with little input into the analysis and interpretation of the results. The lifeline of innovation is the "dirty" field knowledge. Firsthand field knowledge of the customers' needs and desires brings with it the insight into how to fix the problem.

The knowledge gained through the ethnographic process is exceptionally rich, and only a fraction of it can be communicated to people who have not been in the field. When possible, the actual product development team should perform the field research and participate in the analysis. The process helps a team to break away from incremental learnings and obtain new thinking by refocusing the team on the problem as experienced by the customer.

When a research team is put together, it should be a cross-functional team who can commit to work together throughout the project. When the team is created, it is important to balance demographics, job function, and experience in ethnography. After the research team has been selected, it should be broken down into smaller field teams consisting of no more than three people to conduct the individual site visits. Teams of more than three people tend to overwhelm customers and make them uncomfortable. The rules to select field teams should follow the same rules as the selection of those for the research team. For example, in a project on skin care for teenage girls, each field team had no more than three people per team with at least one female member and

HOW TO INVOLVE THE SALES TEAM IN TEAM-BASED ETHNOGRAPHY

The question is often posed, "We have salespeople in that part of the world. Can they do the research?" The irony is that frequently the locations where the sales team conducted the research were later determined to have been the most important. In many cases, local representatives are well connected to the local market with customer contacts, navigation and interpretation skills, and they may see voice of the customer as their job. For example, if an hour is spent with Henry Ford, what color car would you want to buy? After an hour with him, it would be clear that "black" is not only socially responsible but also a nice color. These extraordinary salespeople may find value in the ethnography process but are challenged to be unbiased. Bias is a preconceived notion or belief about a situation that could influence or limit a person's ability to perceive what is really happening. Rules to involve the existing field sales and service team are as follows:

1. Treat the field team as an important part of the value chain. Observe and interview them as a unique type of customer.
2. Orient them to the research goals, objectives, and process before participation.
3. Give them an assignment; a digital camera and photo log are good choices.

Properly directed, the field specialists have the ability to become the voice of today's customer, an insightful researcher, and a powerful company asset because they become the customer representative in the organization.

a mix of those with backgrounds in customer insight and product development.

Ethnography can take as much or as little time as the team is willing to devote to it, but it is imperative that each field team member experiences at least two or three site visits. Participation in only one site visit does not provide enough information to compare to other team members' experiences. However, multiple visits allow a field team member to see emerging patterns and themes.

Part 2. Train and Coach the Field Research Team

For those who participate in ethnographic research, it is imperative that they are coached and instructed by a trained ethnographer on what to expect in the field. An experiential learning format is the best way to train a team to conduct ethnographic research (Rosenau, 2002). Key skills include the following:

- Basic theories and techniques in ethnographic research methods
- How to observe and listen
- How to develop open-ended, story-laden, and non-leading questions and probes
- How to take notes, use other methods of data collection, build a field record, and debrief
- How to sustain disciplined subjectivity to understand and manage personal and corporate biases

CASE STUDY: RUBBERMAID OFFICE PRODUCTS

Rubbermaid manufactures office products such as pencil holders, stackable letter trays, and file folder sorters found on many people's desks. The product development team wanted to explore the use and expansion of their products through ethnographic research.

The team decided it needed to do the work itself to gain the most value. An ethnographer spent extensive time to coach and train the team because ethnography was a new research method to them. All team members participated in practice site visits with their consumer target. The visit was used to practice their interviewing, observation, and note-taking skills prior to the research. The practice allowed the team to be more comfortable with the site visit process and work out any challenges within the methods and tools chosen. It also allowed the team to better understand areas of weakness and identify any biases not previously acknowledged, such as expectations of how customers organized their office, managed work at home, and designated personal space. Outside of the practice site visit, other instruction included ethnography basics, observation, interviewing, data collection, and debriefing methods.

CASE STUDY: BUSINESS-TO-BUSINESS DURABLE GOODS

A research team conducted research on farm equipment in Germany. The team had a three-part process: (1) meet with dealership sales representatives, managers, and service personnel; (2) facilitate an in-dealership focus group with invited farmers; and (3) conduct site visits at the farms, which allowed the team to get insight into value chain requirements.

During a focus group, one farmer said, "[The tractor] is perfect. Don't change a thing on the next model." In his kitchen, with his wife present, the story was much the same. "The best one I have ever had. Please don't change a thing." As they moved into the shed where the equipment was stored, his words were consistent as he showed the design team over 20 modifications he had made to the product to allow the equipment to work more easily. Even as he described the additional modifications, he continued to praise the existing product. Which of the following conclusions would your team draw?

1. We have found a "lead user." Here are 20 possible design changes for the next model.
2. The ability to customize is an important product requirement for the end user.
3. "Boy, he sure invalidated his warranty didn't he?"

The right answer is 1 or 2, because in response 1, there are ideas for design changes and in response 2, the underlying need may be to give the customer the ability to customize their equipment, rather than to predetermine the design features on the product.

COACHING TIPS TO GET STARTED

- Fewer people on the research team are better than more. It is important to have many eyes to look at research, but those eyes need to participate in multiple site visits with the customer in order to compare and contrast findings.
- When a new research team is coached, use experimental learning. Let the team experiment with those things they will experience in the field.

Part 3. Create the Observation and Discussion Guide

The key word here is *guide*, not a script of an interview or conversation, but rather a point of reference to fall back on if the team gets stuck. Think of it as a roadmap of a "guided conversation." Throughout the research, this guide will be shaped and changed as discoveries are made and hypotheses emerge.

Most guides use a semistructured format, which uses a variety of potential probes but is not strictly followed. Jumping around the guide or going off topic is standard. In the creation of the observation and discussion guide, keep in mind the type of information to be gathered. For new product development, the ethnographer is seeking insight in four categories:

- *Discontinuities in system patterns.* Something that breaks up a traditional pattern; for example, a child who lives with grandparents and has visitation with both natural parents.
- *Disequalibrium or lack of balance in a system.* When one partner appears to gain more than other members of the value chain; for

example, a customer who is angry with credit card companies who appear to overcharge for their services.

- *Disintermediation opportunities.* The ability to bring a step of the process in-house; for example, customers who do substantial research online without approaching the dealer or retailer.
- *Compensatory behavior.* A signal that a system member has experimented with a product while waiting for a market innovation; for example, a fan duct-taped to an operator's station (Miller, 2002).

Creation of the guide also ensures the area of inquiry is broadly and holistically defined. The rule of thumb is to go up to the most general and abstract level of understanding for the topic. For example, to learn about cosmetics, the meaning of beauty must be understood; to understand how people view the mortgage process, you want to know the meaning of home. Once the context is understood, then it is appropriate to drill down. In addition to the discussion guide, create an observation guide that focuses on what the team looks for and what the team looks at. To do this, it is helpful to remember there are many possible elements of observation (see Figure 8-6).

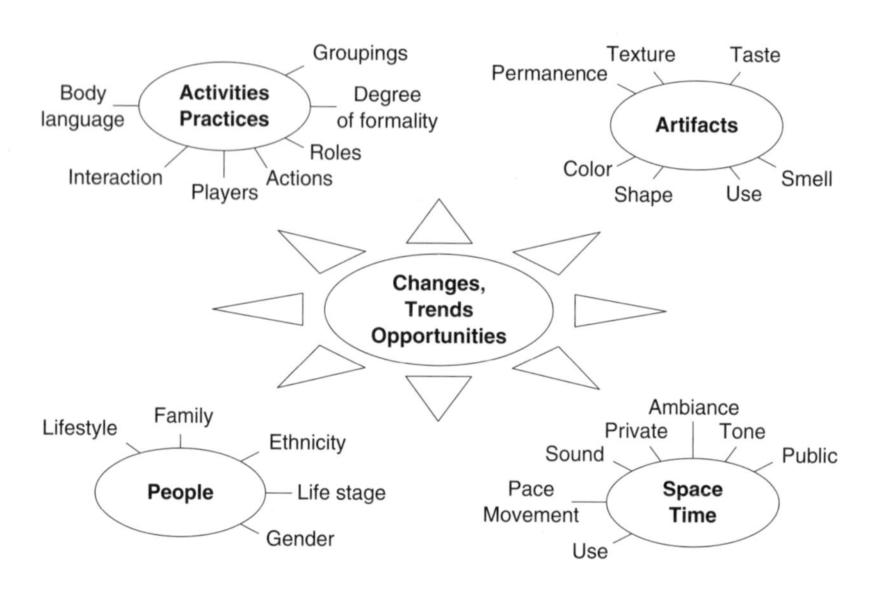

Elements of Observation

Because:	Observation filtered through lens
We must:	Separate actual observations from meanings we attribute
We do this by:	Scrupulous attention to detail Deferring personal judgment Knowing our lens

FIGURE 8-6. Elements of observation.

Following are some rules of thumb when putting together an observation and discussion guide:

- Allow time at the beginning of the conversation to set expectations for the site visit. Introduce the field team members and their roles; review the purpose of the research, how long the site visit will last, and what the customer will be asked to do; and begin to build rapport with the customer.
- Start with a tour of the entire site, home, office, or factory.
- Start the visit with broad topics and then move to more specific topics. Save the sensitive and personal topics until closer to the end of the conversation.
- Ask open-ended questions that allow customers to give as much detail as they want and the researcher the opportunity to ask more specific or probing follow-up questions.
- Do not ask questions that lead or direct the respondent toward a particular answer. Leading questions tend to start with words such as *do, are, can, could,* and *would.*
- Allow time at the end of the site visit for the respondent to ask questions of the field research team. This is a time when the research team can physically stop taking notes but mentally needs to be extremely aware. This is a cue to the respondent that the research is over, and many times the customer will open up.
- At the close of the time together, thank the customers, tell them what is planned for the information collected, and compensate them for their time.
- Have sample and photo notes in your guide, so they are not forgotten.
- After the guide is created, review it for any biases. Ask yourself what hypotheses and filters have already been created in your mind about the research outcomes. Then rework the guide to be more objective.

STEP 3: CONDUCT THE FIELDWORK

Part 1. Prepare for the Field

The research team's expectations should be prepared before a site visit is conducted. Rules are as follows:

- Plan for at least two or three hours per visit.
- If a business-to-business site visit is conducted, get permission from the appropriate on-site authorities before the visit.
- Determine the roles for the team members before the visit. Typical roles are primary moderator, observer/note-taker, and equipment handler. Everyone on the research team should try out all roles.

CASE STUDY: LIPSTICK, DEVELOPING THE DISCUSSION AND OBSERVATION GUIDE

Introduction

Introduce members of field team and thank respondent; state why you are there, define the roles of each person on field team, specify how long you'll be there, ensure confidentiality.

Questions	Observations
Rapport Building	
◆ Go over lipstick log	How often is lipstick used?
◆ What was your first lipstick experience	Note emotions when explaining stories.
◆ What was your best/worst lipstick experience	
◆ Define lip care, lipstick	
◆ Why do you wear lipstick	
Occasions	
◆ What occasions is lipstick worn? Not worn? Why?	
◆ What influences decision?	
◆ How do you decide what shade?	
◆ How do you know how it looks?	
◆ Significance of certain shades?	
Application	
◆ How did you learn to apply?	Where are all the places the process is done?
◆ Describe lipstick application.	Describe the environment. Are there any
◆ When is it applied?	inhibitors (little space, lack of mirrors,
◆ Where is it applied?	multiple people around)?
◆ How much is enough?	What tools are used?
◆ Describe any tricks or rituals in application.	What parts of the body are used?
◆ How do you take it off?	What preparations do you make?
	Where is the process started?
	Kind of packaging?
	What is the shape of the lipstick?
Storage	
◆ Where is it stored?	Note storage locations, temperature, how
◆ How many kinds of lipsticks?	many tubes, what else is stored there.
General	
◆ Describe your ideal lipstick.	
◆ What do you like/hate about lipstick?	
◆ What else do you use lipstick for?	
Closing	
◆ Ask if the field team has any other questions	
Questions for field team	
◆ Thank them, pay them, tell them next steps with the research findings.	
◆ Get homework, nondisclosure agreement, pictures.	

- Make the customer as comfortable as possible.
  - Use words appropriate to the customers rather than company acronyms.
  - Dress in an appropriate manner to the situation. For example, a research team who conducts research with floor installers would wear jeans, boots, and t-shirts.
- The authors prefer full disclosure of the company and its research purposes. In certain circumstances, the research activity can cause concern for an organization. In this case, the customer could sign a nondisclosure agreement or the research team could be more general about its purposes for the research.
- Have unconditional positive regard for the individual and community that is studied. Research teams, who see participation and relationships as a path to understanding, must practice nonjudgmental observation of customers and participation in the field.
- Practice unobtrusive insertion, measurement, and exit from the site and community. The naturalists' motto of "take nothing but pictures and leave nothing but footprints" is a good guide.

Part 2. How to Debrief

A person retains up to 99 percent of unrelated information two hours following an exercise, up to 71 percent for unrelated material two days following an exercise, and less than 14 percent for periods longer than two days (Goldstein, Chance, 1971). Immediately after the site visit, the research team should start to process their learnings, form beginning hypotheses, and take a first pass to analyze the information gathered during the site visit. This debrief period should be facilitated by someone who can capture what is discussed so there is a permanent record of the conversation. The ability to create an archive or permanent record to store the information is crucial to the quality and integrity of the research. As in any research method, an archive provides a long-term reference source and the ability to verify and validate the research. *As an estimate for every hour spent in the field, at least half that amount of*

8:00 a.m.	Kick-off meeting
9:00 a.m.	Travel to site
10:00 a.m.	Conduct site visit, in-home and retail experience
12:30 p.m.	Debrief, lunch
3:00 p.m.	Travel to site
3:30 p.m.	Conduct site visit, in-home, retail experience
6:00 p.m.	Debrief, dinner

FIGURE 8-7. Sample schedule for ethnographic research.

time should be spent debriefing individually, as a team or in a larger group. A three-prong approach to debriefing works as a way to distill the information gathered:

1. *Work alone immediately following each encounter.* This allows individuals time to fill in gaps in their notes while their memory is most clear and to access information in short-term memory before it is discarded. It also allows each person on the research team to sort through, highlight, and begin to process the most important points from the research without the observations of others to sway them.
2. *Work as a research team.* This allows each individual to verbalize and build off each teammate's perspective, knowledge, and experience. An archival record should be created of the team's top findings, issues, insights, and connections from the research (see Figure 8-8). A significant amount of time of the debrief process should be spent in this phase to examine and determine the significance of the data collected.
3. If there are multiple research teams in the field, *a larger group debrief* should be held. This debrief provides the opportunity for all research members to debrief each other on the top issues and insights collected before the next round of site visits. This is a great time to make ethnography happen through the exchange of knowledge and insight gained from others who may have had different or similar experiences.

There are many methods of debriefing as a group. The choice of what technique is most appropriate depends on the audience, size of the group, and the purpose of the debrief, such as whether to inform, explore, analyze, eval-

Team: ______________________________

Date: __________ Place: ______________________ Time: ________

Best quotes, stories, observations.

What has been learned: themes, surprises, and omissions.

So what? Implications and insights.

New questions?

FIGURE 8-8. Rolling debrief insight sheet.

Interview with Amy Brown
Friday, June 14, 2002
6:45 a.m.

Segment Descriptor:

- Married, middle-class, professional
- Full-time job, commuter
- Generation X age bracket
- Laid-back, but busy person with goals in life of house, land, and kids

Brief personal description: Amy lives in downtown Lancaster in a row house. She's married and works full-time. She's cost-conscious about what she buys, including lipstick. She wears makeup every day and feels like she always has to have something on her lips. During the week, she applies her makeup and lipstick in the car on the way to work (at stoplights and when a lot of cars are not around).

Key Quote: "I am rarely seen without lipstick. I think it gives my face color, it makes me feel put together. Lipstick is the core. I use lipstick to pick me up, and it makes me feel polished and put-together."

Other Observations/Quotes: "I don't know if lipstick has a shelf life." "Lipstick is kind of like an instaboost." "I hate when I don't have anything on my lips." "I don't buy a lot of new colors. It costs too much to try them."

Compelling Needs: Cost- and time-conscious. Color-conscious. Portability. Lip products that stay on and keep her lips moist.

FIGURE 8-9. Case study: Lipstick, sample profile sheet.

uate, or present findings. The most common methods to debrief are as follows:

- *Site visit profile sheets.* Each field team fills out a one-page profile sheet of the site visit to give an overview of the site visit and its key findings to those who have not participated in the research (see Figure 8-9).

CASE STUDY: SAMPLE STORYTELLING—THE POPCORN STORY

While an ethnographic research team from General Mills conducted research about snacking behaviors, they realized that popcorn, more than other snacks, carried with it meaning, memories, and ritual. After deeper exploration, the team came back with the following story:

While they sat around a kitchen table and spoke with a mother and daughter about popcorn, it was hard to ignore the father, who was in and out of the room, agitated or perhaps angry. Finally, unable to restrain himself any longer, he went to a far corner of the kitchen, rattled around at the back of the cabinet and emerged, scowling, with a beaten up old aluminum pan. He had come into the marriage with this special pan—the pan in which *his* father had made popcorn. He had seen it as part of his role as father to be the popcorn provider. Now, thanks to General Mills and others, the microwave product had made that role and his pan irrelevant.

The team chewed over the story and came to realize that while they had focused on the functional attributes of popcorn, they had neglected the emotional benefits. Eventually, out of this "kernel" of insight, the phenomenally successful product of Homestyle Popcorn was born.

This moment could not have happened without the team in the kitchen. This story also illustrates the power of a sample of one, if that one gives new ways to see an opportunity.

- *Storytelling.* Each field team chooses a story or series of stories to communicate the key insights from each site visit. The story should be a vibrant reflection of the customer in order to impart a feel for the customer's environment, personality, and needs. This can be an effective method to inform others of the broader context and instill the voice of the customer.
- *Observation, insight, and connection.* Each field team reports back the highlights of what they observed and heard (observation), their interpretation of the observation (insight), and the key needs observed (connection) for each site visit. This provides the rest of the group the opportunity to give their perspective and add other connections and builds to the research team's thoughts. This method is best used as a beginning analysis tool and can be used to build a database of customer insights (see Figure 8-10).
- *Transcription.* Each field team transcribes the site visit with as much detail and context as possible. If video was used, the environmental context is captured in written form. Team members highlight key quotes or environmental cues from the transcript that display vivid visual images and the voice of the customer. Each key quote is discussed, cleansed to remove partiality in the wording, and then reworded to capture the significance. The research team should be the one to use this method because they understand the context of the situation. This method can be used as a beginning analysis tool to highlight key points. (Burchill and Brodie, 1997)
- *Formal presentation.* A presentation of top insights and beginning themes with supporting stories, quotes, and video clips can be made to a broader

	Observation (Raw facts or quotes)	Insight (Your interpretation and twist on the observation)	Connection (Customer need or problem to be solved)
1.	"Ideal lipstick is long-lasting, high on moisture, and will not kiss off."	1. Long-lasting lipstick tends to be cakey and taste bad. This takes away the sexiness of wearing lipstick.	
2.	She applies lip gloss with her finger and said that cleaning it off afterward was a pain.	2. Putting the finger to the mouth when applying lip gloss makes the lip gloss feel as if it is more customized.	2. What if the lip gloss on the finger could be used for another purpose in the makeup process?
		2. Putting the finger to the mouth when applying lip gloss makes the person feel like the product doesn't quite meet their needs.	

FIGURE 8-10. Case study: Lipstick, observation, insight, and connection debriefing form (Woodland, 2002).

audience. This presentation does not include any analysis or meaning of the findings but rather is a presentation of the research process.

STEP 4: ANALYZE THE DATA

The ethnographic research has been conducted; the data has been coded and sorted. Now a deep analysis of the data must occur. Without the analysis, the time and effort in the field will have limited value.

COACHING TIPS TO CONDUCT THE FIELDWORK

- Set expectations with the research team before the field is entered. A research kickoff meeting will help the team to grasp the context of the research in the overall process. Review the profile of the site visits, research schedule, data collection methods, debriefing methods, observation and interviewing coaching tips, and any miscellaneous logistics.
- Stay flexible while the research is conducted. Team members will need to adapt to the customer's schedule and understand that all research cannot take place during normal working hours. There is a balance between the integrity of the research and convenience for the team. The time of the research team is valuable, so if a respondent is not available, have a backup plan for how to effectively use the time to gain valuable customer information. For example, this could include informal visits to a retail environment.
- Do not jump to conclusions during the fieldwork. Beginning hypotheses or reoccurring themes can be noted, but it is important to remain open-minded during the fieldwork.
- It is normal to feel overwhelmed by the amount of data. Trust your research teammates, the process of debriefing, and the next step of analysis.

The field research team must participate in the analysis, but this can also be an opportunity to include key stakeholders, those whose input and blessing on the project can determine its success. Key stakeholders may be the financial sponsors of the project, or the sponsors of the project may be another entity. They provide fresh ears and eyes to ask insightful questions, focus the analysis, and provide the much-needed buy-in to move the research toward implementation. The three parts of ethnographic data transformation are as follows.

Part 1. Description

Great ethnography rests on the quality of the description. Stories and concrete specifics are the heart of this type of analysis. The stories chosen should be vivid depictions of the customer's world and encompass the entire context of the situation. The questions to be answered in this part of the analysis are *what has happened?,* and then, *what has really happened here*? Raw facts, observations, and direct quotes are important, but it is most important to get beyond the surface level.

Part 2. Analysis

Thematic analysis sorts the data into buckets of meaning and answers the question of "How do things work together?" The themes created in the analytic stage are woven together into an actionable framework to show patterns and relationships. The three forms of analysis are as follows:

- *Qualitative cluster analysis.* Participants take the data gathered and pull particularly sticky pieces of information from the mound. Sticky pieces of information are those unavoidable buckets that emerge from the data during the dialogue between informed team members. As these core buckets emerge, insights are linked and categorized together to create qualitative clusters. The key pieces of information are identified and data points continue to be added to grow the current clusters and form new ones until all the data are considered. Clusters are given titles, named for the key element that ties the cluster together. These titles typically become the key theme areas. Clusters are linked and strengthened with stories and illustrations (Miller, 2002).
- *Language processing.* After transcription of the site visits, customer quotes are selected, scrubbed for bias or for information that is not based on fact, and then grouped based on a common theme or thread. Quotes continue to be grouped until each grouping feels like a discrete and valuable insight. Each grouping is then given a title that describes the thread that binds the group together. If group titles have relation-

CASE STUDY: LIPSTICK, QUALITATIVE CLUSTER ANALYSIS

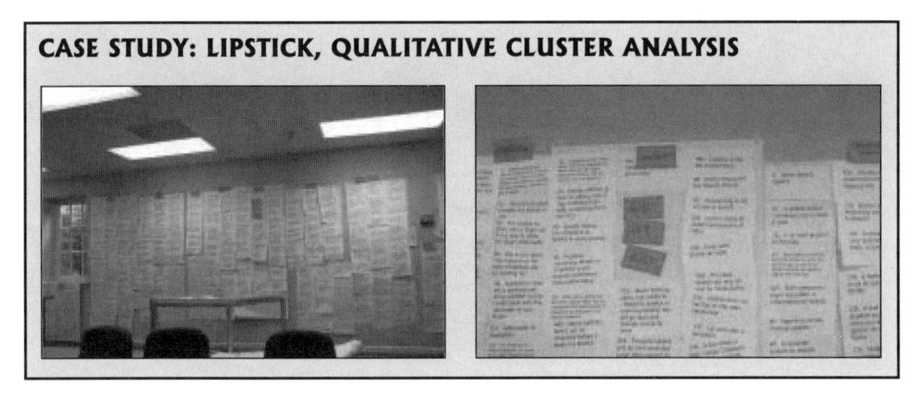

ships between them, it is given a more abstract title to describe their relationship. Connections between titles are mapped to show their relationship and interconnectedness. From the groupings and relationships between them, themes are developed (Burchill and Brodie, 1997).

- *Value pyramid.* This method assumes there is a higher order of benefit to the information gathered. Information from the site visits is mapped onto a hierarchy of customer benefits that range from the most basic to emotional to higher order needs. Pyramid levels are as follows:
  - *Fundamental product attributes.* The physical characteristics and features of a product.
  - *Functional benefits.* These flow directly out of the product's attributes.
  - *Emotional benefits.* How the use of product or service makes the customer feel.
  - *Higher-order benefits.* This is the "something bigger." Although difficult for customers to articulate, they are very real and powerful for many products and tend to address fundamental human values.

  As with any hierarchical structure, conditions must be satisfied at each level before the benefits at the next higher level can be obtained. The top two levels of the pyramids are the battleground of brand distinctiveness. As categories mature and marketing sophistication increases, the ability to establish and communicate higher-order benefits becomes critical (Ward, 2003). Value pyramids can be used to better understand the customer, product, technology, process, and global platforms, but it is built on the objectives, goals, and outcomes of the research (Sawhney, 1998). See Figure 8-11.

Part 3. Interpretation

The interpretative part of the process requires critical teamwork because, according to John Seely Brown, director of Xerox at the Palo Alto Research

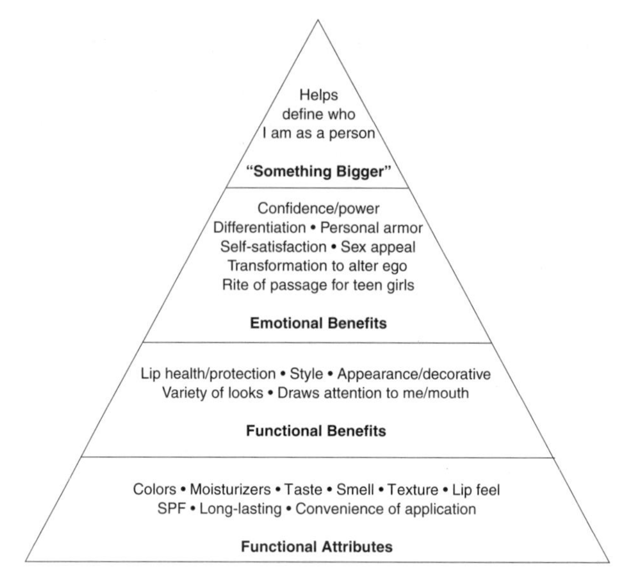

FIGURE 8-11. Case study: Lipstick, value-pyramid example (Ward, 2002).

Center, "it is not shared stories or shared information so much as the shared interpretation that brings people together." Interpretative analysis is a synthesis process that answers the question "so what?" The research team transcends the actual data and makes creative leaps in relevant directions. Each person's perspective and experience with the information is taken into consideration.

Synthesis of diverse perspectives is a learned skill involving dialogue, the ability to listen, and grounding in the tangible data from the research. Robust themes are developed to determine customer insights, needs, and problems worth solving. A theme could be outlined as follows:

Theme Title (title that depicts an overview of the theme)

- *Problem worth solving.* Description of key customer insight and/or need seen in the research; an explanation of what the problem is and why it matters to the customer.
- *Supporting observations, quotes, and facts from the site visits.* From all the data collected and the buckets of themes developed, pick those data points that best represent this theme. These could be quotes, observations, or personal insights.
- *Supporting customer story.* The story could be a real story heard from a site visit or a synthesis of elements of stories from site visits that portray the theme need. This helps give the theme context.
- *Diagram of the theme.* A picture or diagram that visually describes the theme.

- *Pluses and concerns.* What are the benefits and concerns for moving this theme forward as it directly relates to the sponsoring organization?

See Figure 8-12

STEP 5: USE THE OUTCOMES

The research is done; the results have been determined. Now what do you do with it and how is it communicated to the rest of the organization? Ethnography used in the fuzzy front end of product development provides individual, team, and organizational outcomes. It can have a deliberate impact on company strategy and culture: It stimulates team unity, and it helps to mold professional and personal aspirations.

Individual Outcomes

To make a difference, individuals must feel empowered. Ethnography empowers team members because they have a better sense of the customer's needs and the context of use, which makes their intuition more finely tuned and enables more accurate interpretations. There is also an opportunity for personal growth, a chance to see and think about the big picture. As one product developer put it, "As important as the information was, the people for whom we were designing had suddenly become real to us. We continued to think about and refer to them by name as our products took shape."

Team Outcomes

The team that conducted and analyzed the research should feel energized both by the process and its alignment on discoveries and insights. The shared team experience of ethnography will build the team. It will provide the glue to help them find common ground among their diverse backgrounds and experiences. It is also an effective way to create team energy and passion. The customer is now at the center of the innovation process rather than a part of the team's functional agenda. The team will feel ownership of the outcomes, which will pay off in execution.

Organizational Outcomes

Through ethnography, the organization will have learned a new way to listen to its customers. The real gift of ethnography is the ability to get two interconnected kinds of learning: data about "out there," or the customer's world,

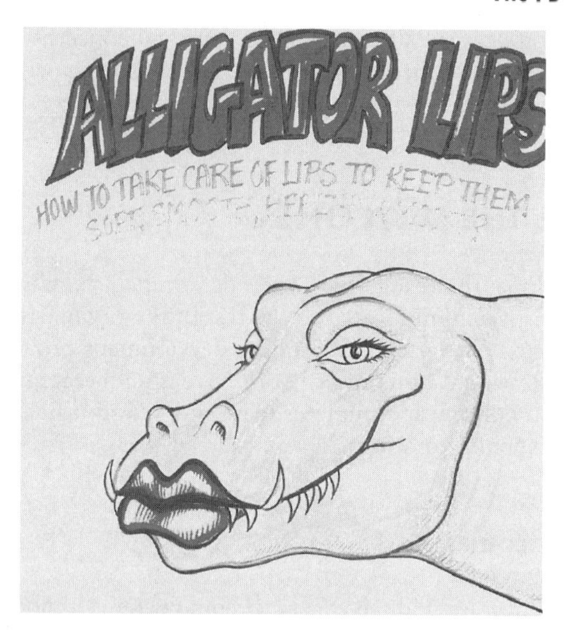

Theme Description Sheet

Theme title: Alligator Lips

Problem worth solving: How to take care of lips to keep them soft, smooth, healthy and moist

Supporting quotes, observations, insights, connections and applications:

- Customers are trying to keep their lips healthy by using lip balm, Chapstick and Blistex because lipstick dries their lips out.
- Lip care is taking care of the lips, keeping them soft, and protecting them from the sun.
- She brushes her lips with her toothbrush to get off the dead skin.
- Chapstick or Blistex in every pocket.
- "I use bag balm on my lips in the winter for heavy-duty chapping."
- Lip maintenance and repairs are done throughout the day and at night.

Ideal customer market and supporting story as an example of the theme:
Rebecca has just awakened and has extremely dry lips from skiing a week before. The lipstick she uses seems to dry out her lips and just accentuates the problem. Lately she's been carrying lip balm with her constantly so she can put it on before reapplying her lipstick. She really wishes there was a product that did both and didn't cake up on her lips later.

Pluses:
+ Meets a customer need
+ Crosses boundaries for reluctant lipstick users

Concerns:
How to make moisturizing new and believable

FIGURE 8-12. Theme description example.

but also about "in here," or the organization's view of the world and which biases may have limited or tainted its view. Many organizations operate business divisions that have been created based on convenience and history to the organization rather than on customer need. These artificial boundaries can make it very difficult to see opportunities in the white space between divisions. For example, customers do not think of orange juice as frozen or concentrate;

COACHING TIPS TO ANALYZE THE DATA

- Many organizations have a quantitative mind-set, and ethnography stretches beyond their comfort zone in terms of what constitutes validity. Be sure to contextualize this research as one piece of a broader learning process.
- Involve key stakeholders early. Provide them with direct primary data such as audio, photography, and video, and then involve them in the interpretation.
- Beware of the "commitment gap." Sometimes people who participate in the fieldwork are then too busy to devote the necessary quality time to analyzing the data. Their input is needed, so make sure they are committed for the long haul.
- Allow time for synthesis and incubation to occur and for stories to emerge.

they just know they want it and buy it. Ethnography provides a holistic view that allows an organization to learn and listen through the customer's eyes, and then they can create compelling value for the customer because of their knowledge. They can ask better questions and create more focused and useful theories.

How to Reframe and Communicate the Outcomes to the Larger Organization

The manner in which the outcomes of the ethnography are presented to the larger organization will frame the effectiveness, acceptance, and implementation of ethnography as a research tool. As an organization, evaluate the type of impact you would like ethnography to have, and then choose the method of communication carefully. Following are some common tools and techniques to communicate the outcomes to the larger organization:

- *The springboard story.* Telling stories that catalyze the organization into action. Noel Tichy, a noted professor of management at the University of Michigan, said, "Leadership is not about change. It's about taking people from where they are now to where they need to be. The best way to get people to venture into unknown terrain is to make it desirable by taking them there in their imagination . . . telling stories." Great stories are sources of energy because they generate momentum, catalyze understanding, are more believable and persuasive than statistics, are memorable, and can easily ripple through the organization (Denning, 2000).
- *The innovation cycle.* Take the themes developed and make them the focus of the next new product development brainstorming and concept development session. Then a platform can be developed around a theme (Miller, 1997).
- *Highlight video.* Create a highlight video of interviews put to music and play it in a well-traveled location.

CASE STUDY: WESTERN UNION—THE PATH TO CORPORATE TRANSFORMATION

Western Union, a well-known telegram and money transfer business, chose ethnography as a way to directly interact with their end-users. They wanted to get an in-depth understanding of their customer, which would help lead them to generate breakthrough new product concepts, develop marketing strategies, and align Western Union with their customers.

After in-field research, the Western Union new products team felt changed by the customer immersion experience and wanted to give a sample of this experience to the larger organization. To do this, customers who participated in the research were invited to present the portfolio of products for implementation to the executive committee. Each end user not only effectively discussed his or her story, needs, and interests with the committee but also wrote a letter addressed to the company. As one customer said, ". . . remember to keep plain people like us and our needs in mind because other companies don't!"

As a result of the research and the presentation to the executive committee, the story of the visiting customers created a buzz within the organization. The change in the research team's mind-set and perspective of its customer had a rippling effect for the company. One research team member described the customer visit as "the best [two-hour] investment our company has ever made." As the field research team spread throughout the organization, the vividness of their customer interactions continued to be at the forefront of their work. As a result, the team was given overwhelming support for the implementation of the research results, and the Western Union team was given permission to focus on a new product development path.

- *Quote and collage board.* Create a theme area within your workspace. Post quotes from the customer, pictures, any homework, and anything else collected from the site visit during the research. Then leave space for those who pass by to record potential applications for the data and themes.

CONCLUSION

"We become slaves to demographics, to market research, to focus groups. We produce what the numbers tell us to produce, and gradually, in this dizzying chase, our senses lose feeling and our instincts dim, corroded with safe action," said Barry Diller, the founder of Fox Broadcasting and a media mogul. Ethnog-

CASE STUDY: LIPSTICK, INNOVATION CYCLE

Following the analysis of the lipstick ethnography, the team conducted ideation sessions to develop potential new products. The outcome of these sessions ultimately led to the launch of single topcoats, fashion topcoats, new shimmery shades, and new wedding copy for Outlast, Procter & Gamble's Cover Girl brand.

raphy is about how to gather and interpret the many signals customers give us from their perspective. Quantitative and evaluative data, though valuable, does not give organizations insight into where the customer will be and why. For any company to create a successful new product, the most important prerequisite is to understand the customers and their world. The objective is to provide meaningful innovation, and this can only be derived from one source, the customer.

Wal-Mart sets the example through their devotion to the customer. *Fortune* writer Jerry Useem explains it best when he says, ". . . and so, you see, there are two types of executives these days: those who have learned to play by Wal-Mart's rules, and those who still haven't learned the right answer to the [company] cheer's closing question: 'Who's number one?' 'The customer! Always! Whoomp!!!' "

BIBLIOGRAPHY

Abrams, Bill. 2000. *Observational Research Handbook: Understanding How Consumers Live with Your Product*. Chicago: NTC Business Books, 2000.

Agar, Michael. 1999–2003.*The Professional Stranger: An Informal Introduction to Ethnography*, 2nd ed. San Diego: Academic Press.

Bernard, Russell H. 1988. *Research Methods in Cultural Anthropology*. Thousand Oaks, CA: Sage Publications.

Bernard, Russell H. 1998. *Handbook of Research Methods in Cultural Anthropology*. Lanham, MD: AltaMira Press.

Boyle, David. 2001. *The Sum of Our Discontent: Why Numbers Make Us Irrational*. New York: Texere.

Burchill, Gary, and Christina Hepner Brodie. 1997. *Voices into Choices: Acting on the Voice of the Customer*. Madison, WI: Joiner Publications (Oriel Inc.).

Denning, Stephen. 2001. *The Springboard: How Storytelling Ignites Action in Knowledge-Era Organizations*. Boston: Butterworth-Heinemann.

Duguid, Paul, and John Seely Brown. *The Social Life of Information*. 2000. Boston: Harvard Business School Press.

Geertz, Clifford. 2000. *Local Knowledge: Further Essays in Interpretative Anthropology*. New York: Basic Books.

Goldstein, A.G., and J.E. Chance. 1971. "Recognition of Complex Visual Stimuli." *Perception and Psychophysics*, 9: 237–241.

Griffin, Abbie, and John R. Hauser. 1992. "The Voice of the Customer." Report No. 92–106. Cambridge, MA: Marketing Science Institute, 1992.

Groves, Robert M., et al. 2001.*Telephone Survey Methodology*. New York: John Wiley & Sons.

Hirshberg, Jerry. 1999. *The Creative Priority: Putting Innovation to Work in Your Business*. New York: HarperCollins.

Kelly, Tom. 2001. *The Art of Innovation*. New York: Doubleday.

Leonard, Dorothy. 1997. "Spark Innovation through Empathic Design." *Harvard Business Review* November. pg.102–113

Martin, Justin. 1995. "Ignore Your Customer." *Fortune*. May 1.

McMath, Robert, et al. *What Were They Thinking?* New York: Times Books, 1999.

Miller, Christopher W. 1997. *Focused Innovation Technique*. Lancaster, PA: Innovation Focus Inc.

Miller, Christopher W. 2002. "Hunting for Hunting Grounds." Chapter 2 in *The PDMA ToolBook for New Product Development*, ed. Paul Belliveau, Abbie Griffin, and Stephen Somermeyer. New York: John Wiley & Sons.

Mishler, Elliot. 1991. *Research Interviewing: Context and Narrative*. Cambridge, MA: Harvard University Press.

Naparstek, Belleruth. 1998. *Your Sixth Sense: Unlocking the Power of Your Intuition*. San Francisco: Harper.

Perry, Barbara. 2002. *Get Ready for Ethnographic Fieldwork*. Barbara Perry Associates.

Perry, Barbara. 2002. "Using Ethnography to Sustain Growth." Paper presented at Tomorrow's Products Today: From Imagination to Implementation, PDMA International Conference. Orlando.

Perry, Barbara. 1998. "Seeing Your Customer in a Whole New Light." *The Journal for Quality and Participation* November–December. 38–43

Rosenau, Milton. 2002. "From Experience: Teaching New Product Development to Employed Adults." *The Journal of Product Innovation Management* 19, 1, January: 81–94.

Sawhney, Mahanbir S. 1998. "Leveraged High-Variety Strategies: From Portfolio Thinking to Platform Thinking." *Journal of the Academy of Marketing Science* 26, 1: 54–61.

Seybold, Patricia. 2001. "Get Inside the Lives of Your Customers." *Harvard Business Review* May. 81–89.

Sherrod, Earle. 2001. "It's All Greek To Me: Translating Data into Concrete Results, Adult Care Ethnographic Study." Paper presented at Ethnographic/Observational Market Research conference, Institute for International Research, February.

Useem, Jerry. 2003. "America's Most Admired Companies: One Nation Under Wal-Mart" *Fortune* March 3.

Ward, Kirk. March, 2003. "Value Pyramid." Paper presented at the Growth Strategies Seminar, Innovation Focus, Inc., Lancaster, PA.

Weiss, Robert. 1994. *Learning from Strangers: The Art and Method of Qualitative Interview Studies*. New York: The Free Press.

Wellner, Alison S. 2002. "Watch Me Now: Special Report on Ethnographic Research." *American Demographics* October. Woodland, Cara. 1998. *Creating the Customer Connection Workbook*. Lancaster, PA: Innovation Focus, Inc..

Woodland, Cara L. 2001. "Lighting a Different Path." *Quirk's Marketing Research Review* December 22

Woodland, Cara L. 2002. "Are You Listening?" *Alert!* April. Woodland, Cara L. 2002. "First Hand Experience or Second-Hand Information." *Quirk's Marketing Research Review* March 22

Shifting Your Customers into "Wish Mode": Tools for Generating New Product Ideas and Breakthroughs

Jason Magidson

You've got to be careful if you don't know where you're going because you might not get there."

—Yogi Berra

This chapter provides guidance to new product development teams on applying an approach that helps generate new product and service ideas and breakthroughs by getting customers into a frame of mind called "wish mode." The wish mode tools discussed here help generate better understanding of customers' needs and desires and help increase the likelihood of new product success. The wish mode approach involves (1) creating an environment in which users are continually encouraged to wish for their ideal product or service, (2) capturing those wishes,; and (3) converting the wishes into a product or service design.

The wish mode approach is particularly useful in bringing some clarity to the "fuzzy front end" of new product development (NPD) that precedes physical development. Although NPD projects will benefit most by using this approach earlier rather than later, it can be helpful later in the NPD process and even over the life cycle of a particular product or service.

The chapter's roadmap is as follows:

- Overview of the wish mode approach
- Application of customer idealized design (CID), the key wish mode tool that constitutes the bulk of the chapter, at IKEA and GlaxoSmithKline
- Planning, setting up, and leading customer idealized design sessions

- The post-CID session iterative design process
- Ongoing wish mode processes for capturing ideas that occur through the "daily usage" of products and services
- Resources required for applying wish mode processes

OVERVIEW OF THE "WISH MODE" APPROACH

Connecting with Users

Shifting customers into wish mode and learning how they would ideally like a product, service, or system to operate can provide great opportunities to NPD teams. Microsoft provides a good example of wish mode. In the 1980s and 1990s, the software giant used multiple channels to generate new product ideas from users. It set up a phone "wish line" through which users could communicate their wishes for new software features. It also created a "wish fax" template in Microsoft Word that people could fill in, print out, and fax in. It even advertised in trade magazines, asking readers to submit wishes for new features in future releases. All of these wishes were captured in central databases. NPD and existing product teams then reviewed the gold mine of ideas and implemented novel, breakthrough functionality in subsequent releases.

Microsoft is not alone in recognizing users' insights as a potent source of ideas for new products and services, as well as continuing enhancements to existing offerings. Cross-industry studies have shown that users are not only the best single source of new ideas for products and services but are also responsible for the majority of *actual innovations*. To illustrate, MIT professor Eric von Hippel traced the origins of commercial innovations in semiconductors and found that users were the source of 63 percent of the major functional improvements (von Hippel et al., 1999).

Unfortunately, many new product initiatives fail to connect with users. Two prominent scholars of new product successes and failures, Robert Cooper and Elko Kleinschmidt (1988), pointed out that proportionately very little time and resources (3.4 percent) in NPD projects are devoted to gaining a detailed understanding of users and the market. Furthermore, the number one reason for failure of new products is inadequate understanding of the market and users—a factor responsible for fully 24 percent of all failures (Cooper, 2001).

Even when new product teams do attempt to connect with users, the way they involve users has a big impact on the results. Three models illustrate the role of users in new product development: designing *for* users, designing *with* users, and design *by* users.

- In designing *for* users, NPD team members typically believe they know best what the users need. They design something, then check it with only a small circle of managers or executives—often with minimal input from actual users. This practice often fails because the NPD team's outputs do not meet the most important needs and desires of those using the product. In general, those who provide a product or service do not, by themselves, possess a sufficient awareness of the nuances and subtleties of the users' needs and ways they use a product nor how they would like the needs to be met. Additionally, those who provide a product are usually self-constrained because of various organizational factors including current technology in place, culture, assumptions about current offerings, and infrastructure.

- In designing *with* users, the NPD team interviews users and gets their input on what is important to them. Still, they only get high-level requirements, not the details of how the users want the product to work, look, and feel. Thereafter, team members retreat to distill and synthesize the requirements, make decisions on the relative importance of various factors, and create the design. If they do eventually come back to users for their reactions to it, it is often in a cursory manner.
- In design *by* users, users are the actual designers of their *ideal* product or service. They not only specify the ideal characteristics they want, but they also design (minimally in concept) what it will look like (its structure) and how they will use it (the process). The users' initial design choices reveal user-valued priorities and opportunities. NPD team members and users then work closely to develop a prototype, modify it iteratively based on feedback from a wider group, and then execute the users' design. Through this approach, the NPD team captures more than the customer's "voice"—it captures the customer's "design."

If an NPD team wants to capture potential breakthrough ideas and deeply understand users' needs, the best of these alternatives is the last—*design by users*.

Finally, there are two modes of connecting with users: (1) "scheduled," planned idea generation sessions/meetings and (2) collection of ideas generated unpredictably and perhaps serendipitously during the simple "daily usage" of a product or service. Both of these dimensions are important to the wish mode approach and are covered in detail. They are also interrelated. Figure 9-1 shows the complementary and cyclical relationship of holding periodic "scheduled ideation" sessions and creating ongoing processes for capturing "daily usage" ideas whose timing is "unscheduled." Typically, a product team will hold a set of CID sessions for a specific product periodically (e.g., every 12 to 24 months) and, in between sessions, maintain robust processes for continuing to encourage and capture users' "daily usage" wishes.

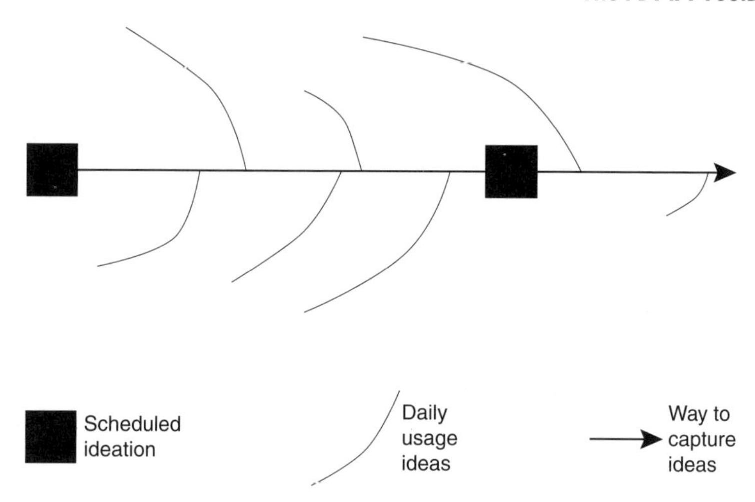

FIGURE 9-1. Capturing "scheduled" and "daily usage" ideas/wishes

CUSTOMER IDEALIZED DESIGN—SCHEDULED IDEATION AND DESIGN BY USERS

Customer idealized design is a wish mode approach that combines scheduled ideation and design *by* users. With CID, which was originally developed by management educator Russell L. Ackoff, users actually design a product, service, or system. Breakthroughs are generated when users pretend that the product or service was destroyed the night before. They are instructed to start from a clean slate and design what they would like if they could have whatever they wanted. This approach is very different from starting with the existing product and making improvements or identifying deficiencies.

Figure 9-2 illustrates why designing an ideal from scratch is so much more effective than starting from the existing product. If improvement efforts start from the existing product, some minor enhancements and additional functionality might result, as indicated on the left-hand side of the figure. In contrast, idealized design greatly expands the users' concept of the product or service because users start from scratch and focus on their ideal. This creates the possibility of major breakthroughs.

CID is generally implemented as small-group qualitative research, but there are ways to engage larger groups. At first glance CID looks similar to a focus group because initial half-day sessions are held that involve 10 to 12 "participants" per group, with additional attendees in an observer role. However, its process is very different because the principal activity is the design from a clean slate, *by users*, of a product or service, in contrast to having users react to predefined concepts/designs.

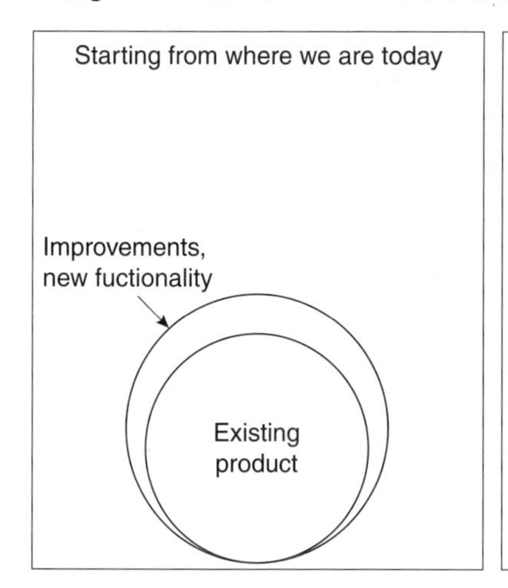

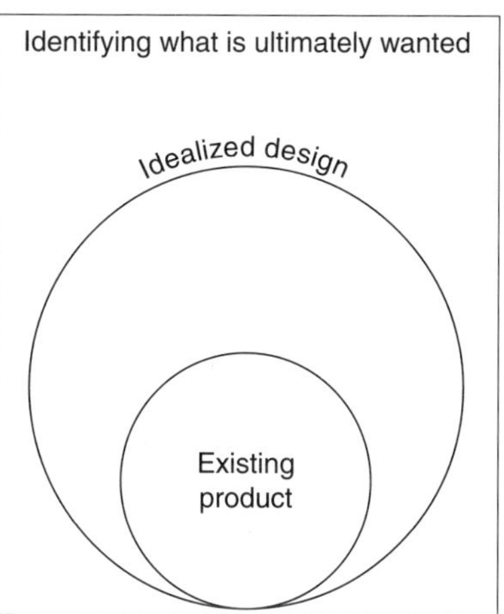

FIGURE 9-2. Enlarging the possibilities

During a CID session, the participants perform two main activities:

- *Generate "specifications."* A specification is a statement of a desired property or characteristic of a function, a process, or input. For example: "When using an elevator, if I push a button for the wrong floor, I could cancel it so the elevator wouldn't stop there" is a specification of desired functionality.
- *Develop a "design."* A design is a structure and a process that will bring about one or more desired specifications. To continue the preceding example, a design to bring about the desired specification of canceling wrong floors on an elevator might be to develop "cancel" buttons that would be placed next to each of the floor buttons.

In addition to generating breakthrough ideas, there are numerous other outcomes from holding CID sessions. For instance, they help NPD team members remove their self-imposed constraints or "manacles of the mind." The ideas generated can open up options and solutions that were not considered before but are in fact doable. And, NPD teams are often surprised by how many of the wishes they are able to realize.

APPLICATION OF CUSTOMER IDEALIZED DESIGN

Application of CID and "Wish Mode" Processes at IKEA

In the mid-1990s, IKEA's chief executive for North America, Goran Carstedt, was looking to grow the home furnishings business by creating an even deeper understanding of and connections with store customers, which he hoped would help generate breakthrough product and service ideas. To that end, IKEA held CID sessions. Nine separate small groups of customers were tasked with designing their "ideal IKEA shopping experience." They were instructed to start by pretending that the existing stores, products, services, and so on had been *destroyed the night before*. For a half day each, groups of customers became "design teams" that would start from scratch and design their ideal shopping experience.

Participants began by developing a list of specifications of the ideal shopping experience. A few of their specifications follow:

- "I could quickly and easily get to and find what I am looking for."
- "I would never feel disoriented—I would always know exactly where I am in the store, where everything else is, where the checkout is, and so on."
- "If I am buying one item, all of the other items that go with it would be in the same place. I would not have to hunt around the store to find related things."
- "For example, if I'm buying a sofa, then pillows, curtains, carpets, lamps, and picture frames would be nearby."
- "Checkout should always be fast, and there should be self-checkout capabilities."
- "Shopping at IKEA would be a relaxing, pleasant experience."

After the teams had developed their specifications, they were then tasked with developing a design that would bring the specifications about. They were asked to draw their designs on flip charts. Participants indicated that they wanted to create a "home base" in the store from which they could orient themselves to the vast array of products and departments and not suffer from information overload or disorientation. They drew an octagon-shaped building layout with an open, airy center area with a cathedral ceiling that would meet the specification quite well. They described and sketched out additional features that they wanted:

- The central area should be open and multistoried so that shoppers can look up or down, into or out of the center, and see where other departments are.
- The departments around the sides of the central area would be clearly labeled.
- There would be a Guggenheim Museum-like circular central area with a natural-light ceiling.
- At the top would be a restaurant providing an "oasis" where shoppers could relax, have Swedish food, converse, and think about their shopping needs.

FIGURE 9-3. Drawing of the new IKEA building design

- People would be able to go up and down levels using conveyors located in the central area.

Figure 9-3 shows a subsequent rendering of the octagonal building.

In 1998 IKEA opened its Chicago store, which is based on the users' design (see Figure 9-4.) The building was 411,000 square feet (roughly seven football fields). Because the store quickly was doing twice as much business as expected, IKEA added 50,000 square feet in 2001.

Many features from the CID session were implemented:

Finding Things Quickly without Getting Disoriented

- The store has the unique octagonal architectural design and three levels.
- The central area, circular in shape with a cathedral ceiling, has departments around the eight sides. From the central area, shoppers can easily see into and identify the departments. For example, standing on the first floor, shoppers can easily see the eight departments on the second floor.
- To get to another department, the shopper can either go back to the central area or follow a circular floor plan that rings the central area and includes wide inner- and outer-loop aisles.
- Six specially designed escalators carry people *and* their shopping carts and ensure that no one has to walk far or wait to change floors.

FIGURE 9-4. Photograph of the new IKEA building

- The many large windows help shoppers know where they are and later find their way back to their automobiles.

Quick Checkout

- To reduce waiting at checkout, the store increased the percentage of large items that shoppers can retrieve themselves (85 percent) from the self-service warehouse.

Related Products Nearby

- Related products are available in the same area. For example, near the beds, shoppers can locate sheets and pillows. Near the sofas, they can buy lamps, pillows, curtains, carpets, picture frames, and CD holders. Near dining room furniture, people can find cookware, cutlery, plates, glasses, and so on.

Relaxing Shopping Experience

- A restaurant at the top of the central area provides a relaxing atmosphere.
- Large windows let in soothing natural light and eliminate claustrophobia.

Large-sample surveys of customers, conducted by IKEA, reflect well on the Chicago store:

- Fully 85 percent of customers rated the shopping experience "excellent or very good," and 15 percent rated it "good." None rated it "fair or poor."
- Fully 93 percent of customers (purchasers and non-purchasers) said they would "definitely or probably shop at IKEA again."
- Return visits to the Chicago store are higher than at other IKEA stores.
- After the store had been open only five months, 25 percent of customers had visited the store six or more times.
- Shoppers spend an average of one hour longer in this store than do customers of other IKEA stores.

In addition to the redesign of the IKEA shopping experience as a whole, CID sessions were held for various components of the shopping experience: for example, the sofa department, the sofas themselves, design services (kitchen, etc.), installation and setup services, ordering methods, problem resolution processes, and delivery services.

IKEA's Sofa Department redesign is illustrative. During the session, the participants said they would like to be able to "borrow" large pieces of fabric (five feet by seven feet)—in contrast to the conventional small swatches—that they could take home and place on their current couch to help them judge whether the new couch pattern and colors will look right alongside everything else in the room. Participants explained that this would help avoid feelings of disappointment and foolishness that occur when they order a nonreturnable custom sofa that doesn't look right in the room. For delivery services related to custom sofas, the users designed several innovations, including a way to get quick updates on manufacturing and order status, and flexible, customizable delivery times, dates, and processes. Improvements included being able to do the following:

- Call a phone number for more precise delivery times.
- Arrange to be called at work to meet the delivery truck.
- Give the address and phone number for a neighbor who could admit the delivery personnel.

APPLICATION OF CID AT GLAXOSMITHKLINE

GlaxoSmithKline (GSK), a leading pharmaceutical company with $30 billion in annual sales, has embraced CID, having held well over 100 CID sessions since 2000. Whereas the IKEA example represents application of CID to new products for "external" users/customers, the GSK example represents application of CID to new products serving nearly 100,000 "internal" users/customers around the world, most of whom are GSK employees.

GSK applied CID to, among other things, creating software systems or "products" that employees use to get their work done. Software products were created for many functions including procurement, R&D, accounts payable, sales and marketing, finance, manufacturing, and travel and events management. For example, one product created using CID was a contracts management system called ConTrak for users in the procurement function. Several CID sessions focused on creating a global, Web-based system to help GSK manage its contracts with suppliers. Users' specifications included the following:

- A standard executive summary page for each contract, to avoid having to read 30 pages to learn key contract terms (e.g., pricing, start date, and expiration date).
- Pre-approved templates for creating new contracts.
- Automatic alerts about upcoming contract expirations or automatic renewals.
- The ability to do focused searching on specific fields—supplier, product/service, contract manager, contract value, expiration date, and so on.

The resulting contracts management software product incorporated all of the preceding specifications. The product offers these concrete benefits:

- It is very easy to locate and understand signed contracts.
- The contract templates enable much faster legal review of contracts.

- Since more contracts are in one place, there is better visibility of contract coverage.
- There is less frequent use of watered-down contracts—the contract manager starts fresh each time with an up-to-date template whose terms protect GSK.
- Automatic alerts about approaching contract expirations help ensure continuity of contract coverage, which protects against business interruptions.

PLANNING AND UNDERTAKING CUSTOMER IDEALIZED DESIGN

This section provides detailed guidance to NPD project leaders and their teams on how to arrange and lead CID sessions. It includes the following:

- Where CID and "daily usage" wish mode processes fit in the NPD process
- Planning and setting up CID sessions
- Leading a CID session

Where CID and "Daily Usage" Wish Mode Methods Fit in the NPD Process

Before going into the planning, setup, and leading of a CID session, you need to first provide NPD teams with some perspective on where CID and daily usage idea generation and capture processes fit in the NPD process. CID is most advantageous in the early stages of the NPD (i.e., the fuzzy front end), when there is likely to be more opportunity to act on breakthrough ideas and designs. At this point, the NPD team generally has resources available and is less likely to be invested in a particular approach or solution.

In addition to identifying when to conduct wish mode processes in an NPD project, it is also helpful to see how wish mode processes relate to conventional customer research activities, both in terms of timing and their contribution to generation of insights. Figure 9-5 shows some of these types of activities and a suggested ordering in relation to the wish mode processes.

One-on-one interviews, if open-ended, could be held in tandem with CID sessions. Open-ended interviews are exploratory and geared at gaining an understanding of "what you don't know you don't know." This type of interview has no questions. The interviewer gains insights by saying to the participant, "Tell me about _____." The interviewer listens carefully to the responses and then, in search of further insight, follows up by saying, "Say more about _____." An open-ended interview is in stark contrast to a closed-ended interview, which has been created with specific questions the organization wants answered. Unfortunately, predefined questions often leave the best insights

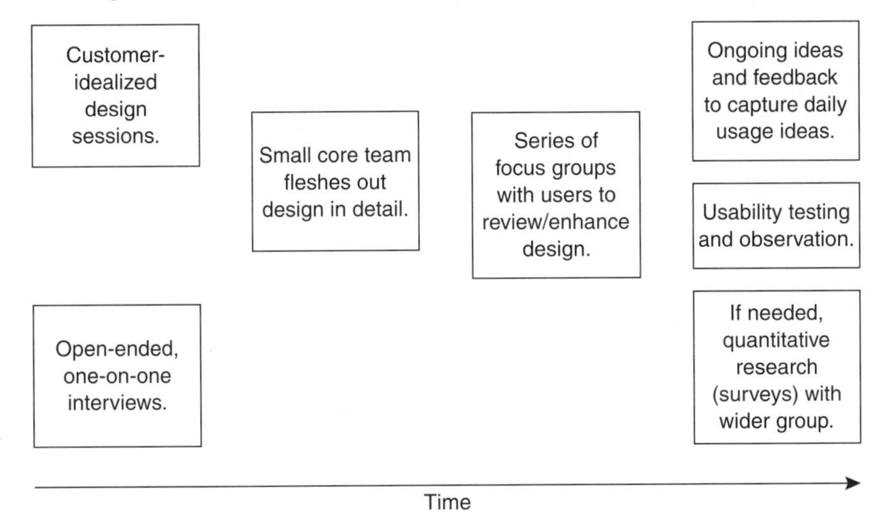

FIGURE 9-5. Ordering of research activities

unrevealed because they "lead" the user rather than letting the user reveal the opportunities for insights.

Focus groups can be very useful but are best held after CID sessions. Focus groups generally solicit a reaction from the user to a concept that has been developed, often by an internal team. These sessions do not generally lead to breakthroughs because users are reacting to the organization's concept or design, rather than reaching beyond to the ideal. For this reason, focus groups are much more useful when their purpose is to enhance and correct designs that have been fleshed out based on users' ideas. They provide an opportunity to get confirmation, disconfirmation, or adjustment to the in-process design. More importantly, they can be used for continuing to iteratively enhance the design concept for the product or service, so by the time development work begins, the product design is on the mark.

The ongoing collection of daily usage wishes (or ideas and feedback) that come on a daily basis from users and team members is another important activity that should continue as the product goes through the development process and even after it has launched (This process is discussed in detail later).

If quantitative research such as large-scale surveys are to be used, they should be undertaken after CID and subsequent qualitative feedback on the design. Surveys are stronger in the evaluation-of-options and confirmation/disconfirmation stages than in the initial stage of generating ideas/wishes and gathering requirements.

Planning and Setting up CID Sessions

The process of planning and setting up CID sessions includes the following key elements:

NUMBER OF CID SESSIONS. The NPD team should try to hold at least two initial CID sessions. Although it is possible to hold just one if time and resources are limited, multiple sessions provide more opportunities for generating ideas and breakthroughs and wider confirmation of the results. Multiple sessions also allow for including more types or segments of users or markets, which is helpful when the NPD team has identified significant differences between types of users or is targeting specific types of customers they want to understand and ultimately attract. Another potential advantage of multiple sessions is that they allow the initial session(s) to have a broader focus, with subsequent sessions that can go into more detail on specific things the users have identified as priorities. Given the preceding considerations, an NPD team might consider holding between two and six CID sessions for a particular product or service when starting to use the process.

SETTING THE CID SESSION SCOPE. In planning one or more CID sessions, the NPD team will need to define what will be the session scope. This is a brief statement—one or two sentences—on what the users will be asked to design. Getting the correct scope is important but can be tricky and requires some thought. This is where multiple sessions can be helpful. The initial session(s) could be broader and subsequent sessions could focus in more detail on specific aspects that users have identified as important. This is what happened at GSK, where, in initial sessions, users sketched out a suite of software products they wanted, and in subsequent sessions they designed specific modules. Table 9-1 provides some illustrative examples of scope.

One final example on CID session scope illustrates the situation where an NPD team sets some scope parameters/assumptions that force the participants to focus on things that are not important to them. One NPD team wanted input from users on a product for setting retirement savings goals. They initially planned to ask participants to design the ideal credit-card-size "calculator" they would carry around with them and into which they would enter information on expenses they incurred during the month. The CID facilitator advised the team to allow the users to design their ideal product without being limited to a calculator representation, and the team took this advice. At the end of the session, the team asked the users if they would use a calculator. The users completely rejected the idea, saying they already had too much stuff to carry around.

In general, a way of keeping things open when a CID session starts is to ask participants to design their ideal product and/or process for a particular activity. Giving them the flexibility of both product and process allows them to design from their perspective.

FACILITIES. A single CID session will require a room plus food and beverages for about 12 participants (i.e., users) and as many of the "providers" (i.e., those who provide a product or service) as can attend. The providers will be there to listen. Sometimes the organization can use its own large conference

TABLE 9-1.
Examples of CID Session Scope

Example	Comments About Scope
"Design your ideal airline overnight flight experience" vs. "design your ideal airline overnight flight cabin."	In initial session(s), an airline company or an airplane manufacturer should start with the broader "flight experience" rather than "cabin" because many aspects of the experience are not specifically tied to the cabin alone (e.g., services, food, and sleeping conditions). Design of the "cabin" might be appropriate for a subsequent session once the larger possibilities have been identified and if the users have indicated that this is important.
"Design your ideal IKEA *shopping experience*" vs. "Design the ideal IKEA *store*"	In the IKEA example, if the users had been asked to focus only on "designing their ideal store" in the initial sessions, it may have limited their thinking about design services, assembly services, online services, the restaurant, returns, products, customer services, and so on.
"Design your ideal residential roof" vs. "Design your ideal residential roof shingles"	If the roofing materials company wants wider opportunities for breakthroughs and/or this is the first session, the first choice would be more appropriate. The latter choice is more appropriate if the company only intends to stay with shingles and not consider other options or has already explored other options in previous sessions and now wants a detailed design.

room. In other cases, it can rent a conference room in a hotel or a commercial market research facility that is set up to host groups. The room should include two flip charts, a big screen, and an LCD projector for use when introducing the process and for displaying the specifications as they are captured. The participants will use the flip charts during the design phase to record diagrams and/or text. Seating should be arranged so that the participants and the facilitator are very close together and yet all of them can easily see each other. Often the facilitator stands or sits at the head of a table or at the top of a U-shaped arrangement.

CID SESSION FACILITATION. The NPD team should identify a facilitator for the CID sessions. It can use an internal facilitator, if available, or engage an outside facilitator, if desired. There are several skills that help a facilitator be successful. The facilitator needs to be a good listener and she needs to be able to quickly, accurately, completely, and unobtrusively record people's statements. She also needs to be able to let go of her own opinions, be able to "be with" the participant who is speaking, and be nonjudgmental. It also helps if she can put people at ease and make the process fun. She can do this if she can set the stage where she lets the participants become the designers while she quietly captures what they are saying, intervening only when needed to keep them following the guidelines for the process. The activities the facilitator will perform are discussed in detail later.

SCREENING AND RECRUITING. The first step is to identify the types of participants desired. The most important criterion is that the participants should include those who actually use the type of product or service of interest (e.g., people who drive sports cars, users of surgical instruments, people who travel frequently, or people who use television program guides). Once usage is established, the NPD team may want to segment the users and hold separate CID sessions to see if different types of users have different needs and wishes. For example, there may be "lead users" who are very experienced and sophisticated in using a product or have advanced needs and special applications. It can be very fruitful to hold separate sessions with them and other sessions with users who have less advanced needs. This can identify opportunities to build a product or service that meets the needs of both. Nevertheless, if resources are limited, it is acceptable to include both types of users in the same session, because this captures the needs of both and helps ensure more users' needs are met.

Another important consideration is that *customers* (those who pay) and *consumers* (the actual users) are not necessarily the same people, so their needs and wishes may be different. If those who pay are different from the users, they should be considered for involvement as well, possibly in a separate session. In some cases, the parties who are in the "chain" between the producing organization and the user (e.g., distributors, retailers, and third-party payers) should be considered for involvement as participants. However, if you must make a choice between involving them and involving actual users, you should usually choose the actual users, because they are the ones who are ultimately being served and it's rare that others can identify what *they* ideally desire.

Demographic characteristics could also be considered (e.g., age group, gender, household income, ethnicity, and level of education). These characteristics are not related to successful user performance in the CID sessions but are relevant to considerations about the target customers the NPD team wants to attract. Again, depending on the time and money available, numerous CID sessions could be held with different segments, or a few sessions could be held with a mix of participant types in the same session.

The next step is to identify how participants will be recruited. They could be identified from active customer lists, former customer lists, third-party vendor lists, third-party phone contact and screening, or, less formally, recruited from personal contacts of people in the organization itself.

The team can either recruit on its own or pay an outside firm to do the recruiting. In recruiting, often "screener" questions are used to screen people in or out to get a particular age group, gender mix, level of product usage, household income, and so on. In seeking "active" participants, some organizations include screener questions that attempt to recruit participants who indicate that they usually express opinions, are considered leaders, are able to verbalize thoughts clearly, and so on. These screeners can be helpful because it is better to get participants who will speak up than those who will be less inclined to speak. Screener questions about people's creativity are not as important for CID because the process itself generates a high level of team creativity.

When the organization-to-user relationship is business-to-business rather than business-to-consumer, it can sometimes make recruitment easier. It may be possible to work out arrangements in which user (i.e., business customer) input is traded for the opportunity to influence the provider's plans for a product or service. Where business relationships are already in place, sessions can sometimes be run at the customers' sites, where it is convenient and inexpensive to assemble users. In these cases the users usually need not be compensated. Likewise, similar recruitment arrangements can be made where there is a multiorganization user group that can be engaged as participants. In cases where the NPD team cannot get access to users through arrangements with businesses, they can simply recruit users directly and compensate them. The users may need to take time off from work but in some cases can get permission to participate as a work-related activity.

One additional factor should be considered when the organizational relationship is business-to-business or the users are part of an industrial firm. As previously mentioned, it is important to understand where the users in the industrial firm fit in the "chain" of customers and users, and whether users from other parts of this chain should be included. For example, if a user in an industrial firm uses a machine to make the ultimate product but the focus is on enhancing the machine for their use, this may not require the involvement of the ultimate consumer of the final product. However, the NPD team should explicitly decide whether to do so.

NUMBER OF PARTICIPANTS. The NPD project team should recruit 12 participants per group (depending on the level of incentive, this can mean over-recruiting by roughly four people). It is best to have 10 to 12 participants per facilitator. In a session with more than 12 participants, people may feel they have to wait too long to speak; in sessions with fewer than six participants, the facilitator risks less of a dynamic through which participants can build on one another's ideas. CID sessions are designed as group sessions rather than one-on-one sessions, because people contribute their individual ideas, build on each other's ideas, and are stimulated to think of things they would not likely have come up with on their own. Having 10 to 12 participants also provides additional insurance just in case some of the participants offer few ideas.

Another advantage of having 10 to 12 participants per facilitator is that they can work together on specifications, then split into two breakout groups that work separately on their designs during the design phase. The use of two groups creates more opportunities for design ideas to emerge. Having five or more people in each breakout design team seems to work well. Fewer than five can make it hard for people to build on each other's design ideas. Incidentally, when there is more than one breakout group, and time permits, it is helpful but not essential to have the groups present their designs to each other at the end of the session. This allows the groups to cross-fertilize ideas and even create a bit of fun, healthy competition.

It is possible to engage 30, 40, or 50 or more participants at one time by using multiple facilitators. In such cases, the lead facilitator could introduce the process to the entire group, and then, when starting specifications, all of the facilitators would lead breakout groups.

OBSERVERS. Invite as many "providers" as possible to the session so they can experience it "live." The broader the involvement of providers, the less need for those who did attend to go back and "sell" the results to others. Certainly, as many members as possible of the NPD team or existing product/service team should attend. It will also be helpful for one or more high-level sponsors and individuals/groups who support the project team to attend so they can be exposed to the ideas. It is remarkable how hearing the users' ideas broadens the observers' consideration of opportunities and their perception of priorities. When inviting observers, it is helpful to give at least one month's notice of the session in order to obtain better attendance.

It is acceptable to include providers in the room (versus behind one-way mirrors) so they are less likely to distract one another and more likely to concentrate on what the users are saying. Providers should sit on the perimeter of the room taking notes, having been advised prior to the session that they are there only to listen, at least until the session nears its end. Experience has shown that observers in the room do not distract participants, because the participants become so focused and energized around designing their ideal that they forget about their surroundings However, if desired, it is possible to use a focus group facility with one-way mirrors.

ANONYMITY. Some organizations may have specific reasons for not revealing their identity. Whether or not a firm remains anonymous, experience has shown that results are very good.

Leading a CID Session

At the beginning of the CID session, the facilitator welcomes the participants and tells them that they will be taking part in a fun process called customer idealized design, in which they become the designers of their ideal product or service. At this point, the facilitator should very briefly review the scope of what they will be designing (i.e., talk about and/or display the one- or two-sentence description of the scope). Next, the participants briefly introduce themselves (providers are usually just acknowledged but not individually identified). This is followed by an introduction to the CID process, which typically ranges from 15 to 30 minutes. This introduction includes guidelines for the

CID session (discussed in the text that follows). It should also include a story about an application of idealized design (e.g., the IKEA store design). Doing so makes things clearer for the participants and gives them a sense of what they will be doing for the remainder of the session. A story also helps people understand what the specifications phase is about and how the design is derived from the specifications.

Guidelines for Facilitating the CID Session

Many ideas or actual innovations are never conceived or executed because the right environment is not created and/or people constrain themselves from thinking creatively. Sometimes teams or individuals just assume they do not have permission or power, and consequently don't give themselves permission or the boldness to explore possibilities.

During the introduction to the CID session, the facilitator creates the right environment by instructing the participants to follow all of these interrelated guidelines:

- *"The current system was destroyed last night."* This assumption is critically important because referring to the existing system holds users back from thinking about what they really want. Their minds tend to get caught up in the constraints of the current situation, and they do not allow themselves to think beyond it, which is the most common constraint to breakthroughs.
- *"Focus on what you would like to have if you could have whatever you wanted: Think of it as your ideal; stay in design-from-scratch mode."* The idea is to get the participants to shoot for the stars, not hold back at all. The facilitator can tell them, "We can all come back to reality tomorrow."
- *"Don't focus on what is not wanted."* Focusing on the negative aspects saps the positive energy and interaction and takes time away from focusing on ideals.
- *"Don't worry about whether resources are available to implement the wishes/ideals or whether it's even possible to implement the wishes."* Imparting this message is critical because worry or skepticism about whether the design will ever be implemented frequently holds people back from a "what-ought-to-be" mind-set and using their imaginations. Furthermore, breakthrough thinking often increases available resources because decision makers will see additional areas that are attractive and worthy of investment.
- *"If you disagree with someone else's specification, simply state an alternative specification."* The facilitator must not allow participants to criticize each other's specifications, as this will destroy the constructive and creative mood. She should tell them that evaluation and prioritization

come after the session. She may need to remind participants of this guideline during the session.

- *"The product's larger containing environment remains intact."* This "containing environment" includes the contexts, processes, and settings in which the product or service is to be used. This assumption encourages participants to focus most of their energy on the product they've been asked to design and less time—less than 10 to–20 percent, as a rule of thumb—on the variables that don't fall under the NPD team's (provider's) purview. The reason for doing this is to ensure that the participants don't spend most of their time designing something that the providers cannot implement because it relies on changing many things they don't control. For example, if the focus is on designing a corkscrew and the NPD team has no involvement in manufacturing corks, the participants should not spend the bulk of their time redesigning the cork. On the other hand, the permission to stray broader for 10 to–20 percent of the time can sometimes identify significant opportunities, which can then be selectively pursued by the NPD team.
- *"Providers* must *agree to remain in listen-only mode."* Providers must remain listeners until the end, when they may ask questions. Following this guideline prevents providers from disrupting the flow of user wishes by asking questions that divert the users from talking about what they want. It also avoids the common problem of providers becoming defensive about their current offerings or disrupting the users' flow by interjecting that they already provide a feature being discussed. This rule also eliminates the providers' tendency to say, "We can't do that because of such and such," or "Our technology is this way, so we can't do that." These types of reactions shut down the creative thinking of users and stop them from wishing.

On occasion, providers want to make sure their concerns are addressed in a session or they want to hear the participants' reactions to topics they feel are important. It is best to delay the introduction of providers' topics until later in the session or even until the end so the creative flow of the users' own ideas is not disrupted. In some cases, the facilitator can get a list of topics from the providers and can, at or near the end of the specifications phase, ask the participants to give their thoughts (and possibly specifications) on the providers' topics that haven't been covered. In the post-CID session design document (covered later), it should be noted that these topics were brought up by the providers so that these can be distinguished from what the users came up with unprompted. It is also possible to build in time for a question-and answer period near the end of the session, when the providers can talk with the participants about topics of interest.

As the facilitator is finishing the introduction to the CID process, it is sometimes helpful for her to hand out a checklist of guidelines to the participants and observers; it helps remind people to follow them. Additionally, just as the specifications phase is beginning, she can help get the participants' creative juices flowing by reminding them to "shoot for the stars" in what they are about to design so that the organization can "hit the moon in implementation."

Specifications

The facilitator should start the specifications phase of the session by asking the participants to give their bullet statements about what they would like in their ideal. He should do the following:

- Instruct participants that their specification statements should begin with phrases such as "There ought to be . . ." or "I would like . . ." This helps keep things positive. If participants start talking about what they don't want, they should be asked to convert that into a specification of what "ought to be."
- Record *all* specifications so everybody can see them being written in real time (e.g., via LCD projector and screen or flip chart). This not only captures everything but also shows participants that their input is being taken seriously, which is very important in generating full participation and a genuine sense that what they are saying is not falling on deaf ears. At the outset, the facilitator should ask participants to speak up whenever something has not been captured accurately.
- Remind participants that it is important that they stay in design-from-scratch mode. If participants refer to the existing system or complain about current problems, the facilitator should remind them that the system was destroyed last night and that there are no problems to complain about. The participants should be asked to convert their statements into something relating to their ideal. This will keep the group positive and creative.

There's one aspect of facilitating that appears subtle but is very important. Often, people state specifications that are worth following up in order to learn more about what they want and/or why they want it. Understanding why people want something (e.g., understanding a difficult or frustrating situation in their lives that a product feature would address) can be invaluable. It could point to breakthrough design or even strategic positioning opportunities. Probing further will get at the underlying emotional and rational basis for the specification. The facilitator should probe by simply saying, "Tell me more about that" or "Say more about that." These phrases keep the discussion open-ended and encourage people to expand on their specifications. This approach is much better than asking "Why?"—which could be interpreted as a judgment of the statement's value and can elicit an incomplete or inaccurate response. Answers to "why?" questions tend to result in rational reasons instead of gut, emotional responses (which play a big role in product or service usage decisions). Another subtle but important point is that the facilitator should not say, "Do you want that because . . ." Saying "Tell me more about that" allows the underlying reasons to come from the *user's* perspective, not the facilitator's.

Knowing when to say "Tell me more about that" is something that comes with practice, but there are a couple of situations where it is clearly needed. One is when it's apparent that to implement what the person is saying, more detail or explanation is needed. For example, someone might say, "The product should be user-friendly." If the facilitator does not follow up, the NPD team will never know what the user had in mind. The other situation is when someone uses words that are emotion-laden. Probing here can shed light on product or service features and/or emotional benefits that may be very important. Probing also shows that the facilitator is interested in what the person has to say, making it safer for other participants to offer what they may personally feel are outrageous ideas but in reality could be major breakthroughs. In some cases, the facilitator can reveal deep insights about a particular statement by following up multiple times to the participant's responses to his probing.

The facilitator should be careful not to judge ideas—either negatively or positively—because this will discourage the participants from offering them or can guide them in the direction they think the facilitator wants them to go. Instead, he should simply write the idea down. There will be plenty of opportunity after the session to evaluate specifications. The facilitator needs to "go" wherever the participants' design takes him. Similarly, the facilitator should not ask questions (e.g., "Do you want . . .") This "leads" the participants rather than lets them lead where the design is going.

It is essential that the facilitator create an environment in which people aren't worried about sticking their necks out. They need space to offer bold ideas that may provide breakthroughs. Sometimes someone will offer a wild idea, and some of the other participants will react with laughter or negative body language. This reaction indicates their uncertainty about the acceptability of any and all ideas. In this situation, the facilitator must support the person who offered the idea. Otherwise, that person and others will hold back on future ideas. To provide support, the facilitator should focus intensely on that person and ask him or her to say more about the idea. It may be necessary to follow up yet again to demonstrate that wild ideas are desired. The facilitator could also remark to the group that it's important to hear bold ideas because they frequently lead to breakthroughs.

If there is a lull in the offering of specifications, it's important for the facilitator to sometimes say, "What else?" When doing this, the facilitator should continue to make eye contact with everyone in order to send a clear signal that he is eager to hear something they may have held back on. It is important for the facilitator to be totally comfortable and patient, waiting through any awkward silences, because someone will eventually speak up. Often, great new ideas and/or totally new areas of discussion come out that someone might have been hesitant to raise before. It is worth noting that saying "What else?" is much better than saying "Is there anything else?" The latter allows for an answer of "no," which can end the process prematurely, leaving fruitful ideas unexplored.

Often, the specifications phase will run about two hours, but this time frame is not set in stone. When the group runs out of new specifications to offer, the facilitator can begin the design phase of the session.

Design

In the design phase, which takes place in the second part of the session, the participants create (or draw on) structures and processes that would bring about the specifications. For example, the IKEA store's octagonal design helped realize the specification of being able to easily find departments, products, and so on. Design is essentially about showing possible ways that selected specifications could be brought about.

There are a few things the facilitator can do to get the design process moving efficiently. A way to begin is to ask the participants to think about the specifications and start with what is most important to them. It is not necessary to go back through the specifications to see or rank what was discussed; experience shows that this just takes up time that could be spent working on the design. If the participants cannot collectively remember something that was stated, then it's not important enough to be covered in the design phase. Instead, the participants should be instructed to start with what they collectively feel was most important. To determine this, sometimes a group will have a brief discussion and make a short list of candidate items they might work on, and then choose one or more for which they will create a design. It is better for the group to make its choices by consensus (i.e., everyone can live with the choices selected) than by voting, because those in the minority may not be as engaged in the process.

Designing is a creative process, so no one ever knows what will emerge or exactly how the groups will work. Some groups "click" and come up with designs right away. Others struggle for a while, coming up with a design only at the end. The design phase is the hardest part of the process, and it is not unusual for some groups to flounder for a while. Reassuringly, there will be many good ideas in the specifications—and accordingly, no shortage of possibilities to develop.

As stated earlier, having 10 to 12 participants allows the group to form two breakout teams during the design phase. Often it's useful at this point for the facilitator to step aside and allow the participants to self-lead the breakout teams. When there are enough people to create more than one breakout team, this frees the facilitator to move between breakout groups and re-explain the process, if necessary, answer questions, and observe progress. Having participants facilitate the design phase also helps solidify the breakout teams' ownership of their designs, which helps promote full participation and leadership rather than passivity and "following." One way to identify participant-facilitators is to ask for volunteers who pass the following test: People they spend time with, whether at work or elsewhere, tell them they are good at leading groups, at listening, and at accurately capturing and restating the ideas of others.

Finally, the facilitator should tell the teams that they should try to capture as much as possible on a flip chart so that all ideas are preserved—whatever is not written down is lost. It is often necessary, as well, to remind the teams of this after they've gotten into the work.

Typically, during an initial half-day CID session, the participants will spend one to two hours working on their design. (Where more than one-half day is available, it can be very productive to take this time to enable participants to go into more detail.) The facilitator should instruct the participants as to how long they should work on their designs, taking into account when the session is scheduled to end. In cases where the breakout teams will be reporting their designs back to the entire group, about ten minutes per group should be factored in. In cases where the "providers" want to ask the users questions at the end, approximately 15 minutes total should be factored in. It is obvious that both reporting back and Q&A take time away from designing, so this trade-off should be taken into account when you are considering whether to include these.

Bringing a CID session to a close is fairly straightforward. The facilitator and other providers who are present should express their appreciation to the participants for their time, effort, and ideas. They should say that their ideas are highly valued and that they will be reviewed and considered for implementation. Interestingly, it is not unusual for some groups of participants to want to keep designing past the time the session was scheduled to end. They've been having so much fun, they don't want to stop. In such cases, it is fruitful for the facilitator to allow them to continue, but she should say that she realizes they are going past the scheduled time and that if anybody needs to leave it is okay. In cases where some form of compensation is to be given to the participants, this is the time to tell them how they will receive it (e.g., pick up an envelope on the way out).

In the unlikely event that a CID session has not produced many new ideas and insights, the NPD team need not be anxious that the session is "bombing." It might just be a sign that the NPD team's concept is already on track and the session is giving confirmation. Additionally, there won't be any awkwardness because the participants will be happy to have been asked for their input. No matter what the outputs, the graceful way to finish is for the facilitator or session hosts to thank the participants for taking the time to share their ideas and designs.

POST-CID SESSION

Preparation of a "Design Document"

The next step following the CID session is to produce a "design document," which categorizes the users' specifications, or "wishes," and includes the beginnings of a design that will bring some of the specifications about. The design document serves to capture the work and helps in sharing the results with others who weren't at the session. It should be written so that someone who wasn't present can pick it up and easily understand it. It is also the starting point for the subsequent process of fleshing out a detailed design.

The person selected to prepare the design document should start by doing two things: (1) go through the specifications statements and turn them into clear, complete sentences (during the session, the goal has been to capture the participants' statements quickly, which does not allow time for full sentences and total clarity/efficiency of words); and (2) put the specifications into meaningful categories so readers will be able to easily find and understand the topics of interest.

Putting the specifications into categories requires judgment, but some guidelines have proven useful:

- The categorization should preferably be done by someone who attended the session. This is because the categories should reflect the interest in and level of conversation around each specification in the session, and someone who was present will have a better sense of this than someone who wasn't.
- The document preparer can start at the top of the bullets and go through each specification, in order, and decide whether it is part of a theme that merits its own category, whether it fits into a category already created for another bullet, or whether it belongs in "miscellaneous."
- There should generally be eight or fewer bullets per category, which is roughly the limit on how many items a person can remember at once.
- Once the categories have been finalized, they should be alphabetized so they can be located more easily.
- The category headings should be copied to the front page to create a "quick index."
- It's not practical to combine design documents from separate sessions because this will create a document whose categories have too many bullets to keep track of.

The outputs from the design phase—drawings and/or text—should also be included and should be placed underneath the categorized specifications. This material can simply be transcribed from the session outputs.

The design document should be forwarded to other relevant stakeholders who are involved in or impact the NPD project so they have an opportunity to review, comment, and take ownership of the process and outputs. Throughout the rest of the process, the design document can serve as a reference that a design team (discussed in the text that follows) can draw on.

Iterative Development of the Design

As the process of developing the design further in post-CID sessions begin, in some cases the facilitator may be aware of design alternatives that could help the NPD team realize some of the users' specifications. In such a case the facilitator could potentially have a role as an educator on these design alternatives.

The next step is to establish a relatively small design "core team" (often three to five people) that will lead the process of iteratively fleshing out the design further. It is possible to have a larger group, but that can be more difficult to manage, which makes it even more important to have one person clearly designated as the leader and made accountable for moving the process forward. The core team often includes some or all of the NPD project team members if the team has already been formed. The time commitment required of each of the team members varies widely depending on the scope, complexity, priority, and timeline of the NPD initiative, but frequently most or all members will devote only part of their time to the iteration of the design.

Criteria to consider for membership on the core team include the following:

- Members each "bring something to the table," such as a skill, knowledge, or the ability to lead and make the process fun.
- Members embrace the philosophy of continuing to engage users.
- There is involvement by the key organizational functions that will enable the design's eventual implementation.

The core team will draw on ad hoc input as well as the periodic participation of an extended team, while continually reconnecting with users in the process of iteratively enhancing the design. The core team continues the design process by mocking up design possibilities and working with users and other providers to refine and enhance the design over the following weeks or months. The more complex the product and the more diverse the customer base, the longer the process will take.

Experience shows that it is efficient for the core team to begin by sketching ideas out on large flip chart paper. In the early stages, when there are many ideas and the design goes through many refinements, paper allows for quick capture and frequent changes. Figure 9-6 shows an example of flip chart paper divided into "pages" that allow a small team to sketch out initial design ideas while viewing many pages at once. This technique is especially useful for visual

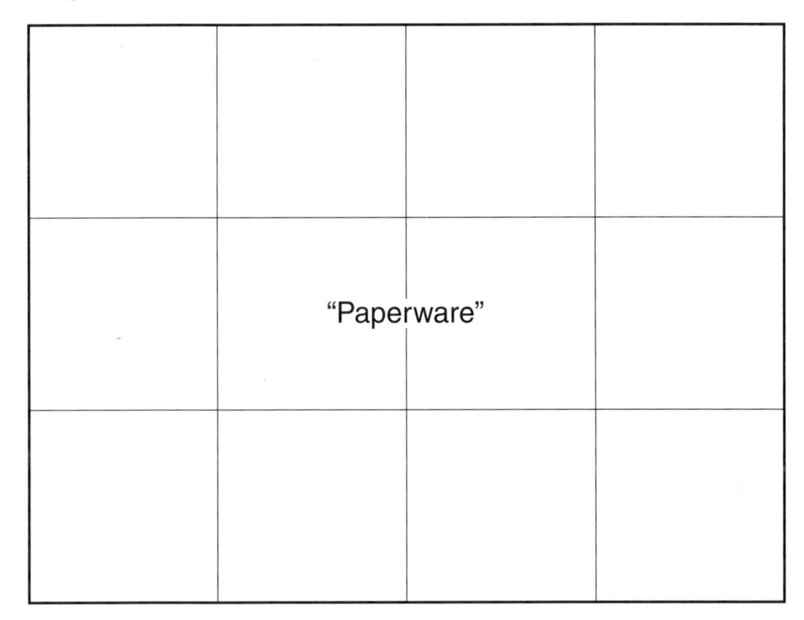

FIGURE 9-6. "Paperware"

representations. In some cases the design ideas are accompanied by textual descriptions of associated processes (e.g., services). Hand-drawn sketches should eventually be converted into an electronic format. As the design starts to take shape, electronic format allows the design to be easily shared with others for additional input.

In the development of software products at GSK, NPD teams called the various iterations of the electronic design mock-ups "Powerpointware." Figure 9-7 provides an example of two mock-up screens for GSK's contracts management product. The left side shows the cursory "search" feature that had been sketched out in the initial CID session. The right side shows the fifth iteration of an eventual ten iterations of Powerpointware.

Visual representations such as these help generate clarity and consensus on the basic design that will be created and are also very important in obtaining user reaction/suggestions. The visual representations are usually complemented by a textual "requirements" document that is also iteratively refined as the development and feedback process continues. This document includes information on what is going on behind the mock-ups.

The design mock-ups typically go through about ten or more iterations of additions, refinements, enhancements, and deletions before the design is sufficiently fleshed out. In some cases the mock-ups are two-dimensional (e.g., for software) and in others is it very helpful to have three-dimensional models, electronic and/or physical. As an example, a construction materials company using CID involved users in a process of assembling 3-D physical models of roofing materials.

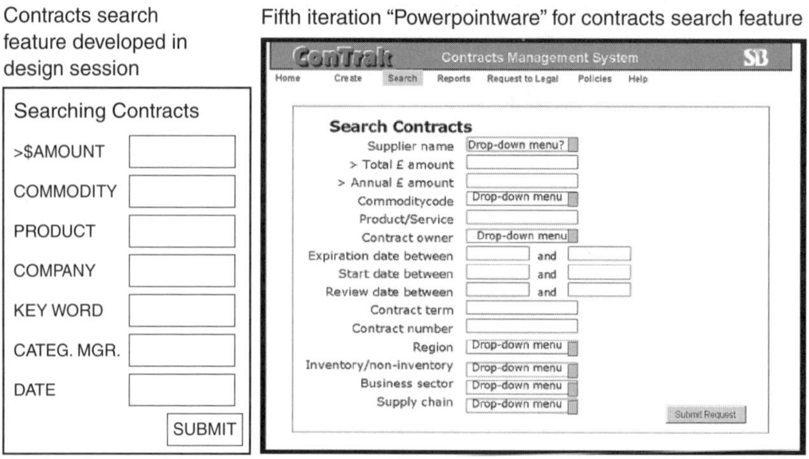

FIGURE 9-7. "Powerpointware"

The mock-up process involves not only continual fleshing out by the design core team members and others in the NPD project team but also periodic input from several sets of users, senior management, functional heads, and other key stakeholders in the design. The NPD project team should be sure to get the input of not just those directly affected by the design but also those who are likely to influence the new product's launch. This allows relevant personnel to shape the new product, which in turn helps generate buy-in.

Additional noteworthy benefits of thorough and iterative prototyping include the following:

- It allows the team to make mistakes and corrections before the product, service, or system is physically designed. This reduces the incidence of having to redo products (or worse).
- It helps reveal potentially flawed assumptions about the needed solution, helps identify details about what is needed, and points to how the solution would work in practice.
- It shows the NPD team it is making progress, which gives them satisfaction and momentum. This further promotes buy-in within the team itself and gives members confidence that the product or service will be aligned with user needs.

Iterative Implementation

In addition to developing iteratively in customer idealized design, implementing iteratively also works well for many types of products and services. Although idealized design is about "thinking big," the range of ideas produced in a CID session presents opportunities for "quick wins" as well as for longer-term initiatives. However, for many but not all products and services it is often helpful to "think big but implement small." In other words, in many industries, trying to implement a grand design that would take many months or even years before yielding any benefits would likely mean that nothing would get implemented, and the proponents of the approach might be written off as "in the clouds." Therefore, it is helpful to handle the implementation in phases, such that actual benefits can be continually delivered to customers, often in a series of releases over months or years. As phase one is being developed, ideas can continue to be collected and considered for future phases or releases. In other words, numerous features and functions can be identified and pursued in the context of successively approximating the overall idealized design.

Approaching both design and implementation iteratively also helps avoid the situation where products become overengineered, offering more features, functions, and complexity than most users want. This process allows for ongoing reality checking and refinement of the design.

The GSK contracts management software illustrates both the phased implementation and the iterative enhancement process. It went through a number of key enhancements over several phases of release. At the end of the CID session, the design was very sketchy. Ultimately, the NPD team decided that a phase-one launch should include only contract templates prepared by the legal department, which would enable procurement department negotiators to use boilerplate language to quickly create high-quality contracts. The NPD team felt they needed a quick win with the customers, who received useful functionality within five months. And, while the preparations for the phase-one launch were still being made, work began on the phase-two product that was released several months after that.

Avoiding Potential Customer Idealized Design Pitfalls

The following caveats can help avoid failure in implementation:

- The finished product or service can be way off the mark if users are involved only in generating specifications but not involved in converting them into a *design*. Get the users to draw pictures of what they want, to build models, to indicate how it would work, to determine the steps involved, to indicate how they want to interact with it, and so on. Continue going back to the users.
- The ideas and design generated in a CID session are like a big wish list. It *does not* mean the entire design should be implemented. Recall that the design process has asked the users not to worry about resource requirements. The importance of implementing only what users are willing and able to pay for cannot be overstated. The final design that is actually implemented must be financially viable. Various other NPD

processes that help identify what will be financially viable should complement the post-CID session design process.

- There are some things customers actually want, but in practice, many of the customers will not use them. In some cases they are not willing to put in the time that may be required to learn how to use a feature (e.g., features on a computer). In deciding what to implement, it is very important to understand human behavior, people's circumstances, people's other responsibilities and commitments, and their incentives—in short, the dynamics of people and organizations. Use judgment about whether people would actually use the product or service features as designed or whether they need further refinement or even elimination.
- During idealizations, users stretch their concept of what they *want*. This broadens the set of ideas the providers can draw on when identifying what the users will use and are willing and able to pay for. However, there is some wisdom in Microsoft's saying, "Build what people need, not what they want." Be careful to discern what people will actually use or do, and consider this in deciding what to build.
- CID is not appropriate in some situations: for example, when there is virtually no chance that an NPD team will get resources or when it is certain that the product or service would not be implemented. These situations would only lead to frustration.

ONGOING "DAILY USAGE" IDEA GENERATION AND COLLECTION PROCESSES

"Scheduled ideation" sessions are not the only way to generate ideas and insights from users. Many opportunities present themselves on a continuing basis, especially in the moment that a person is using a product or service. For example, 3M's Art Fry conceived Post-it Notes while attending church, when some bookmarks fell out of his hymnal. Mr. Fry had a "daily usage" idea and was in a frame of mind and in a position to do something about it. Mr. Fry was aware of an adhesive (for which no use had yet been identified), developed by his colleague at 3M, Spencer Silver, that he realized could be used to hold the papers in place.

Like the 3M example, NPD teams should create an environment in which users' unscheduled daily usage ideas and wishes *come forth*, where users are encouraged to submit them and where there are easy processes for submitting and capturing ideas. A good example of a company that does these is USAA, an insurance, mutual fund, banking, and services business. Highly courteous phone agents are trained to listen for unmet customer needs; when they hear them, they enter them into a database that is reviewed by product managers. Based on this process, numerous successful new products have been developed and substantial new business has been generated over the years. This practice of listening to customers has reduced the need for expensive market research studies because the employees remain closest to the users.

Here are some suggestions for how NPD teams and existing product teams can create a fertile environment for capturing daily usage ideas/wishes:

- Ask the users—frequently—how the product ought to be. Even say, "If you could have whatever you wanted, what would it be?" Use personal visits, telephone calls, meetings, e-mails, Web site idea/feedback buttons, and so on.
- Whenever someone submits a wish, express appreciation and enthusiasm. Doing so encourages future submissions.
- When people submit ideas online, the system should give them confidence that it was received by thanking them for the submission and presenting their input back to them. Web idea pages should explicitly state that users are the single best source for improvement ideas. They should also let the submitter know that all ideas are to be stored in a database that will be used as the basis for new releases, and that some ideas may take longer to implement than others.
- Depending on the number of users, all of the electronic ideas or a sampling should also be e-mailed to one or more team members. This helps create deeper awareness of and self-commitment to the opportunities. Over time, more NPD team members will be more connected to the users' wishes and will become champions for their ideas without prodding from the NPD team leader. Additionally, patterns in user ideas will emerge and help shift priorities to be closer to user wishes.
- Embrace all ideas as possibilities and never immediately judge any idea. Don't say, "We can't do that because . . ." Instead, say, "Thank you very much; we'll add that to our database of new ideas we're exploring for the next release."
- Sometimes a daily usage idea comes in that a manager believes is already available. Some write this off, thinking, "We have that already." Some blame the user. They miss the opportunity to take a closer look at how they might change/redesign their functionality, process, and so on so that the issue is "dissolved." When hearing things he or she believes are already provided, the manager should assume something is missing or wrong in the current design/solution and look for ways to improve it. Again, what it boils down to is whether the listener is in wish mode and open to the idea that there is always room for improving the product or service.

As an example of an electronic form for encouraging users to be in wish mode, Figure 9-8 shows one of the means by which daily usage ideas are captured at GSK. This is the electronic "ideas and feedback" form available on the GSK contracts management product. On the form is a thank-you and a statement of the importance of user wishes to improving the product.

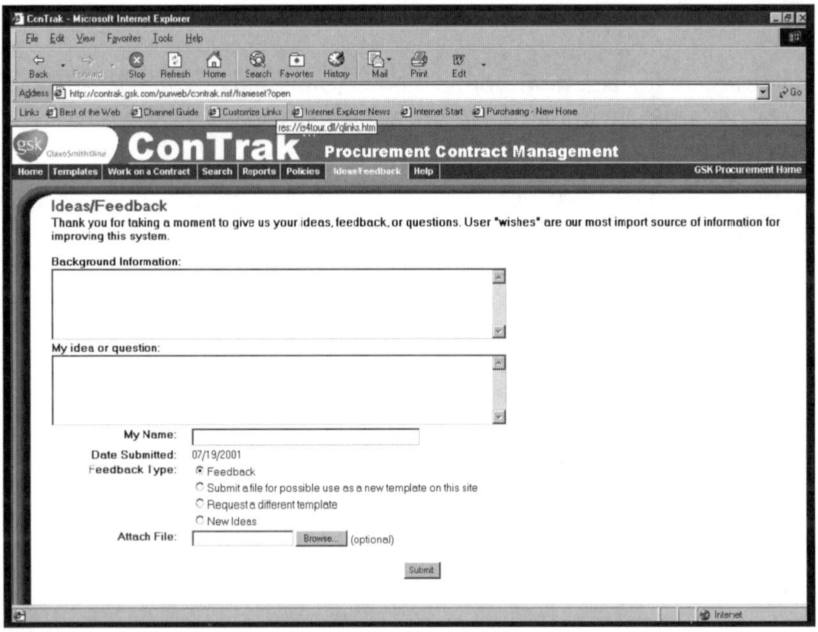

FIGURE 9-8. Ideas and feedback form

Figure 9-9 shows the database that automatically captures the wishes that are submitted, enabling easy review by the NPD teams. Wishes and ideas are reviewed frequently by the product teams, which—depending on the product—could mean daily, weekly, or monthly. Typically, one person on the NPD team is made accountable for managing a process of reviewing the ideas, looking for and highlighting opportunities, and working with other team members to identify priorities.

Observation and Usability Testing

It is worth mentioning another valuable source of ongoing learning about users' needs and desires: observing people using a team's product or service. Many processes for observation and "usability testing" reveal opportunities for improvements and breakthroughs, both for mocked-up/prototyped new products and for existing offerings. In some cases NPD team members simply observe people using a current product or service in their natural environment. In other cases they get input and wishes from the users as they are using the product or service. The key point here is that NPD teams can benefit greatly when observing customers using products or services and encouraging them to state their wishes.

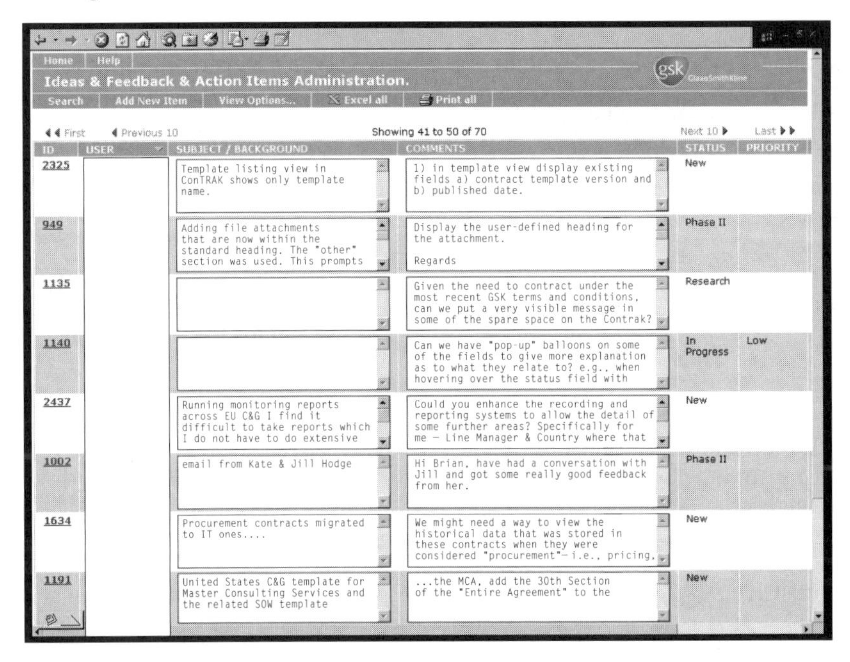

FIGURE 9-9. Database that captures submitted wishes

Microsoft, which does extensive usability testing, provides a good example of the powerful ideas that can be generated by observing how people use a product. For example, company representatives noticed that saving or printing a document took multiple mouse clicks and menu selections. They reduced the number of steps by creating "save" and "print" icons right on the toolbar.

When observing users working with a product or service, you should do the following:

- Notice where they get stuck.
- Observe where they seem confused.
- Encourage them to say what they're feeling as they're using the product or service.
- Perceive opportunities to reduce steps, time and effort required, and so forth.
- Think about possible new features.

To be effective, the person or team that goes out to observe and speak with the users must be in a "what-ought-to-be" frame of mind themselves and should encourage the users to say what they would ideally like. The observers should also be prepared to listen; otherwise, they won't pick up on the potential ideas, problems, and opportunities a user presents. Since talking often gets

TABLE 9-2.
Illustrative Resource Requirements for a New Product Development Project

Activity	How Long Takes	People Required	Cost
Planning CID session	½ day	2 (NPD team lead and facilitator)	Variable
CID session setup ◆ Arrange facility	½ day	1	Facility cost—if hosted internally in conference room, only $250 to $500 for food. If externally hosted at hotel or market research facility—$1,000 to $3,000 including food
CID session setup ◆ Identify and recruit user participants	Depending on selectivity and avail. of candidates, 1 to 4 days	1	Variable—If arranged internally, possibly no variable cost.
Leading CID session	½ day for session itself	1 facilitator per 10 to 12 user participants Variable number of participants	Variable—If internal facilitator, possibly no variable cost. An external facilitator would charge for planning session, conducting session, and writing up the outputs. Costs can vary quite widely. A rough estimate might be $3,000 to $8,000 for all of these activities for one session. This wide range depends largely on how much time it takes the facilitator to work with the project team. Cost can be reduced if someone from the organization writes up the design document. Cost per session can be reduced when there are multiple sessions.
Screening and recruiting	2 to 4 days	1 or 2	Using internal employees to recruit potentially incurs zero variable cost. The costs of using an external firm to recruit can vary widely based on a number of factors, including how narrowly the targeted participants are defined. This may range from $50 to $150 per successful recruit.

TABLE 9-2. *(continued)*

Activity	How Long Takes	People Required	Cost
Participant incentive	N/A	N/A	Incentive for user participants depends on local market conditions and the type of participants being recruited. For example, for an evening session lasting from 6:00 to 9:30, a monetary incentive for users might range from $50 (widely avail.) to $300 (expert/scarce) per user.
Travel costs	Variable	Variable	If CID sessions are held in multiple cities, travel expense and time may need to be factored in.
Completing design document	1 to 2 days	1	Variable—If internal facilitator, possibly no marginal cost; if external could be $1,000 to $3,000.
Iterative design process ◆ Development of mockup ◆ Iterative review with users	2 to 5 months, but may vary	Core team—four × approx. ¼ to ⅓ time. Extended team—6 × 1/20 time. Users—2 to 6 groups × 2 to 4 hrs. and/or one-on-one reviews × 1 hr.	Variable
Usability testing ◆ Prelaunch ◆ Post launch	From first prototype to final product. Ongoing	0.1 team member for each product. Users—1 hr. × number of users	Variable
User ideas and feedback mgmt. (post-launch)	Ongoing	0.1 team member for each product	Variable

in the way of listening, this means minimizing talking and encouraging users to talk while they are using the product or service.

Resources Required for Wish Mode Processes

Table 9-2 gives NPD teams a sense of what time and resources they can expect to devote to wish mode processes. These resource requirements can, of course, vary widely depending on a number of factors, such as product complexity, organizational culture, and variety of customers.

SUMMING UP

Customer idealized design and the capture of daily–usage ideas are versatile approaches because they can be applied to virtually any kind of product or service. It is easy to start applying wish mode processes, and there are many advantages. To review, here are just a few:

- The wish mode approach helps create connections with users and better understanding of their needs, thus increasing the likelihood of new product or service success.
- This frame of mind creates and opens access to a gold mine of ideas that can help propel the organization to the forefront of competitiveness.
- Providers frequently abandon deeply held assumptions, self-imposed constraints, and rigid positions.
- Customer satisfaction rises dramatically.

This chapter provided a basic wish mode process guidance "template" that NPD teams can use to apply and adapt to their specific situations. When a new product development team drives these processes, it gains access to a wealth of possibilities.

REFERENCES

Cooper, Robert G. 2001. *Winning at New Products.* Cambridge, MA: Perseus Publishing.

Cooper, Robert G., and Elko J. Kleinschmidt. 1998. "Resource Allocation in the New Product Process." *Industrial Marketing Management* 17, 3: 249–262.

von Hippel, Eric, et al. 1999. "Creating Breakthroughs at 3M." *Harvard Business Review* September–October: 47–57

10 The Birth of Novelty: Ensuring New Ideas Get a Fighting Chance

K. Brian Dorval and Kenneth J. Lauer

Developing successful new products and services is the lifeblood of today's acknowledged industry leaders. This development enables them to commercialize new approaches that create or meet customer needs while at the same time generating revenue and financial growth for the organization. However, maintaining a dynamic product and service portfolio is not an easy task, as market conditions and consumer behaviors continually change in ways that produce unexpected opportunities to emerge and then disappear. Success today requires companies to produce new products and services that create or meet opportunities at a faster pace, with more quality, while using fewer and fewer resources. It also requires companies to try to have more ideas make it through the NPD process and be successful than they have ever done in the past.

The best practices of industry leaders around the world provide some insights into what's working and what's not when it comes to NPD. NPD in these organizations is strongest in mapping and managing the process steps and stages that ideas travel through as they are screened, analyzed, tested, and commercialized. These industry leaders use classic Stage-Gate™-style approaches to managing there NPD process. This has helped them decrease the time and cost of getting new ideas to market. It has further helped increase the overall quality of those product or service ideas.

These NPD processes work best when promising ideas are put into them, but where do these "new," "good," or "creative" ideas come from? The answer for most of today's industry leaders is that they come from many sources, both internal (e.g., idea suggestion systems, R&D labs, skunk works, or consumer research) and external (e.g., purchasing concepts, patents, products or even entire businesses) to the organization.

The creation and development of internal ideas into concepts that can be brought to the NPD process occurs in so many different, unmanaged, and often chaotic ways that it is often characterized as "fuzzy." NPD approaches have been exploring this fuzziness, and now it is included in many NPD models as a stage or even a self-contained model called the *fuzzy front end.* These models and ensuing practical processes have helped make the development of ideas into concepts more manageable and repeatable. Many of these fuzzy front end approaches also promote the idea of trying to get closer to consumers and understanding what drives their behavior so that new ideas can be found and developed into products and services consumers are more likely to need (and acquire).

Managing the fuzzy front end creates opportunities to develop entirely new concepts and products that create or meet customer needs and generate financial profits. Insights about customers provide NPD teams with triggers for new thinking that results in new ideas and concepts. However, as important as generating new ideas are, organizations are learning that consumer insights and idea-generation efforts are wasted when standard analytic approaches are used to evaluate the ideas generated. Idea screening approaches (e.g., criteria, business cases, etc.) typically used later in the NPD process are less effective when applied to ideas in the fuzzy front end. They systematically remove novelty from concepts to the point that new ideas become old ideas in new wrappings. This is a critical problem at the fuzzy front end, because if old ideas enter the NPD process, organizations end up spending time, energy, and money developing products and services that look remarkably similar to what they already have in their product portfolios. Although standard analytic approaches to evaluate new ideas do not work well in the fuzzy front end, many people do see them as a better alternative to relying on their personal judgment and intuition.

The purpose of this chapter is to describe and illustrate a fuzzy front end approach called SWIFT. SWIFT is a method that NPD teams can use to help ensure new ideas have a fighting chance to make it through the screening activities of the fuzzy front end. SWIFT is designed for use after many, varied, and unusual ideas had been generated, or consumer insights have been obtained, and some initial screening of ideas has taken place. It helps teams elaborate on promising ideas in a way that keeps novelty alive and enables teams to more effectively support new ideas later in the more formal NPD process.

WHAT TYPICALLY HAPPENS WHEN IDEAS ARE EVALUATED IN THE FUZZY FRONT END?

To understand the normal idea development and evaluation process, consider the typical activities in a concept development session taking place at the fuzzy front end. The process usually involves understanding the consumer/customer needs to be addressed. It involves a team generating or brainstorming many, varied, and unusual ideas that can address a consumer/customer need. Ideas generated are reviewed, and the more promising ideas are identified and separated from the rest. The most promising ideas may then be organized or compressed into groups or themes. These clusters are then described to form concept descriptions that are evaluated for promise or potential before being moved into the organization's formal NPD process. Sounds like a simple and straightforward process, doesn't it?

Now consider your real-world experiences. Think back to the last NPD meeting in which your goal was to produce highly novel concepts. How much time was spent during the session generating many, varied, and new ideas, and how much time was spent evaluating or developing the ideas? What was the balance between generating and focusing? What did you notice about ideas selected from brainstormed lists? Were the ideas highly novel, original, or even slightly silly? Or, were they close to the norm, slightly familiar, or safe? What happened to the ideas not selected?

Some common responses we have heard to these questions are paraphrased in the following:

- We spent most of our time in the concept development sessions generating ideas.
- We came up with crazy and interesting ideas—which was fun, but we actually went past the time on the agenda and had to work with the more "practical" ideas.
- We spent little time, mostly at the end of the sessions, choosing the ideas we wanted to develop.
- Being behind schedule and tired, we ended up choosing ideas from the reams of flip charts that are pretty close to the concepts we came with to the sessions.
- We don't have the time to think, discuss, or evaluate all the ideas.
- Most of the time, the interesting ideas we generate are left on the flip charts.
- It is rare for us to actually choose a silly idea because our time does not allow for the development.
- I wonder what will happen to the ideas we did not select, because we don't want any good ideas left behind.

If your experiences are similar, there is an opportunity to productively impact your NPD work by *getting the interesting ideas off the flip charts and into the development process.*

Getting ideas off the flip chart and into the process requires NPD teams to have time to wrestle with the novelty they created in the ideas in order to make sense of it, because the idea's usefulness probably will not be obvious at first. Teams need to deliberately consider the risks and rewards of pursuing novel concepts for the team, the function, and the organization. Team members need to be protected from the realities, time pressures, and constraints of the organization (at least long enough to figure out the true nature of the novel concept and its implications). They need to be given structures and language to help change their thinking from "what is" to "what could be" as a means to avoid the regression of new concepts to what is currently acceptable or normal. These are the purposes that the SWIFT approach can help you accomplish.

WHAT IS SWIFT?

SWIFT is a creative thinking approach that helps NPD teams evaluate, develop, and strengthen highly novel concepts generated in the fuzzy front end of the NPD process. SWIFT stands for Strengths, Weaknesses, Individuality, Fixes, and Transformation. Since it is a disciplined structure, it also helps teams plan the use of their time more productively to evaluate novel concepts and turn them into something useful. It is based on how highly creative individuals naturally seek feedback on their ideas. When asked what they want from others, people frequently say they want to be told what is good about the idea (its strengths). However, they also want to be challenged to think about the problems with the idea (its weaknesses). They also want input on what it would take to make the idea work (how to fix it).

SWIFT helps NPD teams accomplish three very important activities. First, it helps teams understand exactly what is new about the concept they generated. The Individuality step in the approach forces teams to be explicit about identifying and embracing the new thinking in an idea. This is particularly important because it is often the novel aspects of a concept that scare people away from working on it. Second, SWIFT encourages people to be forward-looking as they evaluate new ideas and concepts. SWIFT helps remove the temptation NPD team members have to use past experience to evaluate new concepts. SWIFT provides a structure for team members to consider what the concept would look like in the future if they *fixed it up* (its transformation). Third, it helps teams make better decisions about and support for the concepts they choose to send through the NPD process. Decisions in the fuzzy front end influence investments made in subsequent stages of the NPD process. Therefore, SWIFT helps teams make better choices about the concepts they choose to invest further resources in to develop.

HOW DOES SWIFT WORK?

Getting the most power from SWIFT requires teams to follow a specific step-by-step sequence of activity in which they systematically evaluate and develop highly novel concepts. The sequence of steps is particularly important because it will mean the difference between novelty having a fighting chance or remaining on the flip charts and stored away in some closet. The sequence of activities will be described in this chapter, but before proceeding with SWIFT, there are a few practical things to consider.

Preparing to use SWIFT

SWIFT structures opportunities in the fuzzy front end for single or multiple teams to engage in creative thinking. To promote success, consider the following points when you are preparing to use SWIFT:

- *Who to involve?* To increase the likelihood of novel concepts being accepted for later development, have a diversity of people and functions present in the group. Also, have key decision makers related to the topic present so that their input is included and decisions about the ideas are made on the spot. A facilitator can be used to manage the application of SWIFT. This gives all the team members the opportunity to focus on the content of the session. However, a member of the group can also manage the discussions and flip charts.
- *How many team members?* Optimal-size teams working with SWIFT are five to seven. When less than five, the team begins to lose the diverse input it needs to stimulate creativity. When there are more than seven people, subgrouping and social loafing begin to take place, as individuals have less airtime to share their ideas. SWIFT can be used with larger groups if they are subdivided into groups of 5 to 7 and results are shared across groups.
- *What are the roles and responsibilities?* Make sure all in the meeting understand their roles and responsibilities.
  - *Client.* A person who "owns" or is responsible for the outcomes of the meeting. The client makes key choices and decisions about the meeting content. However, if the entire group or team is responsible for the topic or outcome of the meeting, then all function as clients and are involved in generating the ideas as well as making decisions about what ideas to carry forward.
  - *Recorder.* Many thoughts will be shared during the use of SWIFT. Establish who will record the ideas, thoughts, and suggestions as they are shared. The recorder needs to write on flip charts all the options exactly as they are shared. It may be necessary to have more than one recorder. The recorder may be a member of the team. However, since recording effectively can be a large task, it may be beneficial to have someone from outside the group to take on this role.
  - *Timekeeper.* Someone on the team should be given the responsibility to monitor time and tell the group when it is time to change to the next topic. A general agenda and time frame for the SWIFT application is provided later.
  - *Team member.* Members of the NPD team are responsible for working in concept development and will share their thoughts and ideas during the meeting.
- *What kind of facilities?* To work on novel concepts, it is best to be removed from the day-to-day reality of the organization. SWIFT works best when people are off-site and less likely to be disturbed by constraints and distractions. The room itself should be large and have plenty of wall space on which to hang flip chart paper. When evaluating a novel concept, the entire team should see all the information discussed and recorded all at once. This allows people to scan information at a moment's notice without leaving their chairs. In the best-case scenario,

the room has windows with shades or curtains to be open or closed, depending on the energy and needs of the team.

- *What are the group norms?* Establish a set of working norms for the group to follow as they participate in the meeting. Talking about novel, ambiguous, or overly complex topics can result in high levels of debate. To help manage this debate, set working norms. The following are examples of effective norms: Keep focused on the meeting objectives, allow everyone an opportunity to speak (one at a time), listen well to each other, seek to understand different points of view, be honest and open-minded, keep to time frames, take responsibility to stay up-to-date with information discussed, and support decisions once they are made.
- *What is the time frame and agenda for the session?* Working with a group of five to seven, plan on taking one to two hours to complete SWIFT for one concept. This will give the group time to apply the approach and interpret its results. If you plan to work with a group larger than seven people, additional time will be needed. The following agenda and time frame can be used in a two-hour format:

ACTIVITY (TIME)

- Select and prepare the concept (15 minutes).
  - Select initial concept.
  - Define the concept.
- Identify concept strengths (15 minutes).
- Identify concept weaknesses (15 minutes).
- Identify concept elements of Individuality (15 minutes).
- Generate ways to fix key weaknesses (30 minutes).
  - Identify top-priority weaknesses.
  - Generate ways to address each one.
- Transform initial concept (30 minutes).
  - Select key information to include into new concept description.
  - Develop updated concept description.

- *What supplies are needed?* In addition to information about the concepts to be evaluated, you will need the following materials to effectively apply SWIFT in a group setting: A flip chart and pads of paper, flip chart markers, 12 medium-point permanent markers, 12 pads of 3-by-5-inch Post-it® Notes and masking tape. When hanging flip chart paper on the walls, make sure they are not covered by each other and be sure to label the top of each page with the information that is contained on it. If the group is over ten people, you should increase the material proportionately.

SWIFT IN ACTION: A CASE EXAMPLE

Learning how to use SWIFT is often best done through experience. Therefore, an example of a SWIFT application will be woven throughout this chapter, in

SWIFT IN ACTION: BACKGROUND AND CONTEXT

A global publishing company was engaged in new product development activities with the purpose of developing and launching new products that would change the way it did business. Its core business focused on developing large collections of information cards that people collected over time. The company needed products that built on efficiencies of its core business but that provided something different to the marketplace.

One division of the organization was considering entering a new and competitive market focused on children. Members of one NPD team conducted research with three groups of consumers containing young mothers with young children. One insight from the research was that mothers needed information to help them play with their children in new and exciting ways—day after day.

This team held a two-day meeting to develop a concept that would meet the consumers' needs. The two-day time frame was selected because the team needed to (a) hear the results of the consumer research, (b) identify insights and key consumer needs, and (c) generate ideas to address core needs. All this needed to be done before being ready to apply SWIFT.

The room was in the office and was selected based on the ability to hang flip charts on walls (so as to see the thinking as it emerged). However, this room was also a challenge in that it was located in the main office where team members could get distracted from interruptions associated with daily business.

The division's editorial director was the client, and the team members were editorial people from the United States, England, France, and Germany (because they wanted the resulting concept to work across cultures). Roles were assigned and operating norms were set at the beginning of the meeting.

the form of sidebars, to help make the concepts more concrete and useful. The example comes from an actual application of the concepts in SWIFT.

Select and Define the Concept

Being ready to apply SWIFT means that certain activities have already taken place. To put this in context, SWIFT is best applied in the focusing activities of the fuzzy front end of NPD. As you can see in Figure 10-1, an NPD team will have already engaged in ideation—typically with tools like brainstorming. There are a number of flip charts hanging on the wall with many ideas recorded on them. This idea generation may have been stimulated with external and internal information about customers or the business (as was the case in the publishing example). Next, the team will have already selected the most promising ideas and turned them into initial concepts. Many of the brainstormed ideas have team member votes indicated on them. The tool used was likely multi-voting, in which each group member got a certain amount of votes (based on the number of ideas under consideration and the number of people in the group). The ideas with the most votes have been clustered into specific themes. Each theme has a short statement describing the essence of the concept

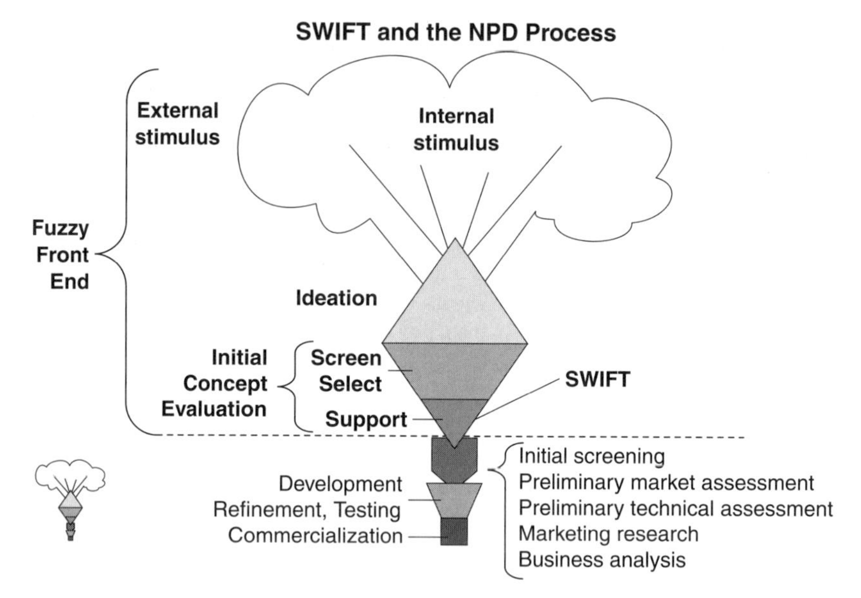

FIGURE 10-1. Locating SWIFT in the NPD process.

that pulls all the individual ideas together. This statement represents the initial concept.

These initial screening and selection steps are important because they have a profound influence on the successful use of SWIFT. As discussed earlier, SWIFT works best to help teams identify and develop the novelty that exists in a concept. It is not designed to put the novelty into a *normal* concept. Therefore, unless the group selects a novel concept, the team will not benefit from the power SWIFT can provide.

There are a number of issues to consider as the team selects and prepares a concept for SWIFT. During the screening and selecting phases, tools like multi-voting can be effective at helping the group agree on what ideas to take forward into concepts. But they do not necessarily ensure that novel ideas are selected. You can fix that by giving each group member a "novelty vote"—that is, a vote in which each person selects a new or unusual idea from the brainstormed list that will automatically be carried forward into the screening and selection process, no matter how many votes on are it. New or unusual means that the idea will result in a product or service that is new to the organization's portfolio or will result in a significant build on an existing product or service. It is not necessary to use an explicit rating scale at this point, as extensive analysis of the ideas has not yet taken place. To ensure novelty votes make it through the clustering and theming activity and onto the concept statement, when writing the initial concept statements, the team is required to use unique language or phrases contained in the new or unusual ideas that were selected by the novelty votes.

SWIFT IN ACTION: SELECTING THE CONCEPT

The NPD team in the publishing company listened to the research and identified key consumer insights and needs. They also generated over 200 ideas for concepts to meet the needs of young mothers and their children. They used brainstorming as the primary creative thinking tool. They also used tools like visually identifying relationships (VIR) and brainwriting (Isaksen, Dorval, and Treffinger, 2000). The challenge was to identify the new and exciting ideas for concepts they could develop into new products for this new market.

The client (also trained as a creative problem-solving facilitator) reminded the group of the core need they were there to address, asked them to select the novel ideas on the list, and asked them to, as much as possible, see the potential in an idea and not to judge it based on how "doable" it was. Each person was given eight votes. Given the strong organizational culture to select ideas that are simply collectable cards, team members were also given two novelty votes each to ensure new ideas with potential were selected.

The biggest challenge the team had was organizing and consolidating the votes. They used a tool called highlighting (Isaksen, Dorval, and Treffinger, 2000) to compress votes into clusters around themes. This resulted in a rough concept statement for each cluster. By underlining the most important words in each idea and using the novel words and phrases to draft the concept statements for each cluster, the team could keep the novel ideas alive.

The result was seven concept statements to choose from. The group used multi-voting again to select which concepts to further develop and refine. It was important to remind the group that those concepts not selected now would be held for later consideration. The initial concept selected was this: *Cards and materials that balance appeal to children with information and directions for ways to play with children for the parent.*

Selecting the concept statement for use with SWIFT is the next activity. The team reads each concept statement and uses multi-voting to make the determination about which concepts to work on.

Define the Concept

Once a concept statement for a new product or service is selected, the team must have a shared understanding of what the statement actually means. Without such clarity, the group will be confused about what is to be evaluated and developed. The confusion then leads to unnecessary tensions and a lack of productivity. The best way to ensure the group has a shared understanding of the concept statement is to conduct a key word definitions activity. Although it sounds simple, it is amazing how often people have different definitions for commonly used words.

To begin the key word definitions activity, write the concept statement on a flip chart and then have the team identify the keywords in the statement. Take each key word identified in the statement and write it on a separate flip chart paper. Now you are ready to define each word with the team. A team member begins by providing an initial definition of the first word. As definitions are given they are recorded on the flipchart. Once the team has run out of definitions, ask the team members to review the definitions and choose the one that is best for the word. The team will use this definition to better understand the concept statement. Be aware that the team will probably want and need to discuss and modify the definition as they attempt to have a shared meaning and to reach consensus.

When the team agrees on what each word means, return to the original concept statement and ask the team members to state what it means to them. Generally, you will find that the team members will have a more clear and shared understanding of the meaning behind the concept statement. If there are still some discrepancies, it will be necessary to work with the team and adjust the concept statement until there is agreement. Once the group understands the concept statement, the group is ready to apply SWIFT. All word definitions must be recorded on flip chart and posted for reference throughout and following the session.

Step 1: Strengths—Identify Strengths or Good Points in the Concept

The first step in SWIFT is to take time to identify the strengths or strong points of the concept. Starting with the strong points helps people open up to the possibility that the concept has some merit. Also, to identify strengths or good points about the concept, you must believe that the concept can be a product or service that can possibly be produced or offered to customers. Shifting to possibility thinking helps limit negative responses to novelty and helps people engage in higher-level thinking processes. The result is higher-quality evaluation.

The following questions help stimulate the team to think about strengths. Record each answer.

- What do you like about the concept?
- Where is the concept strong?
- What does it do that makes it good?

SWIFT IN ACTION: DEFINING THE CONCEPT

Through the word definition activity, the team generated a better understanding of the fuzzy concept it had created. For example, the key words and their definitions are as follows:

- *Materials*. Anything more than cards that could fit in the shipping container.
- *Balance*. Every card would have something for both parent and child.
- *Appeal*. Information and materials are linked to best parenting practices as well as the science of child development.
- *Ways to play*. Not limited to parent-child interaction.

SWIFT IN ACTION: STRENGTHS OF THE CONCEPT

Some of the initial strengths the group identified in the concept included the following:

- It might work because it clearly meets a proven market need for play instructions for parents.
- It might work because it is inexpensive to develop, produce, and sell.
- It might work because potential customers quickly understand the concept; it is easy to convey the idea in marketing material.
- It might work because it can be fast to the market and a low-risk product.
- It might work because it promotes quality time together for parent and child.

- Why will other people like it?
- What is it about the concept that will help it get through later stages of our NPD process?

Invite people to use the phrase "It might work because . . ." at the start of each response. This will help structure their thinking in a productive and positive way. We use the word *might* because we want to create a perception of possibility not certainty. Using the phrase "It will work because . . ." requires a level of understanding that may not be yet available.

Step 2: Weaknesses—Identify the Weaknesses in the Concept

The second step in SWIFT is to identify where the concept does not do well and is weak. This stage identifies what does not appear to work in the concept so that it can be addressed later. It prepares for the developing and strengthening work that comes later in the approach. Identifying weaknesses also helps predict some of the issues or barriers that may emerge as the concept travels through the NPD process. It helps identify reasons why others might reject the concept and prepares the team for addressing them later.

Weaknesses are second because people expect to talk about them. If they had their way, most groups would want to begin with talking about what doesn't work. To test this, share a new concept with a group and see how they respond. Typically, they will begin with why it will not work. The danger with weaknesses is that the group may get carried away with identifying what is bad about the concept. Too much negative energy in the form of weaknesses can remove any hope or perception of potential in the concept.

The following questions help stimulate the team to think about weaknesses. Again, record each answer:

- What problems do you see with the concept?
- Where is the concept weakest?
- What problems will emerge later in the NPD process?
- What problems will other people raise about the concept?

SWIFT IN ACTION: WEAKNESSES OF THE CONCEPT

The team members identified a list of potential weaknesses to the new concept. Some of the key weaknesses included the following:

- It might not work because the children will get bored.
- It might not work because it doesn't feel like a collection.
- It might not work because parents do not perceive value for the money.
- It might not work because the length of series doesn't fit with the age range of the child.

Invite people to use the phrase "It might not work because. . . ." For the same reason, as with strengths, we use the word *might* to help keep thinking less certain and more exploratory.

Step 3: Individuality—Identify What Makes the Concept Novel or Different

Individuality is the part of SWIFT that helps explicitly identify what is new or different about the concept. Individualities are positive aspects about the concept that should be kept, retained, or embellished. What makes individualities different from strengths is that these positive aspects are also characteristics of the concept that are different or new—that are contained only in this concept. Strengths do not have to be unique. This is a particularly important part of the approach because it identifies those aspects of the concept that make it different from all the other ideas being considered. It helps in the future to keep the novelty in the concept visible and observable.

The following questions help stimulate the team to think about strengths. Record each answer.

- What makes this concept different from other concepts?
- What is new about this concept?
- What is the unusual idea in the concept?
- What does this concept have that no other concepts have?

Invite people to begin their responses with the phrase "What makes this concept different is . . ."

Step 4: Fixes—Find Ways to Fix the Key Weaknesses in the Concept

The first three steps in SWIFT focus on evaluating the new concept. Step 4 moves to development. This step is particularly important because new concepts are not often well thought through as to what it would take to make them work. The fixes step provides an opportunity to see what kind of energy

SWIFT IN ACTION: INDIVIDUALITY OF THE CONCEPT

The team members were excited about the concept because they saw a number of key qualities in it that made it different from other products in the portfolio. Some of these qualities included the following:

- What makes this concept different is it is simple enough for older children to teach younger children.
- What makes this concept different is it addresses the needs of parents and can be used by the child.
- What makes this concept different is it uses household items for play and doesn't require a trip to the store.
- What makes this concept different is that it can be used at any time, anywhere.

or effort it would take to make a highly novel concept work. It helps the team understand how much work would be involved to raise the potential of the idea to a degree that makes it worth the effort. In essence, the fixes step helps bring out the usefulness in highly novel concepts. Fixing something makes it more useful or better at meeting a need. It is the combination of high novelty and usefulness that makes new concepts creative and more likely to be accepted by the organization and its customers.

First, identify the three to five weaknesses from the list that would be the most difficult to overcome or have the greatest chance of causing the idea to fail. These are the Weaknesses that need to be addressed for the concept to be considered plausible. To identify priority weaknesses, each team member votes on the top one to two weaknesses using multi-voting. The weaknesses with the most votes are the ones to address. Plan on addressing two or three of the top weaknesses. However, it is possible to address as many of the weaknesses as necessary to make the idea work. Therefore, total consensus on *the* top priority is not necessary. Use the following questions to help the team make their selections about the top priority weaknesses:

- What are the top one or two weaknesses to address?
- What are the greatest weaknesses that stand in the way of success?
- What are the go/no-go weaknesses?

Next, starting with the weakness with the most votes, generate ways to address or overcome the weakness. Use whatever brainstorming or divergent thinking techniques (e.g., Isaksen, Dorval, and Treffinger, 2000; Van Gundy, 1988) are needed to generate the list. The goal is to make sure enough ideas are available to remove the weakness. Repeat this activity until all the priority weaknesses are addressed. Use the following questions to stimulate the generation of many and different ways to address each weakness:

- What might we do to remove this weakness?
- What might be done to turn this weakness into a strength?
- What do we know will work but are afraid to try?

When the group believes it has enough ways to overcome a weakness, turn attention to the next-priority weakness and generate ways to fix it. There is no way to be absolutely certain that enough ideas have been generated to overcome a particular weakness. It is important to remember that working in the fuzzy front end requires the use of hunches, intuition, and even some guesswork. Therefore, decisions can often feel less certain because there is less *data* to consider. More stringent criteria can be applied to the concept later as it passes through different stages of the NPD process. Therefore, the best way to decide about shifting to another weakness is for each person on the team to read through the list of ideas generated and to make an informal rough *estimate* about the question: If we implement each of these ideas, will we remove this particular weakness? Strive for consensus on the answer. If there is big disagreement among team members, spend an additional five to ten minutes generating more ways to address the limitation.

Step 5: Transformation—Describe the New Concept

Up to this point, a lot of work has been done to analyze and evaluate the concept. The group understands its strengths and weaknesses, as well as what makes it different or unique. The team has also identified a variety of ways

SWIFT IN ACTION: FIXING THE CONCEPT

Although the team was excited about the new concept, they realized it needed to be more novel in the product portfolio. Therefore, they truly needed to invest time identifying ways to fix the key weaknesses in order to make the concept stand stronger and be more resistant to suggestions for change others were likely to make that would remove its uniqueness. However, given the amount of time it took to review the research, the team ran out of time at this meeting to explore ways to fix (develop and strengthen) the initial concept they created. This needed to be done in a follow-up meeting.

Two key fixes were chosen from the list previously generated, and ways to ideas to address each were generated. The fixes and their ideas follow. Notice that the team reworded each weakness beginning with the statement "We need to . . ."

FIX 1: WE NEED TO MAKE THE PRODUCT FEEL MORE LIKE A COLLECTION

- Introduce character cards that can be collected over time.
- Use emotional and rational justification for completing the collection in a letter to parents–focus on child development.
- Use numbers, seasons, and days of the week as collection formats.

FIX 2: WE NEED TO ENSURE THE COLLECTION PROVIDES VALUE FOR MONEY

- Add to the instruction cards some that are usable by the child (like color in cards).
- Provide extra supplies children can use when interacting with the cards, such as stickers.
- Make the envelope the product is shipped in part of the "materials" parents and children can use during play.

that the concept can be improved, developed, or strengthened. If the team finishes working on the concept at this point, there is a high risk that each team member will leave the meeting with a different understanding of what the current concept is or how to describe it. Something needs to happen to synthesize all the best thinking from the team and integrate that thinking into one new and improved concept statement. This is the purpose of the transformation step.

The transformation stage of SWIFT is designed to help you integrate the key insights and changes resulting from the evaluation and development work into one new summary of the concept. It is the forward-looking step in the approach, as it encourages teams to look to the future and consider what the concept would be like if all the improvements were implemented. Therefore, it is important that teams do not view this step as simply repeating what has been stated in each of the earlier steps. Transformation is exactly as its title suggests: It is the total integration of new thoughts and ideas about the concept in a way that transform it into something else. These changes may be incremental builds or improvements on the original concept, or they may be fundamental changes to the original concept. In either case, the concept should be transformed as a result of all the thinking that has just taken place, and that transformation should be recognized deliberately and explicitly in this stage of SWIFT. To make the transformation step explicit, team members should do the following:

1. Review the information recorded during the SWI steps to ensure they understand the key thoughts generated about the concept overall.
2. Use multi-voting to identify the most important ideas generated for addressing the priority fixes. Those ideas that receive the most votes will be included in the updated concept statement.
3. Check to confirm that concept changes they recommend do not take away the novelty in the original concept description. Do this by having the group check the recommended changes against the list of Individuality statements (one at a time) to make sure the uniqueness will remain if the change is made. If the recommended change will take away an element of individuality, choose another recommendation from the list of ideas for the priority fixes. This activity is particularly important because the team will be tempted to make recommendations that fix the concept by removing its individuality. If they do this, the result will be concepts that are stronger but not very unique.
4. Redraft the concept description with the integration of the changes. Start with a blank sheet of paper when drafting this new concept description so that the team is not confined to the original thinking. Dividing the group into subgroups of two to three people is the best way to accomplish this. Each subgroup develops a draft concept description on flip chart paper. Beginning with the phrase "The concept is to . . . ," each subgroup writes a five- to seven-sentence para-

graph description of the concept by weaving together the original concept statement, along with the items identified as important to put into the final concept description. This includes the novelty votes identified earlier. Subgroups share their descriptions with each other to prepare for integrating the two statements into one.

5. Once the drafts have been shared, the total group chooses one description that best presents the concept and will function as the base concept. Then the total group identifies unique phrases and elements in the other description and weaves them into the base description. It will be tempting for the team to engage in wordsmithing. This often results in wasted time arguing over words that do not make a significant difference in the overall meaning of the concept. The goal is to get 90 percent of the description and language correct—not to make *the* perfect description.
6. Once the new statement has been created, check it against the list of elements of individuality to ensure the new elements of the concept survived. If for some reason the concept's novelty has been lost in the new description, take the time now to add ingredients into the description that made the original concept unique in the first place.

What Happened to the Children's Play Concept?

The company decided to pursue the concept to the next level. This involved actually developing and testing the product with a small population of customers. The product performance was assessed against the current benchmark product used for all comparisons, and it significantly outperformed that benchmark. In fact, the product's performance was so good that there were some initial challenges with distribution of the product because the company had underestimated the potential response from the test market.

SWIFT IN ACTION: TRANSFORMING THE CONCEPT

As is evident in the description of the transformed concept that follows, the team identified key ingredients in the original concept description, and from the SWIs and Fs it identified, and integrated them within an updated description. In particular, by including new ideas that resulted from fixing the weaknesses, the team has developed the concept further.

> Collectable character cards designed to appeal to children on one side and provide parents with ideas they need to foster play on the opposite side. There is added value in the form of interactive and consumable "treats" providing instant gratification for the child as well as child development information for the parent. The cards and materials are organized by seasonal topics such as winter and summer, with each topic including usable games and activities.

SWIFT was one of the key tools that helped this publishing company create a novel product and test a new marketplace. Today, they have a developing part of product portfolio aimed entirely at the children's market.

TRY SWIFT FOR YOURSELF

One of the best ways to understand how SWIFT works is to try it on something. For this activity to work, you will need to have a novel idea or an unusual concept for some new product or service. Ideally this new concept would be so new that is sounds a bit silly or strange. At the same time, it needs to be a concept that has some form of intrigue or some seed of plausibility. Or, for this demonstration, you can just use the description of a new concept that follows.

Initial Concept

As a scientist in an R&D department, you recently developed a "low-tack" substance made of very tiny spheres that only stick when they are tangent to a given surface, rather than flat up against it. Your hope was to make a substance that would keep objects firmly together over extended periods of time. However, objects held together with the substance can be separated with little force or effort. These nearly indestructible acrylic spheres do give the substance the ability to be repositioned multiple times over. The clear substance can also be distributed through a variety of means, including spray or roll-on processes.

Preparation: Select and Prepare the Concept

On a piece of paper, write a three- to five-sentence description of the concept. Focus the description on what the concept is and provide a high level understanding of how it will work. The description should stimulate a strong mental image about the concept. In this activity, SWIFT will be applied in the same way that it was described previously. To capture thoughts, set up two pieces of paper with the SWIFT structure. Figure 10-2 provides an example of what the worksheets should look like.

STEP 1: IDENTIFY THE STRENGTHS OR GOOD POINTS IN THE CONCEPT Think about the concept from the perspective of what is likable about it. Identify seven to ten strengths, strong points, or places it has value. Write these items in the space labeled Strengths. Remember to start each response with the sentence starter, "It might work because . . ."

STEP 2: IDENTIFY THE WEAKNESSES IN THE CONCEPT. As we said, no idea is perfect. Think about the concept from the negative perspective and identify its

SWIFT TEMPLATE

Strengths. What are the strong points in the concept?

It might work because . . .

-
-
-
-

Weaknesses. What are the weak points or concerns in the concept?

It might not work because . . .

-
-
-
-

Individuality. What makes this concept different or new?

What makes this concept different is . . .

-
-
-
-

Fixes. Identify the top weaknesses and identify ways to address each.

Weakness 1:

-
-
-
-

Weakness 2:

-
-
-
-

Weakness 3:

-
-
-
-

Transformation. Describe the new concept with changes integrated.

The concept is to . . .

FIGURE 10-2. SWIFT worksheet.

major weaknesses. Be sure to generate seven to ten weaknesses statements using the phrase, "It might not work because . . ."

STEP 3: IDENTIFY THE INDIVIDUALITY, WHAT MAKES THIS CONCEPT NOVEL OR DIFFERENT. Reflect on the description of the concept and the images it triggers, and identify what is different or new in the concept. Search for elements that are unique, or aspects of the concept that make it stand out among the pack. Write three to five statements of individuality in the space provided on the SWIFT worksheet. As mentioned before, it is not necessary to come up with large numbers of these.

STEP 4: FIND WAYS TO FIX THE CONCEPT AT ITS WEAKEST POINTS. Look back through the list of identified weaknesses. Select the top three to five that are critical to the potential success of the concept and circle them. Now rank these weaknesses in order from greatest based on the impact addressing each would have on the likelihood of success with the concept.

Next, focusing on the weakness with greatest potential impact, generate seven to ten ways to address it. When there are a sufficient number of ways to fix the weakness, then move to the next weak point and generate seven to ten different ideas to address it. Repeat this for each weak point that has been identified.

STEP 5: TRANSFORM THE CONCEPT. To transform the concept, begin by identifying what can be seen as the key changes or improvements that can be integrated into the concept. These can come from any of the SWIFT categories but will most likely come from the Fixes category. Once identified, circle them.

Next, write a paragraph in the space under the title Transformation describing the concept in a way that integrates these changes. The new statement should be three to five sentences long—longer if it really needs to be. Remember to begin the description with the phrase, "The concept is to . . ."

Finally, check the new description against the items identified in the Individuality section of the SWIFT worksheet. Make sure the new description captures the new aspects of the original concept. If not, rewrite the description to include these new concept ingredients.

REFLECT ON WHAT TOOK PLACE. Now that you have had the chance to apply SWIFT, take a few minutes to make some observations. For example, what have you learned about the concept that you did not already know? What surprises did you have about the concept? Now that you have gone through the evaluation, what might you do differently when talking with others about the concept.

INTERPRETING SWIFT RESULTS

There are some general interpretations that can be made about a concept as a result of using SWIFT. For example, what would it tell you about your concept if, as a result of using SWIFT, it had a large number of strengths, few major weaknesses (with ways to fix each), and a few key elements of individuality? Your interpretation of that concept would be markedly different than if you were able to identify many more weaknesses than strengths and had great difficulty identifying items of Individuality. The former result would suggest your concept is solid with some elements of novelty, along with aspects that need improvement. The latter, on the other hand, would suggest a weak concept that is not very new or different, and therefore it would add no new value to your business.

If you applied SWIFT on the initial concept we provided, you may have recognized what it described. It is the initial concept behind the adhesive used to make Post-it® Notes. This glue was outside the norm of what 3M typically thought about as adhesive. And if it were not for the open minds and persistence of a few key individuals, the novelty in this new kind of adhesive would have been overlooked completely.

- Clarify the frame of reference for evaluating the concept before applying SWIFT.
- Create a template for the team to use as they apply SWIFT.
- Structure each response in seven to ten words.
- Generate Strengths, Weaknesses, Individuality, and Fixes using Post-it® Notes.

FIGURE 10-3. SWIFT techniques

TECHNIQUES FOR MAKING SWIFT WORK

SWIFT is a versatile approach that can be used personally, one-on-one, in small groups or in large groups. To get the best from its use, consider the following four techniques (see Figure 10-3). They will help you more successfully apply the approach based on the number of people involved in its application.

Clarify the Frame of Reference for Evaluating the Concept before Applying SWIFT

Concepts at the fuzzy front end often carry with them novelty, complexity, and ambiguity—three characteristics that make them difficult to effectively evaluate. Teams get in trouble evaluating new concepts when they do not have a shared understanding of the frame of reference to be used to judge the concept. Without a common reference point, there is increased debate about what is a strength or a weakness, what is unique, and what needs to be fixed. However, this debate is caused by inconsistent reference points rather than from differences in opinion about the concept. For example, one team member might identify a particular element of individuality in the concept, while another person disagrees. The disagreement might stem from the fact that one person's frame of reference was *new to the company*, while the other's was *new to the marketplace*.

One way to help ensure effective team evaluation of novel concepts is to make sure there is a common set of reference points before actually engaging in SWIFT. During the Select and Prepare the Concept step, the group needs to talk about the context around the concepts to be addressed. The discussion should include topics like the organizational strategy or reason for developing the concept, the customer need being addressed by the concept, who the target customers are, the general qualities in the concept (radically different concepts or improvements to existing products or services), and the general time frame for working on it. Without such clarity, it will be difficult to have high-level evaluation of the concept or to create consensus about the evaluation provided by SWIFT.

Create a Template for the Team to Use as They Apply SWIFT

It is natural for new ideas to be refined and improved as the work on them continues, since no idea is perfect, particularly highly novel ideas. The challenge is to make sure that the novelty generated in the original concept survives the development and strengthening process. If the group is not careful, "fixing" the concept can result in eliminating its individuality. This is the reason why the categories of SWIFT are sequenced the way they are. The team identifies the individuality in the concept before it generates ideas to fix its weaknesses. Reversing the order could easily result in a team generating ideas to *fix* the concept that actually take its novelty away. If this happens, the team will lose the excitement they felt for the original concept and walk away from a meeting with the same old concepts.

To ensure the best application of SWIFT, develop a template with the categories on prelabeled flip charts and hang them on the wall. Each SWIFT category should have its own flip chart page. Then, when filling in the template, start with the S (strengths) and work through each letter all the way to T (transform). This will help ensure that what is individual or unique is made visible before the group begins fixing the concept.

A template will also help organize the information generated during the meeting. In a one- to two-hour application of SWIFT, a team of five to seven people will generate a large quantity of thoughts, opinions, and recommendations about a concept. If this information is not organized well, a team can get confused, lost in details, or overwhelmed with the sheer quantity of information they generate. Therefore, it is important that the information be recorded, organized, and tracked so that team members can refer back and forth to the information being generated to help with the evaluation and development. The template allows the information to be organized as it is generated.

Structure Each Response in Seven to Ten Words

It will be tempting for team members to elaborate on strengths and weaknesses to different levels and degrees. Some will provide one-word answers, while others provide mini lectures. Both of these situations will make the application of SWIFT less productive. Too little information makes it difficult to remember the original thought. Too much information takes too long to share and will likely bog a team down in unnecessary detail.

To make SWIFT work well in a team setting, team members should share their suggestions in the form of phrases about seven to ten words in length. This will challenge people to be clear and concise in their thinking while at the same time keep the process moving. In some cases, a person will not be able to be that concise unless he or she "talks about it out loud for a moment." This works as long as it does not turn into a soliloquy. However, once he or she has finished, the person should summarize the thought into a phrase of seven to ten words. If the person has difficulty, a team member can offer suggestions to capture the key thought. This will help ensure timely application of SWIFT while ensuring good records of the thinking that took place.

Generate Strengths, Weaknesses, Individuality, and Fixes Using Post-it® Notes

It was mentioned earlier that SWIFT is best used in groups of five to seven but that it can be effectively applied in larger groups as well (between 10 and 15). In situations where the group is large, it will be difficult for one or even two group recorders. People will be waiting to contribute as the thoughts of the last person are being written. It takes a lot of time to capture everyone's input. Also, likely some ideas will be lost as people are waiting for the recorder to finish writing.

Therefore, what can be helpful when using SWIFT in large teams is for team members to record their own thoughts on Post-it® Notes. It works like this. Team members record their suggestions on a Post-it® Note (one thought per note). Then they read it aloud to the group so others can hear their thought and build on it if they choose. They then hand the Post-it® Note to the group recorder, who places it on a flip chart. This enables the team to better use the four guidelines for brainstorming (defer judgment, strive for quantity, freewheel, and seek combinations) to generate large numbers of strengths, weaknesses, individuality, and fixes. However, in these situations, we do not need large numbers of individuality statements, because a concept does not need many elements of individuality to be considered unique or different.

Pitfalls to Avoid with SWIFT

SWIFT was put together to make it as simple to use as possible while maintaining the power of its application. However, there are a number of issues that can emerge that will have a negative impact on its application. For example, SWIFT will not work well in the following scenarios:

- *The concept is not clear or solid enough to be analyzed.* As a result, the team does not have a clear topic on which to apply its creative or analytic thinking. This diffuses the group's energy and attention, resulting in a type of mental wandering, a less concentrated thought process, and a lower quality output. Make sure the concept is clearly articulated to the team and that the team has a solid understanding of what is meant by the concept.

- *You respond to the fixes before the novelty is identified.* SWIFT is designed to help bring novel concepts into existence. One of the key elements in SWIFT that does this is the individuality step. It functions as a safety net to ensure novelty is identified and nurtured. However, people are natural problem solvers. NPD teams may try to jump from the weaknesses step to the fixes step without taking the time to deliberately identify the individuality. It is critically important to make sure teams complete the "I" step before trying to fix its limitations.
- *The team energy is deflated because there are too many weaknesses in the concept.* If a concept turns out to have a massive amount of weaknesses, it can take the energy right out of the group. If this happens, it is important to emphasize the positive aspects of knowing about the quantity of problems with the concept up front. It helps the group make conscious decisions about investing time, energy, and resources. Deciding that a concept is not worth the energy up front is a powerful outcome of using SWIFT.
- *One list of fixes is created to overcome all the weaknesses of the concept.* A team may be tempted to generate ways to fix a concept while thinking about all the weaknesses at once. Although this may be quick, it typically results in less focused and less specific outputs and makes the development of the concept transformation statement more difficult. It is important that the group generate ways to overcome the weaknesses, one at a time.

BEWARE OF CHALLENGING SITUATIONS

SWIFT is a powerful approach, but it is not magic. It requires certain conditions that need to be in place for it to deliver value to NPD activities. To get the full benefit from SWIFT, avoid its use in the following situations:

- *There is no clear ownership for the concept under consideration.* Applying SWIFT on a concept that no one owns will not only waste the time and energy of the team and the organization, it will frustrate people involved and make them less eager to participate in future NPD activities. Establish clear ownership for the concept before convening a team to work on it.
- *There is no clear need being addressed by the concept.* It is difficult to analyze a concept when the problem it was created to address is unknown, unclear, or uncertain. The concept becomes an idea in search of a problem and therefore makes it difficult for the team to have an effective frame of reference when analyzing and developing the concept. Avoid using SWIFT, or delay its use until the need for the concept can be identified or clarified.

- *The team climate is not supportive of creative thinking.* SWIFT involves the use of creativity, and creativity requires people to be engaged. If the context is such that people are tense, frustrated, or mentally distracted in some way, delay the use of SWIFT until these issues have been resolved or at least temporally relieved. Or engage in some form of conflict resolution activity prior to applying SWIFT.
- *There is no actual need for novelty.* SWIFT is designed to help teams develop novel concepts in situations that are ambiguous and complex (as is the fuzzy front end of NPD). There are situations were the problem is clear, the concept to address it is clear, and the pathway to implementation is known. In these situations, where there is no clear need for high levels of novelty, it is unnecessary to apply the power of SWIFT.

CONCLUSION

Bringing novel ideas into existence can be an exciting and rewarding process. It can be a challenging one as well. It takes deliberate and explicit attention focused on managing our natural reactions to novelty to give new ideas a fighting chance. SWIFT is an effective approach for NPD teams working in the fuzzy front end. In addition to the benefits identified in Figure 10-4, SWIFT can be used to create an incubator for developing and strengthening novel ideas, as they are fragile when first conceived. And because they are novel concepts, they may not fit within an organization's existing structures and systems. They need time to grow and mature, because they are likely to have a bumpy ride as they pass through the remainder of the NPD process itself.

SWIFT provides an opportunity to refrain from letting ideas enter the NPD process too early or in a fragile state, since the process will generally chip away all the rough edges of the concept (the edges that made it unique) so that it "fits." For some teams, this will feel like putting a square peg in a round hole. SWIFT will help ensure the "square peg" has enough strength and integrity to survive.

As novel concepts gain strength and momentum, they begin to take on a life of their own. NPD teams may need to be careful what they ask for, because truly novel concepts become contagious, attracting interest and attention of

- Provides a fast way to evaluate, develop, and strengthen novel concepts.
- Manages premature judgment that often kills new ideas.
- Increases buy-in and acceptance of new ideas.
- Helps develop a climate for creativity.
- Provides a common framework for people to evaluate concepts in groups.
- Increases the likelihood that novelty survives as it passes through different NPD process stages.

FIGURE 10-4. Benefits of using SWIFT

more and more people as they progress through the NPD process. It is an exciting process. People have been known to delay their retirement in order to stay involved in the development of a highly new concept because they want to make sure that their new idea does indeed have a fighting chance.

REFERENCES

Engelbaart, D. C. 1962. *Augmenting Human Intellect: A Conceptual Framework* (SRI Project No. 3578). Menlo Park, CA: Stanford Research Institute.

Griffin, A. 1997. "PDMA Research on New Product Development Practices: Updating Trends and Benchmarking Best Practices." *Journal of Product Innovation Management* 14, November 6: 429–458.

Haman, G. 1996. "Techniques and Tools to Generate Breakthrough New Product Ideas." In *The PDMA Handbook of New Product Development* by Milton Rosenau et al. New York: John Wiley & Sons.

Isaksen, S. G., and D. J. Treffinger. 1985. *Creative Problem Solving: The Basic Course.* Buffalo, NY: Bearly Limited.

Isaksen, S. G., K. B. Dorval, and D. J. Treffinger. 1998. *Toolbox for Creative Problem Solving: Basic Tools and Resources.* Dubuque, IO: Kendall/Hunt.

Isaksen, S. G., K. B. Dorval, and D. J. Treffinger. 2000. *Creative Approaches to Problem Solving: A Framework for Change.* Dubuque, IO: Kendall/Hunt.

Isaksen, S. G., K. J. Lauer, G. Ekvall, and A. Britz. 2000–2001. "Perceptions of the Best and Worst Climate for Creativity: Preliminary Validation Evidence for the Situational Outlook Questionnaire." *Creativity Research Journal 13*, 2: 171–184.

Langer, E. J. 1989. *Mindfulness.* Reading, MA: Addison-Wesley.

Osborn, A. 1953. *Applied Imagination.* New York: Charles Scribner's Sons.

Rogers, E. M. 1995. *Diffusion of Innovations*, 4th ed. New York: The Free Press.

Van Gundy, A. B. 1988. *Techniques for Structured Problem Solving*, 2nd ed. New York: Van Nostrand Reinhold Company.

Part 3

Tools for Managing the NPD Process

This ToolBook 2 part provides tools to be applied by project teams or their leaders to improve the performance and eventual outcome of NPD projects. The four chapters in Part 3 offer a diverse set of effective practice NPD tools. The part begins with tools for better quantifying the value of product features. Effective practices are presented to better integrate product requirements into NPD projects, followed by a tool to enable customers to design their own products through user toolkits. The part ends with a wealth of effective practices to successfully implement IT solutions in NPD.

While much of the NPD literature is focused on consumer markets, Chapter 11 is focused on the business market, which is much more focused on economic factors than consumers generally are. This chapter provides a step-by-step methodology for the construction of a logical, convincing economic business case for a new product.

Chapter 12 offers a number of proven techniques to enable an NPD project team to develop accurate, better-understood product requirements, resulting in a more focused product development project. From obtaining top management buy-in to validating customer requirements to generating product requirements and then testing them, this 13-step process will guide a team through this frequently clumsy and misunderstood NPD activity.

An emerging technique in NPD is the active involvement of the end customer in the NPD process itself. Chapter 13 introduces the concept of user toolkits, which allow the actual end users of a new product to design their own new products. This is not an insignificant exercise, as an organization must provide access to its capabilities in an economical means while simultaneously protecting its proprietary assets. Using a number of case study examples, the author provides excellent insight into this evolving practice of NPD. For those organizations already involved in these activities, this chapter offers plenty of effective practices. For organizations that have not tried this practice, the chapter offers enough insight to allow them to get started.

The last chapter in Part 3 addresses an ongoing hot topic: how to successfully implement IT solutions in NPD. Chapter 14 first offers a framework to define and evaluate IT-enabling NPD solutions. It then offers many practical tips for successful implementation. This chapter stresses the importance and resulting value of considering the NPD process in its entirety when you are assessing potential IT solutions. IT solutions for subprocesses rarely deliver their promised value unless the impact on the overall NPD process is considered first. This chapter is a must-read for those looking to better manage the NPD process and all of the data it generates.

11 Establishing Quantitative Economic Value for Product and Service Features: A Method for Customer Case Studies

Kevin Otto, Victor Tang, and Warren Seering

This chapter presents a method to establish the quantitative monetary value for new product features and performance required for prioritization, selection, and trade-off decisions. The method is targeted at product development managers and engineers engaged at the front end of the product development process. The front end is when many critical product development decisions are made about the relative importance of product development activities, what to do first, and what can be traded off. Using a customer case study approach, the method presented in this chapter focuses on a customer's business operations and essentially establishes a business case for a product and engineering effort down to the feature and performance levels. It is shown that this approach is equally effective for high-technology services. This chapter discusses the why, the what, and the how of the method. The why is addressed in the next section. The what is shown with two examples from industry. These examples help develop an intuition and understanding of the method. Finally the how, which is the implementation of the method, is discussed in detail.

WHY THIS METHOD?

First, the method is designed for products targeted at business markets rather than consumer markets (Anderson and Narus, 1999). The distinction is important. Consumer buying behavior is affected by psychological factors. On the other hand, in business markets, the fundamental criteria to commit to a product or a service are grounded on economic factors. For customers in these markets, the question is this: "Will this product or service help my business or organization make money and/or save money?" If so, the product or service is one to consider as a candidate for a procurement decision. In these markets, companies have formal procedures and groups whose mission is to evaluate the value of a product or service, where value is measured in quantitative monetary terms. Emotional satisfaction and impulse buying, characteristics of consumer markets, are not widely accepted practices for businesses. A focus in this chapter is the monetary value of products and services for business markets, because they exert such a strong influence on engineering decisions.

Second, this chapter clarifies the distinction between price and economic value, which is of critical importance in business markets. The approach presented in this chapter uses economic value, which is quantifiable in monetary terms, as a key factor for product development decisions. Price and economic value are not synonymous. It is entirely possible that for some products a low price can mean high value. And a high price can mean more value than a lower-priced product. Having an approach that can put these ideas into practice by means of actionable processes is a key contribution of this chapter. The significance of this approach is that engineering managers can have economic value, in addition to price or costs, as decision variables. Because economic value is expressed in monetary terms, the approach presented in chapter can be easily used for an adapted return on investment (ROI) calculation. Engineering managers now have another approach and evaluation method they can use in development decisions.

Third, the method presented in this chapter is both an engineering manager's product planning tool and a business case development tool. The method very directly links the product's features, functions, and performance to the economic value that a customer can derive by using the product. Although many business case analysis methods for new products provide the means to estimate monetary value, their focus is on estimates of sales, cost, expenses, and profit, not on economic value. Moreover, most approaches are divorced from technical feature levels. This chapter addresses this void. Current approaches to understanding the voice of the customer are weak on quantifying the benefits of new product features. House of Quality (Cohen 1995) and KJ approaches (Language Processing Method 1995) are good for identifying what customers seek in a product. They do not, however, assess the monetary value of these needs. Conjoint approaches (Green and Srinivasan 1990) do provide quantitative value for features but are complex to administer. Cook and Wu (2001) use a price and a product attribute valuation function. We have chosen to focus on the economic benefits that a product attribute can deliver to the customer rather than to the producer.

Fourth, the method presented also helps identify whitespace opportunities. A *whitespace* is an opportunity where needs for a new product and/or service are not being served by any current product. Because the methodology presented in this chapter is fine-grained, the whitespace opportunities are resolved into clear and actionable product development projects. The approach is particularly well suited to uncovering new whitespace opportunities. Whitespace opportunities are simply difficult to identify. This chapter will help in this problem, particularly at valuation. Once identified, whitespace opportunities are also difficult to exploit, even in the best of situations, without internal and external organizational changes to deliver the new solutions. Most R&D organizations are not capable of making required changes of technology. Most businesses are not capable of making required changes in services. Even more difficult are solutions requiring organizational changes by your customer to make use of the efficiency gains enabled by the proposed whitespace solution. Nonetheless, driven by the market, all such changes are often ultimately inevitable. The issue is how soon they can be determined relative to the competition. The approach here is well suited to identify and develop these opportunities early.

Fifth, this method is also a strategic tool. Rather than competing on price, a firm can elect to compete on value (Kotler and Armstrong, 2001). For example, Intel competes on value by increasing functionality and computational speed, although the price of its processors is continually increasing. Intel's customers obtain more value because their gain in productivity, measured in monetary terms, overwhelms any price increment. Although it is the "high-priced product," Caterpillar competes successfully because it produces more value for customers than its lower-priced competitors. The approach presented in this chapter shows how to uncover and quantify which benefits a customer will additionally pay for. How to develop a quantified value proposition is shown in detail with two product and services examples in the context of customer applications. Thus, engineering managers can avoid the siren song of spending R&D dollars on features, functions, or performance, which may be technically elegant but have low economic value when put into use by customers.

Sixth, the method presented in this chapter works equally well for physical products as well as services—that is, nonphysical products. This chapter demonstrates this case and discusses why this is significant.

VALUE IS NOT PRICE

During the course of this work, customers had this to say:

"More expensive can be cheaper in the long run."

"For something more, we'll pay more."

"We always want to meet with the highest bidder to understand the business benefits that explain the higher price."

How can more expensive be cheaper? Why would a customer be motivated to meet with a highest bidder? Customers see a sharp distinction between value and price. The producer of the product sets the price. The customer determines the value. Price is the sum of the product cost and a profit margin for the producer, all expressed in monetary units. Customer value, on the other hand, is the monetary worth of the benefit stream accrued by the customer that is

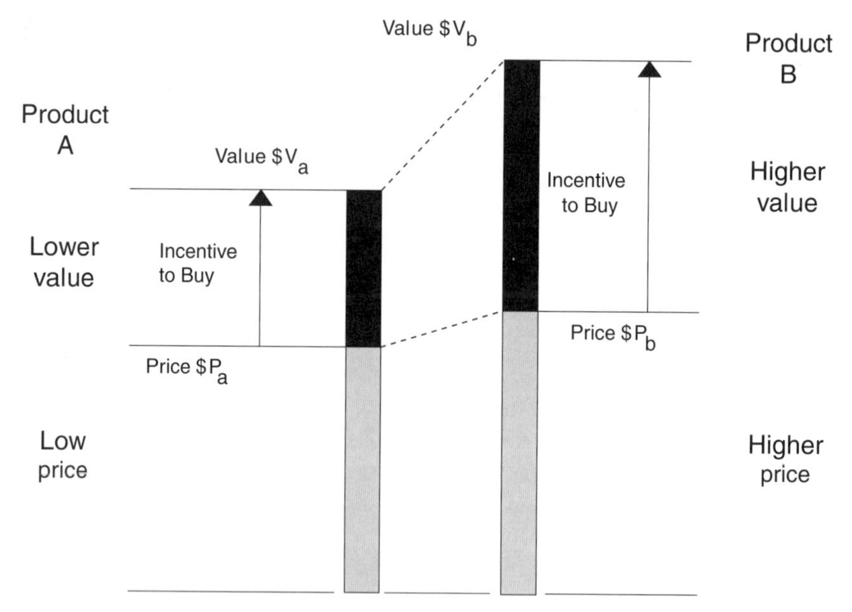

FIGURE 11-1. Value is not price.

enabled by a product's features and performance. The difference between customer value and price determines the customer's incentive to buy (Anderson and Narus, 1999). Given two competing products, the one with the larger incentive to buy will be the winning product.

Consider Figure 11-1. Product A, given its features, can deliver value V_a to the customer. Product B will deliver a different value V_b to the customer. Note $V_b > V_a$; B's incentive to buy exceeds that of product A, even though B has a higher price. Customers will prefer product B in spite of its higher price[1]. This is the meaning of "expensive can be cheaper" and "for something more, we will pay more." The lessons to product engineers are these: (1) Do not confuse price with value, (2) design for features and performance that will generate a value stream that can reduced to monetary terms, and (3) customer value must be quantifiable in monetary terms; otherwise, it is a vacuous sales argument.

As an illustration of the ideas presented, consider an example adapted from Kotler (1991) for Caterpillar. As shown in Figure 11-2, its price is $30,000 higher than its leading competitor, but it has a larger incentive to buy. Relative to its competitor, Caterpillar is able to produce the benefits listed in Figure 11-2 that have the estimated monetary value shown. For example, its durability, which results in a longer product life for the customer, produces a savings of $15,000. Its superior design and product quality requires less repair and purchase of parts; this produces a savings of $5,000. Altogether, as shown in the figure, the benefit stream produces a value of $40,000. On the competitor's base price of $100,000, Caterpillar offers an incentive to buy, which has an incremental value of $10,000. This explains why Caterpillar consistently wins in the market.

Caterpillar Tractors	
Relative Price	
◆ Caterpillar	$130,000
◆ Competitor	$100,000
Difference	**$30,000**
Relative Incremental Value	
◆ Longer usage [durability]	$15,000
◆ Less maintenance [higher quality]	$10,000
◆ Less downtime [superior quality]	$10,000
◆ Fewer replacements [superior parts]	$5,000
Difference	**$40,000**

FIGURE 11-2. Caterpillar tractor value.

EXAMPLES: FORTUNE 500 TELECOM AND FORTUNE 500 SERVICES

To demonstrate the approach, two examples are presented. The first is a Fortune 500 telecommunications systems provider of telephone and IT networks. The second is a world-class information technology (IT) services provider.

Consider the Fortune 500 telecommunications systems provider of telephone and IT networks. The company has a long history of providing communications equipment and network infrastructure, traditionally as a hardware manufacturer and supplier, where, historically, technical performance and reliability allowed the equipment to sell itself. The company was organized horizontally around technical hardware, from small to large systems. This strategy then gradually divorced the end users from the company, with value-added resellers purchasing a bulk of the equipment. With increased competition, the company found itself disconnected from the users in the market, and it lacked a strategy for innovation.

It was not clear, for example, how important reliability really was to the customer. Everyone wanted reliability, but the market was becoming much more focused on price at the expense of reliability. Was 99.999 percent reliability really that important? The same question arose on technical features. Was caller ID important? How is it really used in a business context, and how could it be improved? Was wireless data capability important? Was integrated Internet Protocol (IP) and voice telephone services what the customers wanted, and if so, why? What other R&D opportunities existed for telecommunications that were not being explored—whitespace not being addressed in the market? What level of R&D investments should be made where?

Consider the Fortune 500 services provider of IT consulting and implementation. A similar situation prevailed with the IT services provider presented in this chapter. This company has a long history as one of the best engineering companies of computer hardware products. Recognizing the importance of services, the company created a division for IT services. However, given this company's product culture, it developed a bias to hardware installation, maintenance, and repair services. It acquired an image as very competent "plumbers" that were unable to see how "good plumbing" and excellent products made a difference to customer's business. Why is a new IT service procedure in the organization important for a customer's core business? Unable to explain the business value of their services to their customers, they found themselves competing on price. They became increasingly disconnected with the customer's executives and decision makers. They found themselves consistently excluded from key customer planning meetings where it was important to explain the benefits of excellent products and services, not only in engineering terms, but also in monetary terms. In spite of these weaknesses, this company learned how to bridge technology and customer value. They established formal communications with key business executives to understand the business problems they were trying to solve. They used the economic valuing method presented in this chapter. This process also provided additional reinforcement of the importance of product development to concentrate on features and performance that very directly produce customer value.

A dominant theme of this chapter is this: When making R&D investments, whether extensions to current products or whitespace opportunities, it is key to quantify the value that any new features will provide to a customer as demonstrated by measurable operational improvements. This approach is useful for determining what new white space opportunities exist within a market segment.

Approach and Methodology

The approach can be summarized in six steps, shown in Figure 11-3. The first step process is to identify the market in order to select the customers to serve. The customer's business processes are then modeled in detail as to their business operational activities and flow of value-added goods and information. This result is then analyzed to identify the business difficulties the customer is experiencing and the pain these difficulties inflict at various levels in the customer's organization. The general solutions to the pains in the customer's business activities are generated, as a concept generation activity. The impact these solutions have on the customer's business activities is analyzed. From this, a business case that the customer would make is generated, thereby establishing the value of the solution to the customer. Each of these steps is now presented in detail and illustrated. The approach and methodology is applied first using a case study of a retail customer that uses products from a Fortune 500 telecom provider.

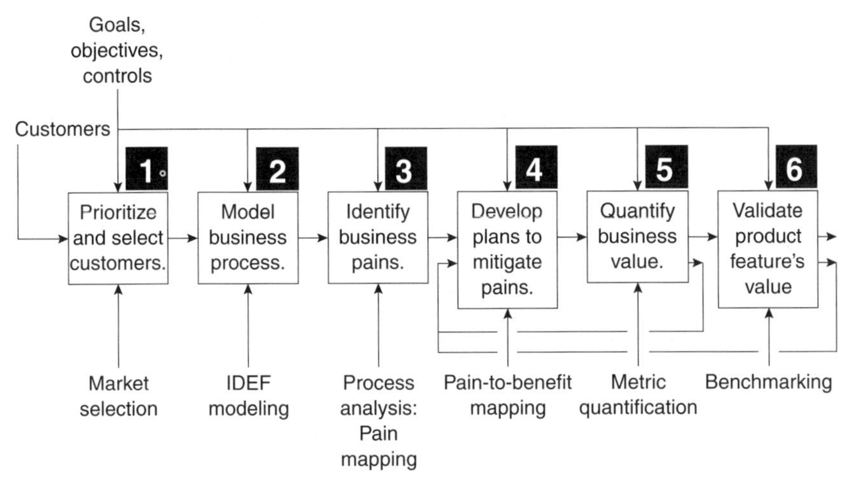

FIGURE 11-3. Developing the value for new products and services.

Step 1: Prioritize and Select Customers

This step is the well-known process of market selection—that is, market segmentation, targeting, and positioning. The marketing group that is part of the product development organization performs this step. The conceptual foundations and procedures for market selection and segmentation are presented in Kotler (2001) and Urban and Hauser (1993). Bauer, Tang and Collar (1992) show an example of this process in the development of a family of IBM computers targeting a business market.

This step is important because (1) not every customer will buy on value and (2) our approach is targeted at customers that can obtain the benefits and the concomitant value that the product can produce in their usage environment. The market will segment into those who must buy on price because they are struggling financially and those who buy on price simply because manufacturers must compete on price—their products provide imperceptible or no distinguishing features. You should examine the customer base for those who could benefit from higher-value products and services. These include customers who have relatively complex or performance-sensitive operations. That is, they may require rapid execution, or intensive capacities, high reliability, or some other distinguishing output feature of their offerings relative to their competition. These are the customers who can understand and appreciate the higher value that unique product features and capabilities can provide.

Step 2: Model the Customer's Business Process

The next step is to understand the customer's business and its operational processes. It is impossible to compete on value unless the engineering, marketing, and sales groups are intimately connected with customers in the targeted market. Without this customer and market orientation, it is nearly impossible to gain insight into the sources of inefficiencies and ineffectiveness in their business operations. A firm cannot compete on value and be a disconnected supplier who sells to value-added resellers without contact with the end users of the product or service the firm is offering. What to prioritize in product development will be largely unvalidated. Corporate guesses on what customers want and value will most likely be wrong.

To understand the customer's business, the best approach is to construct a flow model of goods and services of the customer's business operations. Creating a business process model is well researched (Damelio, 1996). A business process model describes how goods and services are developed and transformed by activities within the organization. The Identification-of-Function (IDEF) process modeling methods (IEEE 1999) are the common and effective approaches to modeling business process, though typically for process improvement through forming metrics (Frost, 2000; Harbour, 1997) or reengineering practices such as lean manufacturing (Womack and Jones, 1996).

The heart of IDEF modeling is constructing activity diagrams of required tasks. In IDEF, process activities are represented as individual black boxes with explicitly identified inputs, outputs, controls, and mechanisms, represented by links between activity boxes. *Outputs* are the value-added results of the activity. *Inputs* are what are needed to complete the activity. *Controls* regulate the activity. Finally, *mechanisms* are tools needed to convert the inputs to outputs. (See also Figure 11-18).

To construct an activity diagram for business operations, start with outputs of the entire process—what goods and services do they provide their customers? Generally, they are flows of information or materials to their customers. For example, consider the Fortune 500 telecommunications hardware provider, whose customer is a retail store chain. The retail store operations needed to be modeled with an IDEF activity diagram. To construct this, the outputs of the retail store are examined in detail. The outputs are the goods in the retail shopper's hands as they leave the store. From these outputs, you trace the flows back into the organization and examine what activities are done as a part of transforming these flows. This tracing goes back completely to the inputs of the customer's organization. This is done for each output flow.

For example, again consider the Fortune 500 telecommunications hardware provider and their retail store chain customer. The objective is to examine the retail store's chain operations for their business value of IT hardware. To do this, you need to look at the flow of retail goods and information between the corporate headquarters, warehouses, and retail outlets. Within the retail store, the primary output flows leaving the store are goods sold to consumers and various forms of payments. These flows can be back-traced into the store, chaining through the front point of sale, the shelves, restocking, back-store operations, and delivery from the warehouse. And so on. The results are chains of activities linked by inputs and outputs. For example, there is a chain of activities that are linked by the flow of goods from shipping to store shelves to the buying customer. There are also chains of cash flows.

There are also chains of order information flows. Each of these are constructed as independent chains.

The next step is to examine each activity in the chains constructed so far and to determine what is needed as mechanisms to complete the activity and what is used to initiate, stop, or regulate the activity as controls. The customer-facing point of sale needed the networked terminals (a mechanism), for example, and the store manager was a control when problems arise (a control).

The last step is to merge the flow chains into a network of interlinked activities that represent the customer's total business process. The independent chains will have unconnected inputs, outputs, mechanisms, and controls, which when merged will link with the activities of other flow chains. Missing activities or flows may be included as a part of this step. The result is a complete process map for the customer's business operations, from inputs through value-added activities to the sold outputs. As an example, Figure 11-4 shows a simplified retail store's business process.

Step 3: Identify the Business Pains

The next step is to explore, through customer interviews, where in the customer's business process the customer is experiencing problems. These problems identify the source of business pains, such as shown in column 1 of Figure 11-5. Uncovering these pains is naturally accomplished as a part of the previous step. The point here is to establish a working document of these pains in a systematic and structured way. The interviews will help pinpoint the exact source of the business pain and the way it is expressed. The critical issue about business pains is that its mitigation must be actionable.

For each pain, it is necessary to estimate how important it is to alleviate this pain. This begins by developing a metric for a solution to this pain, a benefit metric, as shown in column 2 of Figure 11-5. This is a nonmonetary, qualitative, but subsequently quantifiable number that will be used to assess each pain, in units that are sensible and directly measure the pain in its units. Because the pains express themselves in operational dysfunctions, it follows that benefit metrics are typically in units of speed, volume, or quality. Benefit metrics must be quantified so that they can then be transformed into a monetary assessment.

The logical next step is to examine the modeled business system at a higher level of abstraction than the individual activities. Rather than improving any particular internal activity in isolation devoid of context, you should pose questions on replacing parts of the entire business system. For example, in the retail store example, customer-purchasing queues can be entirely eliminated with home delivery. This, however, would be a new service offering for the retail chain that they may or may not be capable of providing.

An important point in completing these tasks is to not focus solely on cost reductions for the customer. That is, benefits generally fall into two categories: revenue increases and cost reductions. Cost reductions are easier to understand and for a customer to accept, but you should also look for ways to increase the customer's overall business volumes. For example, with the retail

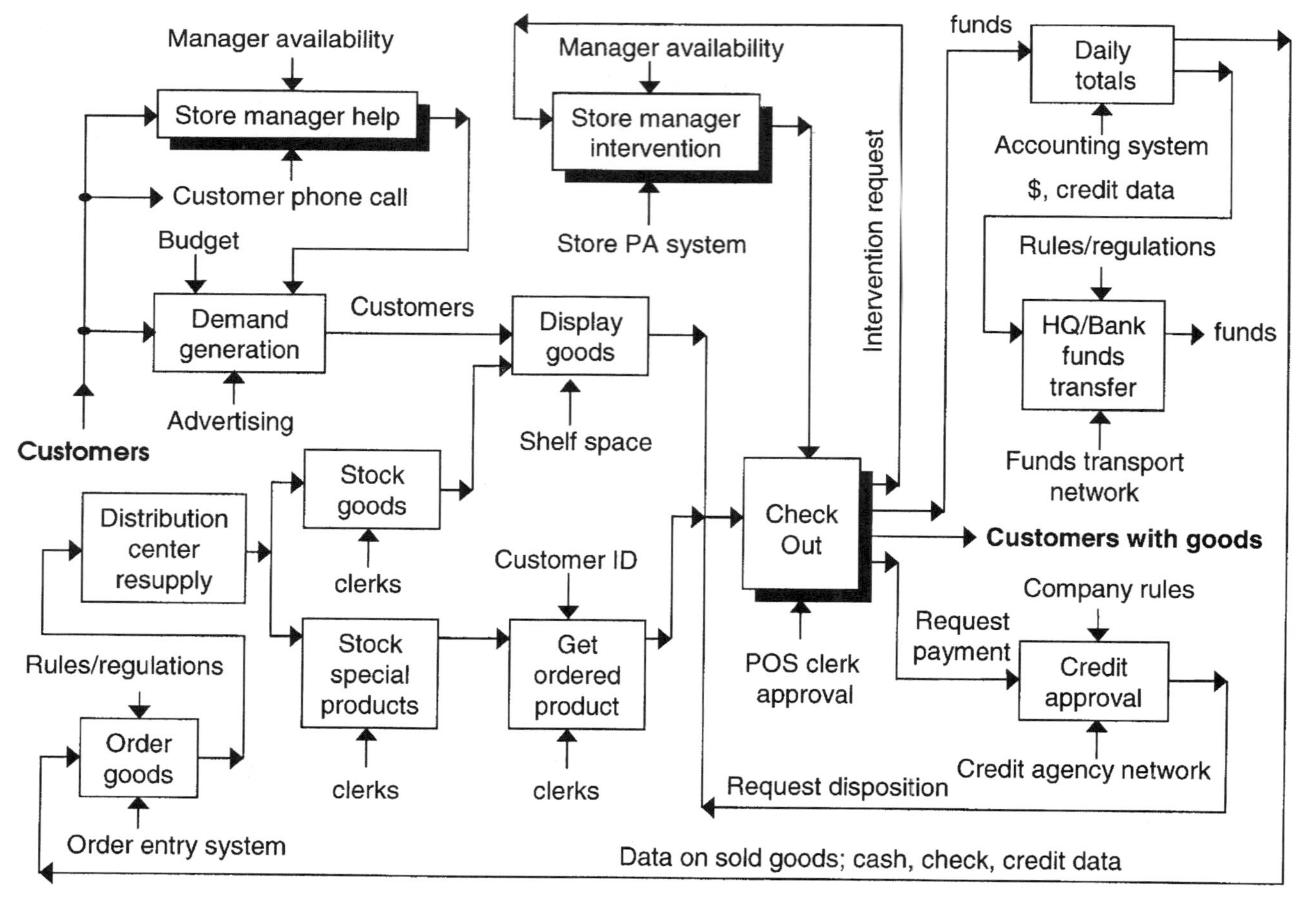

FIGURE 11-4. IDEF diagram of store retail activities.

Pain	Benefits of pain removal
Electronic transaction hardware down. Electronic transaction software buggy. Manager not available during crises at checkout. Checkout delayed and customers leave the store. Lines too long and customers leave the store during peak season.	Speed and reliability of transaction approvals. Security of transaction approvals. Speed of managerial access. Reliability of connections during checkout. Ease and flexibility expanding checkout during peak seasons.

FIGURE 11-5. Retail store process pains, and benefits.

store, installing wireless Internet will reduce installation wiring costs of terminals, and Voice over IP will reduce the costs of connectivity to the store. These are all attractive opportunities to provide new IT products and features that can reduce a customer's costs. However, there are also opportunities to provide IT products that have increased value to the customer through revenue-producing results. Providing prioritized calls to the store manager increases the company's ability to respond to crises and thereby reduces the number of upset customers. Fewer upset customers translate into an improved service reputation and, in turn, into fewer customers defecting to the competition, both of which help improve the retail store's business volumes. In the sections that follow, the examples demonstrate the substantial difference in monetary value that these different product features provide to the retail customer.

Step 4: Develop Pain Mitigation Ideas

The next step is to develop new ideas for products or product features that can improve the attainment level of these benefit metrics, thereby alleviating the customer's business pains. This is illustrated in column 2 of Figure 11-6. This is often an internal concept-generation activity with participation from several diverse functional groups of the customer's organization. As many concepts as possible should be explored. Given a large candidate set of concepts, culling should follow a disciplined process with reproducible results. Technical research and development teams can examine the analysis and generate new product concepts (Otto and Wood, 2001). Marketing, sales, and services teams can examine the analysis and generate new service plans and activities. The descriptions are in a form that traditional product development activities can readily digest: A customer requirement is well formed as a benefit metric, and the context of the metric is well formed as a pain at a customer activity. As an example, Figure 11-6 shows various telecommunication system ideas for new products and services to improve the various benefit metrics of a retail store. Note the column that identifies the whitespace opportunities.

This table identifies the new product features that the product develop-

Benefit of pian removal	Operational Metric	New Product Features	White Space
Speed and reliability of transaction approval.	Increased credit approval access.	Highly reliable external network.	—
Security of transaction approval.	Transactions lost due to system being down.	Highly secure software. Rapidly easily updated software. Service team and help desk.	Yes — —
Speed of managerial access.	Increased POS throughput from reduced time. processing exceptions	Caller ID with call prioritizing software.	Yes
Reliability of connection with back office.	Interest income lost when connection down.	Highly reliable internal network.	—
Ease of peak season expansion.	Increased sales with decreased queues.	Field upgradeable internal network. Wireless POS. Service team and help desk.	Yes Yes —
Reliability of nightly bank transfers.	Interest income lost when connection down.	Highly reliable external network.	—
Security of nightly bank transfers.	Interest income lost because of system being down.	Highly secure software. Rapidly easily updated software. Service team and help desk.	Yes — —
Reliability of restock order processing connection.	Out of stocks because of system down.	Highly reliable external network.	—
Ease of establishing connection to vendors on heterogeneous systems.	Out of stocks and misorders because of communication errors.	Interoperability standards.	—

FIGURE 11-6. Product benefits, metrics, and features to supply the benefit.

ment team can potentially design into the product. However, before engineering and management can commit to design and development, it is necessary to determine the economic value to customers and the development cost to the developers. Then management can decide whether to commit resources to developing these new features and functions. The quantification of economic value of product features, to the customer, is the subject of the next section.

Step 5: Quantify the Business Value: Develop Quantified Value Propositions

Having a list of pains, benefit metrics, and ideas, you can now assess the value of the benefits that each idea provides. Much as a business customer will do, you can determine how much improvement on the benefit metric any new product or service will provide. Improving this metric will then directly translate into improved or increased business operations for the customer that can be valued in monetary terms.

Forming the monetary equivalent of the benefit metric improvement is not generally simple or direct. Data are required from actual customer operations to ascertain required activity times, uptime of systems, and flow rates of customers and goods. This is a joint activity between the vendor and the customer, led by the product vendor who is motivated to demonstrate the superior value of its product. It must have active customer participation for the data and value derivation to be credible and convincing.

If a new idea can reduce the customer's operational business costs, this is a direct benefit they would want to accrue, and it offers you the direct ability to charge a higher price. This higher price is justified by the increased benefits. If, on the other hand, a new idea can generate more revenue for the customer, such as offering higher business traffic or reduced lost sales, then this also provides benefits that the customer would like to accrue. Again, this offers you the ability to charge the customer a higher price for your product and services that produce for them higher revenues.

Generally, cost reduction benefits speak louder to customers than revenue increase capabilities. Customers can visualize increased efficiency at providing the same goods and services, whereas they often have more difficulty visualizing increased business. Therefore, the revenue increase numbers generally get discounted until proven by an actual demonstration in the field.

In the telecommunication equipment provider example, the company was looking for new product or services offerings for retail. One benefit often discussed in the telecommunications industry is system reliability. Many equipment providers advertise their equipment having over five-nines reliability (99.999 percent) per year, for instance, Business Week (2003). Yet, when you examine the business operations of retail, you'll see that very little benefit is gained from this reliability. In fact, three-nines reliability is more than adequate, especially given the typical three-year replacement rates of telecommunication equipment because of technology capability improvements. Figure 11-7 details this calculation for storefront point-of-sale terminals, where the value of five-nines reliability over three-nines is merely $501 per store. Note that the incremental revenue is $297; the expenses avoided are $204. The cost benefit ratio is not sufficiently large. Five-nines reliability does not pay for the retail store. The key question for those responsible for making product development decisions is this: Is there a more effective use of resources than for developing five-nines?

Another example of quantifying the value of telecommunication benefits is shown in Figure 11-8. Referring to the figure, consider the benefit of increased speed of managerial access. The store manager is a critical person whose job is primarily to take care of all exceptions to the standard procedures of the store. Improving his or her efficiency is critical to eliminating problems that impact a store's operational efficiency and effectiveness.

Added Revenue = $C*N*f_1*f_2*(r_1-r_2)$			
C	Average customer transaction	Dollars	$10.00
N	Number of customers transactions per year	Count	500,000
f_1	Fraction of customer paying via e-payment	Percent	30%
f_2	Fraction of customer paying via e-payment who leave the store when a problem occurs	Percent	20%
r_2	System reliability of ban-to-checkout using highly reliable equipment	Percent	99.999%
r_1	System reliability of bank-to-checkout using less reliable equipment	Percent	99.000%
			$297
Reduced Expense = $(n_1-n_2)*R*t$			
n_1	Number of failures per year using less reliable equipment	Count	3
n_2	Number of failures per year using highly reliable equipment	Count	0
R	Labor rate	$ per hour	$17.00
t	Time required to fix each failure	Hours	4
			$204
Total Quantified Value to Customer			$501

FIGURE 11-7. Value proposition for increased retail reliability.

Added Revenue = $C*N*f_1*f_2*f_3$			
C	Average customer transaction	Dollars	$10.00
N	Number of customer transactions per year	Count	500,000
f_1	Fraction of customer transactions requiring a supervisor	Percent	2%
f_2	Fraction of f_1 not processed because the supervisor did not show up in time	Percent	35%
f_3	Fraction of f_2 that is processed because new equipment speeded supervisor access	Percent	20%
			$7,000
Reduced Expense			
None			$0.00
Total Quantified Value to Customer			$7,000

FIGURE 11-8. Value proposition for improved managerial access.

Figure 11-8 details the calculation of the value proposition for the benefit of a store manager's availability to address exceptional situations. The value of more quickly accessing the store manager equates to $7,000 per store, which would easily pay for the cost of a mobile telecommunications device and in-store system.

An important facet of this managerial access benefit, though, is what it will take to fulfill it. This benefit cannot be delivered alone by a telecom hardware solution such as a cell phone or pager. A set of customizable prioritization software embedded into the telecom systems is also required, to permit front-store calls[2] to be always passed through to the manager. Because of this combined requirement of both telecom hardware that the provider is familiar with and software that the provider has no experience with, this benefit to the retail store chain customer is unfulfilled. In fact, it is a benefit not even considered by the telecommunications company and was not understood until the retail store operations were examined as described. This was a whitespace opportunity unmet in the market.

While telecommunications systems can provide this benefit through new innovations such as in-store cell phones, caller ID, pagers, or some form of technology, there is a fundamental reason why the whitespace remains unsolved. The notifications must be customizable and offer prioritization of notification or the managers will not use them. That is, simply providing a telecommunications piece of hardware will not suffice, and providing a set of software code will not suffice. A new solution combining telecommunications hardware and software is required. The Fortune 500 telecom company needed to expand its R&D capability beyond what it currently understood. That is typical of whitespace opportunities.

Nonetheless, the value benefit calculations need to be constructed and calculated. There are assumptions built into the value equations that need to be calibrated at the customer sites. But the basic approach is to understand the customer's business operations, identify the difficulties within those processes, explore a wide range of conceptual solutions, select the solution that can most effectively and efficiently improve the business operations, and then to quantify the value of these improvements.

Step 6: Validate Numerical Worth of Your Product's Features

After constructing these value propositions, they need to be validated in the field. This validation comes in four forms. You need to compare the results against known summary financial statistics, and against customer field measurements, and to compare against benchmarks such as industry sector specific metrics.

The first validation is to sanity-check the summary numbers. The cost savings results should sum up to a reasonable fraction of total customer expenses. The revenue increases should sum to a reasonable fraction of existing sales. The second validation is to sanity-check the inputs against the client operations. Time constants, failure rates, costs of equipment and labor, and similar inputs should be realistic to the actual operations. The third validation is to determine whether the improvements are sufficient to compete against leading competitors. Note that you should be vigilant to avoid double counting.

Lastly, the final validation is to complete a demonstration prototype implementation and measure the actual value brought to the customer. This needs to be set up as a controlled experiment, to fully understand cause and effect. It must be said, though, that this can be problematic, as revenue changes can also be due to exogenous factors beyond the control of the experiment. Nonetheless, some form of such a prototype validation should be an objective and is most effectively performed as a team with the customer.

Summary of the Telecom Example

This field study of a Fortune 500 telecom provider's products in a retail store has illustrated how the approach to value quantification works. The six steps of the value quantification method were presented. This presentation included detailed descriptions of how to derive the monetary value of a product's features and performance for a customer in the customer's actual usage environment. We have shown that value is obtained through a series of successive mappings, which begin with the customer's pains and end with a derived monetary value. This methodology also unambiguously identifies whitespace product opportunities that product managers and engineers can capture for a competitive advantage.

FORTUNE 500 SERVICES COMPANY

Unlike physical products, one cannot touch or weigh a service. How do you demonstrate value for something you can't touch or weigh, that has to be experienced, and that makes demands on customers to succeed? This section of the chapter shows how. For the next example, we use a Fortune 500 services company in a financial retail institution. This example illustrates that the approach and method described in the previous section applies equally well for services in a business market. That this is so is not simply an issue of making this claim. It is important to understand the unique challenges services present to product development and quantifying customer value.

Services are a unique kind of product. Services, apart from being nonphysical, differ from physical products in very fundamental ways (Bitran, 2000). Services' unique defining characteristics are intangibility, heterogeneity, simultaneity, perishability, and mutuality. Unlike products, which can be described by physical properties, services must be experienced. Products are tangible and physical; services are intangible and experiential. This is services' intangibility property. Key aspects of products are separable; in services, they are inextricably fused. The inseparability of function and delivery is one of the most salient properties of services. Whereas products can be inventoried, services cannot. This is the perishability property of services. Closely tied to perishability is simultaneity. Services are produced and consumed at the same time. It is impossible to inspect a service before it is delivered. The customer buys a service as well as its delivery; they are inseparable. People deliver services. Despite best efforts, skills and attitudes vary widely among individuals. This makes services heterogeneous. The effectiveness of a service also depends on customers fulfilling their roles and responsibilities. This customer dependency of services is the property of mutuality.

Customer Services

Step 1: Prioritize and Select Customer

The customer of this example is a retail securities trading business unit of a very large financial institution that places it among the Fortune 100. The business unit has a customer base of approximately 350,000 clients for securities and over a million clients for mutual funds. This customer came to the attention of the IT services provider of this example, when executives of the business unit expressed dissatisfaction with the growth in sales volume and in the apparent productivity decline of its securities workforce. They were looking for ways to improve. Because of the market potential of the securities industry, the IT services provider elected to respond to this customer.

Step 2: Model Business Process

Traders at their desks take orders for securities transactions. Customer orders are taken over the phone, recorded, and subsequently executed electronically with an online system that is connected to major securities exchanges. A detailed audit of the business processes and operations was performed by IT systems engineers with substantial experience in the securities industry. A simplified IDEF model of their business processes for securities transactions is illustrated in Figure 11-9. As in the telecom example, the place to start is with the output of the process—for example, Placing-an-Order-to-a-Stock Exchange. Working backward from that output, trace and identify all the activities until you reach the input of this business process—a Customer-Call. The chain of activities, Accept-Customer-Call, Authenticate-the-Customer, Document-the-Order, Evaluate-the-Order, Finalize-Order-Status, and Place-Order-on-Stock-Exchange, form the backbone of the business process. Each activity has a supporting mechanism—for instance, Customer-IT-System

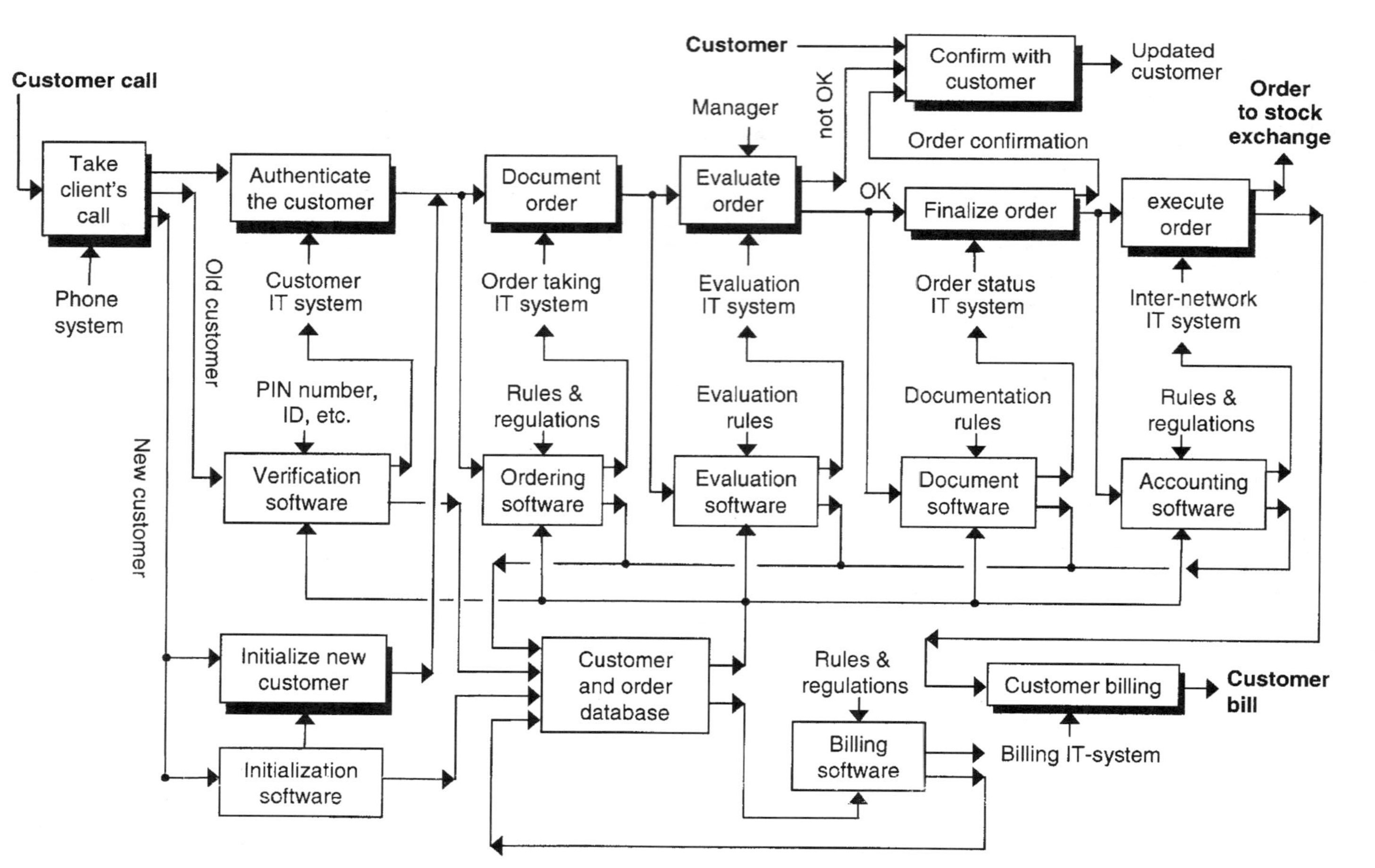

FIGURE 11-9. IDEF diagram for securities retailer.

(mechanism) for the Authenticate-the-Customer (activity). The supporting mechanism is, in turn, enabled by Authentication-Software, which is regulated by PIN number, ID, etc. (control). Another output of the business process is Customer-Bill, which is produced by the activity Customer-Billing. Customer-Billing's mechanism is the Billing-IT-System, and the Billing-Software is regulated by Rules-and-Regulations (control).

Steps 3 and 4: Identify the Business Pains and Develop Pain Mitigation Ideas

Recall that business pain was expressed by the executives of the business unit as their dissatisfaction with the growth of sales and productivity of its securities workforce. Careful observation and measurements of the business processes' activities found that its workforce did meet its productivity targets. But it found that the overall productivity of its workforce was inhibited by frequent system outages. The excessive downtime during working hours reduced the overall operational efficiency of its workforce. The system was frequently not available, during peak business hours, when it was needed to execute and complete a transaction. Untimely execution of customer orders made many customers sufficiently dissatisfied that they took their business elsewhere. To raise the availability level of the IT system, the proposed solution was to improve key hardware elements of the IT system and many of its maintenance and crises management procedures. Figure 11-10 summarizes the pains, the potential benefits, and the operational metrics that were used to calculate the monetary value of the benefits. Note that the business pain of productivity was reduced to actionable activities that are measurable and quantifiable in monetary terms.

As examples of the quantification of operational metrics that are derived from the sequence of pains, benefits, it is now shown how the monetary value is obtained from the knowledge of the customer's businesses process activities. Figure 11-11 shows the calculation of the operational benefit of Increased Sales by Increasing Systems Availability. Increased systems availability will make it possible for securities traders to take orders and execute them on demand without any delays at the time the customer calls.

As another example, consider the business pain of customer defections. Customer surveys undertaken by this securities retailer indicate 0.4 percent of their customer elect to go to a competitor because they consider untimely execution of their transactions to be unacceptable. What is the value of customer loyalty? This value calculation is shown in Figure 11-12.

Summary of Services Example

In contrast with the telecom example of the previous section, another example, which is qualitatively different because it is applied to services, has been presented. The six steps that compose the core of this chapter's method have been illustrated in detail. This included the derivation of the monetary

Pain	Benefit of pain removal	Operational Metric	New Services offerings	White Space
System crashes too frequently, impacting sales maintenance sales expenses.	Sales productivity and customer satisfaction improvements.	Increased sales by increasing system availability.	IT system design & reengineering.	—
			Reporting software for itemized business value of availability.	yes
Customers defecting to competitors.	Improved customer loyalty.	Fewer customer defections increase lifetime value of customer.	Software to monitor customer loyalty, analysis, reporting value analysis.	yes
Sales productivity below competitive benchmarks.	Increased effective working hours through reduced dead time.	Reducing cost of non-revenue-generating time intervals. Improvement of end-user productivity.	System reliability services.	—
			Software for end-user productivity monitoring, analysis, and reporting.	yes
IT maintenance and service expenditures exceed industry benchmarks.	Fewer skilled resources required and less overtime during crisis outages.	Skilled resources applied to revenue producing applications, not solving crises.	Crisis management procedures.	yes

FIGURE 11-10. Customer business process pains, and operational metric.

Increased Sales = $T*S*D*P*(m_2 - m_1)*r$			
T	Average number of trades per minute	Count	4
S	Number of securities per trade	Count	150
D	Number of trading days per year	Count	200
P	Average price per security	Dollar	$35.00
m_1	Original availability per day	Minutes	m_1
m_2	Improved availability per day	Minutes	m_1+10
r	Fee rate=0.04 of the dollar value of trade	Percent	4%
Total Quantified Value to Customer			$1,680,000

FIGURE 11-11. Value proposition for system availability.

Customer Loyalty = C*L*r			
C	Total number of customers	Count	350,000
L	Average lifetime value of a customer	Dollar	$16,000
r	customer defection rate	Per cent	0.004
Total Quantified Value to Customer			$22,400,000

FIGURE 11-12. Value proposition for customer loyalty.

value of service features for a large financial institution. We have discussed in detail the process for the monetary value derivation. This derivation is obtained through a series of successive mappings, which begin with the customer's pains and end with quantified value propositions. As in the telecom example, the methodology presented identifies whitespace that represent new service offerings opportunities.

Implementation

In this section, the "how" of this method is presented. The "how" procedure of implementation follows the template outlined in Figure 11-3 with the telecom and the F500 services examples. In any implementation the first step is always preparation.

The first step in preparation is to obtain senior management and customer buy-in. Senior management is whoever must review and approve the results and conclusions of this effort. Senior management also will approve or make the funding decisions that will be required as a result of this work. It is important that senior management also exert leadership on this work. New potential business activities will arise that the company currently does not address. They will require a business management decision. It is fundamental that senior management set appropriate standards of objectivity and suspend judgment while developing the process models, benefit metrics, new potential solutions, and their economic value. The operations leader of this work is appointed by the senior manager, and the operations leader's jurisdiction lasts as long as the work is being done. Cross-functional teams are mandatory, which include, but are not limited to, engineering, marketing, service, finance, and sales. Each function should have a team leader that works under the guidance and supervisor of the operations leader of the work. Customer buy-in from multiple management levels of the organization that will participate in the work is also a fundamental requirement. Also critical is an overall work plan, with schedules and checkpoints that are established, discussed, and reviewed by all the team before starting.

For each of the implementation steps, the specifications in Figure 11-13 will be used to focus attention on what are the objectives, what is to be accomplished, who leads the work step, and what are the deliverables. Note

Process name	
Process objectives	What this process step is supposed to do
Who does it	Who leads this process Who are in the team
Input	Input for this process
Control	Parameters and constraints that guide the process
Mechanism	Means to effect the process
Output	Output of the process These are the deliverables of the process

FIGURE 11-13. Process specifications.

that leadership for each step changes hand based on the functional intensity and expertise required. Nevertheless the overall leadership of the work always remains in the hands of the operations leader. The structural spirit of the specifications follows the practice of IDEF process definition.

Step 1: Prioritize and Select Customers

The purpose of this work is to select the customers with whom to do the work. These customers should be representative of the target market for which the product is intended (see Figure 11-14).

What makes a customer in a targeted market a good one to work with? Figure 11-15 is useful as a checklist to consider in customer selection. Customers and firms with incompatibilities revealed by this checklist should be avoided. In a business markets, the "customer" is an organization, potentially distributed across many sites. The end users of your products are one group that you can make more efficient. However, the purchasing agents and their view of the process benefits of new products are also clearly important. Senior executives opinions are also important, as well as their views of the business impact of your products. All of their views of the benefits, metrics, and valuation should be factored into the process model and the benefit valuation. The industry of which the customer is a player is another consideration. The industry should be one that the firm has targeted as important. The reasons could be because of potential growth, because it has a large install base to protect, because it wishes to establish a foothold in new markets, and so on. The reasons will be company-specific.

Step 2: Model the Business Process

The purpose of this step is to understand the customer's business process under consideration and to have customer participation in the work. See Figure 11-16. Obtain an agreement from the customer to have the product

Prioritize and select customers	
Process objectives	To identify customers that will buy on value. Select customers that can obtain the benefits and recognize the economic value that the product can produce in their usage environment.
Who does it	Marketing leads for targeting markets for the product. Sales function leads for selecting customers to study. Engineering reviews and approves the markets and customers selection. Participation and support from all the remaining functions of the work group. Participation from relevant groups from the selected customer.
Input	Business and market opportunity studies. Market segment studies. Product positioning studies.
Control	Business unit's goals and objectives articulated by senior management. Customer expectations and agreements reached.
Mechanism	Market segmentation, customer targeting, and product positioning processes. This is a marketing planning process that is well known to marketing professionals.
Output	A list of targeted customers.

FIGURE 11-14. Process specifications for prioritizing and selecting customers.

development teams from engineering, marketing, sales, and so on work on-site with the customer.

The implementation of this process begins with a conceptual big picture of the overall business process. For the telecom retail store example, a simple graphic as shown in Figure 11-17 illustrates the concept of the business process.

With this big picture as a base, the implementation follows the algorithm that is presented next.

1. *Identify the all the outputs of the overall process.* In the telecom example, shown in Figure 11-4, the store's retail activities have two outputs: Customers-with-Goods, and $$-Credit.Data-Daily Totals. In the Fortune 500 example, the security retailer's activities have three outputs: Customer-Bill, Order-to-Stock-Exchange, and Updated-Customer. Note that each of the outputs is shown as an arrow pointing to the right
2. *Identify all the inputs of the overall process.* In the telecom example, in Figure 11-4, there are two inputs: Goods and Customers. In the For-

Factors to Consider		Who is Responsible
Customer needs	The customer has a clear need for the product. The fit between their business problem and the product's functions, features, and performance are self-evident.	Sales and marketing
Your commitment	Your management has the budget and staff to execute the process; it has the will and tenacity that will prevail under competitive measures.	Engineering with approval from senior management
Customer relationship	There is trust and confidence such that the customer will disclose details of its operations, sources of inefficiencies, and other key problems. Trust and confidence permeates the organization.	Sales
Customer situation	Determine whether the customer is ready or receptive to what the product will be able to do. Also determine whether the problem is resulting in significant economic impact to the customer or whether the status quo is intolerable.	Sales
Competitive intensity	There is minimal competitive activity that will disrupt the study. There are contingency plans in place in case competitive activity heats up.	Sales
Your capabilities	You have done it before, or are prepared to learn and bear the burden of learning.	Engineering
Work product: Customer report	A report that identifies targeted customers. For each customer it describes the preceding customer elements. Attached to the report is a memo committing sales to the approach.	Sales with input from product development
Work product: PD Commitment letter	A memo from the product manager committing to support the approach with the identified customers.	Product development manager

FIGURE 11-15. Key factors for customer prioritization and selection.

tune 500 services example, there are two inputs: Customer-Call and Customer. Each of the inputs is shown as an arrow pointing to the right and into a box in Figure 11-9.

3. *For each output, identify the subprocess that produces this output.* For the telecom example, in Figure 11-4, the output Customer-with-Goods is produced by the subprocess called Check-Out. In the Fortune 500 security retailer example, the subprocess that produces the output Order-to-Stock-Exchange is called Place-Order-on-Stock-Exchange. Each of these subprocesses is also presented as a box.

Model Business Process	
Process objectives	Understand the customer's business and operational processes. Develop a flow of the major functions of the business and operational processes. Integrate the material and information flow that make the process operational.
Who does it	Lead the process: Sales Mandatory active participation: Engineering, marketing, and finance function Support and participation from the remaining functions. Participation from customer relevant groups.
Input	A key customer business process or operation is identified Present this business process in the context of the customer's business, i.e., what is the big picture in which this process is embedded?
Control	Description of the scope of the project and customer's participation (the issue is to avoid scope creep).
Mechanism	IDEF process description. Work teams composed of the firm and customers at the customer's site.
Output	An IDEF model of the business process or operation.

FIGURE 11-16. Process specifications for business process modeling.

4. *Identify the inputs, controls, and mechanisms for each of these subprocesses.* For example, in the telecom retailer example, the Check-Out subprocess has three inputs: Goods-Displayed, Goods-Ordered, and Credit-Approval. These are shown by arrows pointing into the Check-Out subprocess box. Control-for-Checkout is shown by an arrow pointing downward into the process box. This control is the output of

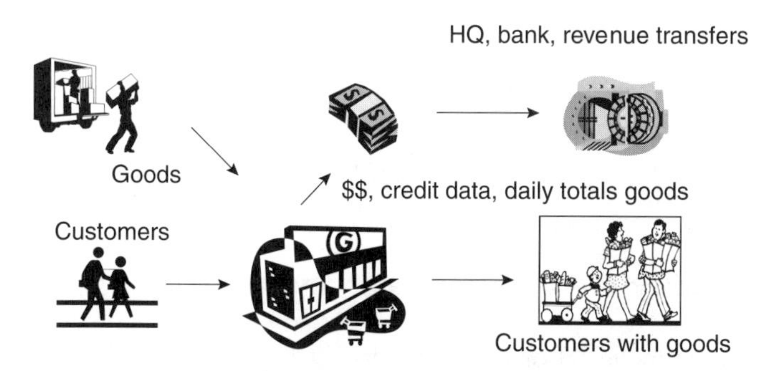

FIGURE 11-17. Schematic of the overall business process of retail store.

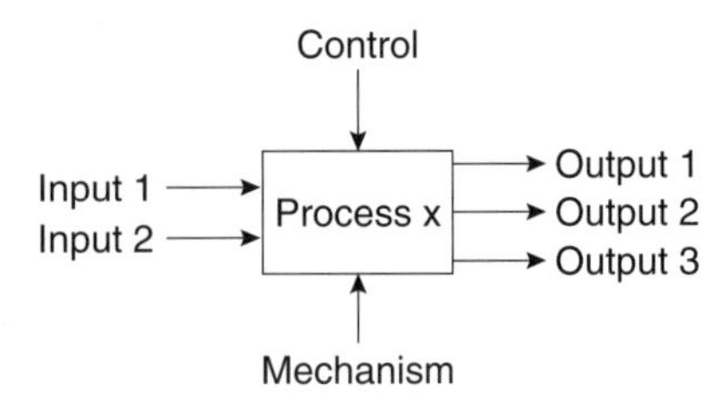

FIGURE 11-18. Schematic of a business operation.

the Supervisor-Intervention subprocess. The mechanism for Check-Out is an upward pointing arrow into the process. The mechanism is simply POS-Clerk-Approval.

5. *Draw a diagram as shown, using the basic building block shown in Figure 11-18.*
6. *Trace backwards the complete flow of inputs and outputs.* The input to each subprocess is the output of another subprocess. Repeat steps 2, 3, and 4 until the backward tracings have reached an overall input of the overall process. For example, for the Fortune 500 securities retailer, the initial input to the overall process is simply Customer-Call. For the retail store example, the initial input to the overall process, as shown in Figure 11-4, are Goods that are delivered to the retail store and Customers arriving at the store.
7. Link all process flows together into a visually appealing and understandable picture. We recommend that during the first pass of this process, the controls and mechanisms can be ignored and suppressed from consideration. During the second pass, consider the mechanisms of the subprocesses while suppressing consideration of the controls. Now that you've completed the identification and location of the mechanisms into the process, it is now time to consider the controls. At the completion of these iterations, a complete process illustration is obtained. Make sure that the graphic illustration is still visually appealing, easy to understand and to follow.

In the execution of this algorithm, think of flows as either function flows, material flows, or information flows. A business operation can be simply expressed as a verb followed by a function, material, or information. It becomes clear that a business process flow is simply subprocesses of functions, materials, or information strung together to handle the inputs to the overall business process in order to produce the desired output.

Step 3: Identify the business Pains and Benefits

A *pain* is any managerial or executive dissatisfaction driven by unsatisfactory performance of its business processes and business operations (see Figure 11-

Identify the Business Pains	
Process objectives	Identify key sources of inefficiencies and ineffectiveness that inhibit the customer's ability to do business or compete effectively and efficiently.
Who does it	Lead the process: Sales Mandatory active participation: Engineering, marketing, and finance function Support and participation from the remaining functions. Participation from customer relevant groups.
Input	IDEF model that describes the overall business process. This is obtained from the previous step. Notes taken during the business process model description process that hints or points (directly or indirectly) to symptoms of operational problems.
Control	Participation, review, and concurrence of the observations made in the previous step. Management reviews of the firm and of the customer.
Mechanism	Interviews and surveys of the customer executives, operations personnel at multiple levels of the organization. On-site working teams from the firm and the customer.
Output	Pain-benefit table. (Figure 11.5 is an example.)

FIGURE 11-19. Process specification for identification of business pains.

19). The implementation of this process follows the algorithmic steps in order to produce the output, a pain table, shown in Figure 11-20.

1. *Preparation.* Have all members of the task group, including customers, review all the notes on the observed operational dysfunctions. Review these notes and observations with various levels of the customer's organization. Make the appropriate adjustments as needed.
2. *Pain table development.* A "Pain" entry identifies a problem that is causing a malfunction or a dysfunctional performance in the business operations. As in the telecom example, when the purchase transaction hardware is down, a sale cannot be consummated. Clearly this is an operational problem; impatient customers leave the retail store, a cus-

Process Element	Description
Pain 1	Root-cause obtained from surveys and interviews
. . .. . .	. . .. . .
Pain i	. . .. . .

FIGURE 11-20. Pain table.

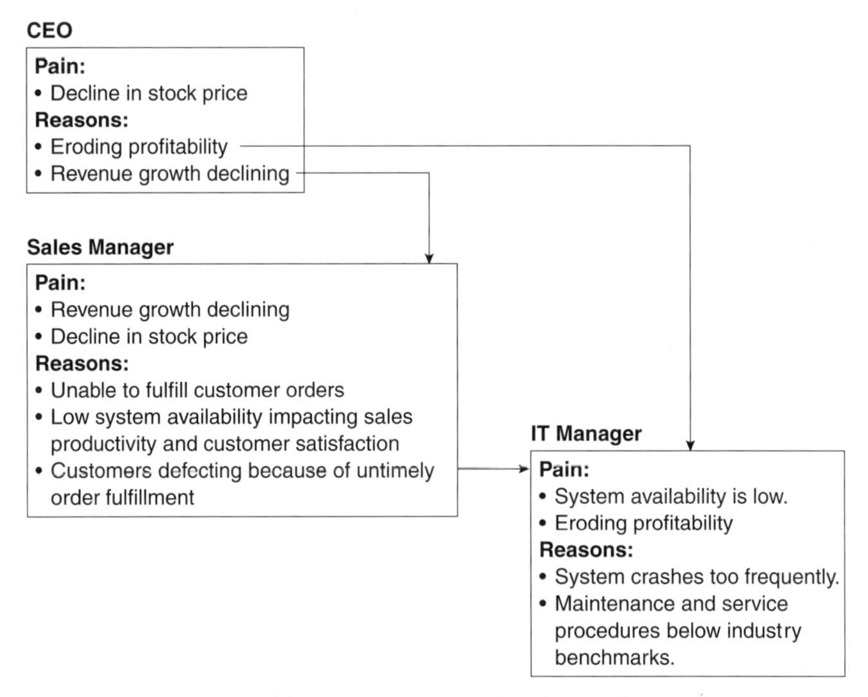

FIGURE 11-21. Decomposition and flow down of business pain.

tomer's buying experience is made more unpleasant, and so forth. How are these pains identified? They are obtained from surveys and interviews of executives, managers, and personnel who have responsibility and visibility into the operations, its efficiencies, and its problems. It is necessary to interview many levels of management and operational personnel, in a variety of positions in the customer's organization, to obtain a complete picture of the pains. Knowledge of these pains is also obtained by working together with the customer on-site over extended periods of time. Occasional visits and conference room meetings are not sufficient. The pain can always be traced to a business operation that is the root cause of the pain. Note that the pain becomes more abstract the higher the position of the respondent. The abstraction can be reduced and made concrete through the multiple-level interviews. How this is done is illustrated in Figure 11-21.

3. *Identify the benefits.* Under the supposition that a specific pain is removed or reduced, state the benefit of these actions. A *benefit* is an advantage or an improvement that can be obtained as a result of pain removal or pain reduction actions that the customer is prepared to take. Clearly, the "benefits" are the inverse of the specified "reason" in the "pain chain." For example, at the end of the chain, one of the reasons is "products frequently unavailable." The corresponding benefit is

then "products are available." Note that the benefit derived from the removal of the root cause propagates upward in the pain chain. Furthermore, note that a many-to-many mapping between pains and benefits is not unusual. In other words, a benefit may apply to many pains and vice versa. Benefits identification is a process of derivation requiring knowledge and insight into the business operations of the customer. The result is to produce a table that looks like Figure 11-5, from the telecom example.

Step 4: Develop Plans to Mitigate the Business Pains

From an engineering point of view, this step and the next one are the most engineering-intensive of the method (see Figure 11-22).

1. *Identify the operational metric.* Using the Pain-Benefit table in the previous section, identify the operational metric that can be used to measure the benefit or the pain in concrete terms.
2. *Specify new product features.* This is an engineering design activity. The questions engineers should be asking themselves are (1) what product features or functions will remove the identified pain and benefit? (2) what is the metric that should be specified to measure the obtained benefit? and (3) is this a white space opportunity (i.e., is a new product required or can new features and functions be designed into the product?)?

Identify Strategies to Mitigate Business Pains	
Process objectives	Identity engineering solutions to the product in terms of features, functions, or performance improvements that will reduce or remove the customer's pain.
Who does it	Lead the process: Engineering Mandatory active participation: Engineering, marketing, and finance function. Support and participation from the remaining functions. Participation from customer relevant groups.
Input	Pain table from previous process step.
Control	Participation, review, and concurrence of the observations made in the previous step by the firm and customer team. Review of ideas and qualitative evaluation of their merit.
Mechanism	Product design methods, concept evaluation methods.
Output	A table as shown in Figure 11-23.

FIGURE 11-22. Process specification of pain mitigation plans.

Pain	Benefit	Operational Metric	New Product Features, Functions,	White Space
What is the pain?	What is the benefit by removal of the pain?	What is the metric to measure the extent of the benefits?	What engineering design is needed for the product?	Yes or No: ◆ Need product enhancements. ◆ Overengineered function or feature. ◆ A new product is needed.

FIGURE 11-23. Product solution guidance table.

3. Construct a product solution guidance table. The schema for that table is shown in Figure 11-23. Figure 11-10 of the Fortune 500 services company is a completed example of such a table.

Step 5: Quantify the Value of the Product's New Capabilities

This step of the process gets to the core of the method: the quantification of a product's capabilities when used in a customer's environment (see Figure 11.24). The implementation of this process follows the algorithm depicted the flowchart in Figure 11-25 and is described in the steps that follow.

1. *Establish the operational metrics' baseline values.* Establish the operational metrics' baseline values under the current situations and conditions. For example, in the table that follows, which shows the Fortune 500 example, these are the rows with the variables C, N, f_1, f_2, and f_3.

Value Quantification	
Process objectives	Obtain a quantification of the value of engineering features when used in the customer's environment.
Who does it	Lead the process: Engineering Mandatory active participation: Marketing, finance, sales Support and participation from the remaining functions. Participation from customer relevant groups.
Input	The table that is the output from the previous step.
Control	Customers goals and objectives in their business operations.
Mechanism	Business process analysis and value quantification as shown in this step.
Output	Quantified value statements for product's features, functions, and capabilities.

FIGURE 11-24. Process specification for value quantification.

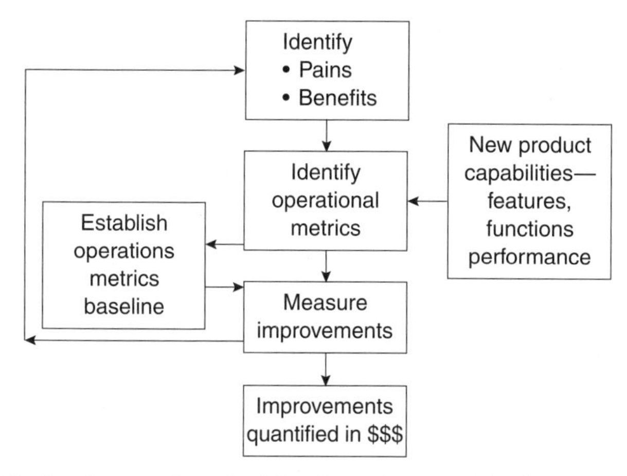

FIGURE 11-25. Implementation algorithm for value quantification.

The units and base values are shown in the next two columns. For example, f_2, the fraction of customers that left the store because the supervisor did not show up, is measured in percentage terms and its value is 35 percent.

Added Revenue = C*N*f_1*f_2*f_3			
C	Average customer transaction	Dollars	$10.00
N	Number of customer transactions per year	Count	500,000
f_1	Fraction of customer transactions requiring a supervisor	Percent	2%
f_2	Fraction of f_1 not processed because the supervisor did not show up in time	Percent	35%
f_3	Fraction of f_2 that is processed because new equipment speeded supervisor access	Percent	20%
			$7,000
Reduced Expense			
None			$0.00
Total Quantified Value to Customer			**$7,000**

2. *Review and confirm the analysis.* Review and confirm the analysis to make sure that the new products capabilities improve the operations as measured by the operational metrics. This entails a review of the tables already discussed in the previous section.
3. *Form a formula that shows the revenue increase and the cost reduction that is obtained by the product's capabilities and its ability to impact the operational metric.*
4. *Calculate the results.*

Step 6: Validate the Results

Lastly, it is important that results are continuously validated throughout this effort from step 1 onward (see the table that follows). In particular, once a process model of the customer is constructed, you should show it to the customer for review and approval. It is important to continuously calibrate the gains in the operational metrics with the customer One of the most important results of the work that needs customer review and validation are the monetary valuation of various improvements. It is key that the engineering group obtain an unambiguous valuation of increasing the efficiency and effectiveness of their business operations as a result of the product's new capabilities.

Validation of Results	
Process objectives	Obtain senior management concurrence from the firm and the customer of the findings, conclusions, and potential engineering solutions. Obtain a senior executive approval for product development. Obtain funding for engineering.
Who does it	Lead the process: Engineering. Mandatory active participation: Engineering, marketing, and finance function. Support and participation from the remaining functions.
Input	Engineering concepts and early design of the product's new features, functions, and performance. Quantified value benefits for customers in the targeted market.
Control	Revisit original goals and objectives of this work. Customer presentations to senior management of the firm presenting their views of the work.
Mechanism	Senior management decision meeting.
Output	A decision by senior management to fund engineering or not.

CONCLUSIONS

A list of customer needs is nice, but it is insufficient for many development decisions. Establishing a quantified, dollar value for each requirement is more helpful. This approach provides this by examining the customer's practice and essentially establishing the customer's business case for your product down to the feature and performance levels. This provides for much better trade-off decisions in new product development. This approach also helps to identify whitespace opportunities. Moreover, because the methodology is fine-grained, the whitespace opportunities are resolved into clear and actionable product development projects. Finally, we pointed out that in spite of fundamental differences between physical products and professional services, our approach applies to products as well as services.

NOTES

1. The graphically illustrated price-value relationship shown in Figure 11-1 can be shown algebraically (Anderson and Narus, 1999). Given that the incentive to buy for product B is larger than for product A, it follows that:

$$V_b - P_b > V_a - P_a$$

and rearranging terms,

$$V_b - V_a > P_b - P_a$$

which shows that the central issue is designing products which produce a value differential for customers that is larger than the price differential.

2. Telephone calls from customers to the retail store are called "front store calls." Many of these calls are customer complaints, purchase inquiries, or disputes that require retail store management or supervisors to handle.

REFERENCES

Anderson, J., and J. Narus. 1999. *Business Market Management.* Upper Saddle River, NJ: Prentice Hall.

Bauer, R., V. Tang, E. Collar. 1992. *The Silverlake Project: Transformation at IBM.* New York, NY: Oxford University Press.

Bitran, G. 2000. *Personal Communications.* Cambridge, MA: MIT Sloan School of Management.

BusinessWeek. 2003. "Broadband Telephony." *BusinessWeek.* Special Annual Issue. SpringCenter for Quality Management. 1995. *The Language Processing Method.* Cambridge, MA: Center for Quality Management.

Cohen, L. 1995. *Quality Function Deployment.* Reading, MA: Addison-Wesley.

Damelio, R. 1996. *The Basics of Process Mapping.* University Park, IL: Productivity Press.

Cook, H. E., Wu, A. 2001. On the valuation of goods and selection of the best design alternative. *Research in Engineering Design—Theory, Applications, and Concurrent Engineering,* 13(1), p42–54.

Frost, B. 2000. *Measuring Performance.* Dallas: Measurement International.

Green, P., and V. Srinivasan. 1990. "Conjoint Analysis in Marketing: New Developments with Implications for Research and Practice." *Journal of Marketing* October: 3–19.

Harbour, J. 1997. *The Basics of Performance Measurement.* University Park, IL: Productivity Press.

IEEE. 1999. *IEEE Standard for Conceptual Modeling Language Syntax and Semantics for IDEFX97 (IDEF Object).* July.

Kotler, P. 1991. *Marketing Management: Analysis, Planning, Implementation, and Control.* Upper Saddle River, NJ: Prentice Hall.

Kotler, P., and G. Armstrong. 2001. *Principles of Marketing.* Upper Saddle River, NJ: Prentice Hall.

Otto, K., and K. Wood. 2001. *Product Design.* Upper Saddle River, NJ: Prentice Hall.

Thurow, L. 1999. *Building Wealth.* New York: HarperCollins.

Urban, G., and J. Hauser. 1993. *Design and Marketing of New Products.* Upper Saddle River, NJ: Prentice Hall.

Womack, J., and D. Jones. 1996. *Lean Thinking: Banish Waste and Create Wealth in Your Corporation.* New York: Simon & Schuster.

12 Integrating a Requirements Process into New Product Development

Christina Hepner Brodie

Past experience has shown that establishing a robust set of requirements can be a challenge. Too many product development teams make the startling discovery at their beta site tests—or even worse, at launch—that their solution misses the mark or is not needed at all. Your teams can improve the success rate of their solutions by integrating a process that supports the proactive generation of requirements prior to concept development, and then tracks the evolution of those requirements throughout product development.

TYPICAL CHALLENGES IN ESTABLISHING A REQUIREMENTS PROCESS

Although many development groups pay lip service to the need for requirements management, problems persist. In some cases, product development leaders fail to prioritize this important dimension of their process. In other situations, developers don't fully understand the customer problem they are attempting to solve. They may believe that what customers ask for (solutions or specifications) are the requirements. Or they may believe that it's pointless to engage customers, because customers don't know or can't tell you what they really need.

In some companies, only the marketing department defines requirements, but R&D either doesn't really believe them or perceives them as too abstract to be useful. In highly technical companies, especially those where R&D drives development decisions, some engineers and senior managers believe that a product's failure is due only to a failure on the part of marketing and sales to properly convey its benefits. What the engineers and senior managers do not recognize is the possibility that requirements were poorly defined at the outset. Where the work of marketing and R&D is mostly separate, misunderstandings around requirements are more likely, and the atmosphere may be contentious or blameful.

An integrated approach—one that facilitates the cross-functional ownership and monitoring of requirements throughout product development—is often missing. By defining or improving the process of generating and tracking requirements, a company can integrate all requirement stages for a given project and measure overall effectiveness in fulfilling requirements. This chapter helps you assess your organization's requirements process and, if needed, provides guidance for improving it

I. CREATE YOUR PROCESS

Should you determine that your process needs improvement and are in a position to influence a decision to act, you will find here an outline of sequential steps that should help you shape and pilot a viable approach.

A. Ensure Buy-in at the Top

Without executive leadership, investing time and resources in a requirements management process and building compliance may prove futile. Leadership for robust requirements work comes from the team of executives or senior managers who oversee the product development process. If a requirements process is perceived to add value, this leadership team must establish, inspire, and guide the process design team in creating the right approach for their organization.

The design team can then establish a system to create, capture, and track the requirements for each project. This system will both expedite the process and create a way to measure how well you meet customer requirements and the impact of the process on product success.

B. Define Requirements Terminology for Your Organization

The initial challenge is to agree on related terminology. The dictionary defines a requirement as something that is necessary or obligatory—which is not always the case with requirements in product development. To further complicate matters, many companies use the terms "customer requirements," "product requirements," and "product specifications" interchangeably. Other companies have created their own distinctive terminology. This chapter sets out current working definitions and related examples as a starting point. You can create examples that are more relevant to your industry or company as needed.

C. Define Your Process

Once executive leaders have sanctioned a new requirements process and key participants have aligned on terminology, your next step is to create a cross-functional design team to increase organizational buy-in. The design team's initial task is to map and evaluate any existing parts of a requirements process and identify areas for improvement. Figure 12-1 is a simple self-assessment to help you begin. Your team's next step is to create a first draft of the sequence of steps involved in generating requirements. What follows is an illustrative, generic set of steps to give those responsible for design a place to start. Figure 12-2 illustrates a typical product development perspective of the flow of requirements generation.

For the following practices, please use this rating scale to answer the questions:

NA—The practice is not applicable to our industry
0—The practice is not evident
1—The practice is in place, but not really used
2—The practice is sometimes used
3—The practice is consistently used

1. We have a documented product development process.
2. We have a requirements process in place that is a documented part of our product development process.
3. Our requirements process scales to accommodate the complexity of the product development challenge.
4. Customer requirements are gathered by a multifunctional team.
5. Customer requirements are explicitly articulated by a multifunctional team.
6. Customer requirements are one key stimulus for idea generation and concept development.
7. Potential fulfillment of customer requirements are one set of criteria for internally screening solution concepts.
8. Once a new product concept is finalized, product requirements are explicitly documented.
9. Following concept development, the resulting product [and functional] requirements are reliably linked to the related customer requirements.
10. Functional groups align on the product/functional requirements set.
11. Product/functional specifications are explicitly written.
12. Performance targets and/or ranges are clearly established for each specification.
13. Product/functional specifications remain clearly linked to the product/functional requirements for which they were created.
14. Test requirements remain clearly linked to the product/functional specifications for which they were created.
15. Requirements are collected for continuous improvement throughout the life cycle of the product.

Looking at your individual item scores will enable you to see opportunities for process improvement.

FIGURE 12-1. Simple Requirements Process Assessment.

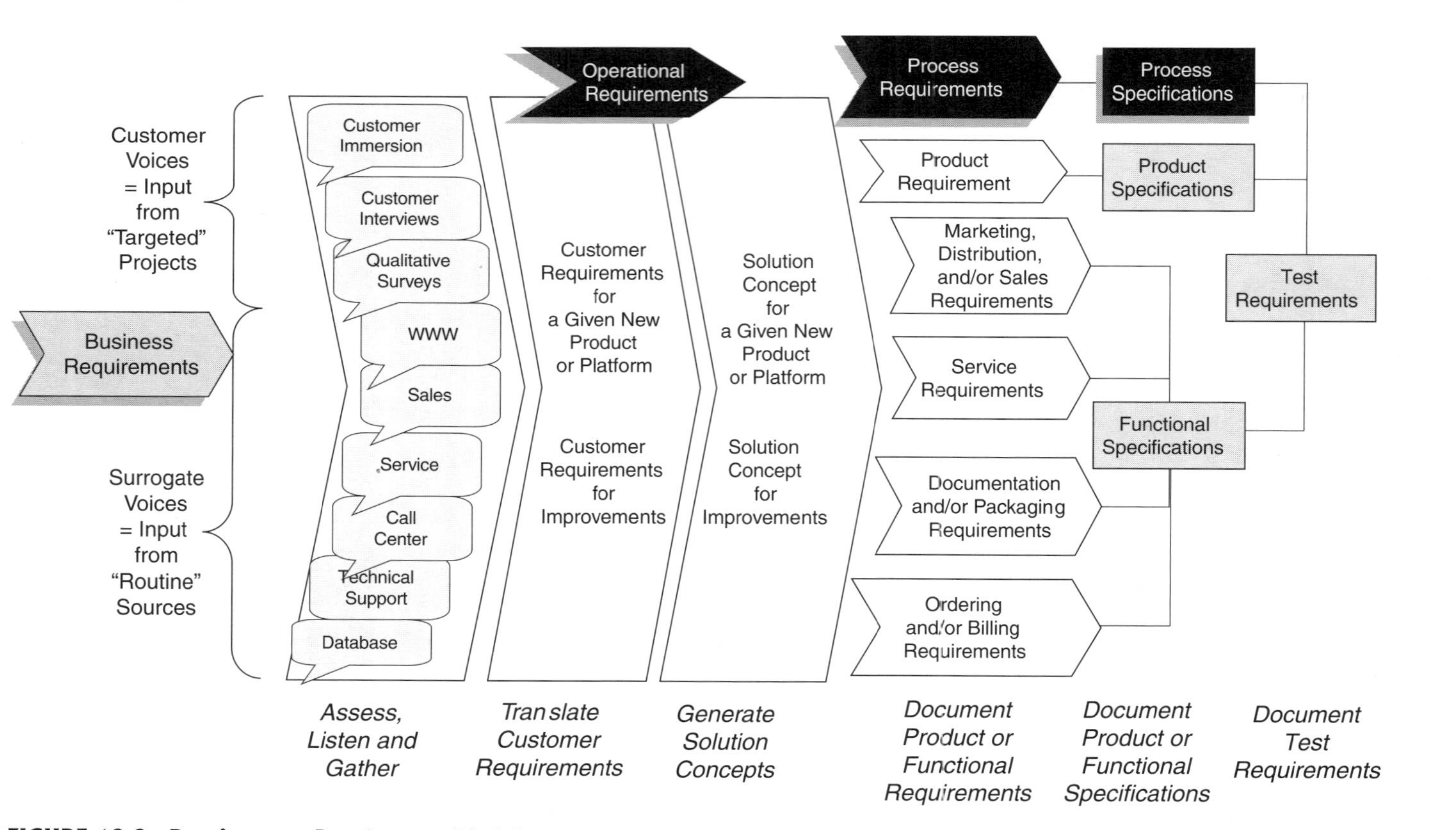

FIGURE 12-2. Requirements Development Model.
Copyright 2003 Pittiglio, Rabin, Todd, and McGrath. Used with permission.

STEP 1: ASSESS. Executives who oversee product development (e.g., a product approval committee or PAC) assess the portfolio of product development plans and projects, identify the targeted opportunity space and set the product innovation charter (PIC). This includes the business requirements for the project and establishes the area or solution space of interest.

DEFINITION

Business requirement: *Clearly articulated boundary conditions for the given project focus; these might include strategic goals that drive the project, target markets for the potential solutions, broad investment bounds, financial requirements, and so on.*

Once the domain of exploration or potential solution space for a given project has been defined, the cross-functional team tasked with delivery should assess the quality and quantity of customer input needed.

a. *Continuous improvement projects.* If continuous improvement of an existing product is the goal of a project, then sales, customer service, or other customer-facing personnel may be the best source of input for requirements.
   1. Idea banks, databases, and other resources that capture customer input, ideas, and complaints are often a good source of requirements.
   2. Depending on the scope of the project—how much change or improvement is sought—the team may benefit from spending time in the users' world to understand directly what their needs are.

b. *New product or platform projects.* If a project involves a new platform or new product development, the team should spend time with current users of related solutions after first collecting any existing knowledge from databases or customer-facing personnel. Standing in the cus-

EXAMPLE: BUSINESS REQUIREMENTS

After launching a platform prematurely, a medical products company lost market share and saw no return on investment because of costly rework. For the next-generation platform, executive management launched a cross-functional team charged with defining the customer requirements before creating the new platform concepts. The business requirements included the following:

- *Strategic objectives.* Regain market share with the next-generation solution and anticipate how the related medical practice will evolve over the next ten years.
- *Development focus.* Next-generation platform.
- *Target markets.* Worldwide, with project focus in Asia, Europe, and North America

tomers' shoes helps the team to better understand the problems and issues that need to be resolved. Even when the team's task is to create something new-to-the-world, they will benefit from understanding directly what customer needs their potential solution will address.

STEP 2: LISTEN. In the earliest phase of product or service development (Phase 0, or the Opportunity Phase), the cross-functional team members collect data that they will translate into customer requirements. Whether improving an existing product or creating a new platform or product, a cross-functional team initiates the work. Team members with diverse functional backgrounds are more likely to perceive the full range of needs than a single function such as marketing. (See Burchill and Brodie, 1997, for step-by-step how to plan and execute such an activity.)

One of the best ways to understand latent requirements is to put team members with technical knowledge into the users' environment. Often, users are limited by their experience and don't fully know what is possible. For example, before television sets had split screens that allowed viewing of more than one program at once, consumers would not likely have asked for this capability. Having no experience of it, they would not know it was possible. However, if technical folks spent time in customers' homes, they might have noticed two TVs in the same room. Why? Because someone wanted to watch two sports events at the same time or family members couldn't always agree on what to watch, and conflicts arose because of channel surfing among different programs. This insight would have revealed the unarticulated desire to "view multiple programs simultaneously."

STEP 3: IDENTIFY VOICES. After completing a customer visit or other customer interaction, the team's first task is to comb through the *verbatim* interview transcripts, notes, or observations to identify useful customer or consumer "voices." What is a voice? A voice is a verbatim quote from a customer interaction.

DEFINITION

Customer or consumer voice: *A verbatim quote from a customer interaction, such as an interview, a call center complaint, a qualitative Web interaction, or a quote from a service call. This might be a word, a sentence fragment, a sentence, or a passage.*

A "useful" voice is any passage that offers a clue as to what customers need. This could be a complaint, insight, solution, or any description that captures the context of what the customer is doing, or trying to do. Not everything collected will be useful, so team members must find those passages that

EXAMPLES: CUSTOMER AND CONSUMER VOICES

1. (From a medical market example) "I think one of the big problems is *sifting through all that data and trying to decide what's important. Most of our medical devices today display far too much data, and yet they don't focus on support, and all the numbers are the same size today without a lot of attention being paid to what's important.* Maybe the important stuff should be big and the other stuff should either be small, or maybe it should be hidden until it's needed, or until there's an alarm, or until it's called up by the user, and then it should pop up a window or something like that. So the equipment, I think, needs to become more aware of what sorts of things are important."

2. (From a financial services market example) ". . . [*Now, short-term goals sometimes can be at odds with that overarching objective. For instance, yield, or spread, or margin, or however you want to approach it—sometimes we find we don't deliver the yields that the company wants. And part of that is driven by the fact we continue to play in the upper tier. We focus on high credit and so we end up pricing it a little thinner, and so we're not getting the margins we'd like as a company.*] So, from a short-term perspective, right now we're trying to figure out how to get our margins up without sacrificing the quality. [*The credit quality is the reputation and hallmark of the company, and that's a conflict right now that's at the heart of a lot of debate in this company.*]

3. (From a consumer foods market example) ". . . *Sometimes I don't even feel like eating dinner; I'll just want a sandwich or just something like soup and salad—it really depends. Then other times I want everything—steak and the whole works. That's the other thing. It's kind of hard because sometimes I'll want to have a big dinner and they don't. Then they want a big dinner and I don't want to cook it. So our clocks are a little uneven.*"

help them to identify or intuit the customer needs that lie "in, with, and under" the language or observations.

In these examples, contextual imagery voices are italicized and voices that likely hold the seed of a requirement by themselves are underlined. Contextual imagery focuses on how the environment of use is now, or has been—not how they would like it to be in the future. Some images also hold a clue to a customer requirement. Other voices hold the clue to a requirement (e.g., a complaint, a wish, a suggested solution) but do not depict the context.

STEP 4: TRANSLATE. Once the key voices have been identified, the team members can translate these into explicit customer requirements. One of the benefits of using a cross-functional team to complete this work is that it increases the likelihood that they will discover a range of customer requirements. Customers may respond to open-ended questions with problems or challenges that functions other than product design and development could solve—requirements that could be addressed by sales or marketing strategies, documentation, packaging, service, or regulatory strategies.

DEFINITIONS

Customer or consumer requirement: *A sentence that describes from the customer's (or consumer's) vantage point the need/issue/problem that needs to be solved or solved more effectively. Other terms with similar meaning might be "market requirement" or "customer need."*

- Explicit customer requirement: *A sentence that describes an evident customer need that can be readily identified within a customer voice.*
- Latent customer requirement: *A sentence that describes an unexpressed customer need that may be intuited from a customer voice or observation, knowledge of customer experience, or a deep understanding of the context of (potential) product/service use.*

For example, the team from the preceding medical product project was developing a new piece of medical equipment. The team interviewed physicians, nurses, bio-medical technicians, staff developers, maintenance workers, and purchasing staff—a diverse set of specific people who would be involved in some way with the medical equipment. After visiting a variety of hospitals across three continents, the team was surprised to learn that in some regions, some of the people they interviewed did not take the time to learn how to correctly use this critical, potentially life-saving piece of equipment. The team also discovered that even after the people learned how to use the equipment, it was difficult for them to quickly understand what was important based on what they were seeing on the existing monitor.

There are two dimensions to creating customer requirements: (1) writing them as clearly as possible and as specifically as is appropriate and (2) thinking deeply about what has been learned from the customers' context and experience. The importance of customer context has become increasingly apparent in requirements generation. For example, if told that the next generation of business carrier—such as a briefcase, computer bag, backpack, or some new solution—must be "lightweight," a designer wouldn't have enough information. *Why* should it be lightweight? The contextual details enable the designer to understand what functional issues need to be solved by making the carrier lightweight (e.g., the business professional has back problems, or lifts frequently for overhead storage, or transports heavy content, etc.). The contextual knowledge helps define the true functional issue that needs to be solved—lightweight is one dimension of the solution that might fulfill the user need.

For important projects, teams gather both explicit and latent customer requirements. Explicit requirements often capture issues that haven't been completely resolved—either because resolution wasn't technically possible or cost-effective in the past or because of a lack of attention or creative focus. If a team can create a better solution than existing products offer, the odds of a

EXAMPLE: WRITING AN EXPLICIT CUSTOMER REQUIREMENT

Customer voice: ". . . sifting through all that data [on the monitor] and trying to decide what's important . . ."

A customer requirement is typically written as a sentence that aims to describe what needs to be solved from the customers' vantage point. At its heart is an action verb—the description of the functional issue that needs to be solved. In this case, the key issue is "comprehend," "determine," or "perceive." Next, the translator must address who should be doing the function. Here the obvious subject of the sentence would be "physician." So, the apparent first crafting of the requirement is as follows.

Explicit customer requirement: Physician determines which information is important.

The translator recognizes that while this is a relatively clear requirement, it merits additional thinking. She could make it more specific by qualifying the specialty of the physician and/or by defining the nature of the information. Since many kinds of specialists might use the equipment, she leaves the subject more abstract. She also realizes that, at this point, she is not clear about which information is the "important" information, so she leaves that abstract as well. However, she notices that this is a two-valued requirement—a yes or no. It means that the physician can or cannot determine which information is important. As the translator thinks about this requirement, she realizes that such physicians are trained to know what's important now, and that by considering the context of the voice, she can think a little more deeply about the issue:

Context of voice: "Most of our medical devices today display far too much data, and yet, they don't focus on support, and all the numbers are the same size today without a lot of attention being paid to what's important . . ."

Explicit customer requirement: Physician determines *readily* which information is important.

By adding "readily," she now has created a multivalued requirement, capable of being measured on a scale.

successful product are good. In mature product categories, identifying the latent issues becomes much more important, as solving those offers greater opportunity for differentiation.

STEP 5: SELECT. A team that has thoroughly explored a range of sources will likely identify more requirements than can realistically be solved. To choose which ones to address, some teams ask a select group of customers which issues are the most important. A reasonable number is 25 to 35—but again, the team must do some relevant thinking. Given the complexity of the product, the medical equipment team in the preceding example chose 28 requirements that the equipment itself would solve, another 20 that would be addressed by marketing strategies and service, and 25 that related to how the product would fit within the hospital environment.

The Center for Quality of Management (Burchill and Brodie, 1997) developed an approach to selection in which team members act as surrogates for the

EXAMPLE: WRITING A LATENT REQUIREMENT

Context plays a key role in illuminating unarticulated or latent issues. If the equipment team mentioned previously is interested in identifying latent customer requirements, intuition and technical knowledge can help. Knowing the context facilitates these intuitive connections. In this case, by linking a voice like, "I need three hands in the midst of a procedure—one for the patient and two for the equipment" with contextual understanding, "Multiple knobs and levers used together to control the flow," plus technical knowledge of what's possible, more subtle thinking can evolve:

Latent customer requirement: Physician controls the flow of *x* to the patient, hands-free yet safely.

In this case, users of current equipment who were interviewed might not suggest hands-free operation—as they might not realize such a need could be met. But by thinking about the voice (three hands) in context (knowledge of current interface explains the why "three hands"), the team was able to intuitively understand the latent requirement. The words "flow of *x*" imply a dimension of the solution. If the team knew they could recommend a far-in-the-future generation of solution beyond current practice, they would be careful not to include even this slight implication. They knew, however, the solution had to fit into the existing practices and hospital infrastructure. If they could solve safe, hands-free control at this juncture, it would be significant—even though years later, new scientific knowledge might eliminate the need for any equipment at all.

customers they visited. They copy each requirement and its source onto a Post-it Note®. Then they post the requirements into logical categories on chart paper on the wall. The selection process proceeds as follows:

- Each team member takes a red pen and places a dot on the requirements that he or she deems most critical, based on customer input.
- A requirement needs only one dot in each round, and dots shouldn't be added when one already exists. (This is a choosing process, not a voting process.)
- At the end of the round, the team removes the requirements with no dots.
- Team members repeat the process in round two, adding a second dot to the requirements they believe should go forward and removing those with only one dot at the end of the round.
- A team member counts the total number of requirements at the start of the process and at the end of each round.
- The process is repeated until the team is within 30 percent of the target number of requirements.

When the team reaches that point, the approach is shifted:

- A team member divides the target number of requirements (25 to 35) by the number of team members participating. This yields the number of final choices each person will have.

- Taking turns, each person announces his or her first choice and explains why it's an important requirement. Hearing the selections aloud ensures that the team covers the full range of important issues.
- The team continues with additional rounds until the target number of requirements is selected.

STEP 6: PREPARE TO TEST AND MEASURE. To measure how well your product or service meets customer requirements, set up customer-oriented metrics after generating the requirements but before developing concepts. Make these metrics a customer-focused component of the testing that comes later in product development—testing designed to objectively evaluate the final concepts, models, or prototypes of the solution. Include a description of the tests and measures to be used, and what is known to date about the performance targets for each customer requirement. This exercise often helps the team to clarify requirements even more fully and sometimes refine the language of the requirements. Doing this work before validating the requirements gives you the potential of collecting additional target performance information during that step.

STEP 7: VALIDATE CUSTOMER REQUIREMENTS. While qualitative techniques deepen understanding of customers' needs, it is also critical to validate this data with a quantitative sample of customers before making significant investment decisions. A common tool to validate and prioritize customer requirements is the Self-Stated Importance Survey. Some teams build in comparable "satisfaction" questions to determine how satisfied customers are with how current products meet specific customer requirements. This allows the team to analyze the satisfaction gap relative to the importance of the requirement. For identifying latent requirements, teams can use a Kano survey instead of—or in conjunction with—the Importance Survey. The Kano survey looks at four types of requirements: "must be," "adds value," "delighters," or "indifferent." Customers rarely mention their "must-be" requirements, assuming they are obvious and will be solved. Moreover, since "delighters" are latent and not clearly articulated in their minds, customers rarely mention them either. Requirements that are latent *and* important—and can be solved—offer a major opportunity for product differentiation.

STEP 8: DEVELOP SOLUTION CONCEPTS. The next step is idea generation and concept development for the selected customer requirements. Focusing creative thinking on the customer issues increases the likelihood of product or service concepts that are truly responsive to what customers need. As noted earlier, the multifunctionality of the team is critical here, since customer requirements may be solved by such nonproduct solutions as marketing, service, packaging, sales strategies, and/or documentation.

STEP 9: ESTABLISH OPERATIONAL REQUIREMENTS. Prior to final specification, a team will be well served to collect the operational requirements—those internal factors that are needed to deliver a product to market.

DEFINITION

Operational requirement: *A sentence that describes from the internal customer's vantage point the need/issue/problem that needs to be solved or solved more effectively.*

It is important to define potential development constraints (e.g., Do we have the capabilities to actually develop the functionalities we've defined?), potential marketing or sales constraints (e.g., Do we have the resources to complete the needed market research later in the development cycle—and for preparing effectively for launch? Can our sales organization prepare sufficiently to approach this new channel?), and potential manufacturing challenges (e.g., Do we have the capability and capacity to scale up as needed? Do we have the suppliers and other supply chain partners we need? Will the timing work for our projected schedule?). If your team and its sponsors can't fulfill these internal, operational requirements, you will have to change your strategy, adapt your final concepts accordingly, or creatively resolve the internal need. If your company produces its own products, some of the operational requirements may lead to process requirements.

STEP 10: SCREEN CONCEPTS INTERNALLY. The next step is to evaluate the team's concepts against all three sets of requirements—business, customer, and operational. With cross-functional alignment on customer and operational requirements, plus the knowledge of business requirements that comes with

EXAMPLE: TRANSLATION OF AN INTERNAL VOICE INTO AN OPERATIONAL REQUIREMENT

Internal operations customer voice: "Our development strategy should differentiate by product performance but take advantage of our existing manufacturing processes." Translates to:

Operational requirements: 1. Build image intensity calibration into standard product quality testing (PQ). 2. Sterilization uses standard company processes.

Process requirement: Device will be sterilized using a high-volume low-cost process and must house pallet size of 4 by 4 by 5 feet.

Process specification: Use 100 percent EtO cycle #73 (SOP ST-010 at 125F, for 4 hours with pressure < 3 atm).

the team's charter, you will have the basic criteria needed to decide how best to meet all three sets of requirements. Common tools for focusing screening discussions include Pugh's Concept Selection Process and PRTM's Simplified Quality Functional Deployment (S-QFD).

STEP 11: GENERATE PRODUCT/FUNCTIONAL REQUIREMENTS. Following concept screening and selection, it is important to test concepts, at least with key customers, so that you can refine the concepts prior to development if necessary. This step typically occurs in Phase 1, or the definition phase of your product development process. Once the multifunctional team has aligned on its final concept(s), product (or other functional) requirements can be written. In the solution space, one customer requirement may turn into multiple product or functional requirements—statements about what the solution will do to solve the customer need. While the customer requirements state *what* needs to be solved, the product requirements state *how* to solve the problem. In turn, these "hows" become "whats" in the next iteration of statements—a classic, QFD way of thinking about the relationship between one phase and the next.

DEFINITION

Product or functional requirement: *A sentence that describes the functionality of the solution that will solve the customer's problem. That functionality might be solved by product design or by other functional solutions like marketing programs, packaging, sales strategies, service support, documentation, and so on.*

EXAMPLE: WRITING PRODUCT AND FUNCTIONAL REQUIREMENTS

Customer requirement: Physician determines readily which information is important.

Product requirements (a few of many that would address the customer requirement):

Here the team decided that part of the solution would be to use graphic design and the underlying software algorithms to automatically differentiate among types of data.

- Data display uses consistent color and fonts across multiple screens to distinguish critical information.
- Data display uses consistent fields across multiple screens for similar kinds of data.

Functional requirements (a few of many that would address the customer requirement):

- Promotional literature clearly defines the distinctions among the important data relative to each key monitor function.
- Sales process includes demonstration of features that address quick comprehension of important data by professional staff.
- Sales process invites professional staff members to experience the ease of using a typical equipment scenario and encourages assessment of ready determination of what is important.

During this step the team documents additional product requirements—beyond those that flow from the solutions to customer requirements. For instance, customers rarely acknowledge their "must-be" requirements, as noted earlier. But many of these are fundamental to the overall concept. If the solution is new or radical, it may not address any of the "must-be" requirements, because it is solving the need in a totally different way.

STEP 12: GENERATE PRODUCT/FUNCTIONAL SPECIFICATIONS. At the beginning of development (Phase 2, or the Development Phase of your product development process), product or functional specifications are set for each of the product or functional requirements. These are the technical details of how each dimension of the solution will be executed, as well as the performance targets for the related functionality. If your team created customer-oriented metrics earlier, you may have already identified some of these performance targets and even validated them with customers during the quantitative survey.

DEFINITION

Product or functional specification: *A description of a specific technical solution and related performance targets for a given product or functional requirement.*

STEP 13: DEVELOP TEST REQUIREMENTS. During Phase 3, or the test phase, customer, product, and operational requirements may be factored into creating the final test requirements. These will be used to evaluate how closely the developers have come to fulfilling their requirements and performance targets. The nature of the original requirement determines whether lab tests are suffi-

EXAMPLE: PRODUCT OR FUNCTIONAL SPECIFICATIONS

Product requirement: Data display uses consistent color and fonts across multiple screens to distinguish critical information.

Product specifications:

- Color luminance for critical information areas is at least 50 percent brighter than for noncritical areas.
- Font size for critical information is at least six points larger than noncritical data.
- Critical information uses at least two indicators (e.g., color plus texture) to accommodate for color blindness.

Functional specification:

For those requirements that market research determines are the most important to medical personnel (e.g., finding critical data readily), medical professionals in the sales process will learn of the solutions to these requirements in at least three different ways (including visual, auditory, and experiential modes).

EXAMPLE: TEST REQUIREMENTS

Customer requirement: Physician determines readily which information is important.

Product requirement: Data display uses consistent color and fonts across multiple screens to distinguish critical information.

Product specifications:

- ◆ Color luminance for critical information areas is at least 50 percent brighter than for noncritical areas.
- ◆ Font size for critical information is at least six points larger than for noncritical data.
- ◆ Critical information uses at least two indicators (e.g., color plus texture) to accommodate for color blindness.

Lab Test:

- ◆ Systematically check each field of critical data in prototype.

Measure:

- ◆ Luminosity.
- ◆ Count number of alternative means to distinguish target data.

Beta Test (part of larger beta test):

- ◆ Invite an uninitiated color-blind user to identify the critical data.

Measure:

- ◆ Time to identify which data are critical.
- ◆ Success in distinguishing critical data by other than color (yes/no).
- ◆ Satisfaction on a given scale with potential ability to determine critical data during a crisis.

cient or whether user and/or field tests are necessary. Lab tests and beta site tests are conducted for validation of prototypes or early products—and to identify necessary refinements to the solution. Once again, the cross-functional team establishes the test requirements based on all of the requirements that precede this step.

DEFINITION

Test requirement: *A description of the tests, measures, and performance targets that will enable objective evaluation of how well the solution fulfills a given requirement*

D. Establish a Way to Track Requirements Evolution

A key to success in the requirements process is the ability to track the development of requirements from one phase to the next and the impact of any changes made throughout development. For tangible solutions—products that need to be manufactured—the further into development a project is, the costlier it is to change a product requirement or specification. But sometimes it is necessary to make a change. Certain technical aspects may not be as feasible as assumed or may prove too costly, a competitor may launch an unexpected product that changes the game, or the market may change in other ways. The key is to have visibility into the changes and understand all of their implications.

STEP 1: ESTABLISH A REQUIREMENTS MANAGEMENT POLICY. Depending on the variability of development scope in your organization, consistent management policies may work. However, in some companies, policies may need to adjust relative to the scope of a project (e.g., new platform vs. new product vs. improvement to an existing product). Policies clarify roles and responsibilities—who will develop the requirements for each phase, who will maintain them, and who has authority to change them. Even the "who" might vary. For instance, a person of authority might manage the high-risk product requirements or specifications, or those that represent critical elements of the solution, while someone else manages the less critical requirements.

While all team members involved with developing the solution will need access to the requirements—and may well have played a role in creating them—policy sets the parameters for changing them, how to get buy-in for a change, and how to make the change. The more specific levels of requirements may be more flexible early in development than their more abstract antecedents. The management policy also clarifies when requirements will freeze at all levels of specificity.

STEP 2: CREATE A WAY TO UNIQUELY IDENTIFY A REQUIREMENT AND ITS RELATED SUBSETS. To see the relationships between each requirements phase, you need a way to link subsequent requirements to their predecessors. Typically you would establish a numbering system or other coding protocol that indicates these relationships.

Besides uniquely identifying each individual requirement, creating such a system enables teams to group the requirements by logical, related batches at the more abstract levels and make visible the linkages between and among them to the finest degree of granularity. Managing requirements in this way can be costly, but it's worth it. Making the investment of time and resources reduces your risk. Any time a change is made during development, it will likely have a ripple effect on other elements or dimensions of the solution—with potentially expensive consequences. That's why it makes sense to clearly identify the relationships and the impact of a change. Another benefit to having a well-managed set of requirements is that your teams can see very clearly where they are in their process and when their development is complete.

STEP 3: ESTABLISH REQUIREMENTS DOCUMENTATION. Organizations often track the evolution of requirements with a spreadsheet or other customized tool. In addition to the requirements being documented as outlined, the original sources of the requirements should be noted as well. Given the richness of

EXAMPLE: UNIQUELY IDENTIFYING REQUIREMENTS

1. Physician determines readily which information is important.
   - 1.1 Data display uses consistent color and fonts across multiple screens to distinguish critical information.
     - 1.1.1 Color luminance for critical information areas is at least 50 percent brighter than for noncritical areas.
       - 1.1.1.1 Design detail one.
       - 1.1.1.2 Design detail two.
     - 1.1.2 Font size for critical information is at least six points larger than noncritical data.
     - 1.1.3 Etc.
   - 1.2 Data display uses consistent fields across multiple screens for similar kinds of data.
     - 1.2.1 Critical information uses at least two indicators (e.g., color plus texture) to accommodate for color blindness.
     - 1.2.2 Etc.
   - 1.3 Etc.

this information, and the need for organization and linkage, a database solution or a systems approach can make documentation far more efficient. Software now exists that helps capture and monitor requirements—especially during the development and test phases of the process. At this writing, companies are finding that off-the-shelf solutions are suboptimal but can be adapted.

E. Establish Roles and Responsibilities

OVERSIGHT. Once you have set up clear expectations for requirements work and deliverables, the review committee or PAC must ask the development team questions related to the requirements dimension of their process at phase reviews or design reviews as appropriate (see Belliveau et al., 2002) This will ensure that the teams have completed sound requirements work. The executives on the review committee may need to be oriented to this added dimension and coached if they don't have experience themselves with writing clear requirements.

A consistent review committee that tracks the evolution of requirements for each project minimizes the odds that feature creep will derail a product's efficient journey through development. Furthermore, when management can trace the decision making about changes in requirements, they can better understand the effect on process cycle times.

DOCUMENTATION. Assign one team member to document the evolution of requirements in each phase of product development. As noted previously, policy may establish responsibility for documentation. It could be part of the team leader's role, or it may be more appropriate to rotate the responsibility, depending on the phase of requirements a team is developing. For instance, marketing might document the customer and product requirements, while R&D documents the specifications. Alternatively, in a highly technical project, a marketing team member might document the customer requirements; then each functional member might be accountable for documenting the product and functional requirements—and in turn the related specifications and test requirements. The key question to ask is this: What's the best way for *our* organization to accurately and thoroughly document requirements during each phase of development?

TRACKING. Once you have set rules for changing requirements, you need controls to ensure that the rules are followed. Make someone accountable for tracking the status of requirements and any changes made to them. As noted previously, while changing requirements is sometimes necessary (e.g., in response to a change in the market, a technical breakthrough, or a failed experiment), keeping changes to a minimum will make your process more efficient and get your products and services to market more quickly.

F. Use the Process to Improve Products throughout Their Life Cycle

Once a new product has been launched (Phase 4, or launch and maintain), customer satisfaction metrics based on product or service use are the ultimate measure of how well requirements have been fulfilled. These metrics, along with customer feedback, are also a source of ideas for continuous improvement, enhancements, or variations. Thus, the cycle begins again.

II. PILOT YOUR PROCESS

Test the new requirements process and the value it delivers before rolling it out to all development teams. The scope of your pilot will vary depending on your project portfolio and how often you start new projects. If your organization develops few new products and each takes several years of development, you may not be able to conduct a full pilot. Your goal is to identify up to three projects that are about to begin. Have the teams follow the new steps and guidelines and provide feedback about what is clear, what works, and so on. If the pilot teams can get to the specification stage of evolving their requirements, you should be able to learn enough to refine your process prior to rolling it out.

III. REFINE YOUR PROCESS

Feedback and insight from the pilot teams' experiences will be valuable for refining your process. What's described in this chapter is necessarily generic, given that companies in a wide range of industries will read it. Your process must be tailored to your specific situation. For instance, if your company has created many generations of a product, you may have preset product requirements that you have built into your documentation in the product requirements step—requirements having to do with durability, safety, or regulatory issues. As noted previously, you may have document-specific "must-be" requirements for a given concept as well. Your process must define when and how new product or functional requirements are established and integrated.

Your rules for complex solutions may be different than those for less complex solutions. For many products or services, 30 customer requirements are sufficient to define a new opportunity. But in a complex product, a team may begin with multiple sets of customer requirements for various dimensions of the solution. For example, a system-on-a-chip (SOC) chip-testing equipment team determined it was important to have a distinct set of customer requirements that would be solved by hardware and an additional set that would be solved by software. That team identified about 40 of each before it was satisfied it had the "key customer requirements."

Service and software solutions present another issue. A solution can have an infinite number of options that, even once established, are readily changeable. The more permutations there are in the solution space, the more difficult it is to track them. One experienced R&D software engineer observed, "Software is difficult for management, because it is difficult to see if it is a success or failure. Code is becoming so complex that human beings can't get their minds around it. The solution components are so variable that one can create an infinite number of combinations in the solution space—how does one measure success against such diversity in customer specifications? Failure is usually easy to spot; success is the one that's hard."

IV. HARD-WIRE YOUR PROCESS

INTEGRATE YOUR REQUIREMENTS PROCESS INTO PRODUCT DEVELOPMENT. Once your requirements process is refined, it must be "hard-wired" into your overall product development process. The related steps and tasks must be built into process documentation and guidelines. This means making the linkages explicit at multiple levels of the product development process—including roles and responsibilities, work flow, templates, decision criteria, metrics, related information systems, and rewards. Some teams may need training on how to gather customer input, how to translate that input into clear requirements, how to document those requirements, and so on. Project oversight committees must ask questions about requirements work at phase reviews and recognize excellence, or when poor work has been done. They must be comfortable confronting poor work and requiring that a project team go back to improve or complete its requirements work. Such actions model to the overall organization that doing a solid job of defining requirements is important to overall product success.

CONFRONT THE CULTURE. In most cultures, people are rarely eager to change or integrate "one more process." Here, you are asking people to show exactly what they are doing. A common process reduces independence and increases accountability. When people feel they are giving up control, they may resist embracing a more deliberate, proactive process.

Seagate became aware of the need to better manage requirements. Scott Warmka, a systems engineer and leader of the company's process design team, explains, "We are driven by the need for global deployment, and we are increasingly using statistics to measure our effectiveness in fulfilling requirements. Not only do we want to assure that we effectively meet a given requirement, we want to be able to measure our success. Because of our geographically dispersed organization, we need common requirements language that works globally. A designer in Oklahoma should be able to log on and clearly understand the requirements for a project in Singapore." Warmka has encountered organizational challenges: "Right now we're focusing on infrastructure; we're building key organizational linkages. For us, managing our requirements process also represents a fundamental change culturally—and such changes always take time."

MEASURE THE RESULTS. Establish the criteria against which your organization will measure effectiveness. For instance, qualitatively: How well do team members perceive themselves to be aligned on the meaning of requirements and expectations for fulfilling them? Or quantitatively: What is the reduction in number of engineering change notices (ECNs)? What is the reduction in rework prior to launch? Ultimately, how successful is the solution in the market?

What gets measured gets done.

SUPPORTING THE CASE FOR CHANGE

Why You Should Care about Requirements

Why should a company care about requirements? When product development teams do a solid job of articulating customer requirements prior to concept development, they are more likely to generate product or service concepts that fulfill customer needs. Research shows that "successful products have much sharper definition prior to development" (Cooper, 1987). According to Cooper, projects that have these sharper definitions

- are 3.3 times more likely to succeed;
- have higher market share (by 38 points, on average); and
- are rated 7.6 out of 10 in terms of profitability (versus 3.1 out of 10 for poorly defined projects).

Further support for the importance of customer input to product requirements comes from a 1997 research study that found that "in 80% of the [product] successes, the developers either had a greater than average level of prior market knowledge, and/or collected a greater than average amount of market information and used a greater than average amount to set the product specifications.

"Similarly, in 75% of the [product] failures, the developers knew less than an average amount of market information, or ignored it when setting product specifications. This ability to separate success from failure 75% of the time indicates that while market information processing does not guarantee success, it raises the odds considerably" (Ottum and Moore, 1997).

Some company leaders also are pressing for change as the demand for new product introductions speeds up. One manager at an electronic components company noted that, "When we had only a few products, formally managing our requirements sets was not so critical. Now that the number of our products and their complexity are increasing, and now that we are entering new markets, we must find an effective way to keep track of them."

Launched in January 2003, HP's AlphaServer received rave reviews from customers and analysts alike. It even attracted new customers—in spite of being the final generation of the AlphaServer line. Revenue projections are in the billions of dollars. When asked why this product was so successful, even though the development team survived two company mergers during the project, Subhash Dandage, the program manager, observed, "Our work in collecting the voice of the customer and translating it into a solid set of customer requirements aligned our team. That alignment kept our team motivated and very focused in face of the external distractions. We kept that set of requirements literally on the wall throughout the five years of development. They focused our work and eliminated the usual 'feature creep.' We were able to deliver a solution that customers really like . . . on time and at half the development cost of the previous platform!"

Companies that make requirements a requirement are better able to trace decision making and measure other critical cycle time elements. But the acid test of solid requirements work, followed by the generation of responsive solutions, comes after market launch. A solid requirements process leads to a measurable increase in products that meet or exceed the key product development objectives—market success and profitability.

REFERENCES

Burchill, Gary, and Christina Hepner Brodie. 1997. *Voices into Choices: Acting on the Voice of the Customer*, p. 217. Madison, WI: Joiner Publications (Oriel Inc.).

Cooper, Robert G., and Elko J. Kleinschmidt. 1987. "New Products: What Separates Winners from Losers?" *The Journal of Product Innovation Management* 4, 3: 169–184.

Deck, Mark J. 2002. "Decision Making: The Overlooked Competency in Product Development." Chapter 7 in *The PDMA ToolBook for New Product Development,* ed. Paul Belliveau, Abbie Griffin, and Stephen Somermeyer. New York: John Wiley & Sons.

Ottum, Brian D., and William L. Moore. 1997. "The Role of Market Information in New Product Success/Failure." *The Journal of Product Innovation Management* 14: 258–273.

13 Toolkits for User Innovation

Eric von Hippel

"Listen carefully to what your customers want and then respond with new products that meet or exceed their needs." That mantra has dominated many a business, and it has undoubtedly led to great products and even shaped entire industries. But slavishly obeying that conventional wisdom can also threaten a company's ability to compete. The difficulty is that fully understanding customers' needs is often a costly and inexact process. Even when customers know precisely what they want, they often cannot transfer that information to manufacturers clearly or completely. Today, as the pace of change in many markets accelerates and as some industries move toward serving markets of one, the cost of understanding and responding to customer needs can easily spiral out of control.

Fortunately, an entirely new approach is being developed on the basis of patterns evolving in a few high-tech fields. In this new approach, manufacturers actually *abandon* their increasingly frustrating efforts to understand users' needs accurately and in detail. Instead, they learn to outsource key *need-related* innovation tasks to their users, after equipping them with appropriate toolkits for user innovation.

Toolkits for user innovation are integrated sets of product or service design tools that are specific to a given product or service type and to a specified production system. They are user-friendly enough to be used by customers with minimal training. Customers use the toolkits in conjunction with their own rich understanding of their needs to design a custom product that is just right for them. Customer-developed designs implemented on a toolkit are producible as is—they can be directly transferred to a manufacturer for production.

Toolkits for user innovation give users real freedom to innovate, allowing them to develop their custom product via iterative trial and error. That is, users can create a preliminary design, simulate or prototype it, evaluate its functioning in their own use environment, and then iteratively improve it until satisfied. As the concept is evolving, toolkits guide the user to ensure that the completed design can be produced on the intended production system without change.

A variety of industries have pioneered and shown the value of the toolkits for innovation approach. International Flavors and Fragrances (IFF), a global supplier of specialty flavors to companies like Nestlé and Unilever, has built a toolkit that enables its customers to develop their own flavors, which IFF then manufactures.

In the materials field, GE provides customers with Web-based tools for designing better plastic products. In software, a number of consumer product companies provide toolkits that allow people to add custom-designed modules to their standard products. For example, Westwood Studios provides its video game customers with toolkits enabling them to design important elements of their own games such as the game scenario. Many users engage in this activity and publish what they have developed online for the use of other gamers. Westwood Studios then incorporates the best solutions into the games it sells (Jeppesen, 2002). Open-source software allows users to design, build, distribute, and support their own programs—no manufacturer required. Indeed, the trend toward customers as innovators has the power to completely transform industries. Results to date in the custom semiconductor field show development time cut by two-thirds or more for products of equivalent complexity and development costs cut significantly as well via the use of toolkits. Today, many billions of dollars of custom ICs are sold each year that have been designed by users and produced in the silicon foundries of custom IC manufacturers such as LSI (Chakravarty, 1991; McClean, 1995).

Although now only applied to the development of a few types of products and services, the toolkits-for-user-innovation approach to product design is likely to spread to many producers that create custom products and services for both industrial and consumer markets having heterogeneous customer needs. Toolkits will also provide the design side that is currently missing for users and producers of mass-customized products. In effect, toolkits for user innovation can provide users with true design freedom—as opposed to the mere opportunity to choose from lists of options that is currently offered by mass customizers.[1]

This chapter begins with an explanation of the benefits of shifting need-related design activities to users. It then explores how this can be achieved via toolkits for user innovation and details the elements of such a toolkit should contain. Finally, where and how toolkits can be most effectively applied is discussed. Keep in mind that building a toolkit is a serious undertaking and that the contents of this chapter will not enable you to simply connect the dots and end up with a toolkit. However, you *can* expect to gain a good understanding of toolkits and to gain the information needed to assess whether toolkits are a realistic option for your organization.

BENEFITING FROM SHIFTING DESIGN ACTIVITIES TO USERS

At first glance, it does not seem to make much sense: Why should one be able to develop better products and services faster by transferring need-related work from the manufacturer to the user? After all, the same work is being done in both cases. However, there are, in fact, great advantages having to do with (1) better access to user information on needs and (2) with achieving faster, better, and cheaper learning by doing (Arrow, 1962; Rosenberg, 1982).

In essence, product development is often difficult because the need information about what the customer wants resides with the customer and the "solution" information about how best to provide a solution for those needs often lies with the manufacturer. Traditionally, the burden has been on manufacturers to collect the customer need information through various means, including market research and information gathered from the field (see Chapters 7 through 9). The process can be costly and time-consuming because customer needs are often complex, subtle, and fast changing. Frequently, customers don't fully understand their needs until they try out prototypes and learn exactly what does—and doesn't—work. This process is called learning by doing.

Not surprisingly, traditional product development can be a drawn-out process of trial and error, often ping-ponging between manufacturer and customer. First, the manufacturer develops a prototype based on information from customers that is incomplete and only partially correct. The customer then tries out the product, finds flaws, and requests corrections. The cycle repeats until a satisfactory solution is reached, frequently requiring many costly and time-consuming iterations.

To appreciate the extent of the difficulty, consider product development at International Flavors and Fragrances. In this industry, specialty flavors are needed to bolster and enhance the taste of nearly all processed foods because manufacturing techniques weaken the real flavors. The development of those added flavors requires a high degree of customization and expertise, and the practice remains more of an art than a science.

A traditional product development project at IFF might progress in the following way: A customer requests a meaty flavor for a soy product, and the sample must be delivered within a week. IFF marketing professionals and flavorists jump into action, and the sample is shipped in six days. A frustrating three weeks ensue until the client responds with, "It's good, but we need it less smoky and more gutsy." The client knows precisely what that means, but IFF flavorists find the request difficult to interpret. The result is more frenzied activity as IFF struggles to adjust the flavor in a couple days. Depending on the product, IFF and the client could go back and forth for several more iterations. This represents a huge problem because clients often expect IFF to get the flavor right the first time, or within two or three iterations.

IFF addressed this problem by creating an Internet-based toolkit for innovation to shift need-related innovation activities to customers. The toolkit contains a large database of flavor profiles plus the design rules required to assemble these into new or modified flavors. To protect IFF intellectual property, specific chemical formulations are *not* included in the toolkit. A customer can design a custom flavor on a computer screen and send her new design directly to an automated machine that manufactures a sample within minutes. After tasting the sample, the customer can make any adjustments that are needed. If the flavor is too salty, for instance, she can easily tweak that parameter on the profile and have the machine immediately produce another sample (see Figure 13-1).

Note that outsourcing product development to customers does not eliminate learning by doing—nor should it. What it does is make traditional product development better and faster, for three reasons. First, a company can bypass the expensive and error-prone effort to understand customer needs in detail. Toolkits are significantly better at satisfying subtle aspects of customer need because customers know what they need better than manufacturers do. Second, the trial-and-error cycles that inevitably occur during product development can progress much more quickly because the iterations will be performed solely by the customer. Third, if customers follow the rules embedded in a toolkit (and if all the technological bugs in the toolkit have been worked out), the designs they send to the manufacturer will be suitable for manufacture the first time around.

Toolkits also provide other major types of benefits to a business. For example, supplying toolkits for user innovation to customers can help a company retain the smaller customers that were prohibitively expensive to work with before. This can greatly expand their accessible market and also reduce the pool of unserved, frustrated potential customers that might turn to competitors or to new entrants into the market. Toolkits can also enable compa-

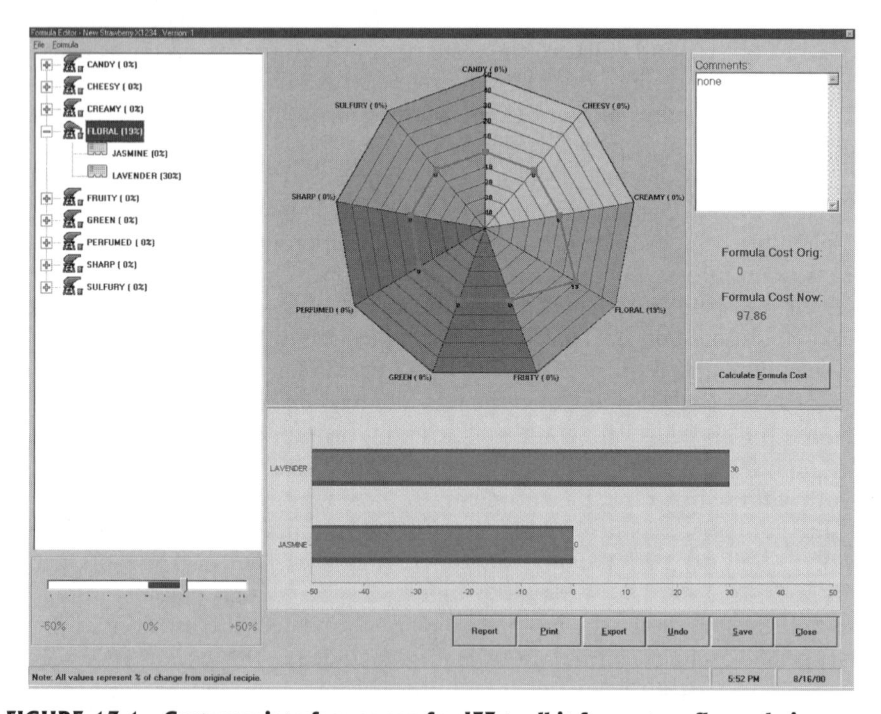

FIGURE 13-1. Customer interface screen for IFF toolkit for custom flavor design.

nies to better serve their larger, preferred customers. That's a benefit most suppliers wouldn't expect, because they'd assume that their bigger customers would prefer the traditional hand-holding to which they're so accustomed. Experience shows, however, that such customers are often eager to use a toolkit, especially when fast product turnaround is crucial.

To illustrate these advantages, consider the case of GE Plastics. This GE division does not design or manufacture plastic products but sells resins to those who do. The properties of those resins must precisely match the requirements of both the end product, such as a cell phone, as well as the process used to manufacture that product. With the formation of the Polymerland division in 1998, GE Plastics gave customers access to company data sheets, engineering expertise, and simulation software to aid customers in both designing plastic parts and designing the molds needed to make those parts accurately. Customers use that knowledge and simulation technology to conduct their own trial-and-error experiments during their design process. For example, they can use GE simulation software to model how a certain grade of plastic with a specific amount of a particular type of reinforcement will flow into and fill a mold they are designing. The approximate cost of bringing such sophisticated tools online: $5 million.

GE Plastics, of course, did not make the investment simply to be magnanimous. Through the Web site, the company identifies and tracks people likely to become customers. That information is then relayed to an e-marketing staff. Today, the Web site attracts about 1 million visitors per year, who are automatically screened for potential sales; that information accounts for nearly one-third of all new customer leads, thus fueling much of GE Plastic's growth. And because the cost of acquiring new business has decreased, GE Plastics can now go after smaller customers it might have ignored in the past. Specifically, the sales threshold at which a potential customer becomes attractive to GE's field marketing has now dropped by more than 60 percent.

The online tools also have enabled GE Plastics to improve customer satisfaction at a lower cost. Before the Web site, GE Plastics received about 500,000 customer calls every year. Today, the availability of online tools has slashed that number in half. In fact, customers use the tools more than 2,000 times a week. To encourage the rapid adoption of its toolkit, GE Plastics runs about 400 e-seminars a year that reach roughly 8,000 customers interested in learning about its tools and products. The company hopes that this effort will help encourage product engineers to design parts made of plastic—and GE resins—when they might otherwise have opted for metal or other materials.

HOW TO DESIGN A TOOLKIT FOR USER INNOVATION

As we have seen, when need-related design tasks are assigned to users, times and costs can be compressed, and learning by doing based on sticky, costly-to-transfer user information can be more seamlessly and effectively integrated

DO TOOLKITS MAKES SENSE FOR YOUR COMPANY?

There are three major signs that your firm can benefit from a customers-as-innovators approach:

1. Your market segments are shrinking, and customers are increasingly asking for customized products. As you try to respond to those demands, your costs increase, and it is difficult for you to pass these costs on to customers.
2. You and your customers often need many iterations before you can satisfy their needs. Some customers are starting to complain that you have gotten the product wrong or that you are responding too slowly. You are tempted to restrict the degree to which your products can be customized, and your smaller customers must make do with standard products or find a better solution elsewhere. As a result, customer loyalty starts to erode.
3. You or your competitors use high-quality computer-based simulation and rapid prototyping tools internally to develop new products. You also have computer-adjustable production processes that can manufacture custom products. (These technologies could form the foundation of a toolkit that customers could use to develop their own designs.)

into the design process. But the user is not a design specialist in the manufacturer's product or service field. So how can one expect users to create sophisticated, producible custom designs efficiently and effectively? Manufacturers pioneering in this field solve the problem by carrying out two major steps: (1) They repartition their traditional product or service development tasks in order to concentrate need-related problem solving within just a few tasks—and then they assign those tasks to users, and (2) they develop the toolkits users will need to carry out the design tasks assigned to them (von Hippel, 2001; von Hippel and Katz, 2002).

Repartitioning Development Tasks

In the conventional product and service development paradigm, problem solving that draws heavily upon *need-related* information has typically been an element within many product and service development tasks. After all, if a manufacturer is to execute all the problem solving in any case, it is irrelevant from the point of view of information transfer costs whether many tasks or few require need-related information. However, if the goal is to transfer only need-related design tasks to users—and to make these tasks as few and simple as possible—then a manufacturer must typically rethink the way its new product and service development tasks are divided up.

This rethinking can involve fundamental changes to the underlying architecture of a product or service. Consider, for example, the repartitioning of tasks that was carried out by semiconductor manufacturers as they shifted to the new toolkits paradigm for custom chip development. Traditionally, manufacturers of custom semiconductors had carried out all chip design tasks themselves, guided only by need specifications from users. And, since manufacturer development engineers were carrying out all design tasks, those engineers had typically incorporated need-related information into the design of both the fundamental elements of a circuit, such as transistors, and the electrical wiring that interconnected those elements into a functioning circuit.

Rethinking the custom design problem led to the insight that circuit elements could be made standard for all custom circuit designs and that all customer need-related information about chip function could be concentrated entirely within the task of designing the unique configuration of the electrical wiring that lay on the top surface of the chip. Chips with an entirely new architecture, called *gate arrays*, were created to allow this repartitioning of tasks, and then the wiring design task *only* was outsourced to users, along with a toolkit that would aid and guide them in performance of those tasks.

The same basic principle can be illustrated in a less technical context: food design. In this field, manufacturer-based designers have traditionally undertaken the entire job of developing a novel food, and so they have freely blended need-specific design into any or all of the recipe-design elements wherever convenient. For example, manufacturer-based developers might find it convenient to create a novel cake by both designing a novel flavor and texture for the cake body *and* designing a complementary novel flavor and texture into the frosting. However, it is possible to repartition these same tasks so that only a few draw upon need-related information, and these can then be more easily transferred to users.

The architecture of the humble pizza illustrates how this can be done. In the case of the pizza, many aspects of the design, such as the design of the dough and the sauce, have been made standard, and user choice has been restricted to a single task only: design of toppings. In other words, all need-related information that is unique to a given user has been linked to the toppings-design task only. Transfer of this single design task to users can still potentially offer creative individuals a very large design space to play in, although pizza shops typically restrict it sharply. Any edible ingredients one can think of—from eye of newt to fruits to edible flowers—are potential topping components. But the fact that need-related information has been concentrated within only a single product design task makes it much easier to transfer design freedom to the user.

Elements of a Toolkit

Toolkits are not new as a general concept—every manufacturer equips its own engineers with a set of tools suitable for developing the type of products or services it wishes to produce. Toolkits for users also are not new; many users have personal tool sets that they have assembled to help them create new items

EXERCISE: IDENTIFY YOUR "NEED-INTENSIVE" PRODUCT DESIGN TASKS

List the basic problem-solving steps your product development group uses to design a customized product. Do some clearly require a lot of user need information and others not? If not, could the steps be repartitioned so that this is true? What tools would a customer need to have to carry out the "need-intensive" product design tasks you have defined without help from your product design engineers?

TWO EXAMPLES

1. A 3M division fabricates enclosures for telecom firms needing to mount equipment outside on poles or elsewhere.
   - *Old task sequence.* Telecom customers like Verizon provide a list of equipment they intend to install in a new, customized enclosure type. 3M engineers then use a CAD program to design a suitable enclosure and interior mounting points. The customer checks the 3M prototype enclosure design and finds it wants to alter its equipment list or other aspects of the specification. 3M redesigns iteratively until the customer is satisfied.
   - *New task sequence*. 3M provides customers with a user-friendly version of its CAD program for enclosure design. (To protect 3M intellectual property, only customer interface aspects of enclosure design are revealed by the toolkit.) A customer inputs a list of equipment to be housed in the enclosure and other need specifications into the toolkit. The toolkit then creates a prototype design. The customer evaluates the design, sees needs for changes, and modifies the design accordingly. When satisfied, the customer sends the completed design to 3M for production.
2. A plastics manufacturer produces custom films for food packagers. Different food products require different forms of protection that are provided by custom film "structures"—successive layers of plastic, each with different properties, laminated into a single film. Specific layers in such a structure may, for example, block the transfer of oxygen to the food within the package, while others block the passage of UV light.
   - *Old task sequence.* Customers in the food packaging business transfer a set of performance specifications and a price target to the plastics manufacturer. Engineers in the firm use a CAD program plus engineering experience to design a film that is as close to the customer specification as is technically possible. The customer revises the specification to achieve a better fit between its needs and what it now knows are the relevant technical limits. The plastics manufacturer redesigns iteratively until customer is satisfied.
   - *New task sequence.* Relevant engineering experience is transferred to a beefed up CAD program that contains the properties of all film layer types used by the firm plus design structure rules such as layer compatibility considerations. The CAD package is equipped with a new, user-friendly program interface, and it is provided to customers as a toolkit for user innovation. To protect firm intellectual property, only the functional properties of each film layer type are included in the program; the chemistry of each layer is kept secret by the plastics manufacturer. Customers insert the specifications they need into the program and are immediately given back the performance characteristics of a practical film structure that matches their specifications as closely as possible, along with pricing information. The customer evaluates the design, sees needs for changes, and modifies the design accordingly. When satisfied, the customer sends the completed design to the producer.

or modify standard ones. For example, some users have woodworking tools ranging from saws to glue that can be used to create or modify furniture—in very novel or very standard ways. Others may have a kit of software tools needed to create or modify software. What is new, however, is integrated toolkits enabling users to *iteratively create and test designs* for custom products or services that can then be produced as-is by manufacturers.

Effective toolkits for user innovation meet five important objectives. First, they enable users to carry out complete cycles of trial-and-error learning. Second, they offer users a solution space that encompasses the designs they want to create. Third, the customers will be able to operate them using their customary design language and skills; they will not require much additional training to use them competently. Fourth, toolkits contain libraries of commonly used modules that the user can incorporate into his or her custom design, thus allowing the user to focus his or her design efforts on the truly unique elements of that design. Fifth, properly designed toolkits ensure that custom products and services designed by users will be producible on manufacturer production equipment *without* requiring revisions by manufacturer-based engineers.

Learning by Doing via Trial and Error

It is crucial that toolkits for user innovation enable users to go through complete trial-and-error cycles as they create their designs—that is how learning by doing is done. For example, suppose that a user is designing a new custom telephone answering system for his firm, using a software-based computer-telephony integration (CTI) design toolkit provided by a vendor. Suppose also that the user decides to include a new rule to "route all calls of X nature to Joe" in his design. A properly designed toolkit would allow him to temporarily place the new rule into the telephone system software, so that he could actually try it out, via a real test or a simulation, and see what happened. He might discover that the solution worked perfectly. Or, he might find that the new rule caused some unexpected form of trouble—for example, Joe might be flooded with too many calls—in which case it would be back to the drawing board for another design and another trial.

In the same way, toolkits for user innovation in the semiconductor design field allow the users to design a circuit that they think will meet their needs and then test the design by "running" it in the form of a computer simulation. This quickly reveals errors that the user can then quickly and cheaply fix using toolkit-supplied diagnostic and design tools (Chakravarty, 1991). For example, a user might discover by testing a simulated circuit design that she had forgotten about a switch to adjust the circuit—and make that discovery simply by trying to make a needed adjustment. The user could then quickly and cheaply design in the needed switch without major cost or delay.

One can appreciate the importance of giving the user the capability for trial-and-error learning by doing in a toolkit by thinking about the consequences of not having it. When users are *not* supplied with toolkits that enable them to draw on their local information about needs and context of use and engage in trial-and-error learning, they must actually order a product and have it built to learn about design errors—typically a very costly and unsatisfactory way to proceed. For example, many custom computer manufacturers offer a Web site that allows users to "design your own computer online." However, these product configurator Web sites do not allow users to engage in trial-and-error design. Instead, they simply allow users to select computer components such as processor chips and disk drives from lists of available options. Once these selections have been made, the design transaction is complete and the computer is built and shipped. The user has no way to test the functional effects of his or her choices before purchase and first field use—followed by celebration or regret. The cost to the customer is unexpected learning that comes too late: "That high-priced memory option I selected did look great. But now that the computer has been delivered, I can't see that it makes any difference in performance."

In contrast, a toolkit-for-user-innovation approach would allow the user to conduct trial-and-error tests to evaluate the effects of initial choices made and to improve upon them. For example, a computer design site could add this capability by enabling users to actually test and evaluate the hardware configuration they specify on their *own* programs and computing tasks before buying. To do this, the site might provide access to a remote computer able to simulate the operation of the computer that the user has specified, and provide performance diagnostics and related choices in terms meaningful to the user. An example of meaningful feedback: "If you add x option at y cost, time to complete your task will drop by z seconds." The user could then modify or confirm initial design choices according to design and preference and trade-off information only he or she knows.

An Appropriate "Solution Space"

Economical production of custom products and services is only achievable when a custom design falls within the preexisting capability and degrees of freedom built into a given manufacturer's production system. We term this the solution space offered by that system. A solution space may vary from very large to small, and if the output of a toolkit is tied to a particular production system, the design freedom that a toolkit can offer a user will be accordingly large or small. For example, the solution space offered by the production process of a custom integrated circuit manufacturer offers a huge solution space to users: It will produce any combination of logic elements interconnected in any way that a user-designer might desire, with the result that the user can invent anything from a novel type of computer processor to a novel silicon organism within that space. However, note that the semiconductor production process also has stringent limits. It will only implement product designs expressed in terms of semiconductor logic; it will not implement designs for bicycles or houses. Also, even within the arena of semiconductors, it will only be able to produce semiconductors that fit within a certain range with respect to size and other properties. Another example of a production system offering a very large solution space to designers—and, potentially to user-designers via toolkits—is the automated machining center. Such a device can basically fashion any shape out of any machinable material that can be created by any combination of basic machining operations, such as drilling and milling. As a consequence, toolkits for user innovation intended to create designs producible on automated machining centers can offer users access to that very large solution space.

Large solution spaces can typically be made available to user-designers when production systems and associated toolkits allow users to manipulate and combine relatively basic and general-purpose building blocks and operations, as in the previous examples. In contrast, small solution spaces typically result when users are only allowed to combine a relatively few special-purpose options. Thus, users who want to design their own custom automobile are restricted to a relatively small solution space: They can only make choices from lists of options regarding such things as engines, transmissions, and paint colors. Similarly, purchasers of eyeglasses produced by mass customization (Pine, 1993) production methods are restricted to combining "any frame from this list" of predesigned frames, with "any hinge from that list" of predesigned hinges, and so on.

The reason producers of custom products or services enforce constraints on the solution space that user-designers may use is that custom products can only be produced at reasonable prices when custom user designs can be implemented by simply making low-cost adjustments to the production process. Costs can go up sharply when requests fall outside of that range. For example, an integrated circuit producer would have to invest many millions of dollars and rework an entire production line to respond to a customer request for a chip larger than its existing production equipment can accommodate.

Manufacturers offering toolkits for user innovation to their customer are freed from having to know the details of their customers' needs for new products and services. On the other hand, manufacturers do have to supply the solution space customers need to be able to design the novel products or services they want. For example, International Flavors and Fragrances has to know which components to put into its flavor design toolkit, even if it does not have to know anything about a specific customer's need, or anything about the attributes of the flavor a customer hopes to make.

Fortunately, determining solution dimensions a toolkit must offer does not take superhuman insight on the part of manufacturer experts. Manufacturer-based developers can create a first-generation toolkit by simply analyzing existing customer products and determining the dimensions that were required to design those. Alternatively, they can simply distribute existing in-house design toolsets as a first-generation toolkit for user innovation, as was done by LSI, a manufacturer of application-specific integrated circuits. Then, they can steadily improve it based on customer feedback.

EXERCISE: THINK ABOUT AN "APPROPRIATE SOLUTION SPACE" FOR YOUR CUSTOMERS

Collect a list of custom products developed for your customers during the past year. What "design dimensions" were involved? What "design elements" were used? If different categories of your customers require very different types of custom designs, you may wish to focus upon only a segment of customers for this exercise.

Example:

- Custom designs for integrated circuits have varying demands with respect to the design dimensions of line width and the density with which elements are packed on chip. Examination of designs developed during the past year show values for these dimensions vary between X and Y.
- Custom integrated circuits incorporate predesigned cells and macrocells as design elements in an overall custom design. A complete list of this type of design element that we used in custom designs during the past year includes X Y and Z.

User-Friendly Toolkits

Toolkits for user innovation are most effective and successful when they are made user-friendly by enabling users to use the skills they already have and work in their own customary and well-practiced design language. This means that users don't have to learn the—typically different—design skills and language customarily used by manufacturer-based designers and so will require much less training to use the toolkit effectively.

For example, in the case of custom integrated circuit design, toolkit users are typically electrical engineers who are designing electronic systems that will incorporate the custom integrated circuit (IC). The digital IC design language normally used by electrical engineers is Boolean algebra. Therefore, user-friendly toolkits for custom IC design will enable toolkit users to design in this language. That is, users can create a design, test how it works, and make improvements all within their own, customary language.

A design toolkit based on a language and skills and tools familiar to the user is only possible, of course, to the extent that the user *has* familiarity with some appropriate and reasonably complete language and set of skills and tools. Interestingly, this is the case more frequently than one might initially suppose, at least in terms of the *function* that a user wants a product or service to perform, because functionality is a face that the product or service presents to the user. Indeed, an expert user of a product or service may be much more familiar with that functional face than manufacturer-based experts. Thus, the user of a custom semiconductor is the expert in what he or she wants that custom chip to *do* and is skilled at making complex trade-offs among familiar functional elements to achieve a desired end: "If I increase chip clock speed, I can reduce the size of my cache memory and . . ."

As a less technical example, consider the matter of designing a custom hairstyle. In this field there is certainly a great deal of information known to hairstylists that even an expert user may not know, such as how to achieve a given look via layer cutting or how to achieve a given streaked color pattern by selectively dying some strands of hair. However, an expert user is often very well practiced at the skill of examining the shape of his or her face and hairstyle as reflected in a mirror and visualizing specific improvements that might be desirable in matters such as curls or shape or color. In addition, the user will be very familiar with the nature and functioning of everyday tools used to shape hair, such as scissors and combs.

A user-friendly toolkit for hairstyling innovation can be built upon on these familiar skills and tools. For example, a user can be invited to sit in front of a computer monitor and study an image of her face and hairstyle as captured by a video camera. Then, she can select from a palette of colors and color patterns offered on the screen, can superimpose the effect on her existing hairstyle, can examine it, and can repeatedly modify it in a process of trial-and-error learning. Similarly, the user can select and manipulate images of familiar tools such as combs and scissors to alter the image of the length and shape of her own hairstyle as projected on the computer screen, can study and further modify the result achieved, and so forth. Note that the user's new design can be as radically new as desired, because the toolkit gives the user access to the most basic hairstyling variables and tools, such as color and scissors. When the user is satisfied, the completed design can be translated into technical hairstyling instructions in the language of a hairstyling specialist—the intended production system in this instance. A few years ago, such a system would have been very expensive to implement. Today, individual hair salons can offer their clients access to a hairstyling toolkit on a midrange personal computer.

In general, steady improvements in computer hardware and software are enabling toolkit designers to provide information to users in increasingly friendly ways. In earlier days, information was often provided to users in the form of specification sheets or books. The user was then required to know when a particular bit of information was relevant to his or her development project, find the book, and look it up. Today, a large range of potentially needed information can be embedded in a computerized toolkit, which is programmed to offer the user items of information only if a development being worked upon makes them relevant (McClean 1995).

Module Libraries

Custom designs are seldom novel in all their parts. Therefore, libraries of standard modules that frequently will be useful elements in custom designs are a valuable part of a toolkit for user innovation. Provision of such standard modules enables users to focus their creative work on those aspects of their design that are truly novel. Thus, a team of architects who are designing a custom office building will find it very useful to have access to a library of standard components, such as a range of standard structural support columns with pre-analyzed structural characteristics, that they can incorporate into their novel building designs. Similarly, designers of custom integrated circuits find it very useful to have access to libraries of predesigned elements ranging from simple operational amplifiers to complete microprocessors—examples of cells and macrocells, respectively—that they can simply insert into their custom circuit designs. Even users who want to design quite unusual hairstyles will often find it helpful to begin by selecting a hairstyle from a toolkit library. The goal is to select a style that has some elements of the desired look. Users can then proceed to develop their own desired style by adding to and subtracting from that starting point.

Translating User Designs for Production

Finally, the language of a toolkit for user innovation must be convertible without error into the language of the intended production system at the conclusion of the user design work. If this is not so, the entire purpose of the toolkit is lost, because a manufacturer receiving a user design essentially has to "do the design over again." Error-free translation need not emerge as a major problem—for example, it was never a major problem during the development of toolkits for integrated circuit design, because both chip designers and integrated circuit component producers already used a language based on digital logic. On the other hand, in some fields, translating from the design language preferred by users to the language required by intended production systems can be *the* problem in toolkit design. To illustrate, consider the case of a recent Nestlé USA's FoodServices Division toolkit test project developed for use in custom food design by the director of Food Product Development, Ernie Gum.

One major business of Nestlé FoodServices is production of custom food products, such as custom Mexican sauces, for major restaurant and take-out food chains. Custom foods of this type traditionally have been developed by or modified by chain executive chefs, using what are in effect design and production toolkits taught by culinary schools: recipe development procedures based on food ingredients available to individuals and restaurants, and processed on restaurant-style equipment. After using their traditional toolkits to develop or modify a recipe for a new menu item, executive chefs call in Nestlé FoodServices or other custom food producers and ask them to manufacture the product they have designed—and this is where the language translation problem rears its head.

There is no error-free way to translate a recipe expressed in the language of a traditional restaurant-style culinary toolkit into the language required by a food manufacturing facility. Food factories can only use ingredients that are obtainable in quantity at a consistent quality. These are not the same as and may not taste quite the same as ingredients used by the executive chef during recipe development. Also, food factories use volume production equipment, such as huge steam-heated retorts. Such equipment is very different from restaurant-style stoves and pots and pans, and it often cannot reproduce the cooking conditions created by the executive chef on his or her stovetop—for example, very rapid heating. Therefore food production factories cannot simply produce a recipe developed by or modified by an executive chef as-is under factory conditions—it will not taste the same.

As a consequence, even though an executive chef creates a prototype product using a traditional chef's toolkit, food manufacturers find most of that information—the information about ingredients and processing conditions—useless because it cannot be straightforwardly translated into factory-relevant terms. The only information that can be salvaged is the information about taste and texture contained in the prototype. And so, production chefs carefully examine and taste the customer's custom food prototype, and then try to make something that tastes the same using factory ingredients and methods. But executive chef taste buds are not necessarily the same as production chef taste buds, and so the initial factory version—and the second and the third—is typically not what the customer wants. So the producer must create variation after variation until the customer is finally satisfied. In the case of Nestlé, this painstaking translation effort means that it often takes 26 weeks to bring a new custom food product from chef's prototype to first factory production.

To solve the translation problem, Gum created a novel toolkit of food "precomponent" ingredients to be used by executive chefs during food development. Each ingredient in the toolkit is the Nestlé *factory* version of an ingredient traditionally used by chefs during recipe development: That is, it is an ingredient commercially available to Nestlé that had been processed as an independent ingredient on Nestlé factory equipment. Thus, a toolkit designed for Mexican sauce development would contain a chili puree ingredient processed on industrial equipment identical to that used to produce food in commercial-sized lots. (Each precomponent also contains traces of materials that will interact during production—for example, traces of tomato are included in the chili puree—so that the taste effects of such interactions are also included in the precomponent.)

Chefs interested in using the Nestlé toolkit to prototype, for example, a novel Mexican sauce would receive a set of 20 to 30 precomponents, each packaged in a separate plastic pouch (see Figure 13-2). They would also be given instructions for proper use.

The chefs find that each component differs slightly from the fresh components he or she is used to. But these differences are discovered immediately via learning by doing, and the chef then immediately adapts and moves to the desired final taste and texture by making trial-and-error adjustments in the

FIGURE 13-2. Pouches of "precomponents" used in Nestlé USA'S FoodServices Division field test of toolkit to enable customer's executive chefs to develop producible custom Mexican sauces.

ingredients and proportions in the recipe being developed. When a recipe based on precomponents is finished, it can be immediately and precisely reproduced by Nestlé factories—because now the user-developer is using the same language as the factory for his or her design work. In the Nestlé case, field testing by Food Product Development Department researchers showed that adding the error-free translation feature to toolkit-based design by users can reduce the time of custom food development from 26 weeks to 3 weeks by eliminating repeated redesign and refinement interactions between Nestlé and its custom food customers.

WHEN AND HOW TO DEPLOY TOOLKITS FOR USER INNOVATION

Toolkits for user innovation are applicable to essentially all types of products and services—both industrial and consumer—where customization is valuable to buyers. In this chapter I've illustrated this point via examples of

> **ELEMENTS OF AN EFFECTIVE, USER-FRIENDLY TOOLKIT.**
>
> Your toolkit should do the following:
>
> - Enable your customers to run repeated trial-and-error experiments and tests rapidly and efficiently on their own.
> - Offer a solution space appropriate to the design interests of your users.
> - Allow customers to work in a design language that is familiar to them.
> - Include a library of standard design modules that enable users to create a complex custom design rapidly.
> - Incorporate information about the characteristics and constraints of your production processes, so that customer designs can be sent directly to your manufacturing operations without extensive tailoring.

toolkits being used to help users design custom integrated circuits, custom telephony services, custom video game modules, custom hairstyles, and custom foods—quite a range. Toolkits can be applied to custom products produced in relatively large volumes, such as custom integrated circuits, or to products designed for single unit production, such as products produced by mass-customization production methods. In thc latter case, toolkits for user innovation provide the "design side" that is missing from today's mass customization practices. Note that toolkits for user innovation can be applied to both physical goods as well as information goods and associated services such as custom telephony software and the services it generates. Also, toolkits can be used to design custom physical and information products or services that are then produced by a manufacturer—*or* are produced directly at the user's site. For example, field programmable logic devices can be programmed into fully functional, user-designed custom integrated circuits right at a user site. Similarly, toolkits for software or software-based services can enable users to create custom software in finished form directly at the user site; there is no need to send it back to the manufacturer to be produced.

Toolkits Are Not for All Users—or All Applications

The design freedom provided by toolkits for user innovation may not be of interest to all or even to many users of a given type of product or service. Users must have a high enough need for something different to offset the costs of putting a toolkit to use. Toolkits may therefore be offered only to the subset of users who have a need for them. Or, in the case of software, toolkits may be provided to all users along with a standard default version of the product or service, because the cost of delivering the extra software is essentially zero. In such a case, the toolkit capability will simply lie unused in the background until a user has sufficient incentive to evoke and employ it.

Also, the toolkit approach cannot be used to satisfy every customer application. For example, toolkits typically will not enable customers to create products that are as technically sophisticated as those an experienced engineer at a manufacturer can develop using conventional methods; the design rules held in a skilled engineer's mind are more subtle than those encoded in today's toolkits. So manufacturers may continue to design certain products that have difficult technical demands, while customers take over the design of others that require quick turnarounds or a detailed and accurate understanding of the customer's need.

Users who do end up using a toolkit will often be lead users, whose present strong need foreshadows a general need in the marketplace. Manufacturers may well find it valuable to somehow acquire the generally useful improvements made by these lead users and supply them to the general market in the default version of the product generally offered. This was the pattern followed by the Technicon Corporation with respect to its clinical chemistry AutoAnalyzer™ products, for example. Information on improvements made by clinician-users was actively collected by that company, and innovations of value to many users were incorporated into analyzers sold to the general market. The pattern is also visible in the case of open-source software products such as Apache server software. Here innovations developed by users are screened in some way, and the best are incorporated into the "official" version of the software, which is then generally distributed.

Competitive Advantages of Toolkits for Manufacturers

Toolkits can create competitive advantages for manufacturers first to offer them. Being first into a marketplace with a toolkit may yield first-mover advantages with respect to setting a standard for a user design language that has a good chance of being generally adopted by the user community in that marketplace. Also, manufacturers tailor the toolkits they offer to allow easy, error-free translations of designs made by users into their own production capabilities. This gives originators a competitive edge even if the toolkit language itself becomes an open standard. For example, in the field of custom food production, customers often try to get a better price by asking a number of firms to quote on producing the prototype product they have designed. If a design has been created on a toolkit based on a Nestlé-developed language of precomponents that can be produced efficiently on Nestlé factory equipment by methods known best to that firm, Nestlé will obviously enter the contest with a competitive edge.

Toolkits can impact existing business models in a field in ways that may or may not be to manufacturers' competitive advantage in the longer run. For example, consider that many manufacturers of products and services appropriate benefit from both their design capabilities and their production capabilities. A switch to user-based customization via toolkits can affect their ability to do this over the long term. Thus, a manufacturer that is early in introducing a toolkit approach to custom product or service design may initially gain an advantage by tying that toolkit to his particular production facility. However, when toolsets are made available to customer designers, this tie often weakens over time. Customers and independent tool developers can eventually learn to design toolkits applicable to the processes of several manufacturers. Indeed, this is precisely what has happened in the application-specific integrated circuit (ASIC) industry. The initial toolsets revealed to users by LSI and rival ASIC producers were producer-specific. Over time, however, specialist tool design firms such as Cadence–developedtoolsets that could be used to make designs producible by a number of vendors. The end result is that manufacturers that previously benefited from selling their product design skills and production skills can be eventually forced by the shifting of design tasks to customers via toolkits to a position of benefiting from production skills only.

Manufacturers that think long-term disadvantages may accrue from a switch to a toolkits for user innovation will not necessarily have the luxury of declining to introduce one. If any manufacturer introduces the toolkits approach into a field favoring its use, customers will tend to migrate to it, forcing competitors to follow. Therefore, a firm's only real choice in a field where conditions are favorable to the introduction of toolkits is the choice of leading or following.

How to Get Started on Toolkit Development

Development and introduction of a toolkit for your customers should be undertaken by a team that has members who are intimately familiar with three important factors: (1) the design "language" and design capabilities of your firm's customers, (2) suites of product or service design tools that can be made appropriate for customer use, and (3) your firm's custom manufacturing system. Given a good understanding of these three factors, toolkit design is not as difficult as it may seem—because it can be approached in an evolutionary way. All that is required for initial success is that your first-generation toolkit offer enough functionality to make it valuable to interested users relative to other existing options. *Note, however, that toolkit development is not a trivial exercise. Both IFF and* Nestlé *invested several months of full-time effort by a skilled team to develop their toolkits to the point of initial field testing.*

With this in mind, following are *k*ey steps your team should take:

1. Your toolkit development team should begin by investigating and richly understanding your customers' natural problem-solving language and the methods they normally use to develop their understanding of what they need. It will be natural for customers to use these same tools for solution development as well—so a toolkit should help them do this. For example, engineers who use integrated circuits think in terms of circuit function when they design, using mathematical techniques like network analysis. They could care less whether the ultimate circuit is made of silicon wafers or green cheese, so long as it delivers the function they devise. In contrast, food designers (chefs) think in terms of desirable combinations of flavors and qualities of natural food ingredients when they design and use trial-and-error experiments involving physical preparing and mixing of ingredients to achieve the result they have in mind.
2. Next, examine the design tools and *solution design language* used by your in-house engineers, and how these are fitted to the input requirements and constraints of your production process. For example, IFF flavor designers think in terms of chemicals that create certain flavor notes in the flavors they develop and produce—and the production

process is designed to precisely measure out and combine those same chemicals.

3. Identify the need-intensive tasks in the overall product design process: These and *only* these are the tasks that you want to enable customers to carry out via toolkits. Design a problem-solving process in which the tasks containing need information are cleanly separated from the ones containing only production-related information. See the example in the text of how this was done by custom semiconductor manufacturers when they created a development process based on gate arrays instead of full custom development procedures.
4. Develop an initial toolkit that will enable customers to carry out trial-and-error product development for need-intensive tasks in *their* product development language—and will translate the completed design into the production language that *your* processes need as input.
5. Begin rollout by carefully selecting the first customers to use the toolkit on a commercial basis. The best prospects are customers who have a strong need to develop custom products quickly and frequently, and who have skilled engineers on staff. These customers will have both the incentive to stick with you during the inevitable shakedown period and the capability to join with you to help work the bugs out of the new system. As these customers begin to apply your initial toolkit to their projects, they will "bump up against the edges" of the solution space and tools on offer and then request (demand!) the additional ingredients and capabilities they need to implement their novel designs. You can then improve the toolkit by responding to these explicit requests for improvement. And/or you can enlist these impatient lead users as codevelopers who will actually create and test and use the toolkit improvements they need.
6. Continue to rapidly evolve your toolkit to satisfy your leading-edge customers. Customers at the forefront of technology will always bump up against the limitations of your toolkit first and will push for improvements. Investments in such advancements will likely pay off, because many of your customers tomorrow will need what leading-edge customers desire today.
7. *Adapt your business practices.* The business implications of toolkits go far beyond the product development department. For example, transferring need-intensive aspects of your product development process to customers may make it economically feasible to produce custom solutions for small, low-volume customers for the first time.

CONCLUSION

I conclude by proposing, as I did at the start of this chapter, that toolkits for user innovation will eventually be adopted by most or all producers creating custom products or services in markets with heterogeneous customer needs. As toolkits are more generally adopted, the organization of innovation-related tasks seen today, especially in the field of custom integrated circuit development, will spread. Firms will be able to help their customers to get *exactly* the products and services they want—by helping them to design them for themselves.

NOTE

1. Mass-customized production systems are systems of computerized process equipment that can be adjusted instantly and at low cost. Such equipment can produce small volumes of a product or even one-of-a-kind products at near mass-production costs. (Pine 1993). Today, producers supplying mass-customized products typically only allow customers to mix and match from predesigned lists of options. Thus they may offer users who want to design their own custom eyeglasses only the possibility of combining "any frame from this list" of predesigned frames, with "any hinge from that list" of predesigned hinges, and so on. Customers who want to stray beyond the proffered options are typically told, "Sorry, we can't supply that. Any new option needs to be carefully designed before it can be manufactured by mass-customized production methods." In other words, in today's practice the cost of *producing* unique items via mass customization has come down, but the cost of *designing* unique items—those not assembled from preexisting design modules—has not. Toolkits for user innovation can provide the design side of mass customization by creating a way to offer user-designers significant design freedom while at the same time insuring that the designs they create can be produced on the intended production system.

REFERENCES

Arrow, Kenneth J. 1962. "The Economic Implications of Learning by Doing." *Review of Economic Studies* 29, June: 155–73.

Chakravarty, Dev. 1991. "Marketing ASICs." In *Application Specific Integrated Circuit (ASIC) Technology,* ed. Norman G. Einspruch and Hilbert, Jeffrey L., p. 31 (Table 1). San Diego, CA: Academic Press.

Jeppesen, Lars Bo. "The Implications of User Toolkits for Innovation.' " Copenhagen Business School working paper. November. Available at http://userinnovation.mit.edu/papers/Jeppesen%20%20Implications%20of%20user%20toolkits%202002.pdf.

McClean, William J., ed. 1995. *Status, 4–13*. Scottsdale: Integrated Circuit Engineering Corporation.

Pine, Joseph B. II. 1993. *Mass Customization: The New Frontier in Business Competition.* Boston: Harvard Business School Press.

Rosenberg, Nathan. 1982. *Inside the Black Box: Technology and Economics,* p. 131. New York: Cambridge University Press.

von Hippel, Eric. 2001. "Perspective: User Toolkits for Innovation." *The Journal of Product Innovation Management* 18: 247–257.

von Hippel, Eric, and Ralph Katz. 2002. "Shifting Innovation to Users Via Toolkits," *Management Science* 48, 7, July: 821–833.

14 IT-Enabling the Product Development Process

Henry Dittmer and Patrick Gordon

Over the past decade we have seen a significant transformation in product development performance as companies adopted best practices and improved their decision making. This has primarily been on the dimensions of speed and quality. Sustaining these performance improvements has created increased demand for accurate and timely information on which to base those decisions. Historically, the information technology (IT) systems that have been deployed to support these transformation programs have been software (SW) applications that support a single subprocess (e.g., project costing, resource tracking) but that do not easily share product and project information with other applications.

This IT enablement has been an improvement over paper-based or manual systems—but, because of the lack of information sharing, it has been local optimization. This has led to problems with duplicate systems, incompatible data, unavailable information, and lost data. The unfortunate outcome is that when an organization learns it cannot rely on the data (or that it is too hard to maintain), it invariably learns to work around the formal IT solutions. Once this occurs, it is very difficult to reestablish the disciplines required for people to rely on the formal system.

With the emergence over the last few years of new enterprise software and middleware, companies have the potential for another jump in performance, with complete management of project and product data across the enterprise and across product development subprocesses. An integrated system is one that facilitates the electronic exchange of product and project information between product development subprocesses. The desired outcome of IT-enabling the product development process is that it allows companies to both institutionalize their hard-won process improvements and build new capabilities leading to improvements in productivity, speed, and innovation.

The first half of the chapter is about IT strategy in new product development (NPD) and focuses on a framework and toolkit an NPD program manager or process manager can use to define, partition, and analyze the opportunity for IT-enabling the product development process. The second half of the chapter provides an implementation approach for IT enablement and tips and techniques for making it work.

DEFINING THE NEW PRODUCT DEVELOPMENT PROCESS BEFORE IT-ENABLING

An effective new product development process consists of several interrelated subprocesses. We have found that the key information-intensive NPD subprocesses and the underlying best practices can be categorized into two dimensions: a business dimension (the management or project layer of product development) and an operations dimension (the operational or product layer of product development). Figure 14-1 shows these major subprocesses in each of those dimensions. Each subprocess has process work flows, organizational roles and responsibilities, data requirements, and metrics.

In the business dimension, portfolio management involves creating the planning and decision-making practices, organizational structures, and information systems that help ensure a winning product development portfolio. Resource management ensures that projects are staffed with the right level and mix of resources based on available skills and capacity. Project planning and management is the application of project management best practices to the execution of each development project. Product/platform planning and management ensure that roadmaps are established and maintained. These processes rely on project data information. The key performance measurements in the business dimension are time to market, revenue from new products (innovation), and productivity.

In the operations dimension, requirements management involves the capture of customer requirements, translating them into product requirements

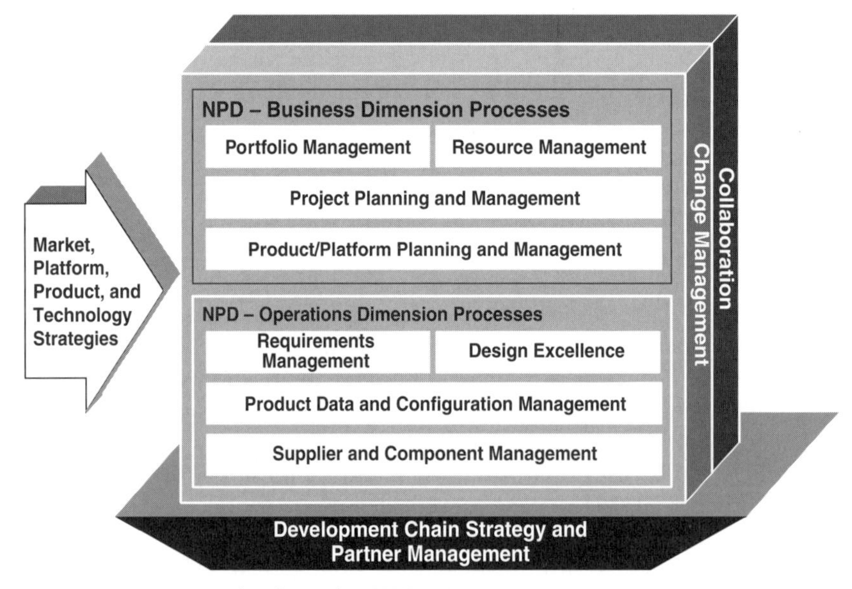

FIGURE 14-1. Information-intensive NPD subprocesses.

and translating those product requirements into product specifications. Design excellence ensures that the product design meets the product specifications, leverages reuse, and can be produced (or delivered in the case of a service) cost-effectively at high levels of quality. Supplier and component management ensure that robust, cost-effective, and reusable parts are designed in to the product. Product data management involves the creation and management of bill-of-material (BOM) information, with linkages to key product design applications, suppliers, and development partners. Key performance measurements in the operations dimension are product quality, productivity, and cost. These processes rely on engineering information for making design decisions, purchasing decisions, production decisions, and product delivery and service decisions. This kind of engineering information typically includes the following:

- Documents (e.g., requirements, cost targets, product specifications, component/software (SW) specifications, CAD files, test routines)
- Drawings, prints, and models
- BOMs, product structures, and rules
- Manufacturing process information
- Engineering change notices and engineering change orders (creation, redlining, approval, and implementation)

Collaboration and engineering change management extend across the two dimensions. For simplicity, we will refer to these two dimensions as NPD-business and NPD-operations for the remainder of this chapter. Note that a system to support NPD-business or NPD-operations can consist of several applications.

Any company doing product development can apply this framework to understand which processes are in place, their maturity relative to industry best practices, and the information intensity and requirements of those processes. The potential impact will vary based on the complexity of the business, the breadth of the portfolio, the volume of project and product information, and the maturity and discipline of its development processes. If business performance can be improved by real-time access or the ability to manage information more effectively—leading to better and faster information exchange and decisions—then that process is a candidate for IT enablement. Even small companies can benefit from this kind of enablement, although a company with simple products may decide that it is not ready for the investment in IT enablement. IT enablement typically requires the implementation of an application or applications that support one or more of the subprocesses shown in the diagram. The starting point for thinking about IT enabling is understanding what performance improvements are possible and what the requirements of the business are. The telltale signs of potential for IT-enablement include the following:

- The information from one system is printed out and then used as input to another system (e.g., sending changes to an engineering drawing to a supplier).

- Information that is required to support a business decision is not readily available (e.g., senior functional manager trying to understand the impact of moving resources to help a troubled project).

A ROADMAP FOR IT-ENABLEMENT

There is no shortage of applications to manage market research and customer requirements, automate and speed the design process, manage product requirements and specifications, manage resources, manage portfolios, maintain project schedules and plans, organize project workflow, and manage documentation and bills of information. What most companies lack is a framework for thinking how their major processes will be IT-enabled and connected. Figure 14-2 shows a simplified framework. The reason why business requirements based on the business process are so critical is that some systems do not easily share information, so the requirements for information sharing (based on the process) need to be known prior to system selection.

Based upon business needs and conditions, a company may elect to IT-enable subprocesses (e.g., resource management) or do something more holistic (e.g., a "full" product life cycle management, or PLM, solution to manage product and project information. Note that IT enablement can run the gamut from Excel spreadsheets to Microsoft Project to multimillion-dollar enterprise systems. The challenge is to both optimize each subprocess and keep these major subprocesses connected.

When companies evaluate their product development process and needs against the framework in Figure 14-1 and their current stage of IT enablement,

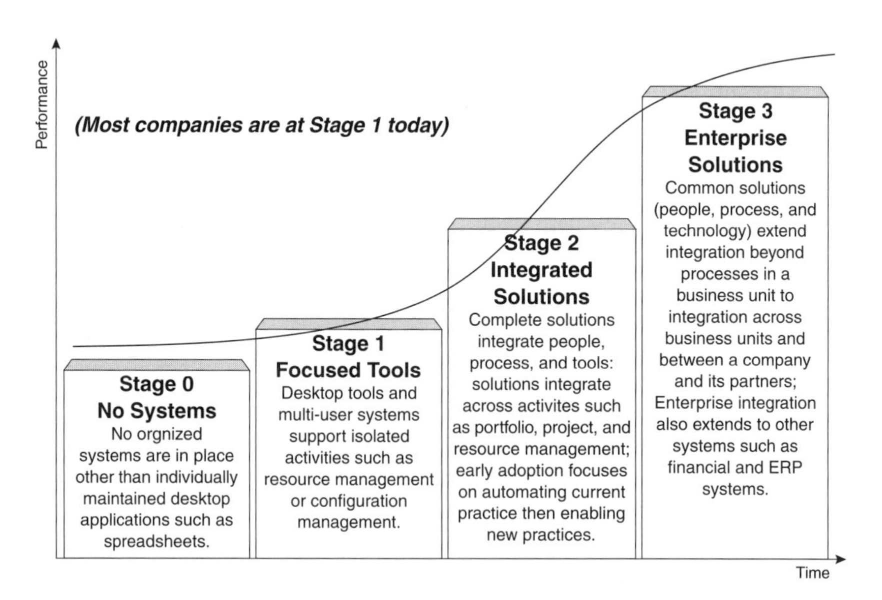

FIGURE 14-2a. Stages of maturity for IT-enabled product development.

Stages / Focus	1. No Systems	2. Focused Tools	3. Integrated Solutions	4. Enterprise Solutions
Project Management	Functional project management with desktop planning applications (e.g. MS Project)	Cross functional project management with limited manual integration of desktop plans via the web	Enterprise project planning and control with networked teams using a common, shared, enterprise application	Advanced, integrated project planning and control linking financial and resource management across the enterprise
Resource Management	Informal resource management with spreadsheets and ad hoc, manual data collection	Utilization focused resource management with systems the focus on tracking assignments	Capacity and demand balanced resource management integrated with project planning systems	Detailed resource requirements planning and enterprise integration with ERP & finance
Portfolio Management	Ad hoc portfolio management with spreadsheets and ad hoc, manual data collections	Periodic portfolio management with point solutions and ongoing, yet manual data collection	Dynamic portfolio management with continuous data collection via project system integration	Enterprise wide portfolio management across multiple portfolios integrated with financial system
Operations Management	No tools or rudimentary tools for PDM, document management, and product design (CAD)	Multiple, siloed applications with manual interfaces for managing product specific data from requirements to BOM	Integration with ERP systems and across key product data applications across an entire business	Enterprise wide integration of product data across the life cycle and integration with PD management systems (above)

FIGURE 14-2b. Practices and tools by stage of maturity.

it helps them prioritize and sequence improvement activities. For example, Johnson & Johnson, because of its process maturity and the strategic importance of information in its portfolio management process, chose to focus on IT-enabling this area first. Nortel focused first on resource and skill management to get better utilization and productivity from its scarce development resources. Avaya applied the framework and, based on a gap analysis against competitive best practices, chose to focus on both project management and resource management first. Faced with increased codevelopment needs, Sun Microsystems focused on improving inter- and intracompany product and component data management and collaboration to increase process effectiveness and efficiency, reduce time to market and product costs, and improve product quality. HP focused on IT-enabling change management to get to a single, common change management system and processes, using integrated information storage systems accessible worldwide.

Increasingly, the product line roadmaps of the leading software vendors address all the subprocesses. However, as of today, no single vendor provides the full spectrum of functionality to IT-enable all of the processes, so some level of company-specific systems integration will likely be required.

IMPLEMENTING IT ENABLEMENT

Many companies fail to achieve the targeted benefits of IT enablement because their return on investment (ROI) projections and timelines are unrealistic or they lack an overarching roadmap. Or perhaps they have no process infrastructure on which to deploy the system, and/or they have failed to address the behavior change (which requires more than just training) required to institutionalizc the benefits. If Figures 14-1 and 14-2 provide a common framework with which to talk about IT enablement of key business processes, the next key tool the product development leader needs is a NPD best-practice approach to seizing the opportunity and implementing successfully.

Developing and driving the standards and disciplines required to develop and maintain the associated information across functions and across enterprises is not easy. As soon as people don't rely on the data within the systems, they will no longer work to maintain the requisite integrity.

While each of the companies mentioned thus far followed a slightly different approach, there is a common implementation approach for IT-enabled product development that can be distilled from the successful elements of their respective approaches. This simplified approach is summarized in Figure 14-3.

A key insight is that an IT-enabled process implementation should be managed likc a phased product development project, with active cross-functional involvement from the very beginning. This means that many of the best-practice principles of good governance, program management, and measurement apply. This includes managing scope carefully and having an issue resolution process and project plans, with deliverables and milestones, in addition to resource plans. The team doing the selection and implementation should not be too low-level; they should have a strong executive-level champion. With multiple perspectives, ownership of data issues, and an entrenched culture, the need for sponsors is critical. Figure 14-3 shows two ongoing boxes for communication and governance/project management; they are key requirements for the success of a process/IT implementation.

Phase 1 Planning	Phase 2 Definition	Phase 3 Pilot Implementation	Phase 4 Rollout
1.1 Current State Assessment	2.1. Requirements and High-Level Design	3.1 Process, System, and Organization Design/Implementation	4.1 Rollout and Improvement
1.2 Product Development Strategy and Vision	2.2 Product Development Capability Roadmap	3.2 IT Vendor(s) Selection	4.2 Training
		3.3 Pilot Solution	
	2.3 Business Case	3.4 Initial Training	

Ongoing Project Management and Governance
Ongoing Communication

FIGURE 14-3. Phased approach to IT-enabllng product development.

TABLE 14-1.

Tip Number	Phase 1: Product Development Capability Planning
1.	Establish an agreed-to, cross-functional vision for product development capability.
2.	Understand the current process and system maturity and what needs to change from a behavioral point of view to achieve the vision.

Phase 1: Product Development Capability Planning

During Phase 1, a company performs a critical assessment of its current product development capability—both qualitative and quantitative (see Table 14-1). The qualitative review is where the process framework, shown in Figure 14-2, can be applied to understand the current level of practices and the gap between the prevailing practices and industry best practices. This requires a fact-based assessment of processes and business rules, organizational roles and responsibilities, existing IT systems, and metrics. The quantitative review entails looking at measurable performance in the areas of innovation, speed, cost, quality, and productivity, and analyzing whether performance levels are consistent with the needs of the organization based on competition and market success requirements.

The other key step in this planning phase is framing the strategic requirements of product development. A company that is trying to be first to market, versus a "fast follower," may have different demands on its product development process and the capabilities of its IT systems. The vision should be cross-functional and should be agreed to, committed to, and communicated by the senior leadership team. The vision should articulate the broad objectives of improved capability. Examples include these: ensure decreased product time to market and "deliver on schedule" within optimal market windows, simplify and enable the use of project planning and management best practices, efficiently plan and utilize project resources, and effectively manage change and coordinate complex dependencies across project participants.

The output of this phase is also a gap analysis (see Figure 14-4) between current capabilities and the strategic requirements of product development.

Phase 2: Product Development Capability Definition

In Phase 2, the vision gets translated into a time-phased roadmap of how that capability will evolve over a three-year or greater time frame. In several companies, the vision and roadmap are done in a single work stream, but we separate them here for clarity. This is the place where guiding principles get

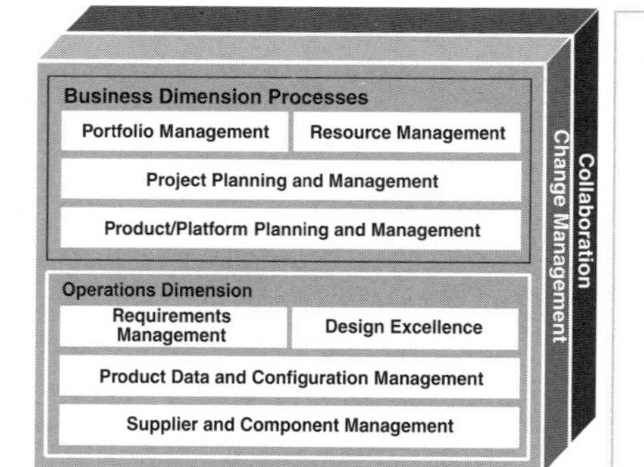

Translation of customer requirements to product requirements to product specifications is ad hoc

- Concept/Engineering/QFD or other formal requirements translation methodologies are not used

Definition of "requirements" limited to customer driven requirements

- Very little consideration of internal requirements (production cost, profitability, resource utilization)
- Goals for Design to Unit Production Cost are not consistently established, tracked, and used to manage design tradeoffs

The DOORS requirement management system is not intergrated with other product development systems

- No linkage to the Product Data Management system which prevent systematic alignment of requirements to engineering BOM
- No access to DOORS for supply chain organization

Requirements reviews/design reviews lack the rigor and necessary discipline to be effective

- Extent and quality of review varies by project and by discipline
- No penalty for non-compliance

FIGURE 14-4. Gap analysis example (requirements management subprocess).

established and ratified by the senior leadership of the company or division. At Avaya, with historically autonomous and increasingly interdependent divisions, a guiding principle was this: The fewer the IT solutions, the better (e.g., one product data management system worldwide). This did not preclude a best-of-breed strategy, but it did establish corporate standards for application types. Otherwise, the various divisions would provide plenty of reasons or obstacles about why this could not be done, and, in the end, the result would be too many applications, too little integration, too many expenses, and—most importantly—insufficient performance. While guiding principles may take years to accomplish, they won't be established at all if they are not part of the roadmap of the future state.

The "Requirements and High-Level Design" component occurs when a cross-functional business team evaluates the current level of each of the major subprocesses (see phase 2 of Figure 14-3) within the organization and articulates what areas need to be improved and over what time frame (using industry best practices as a reference). This is where the team identifies. business requirements such as a requirement that systems should enable cross-enterprise development, a requirement for support of ISO9001 or the Software Engineering Institute's Capability Maturity Model (CMMI) certification, or conformance to an industry standard such as CMII from the Institute of Configuration Management (ICM) (CMII). The "High-Level Design" component consists of "Process Architecture, Applications Architecture, Data Architecture, and IT Standards."

In the process architecture example in Figure 14-5, we define the desired capability of each of the subprocesses based on usage scenarios. The "Process Architecture" component is supported by the "Applications Architecture"—a time-phased view of the (potential) business applications and their relationships.

A "Data Architecture" sets standards regarding the creation, use, maintenance, and dissemination of data given the cross-functional and cross-enterprise nature of the associated processes. The data architecture is underpinned by the information technology standards established by the enterprise. It typically includes all technologies (as-is and to-be) in the environment. It includes server platforms, operating systems, database management systems, network protocols, integration standards, authentication standards, and so on. In many cases this is iterative, since IT tools enable new approaches. A representative example is in the area of resource management.

An often-overlooked element of "High-Level Design" is the basic characterization of the ownership of the business functions, activities, and processes identified in the process design. It must include new activities that may not be formally assigned at present to any function or group of people (e.g., what will be managed by a centralized IT organization).

Flextronics, the contract manufacturer, has done an outstanding job of translating its vision and roadmap into requirements and high-level design. This is necessary given its business model of being the manufacturing and development partner for companies across the globe and in different indus-

Subprocess: Resource Management	IT Enabled Best Practice	Priority and Timing
Scenario: Manage Resource Information		
Nomar Garcia is an R&D functional manager who has just hired a new resource. Nomar wants to add information on this resource to his list of resources—Name, resource category, location, department, etc	The system will allow functional managers to dynamically add, modify, or change resource needs (plans) at any time.	Synchronization with HR systems
	The system will enable functional managers to change assignments for every resource in their group which will trigger a notification to the appropriate Core Team Leaders	High value near term
Based on an anticipated increase in the size and complexity of the project pipeline, Nomar is required to identify additional skill sets needed to support the pipeline in the future. Nomar needs to create and track a new category of resources within his department.	The system will allow hierarchical resource management. The system will allow each manager to manage all resources beneath them.	High value near term
	The system will facilitate planning a group of resources on the same activity. (For example, plan 10 resources for the same training or conference all at one time.)	High value near term
	When comparing long-term demand against the available supply of resources, functional managers will have the ability to use a graphical "what-if" analysis to evaluate the impact of expanding the available resource pool to other functions and departments.	High value near term

FIGURE 14-5. High-level design—process architecture examples.

tries. All of Flextronics's facilities use the same product development systems. Therefore, all of their sites worldwide can share product data; they can move products between sites very quickly; and they have standard points of integration for their customers and business partners.

Integrating the work of this phase is the business case or value proposition. This synthesis document articulates the value to the business of making the changes (process and IT), including the implementation plan and the overall payback (ROI), as well as interim wins (operational improvements). The plan should be realistic: Costs and interim wins should be clearly identified. In several instances where unrealistic payback and timing expectations have been established, the result was a perception that the project failed. The plan should identify the scorecard of key metrics (definition, current value, target, collection responsibility) that will be tracked as part of the implementation and rollout that align with the business case.

During Phase 2, Avaya built its product development roadmap around a stages model. It had benchmarked extensively the performance of its four business units and found they were operating in functional silos, lacking key cross-functional actions, with the corresponding operational performance. Avaya developed a metrics-based value proposition for improvement. It targeted full, cross-functional product development capability in three years as the goal and defined key operational targets. These were reviewed, approved, and actively supported by the CEO and his direct reports.

An often daunting step in this phase is the selection of new applications or systems to meet functional and technical requirements that support increased capability. The selection can sometimes be an extremely short process (selection based on past experience, a preference, or very few vendors), while in some other cases it may be a relatively lengthy process (two to three months, typically), and it may involve evaluation of a multitude of vendors (long list, short list, application demo, request for quote, etc.). For the latter case, Figure 14-6 shows a highly simplified depiction of the system selection process. This process is also best executed by a cross-functional team consisting of key stakeholders including finance, marketing, development, and IT.

The first task in the selection process for application software is to document the functional or solution requirements (see Figure 14.6). The business requirements form a good basis to start. Once the requirements have been documented, the final task in this step is to actually select the application vendor. The first activity during this task consists of looking for any solutions that might meet most of the requirements gathered earlier. This can be accomplished by searching the Web, looking at trade magazines, talking to other companies, and so on. For a start, analysts from research firms like Gartner or AMR Research can be a good source of information (see Table 14.2). The purpose of this activity is to gather a comprehensive, but not exhaustive, list of application vendors and packages that might meet requirements and that need to be considered for evaluation. This helps ensure that there is not an obvious candidate that the team is overlooking.

Once the long list of vendors has been created, an elimination process needs to be undertaken to create the short list of application vendors. This would consist of performing an initial, high-level evaluation of the long list, looking for obvious reasons to eliminate some of the vendors. For example, certain products may not fit within the technology architecture, some may be too new, or some may be too expensive. In some cases, there may be a feature that is absolutely needed but is not available. Based on this short list, the team can now evaluate the vendors against the business requirements. Requests for proposal (RFPs) will be submitted to the agreed-upon list of software vendors to provide an objective, comparable, and repeatable methodology for scoring application vendors against requirements. This removes bias on how well a proposal is answered versus actual system features. It provides a method to

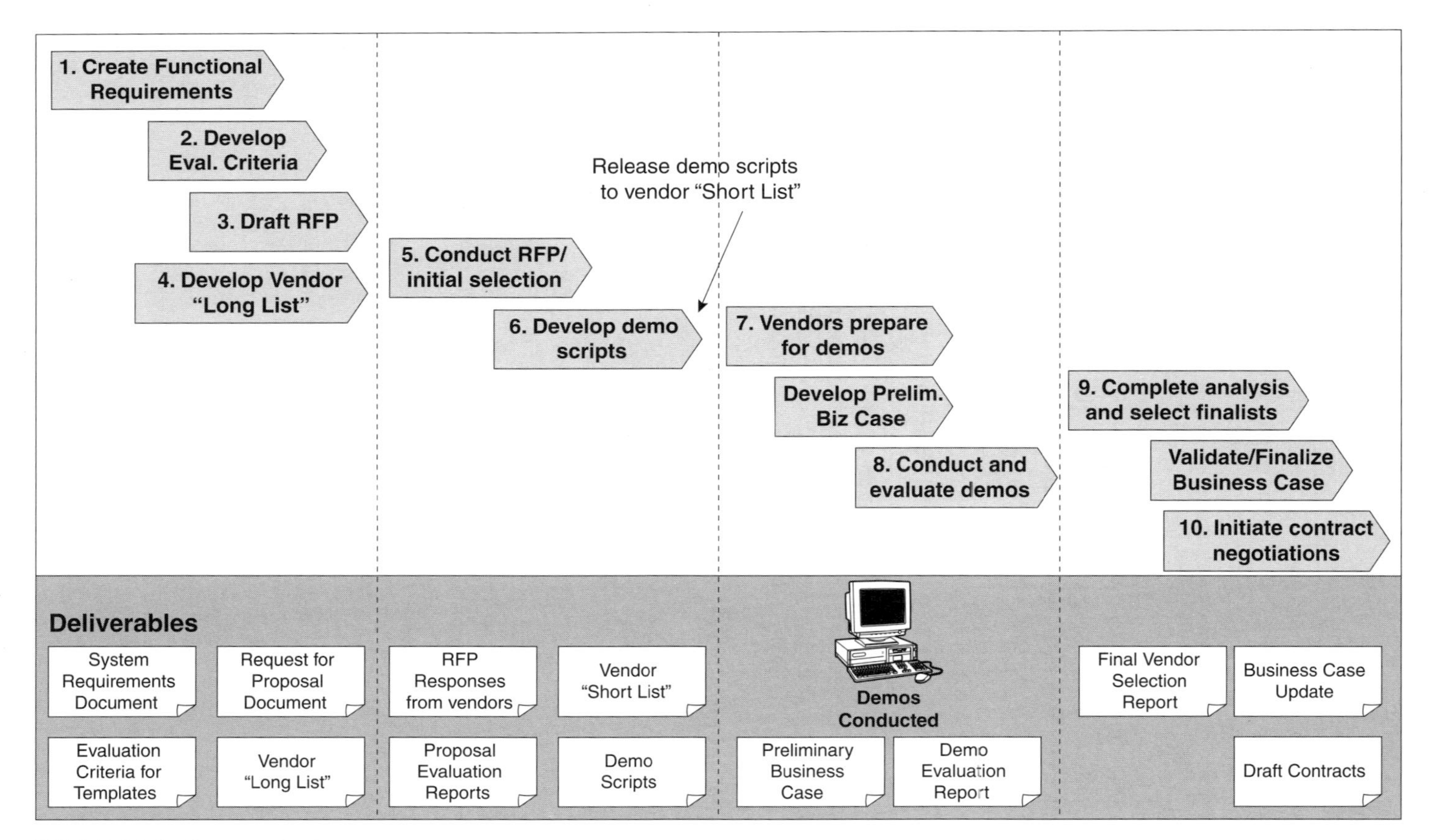

FIGURE 14-6. System selection process.

TABLE 14-2.

Tip Number	Phase 2: Product Development Capability Definition
1.	Define the business requirements and architect the necessary process, applications, and infrastructure required.
2.	Define and measure appropriate improvement metrics and integrate them into an overall business case/value proposition.
3.	Apply a structured systems selection process based on carefully developed usage scenarios scripts: —Cross-functional team helps build business requirements. —Ensure adequate vendor and end user involvement. Use analyst reports as well as references accounts to help you narrow down to a handful of potential vendors.

quantify an RFP response and evaluate vendor solution capabilities and commercial offers.

Next, scripted demos should be conducted that represent how the application will be used in the business process. The scripted demonstration gives the vendor the opportunity to demonstrate how its application can be applied to resolve a representative set of business issues. It gives the evaluation team the information necessary for making the final selection. The demonstration scripts can contain sample data representative of real-life transactions. The vendors must use the data provided to them to demonstrate that their application is capable of accommodating and processing the data structures required by the company.

The final application selection process can sometimes be the most difficult task. The application characteristics must be mapped against requirements, and then weighting factors are applied to determine which application solution most closely meets all the needs. Based on the results of this process, an internal consensus needs to be reached, identifying the application vendor that meets most of the requirements. After the decision has been made, the process is turned over to a formal purchasing or procurement organization for contract negotiations and legal details.

Phase 3: Pilot Implementation

In the pilot implementation phase, the company creates a detailed design of the future business processes and organization. The rough-cut process flow documents that were developed in earlier phases become the starting documents for this step; the high-level process flow diagrams are blown out into the next level of detail. During this step, performance metrics for the business processes (wherever applicable) are developed. Any practices and policies pertaining to the business processes are also developed and documented. In parallel, the company designs the detailed organizational structure to support the defined business processes. Also in this step, the job descriptions and roles and responsibilities of the people who will enable the business processes are determined and documented. Compensation and incentive systems, hiring and employment policies, performance measurement and management policies, career development plans, and spans of influence are the other elements that may be developed during this step.

This step is complete when the detailed organizational structure, job functions (descriptions, roles, and responsibilities), and other organization design-related documents (compensation, incentive, etc.) have been designed and developed. Avaya leveraged an industry standard tool called RACI (Responsible, Accountable, Consulted, and Informed) to develop the roles and responsibilities of its key people for each of the key process activities. An example of RACI being applied is shown in Figure 14-7.

Lastly, the application configuration/integration environment is established and the base application is installed and stabilized. Next, the functionality scope (for the rough-cut prototype), technical scope, and high-level configuration plan are developed—based on the solution design. At this point, the team should know which modules it needs and what it needs to accomplish with them. The application is then configured to create a simulation environment and interfaces. Data are identified, collected, and structured, and a subset of client data or mock data is loaded into the application. The configured application is now ready to demo key functional aspects of the proposed solution.

A useful tool for integrating the preceding activities, refining requirements, and ensuring usability are usage scenarios (sometimes referred to as *use cases*) and conference room pilots. Since product development system usage is voluntary, difficult systems are simply not used, resulting in stalled adoption. Usage scenarios (see Figure 14-8) provide an effective way to model process, system, and organization at the same time. To define usage scenarios, begin with Figure 14-2, which partitions the product development environment into logical work streams. Each work stream consists of one or more solution usage scenarios. A *solution usage scenario* is a grouping of tasks representing a logically complete business event (e.g., assign resources to a new project). Any given usage scenario contains multiple design elements for a logically complete business event:

- Recommended practices and policies
- Process design flows
- Roles and responsibilities
- IT requirements and IT systems configuration elements
- Master data setup rules

The list of solution usage scenarios should be as exhaustive as possible and cover 80 to 90 percent of all possible scenarios in the work stream. In the beginning, avoid focusing too much on exception rules and processes that can

Portfolio Management Process

Note: See appendix for descriptions →

Legend
Accountable
Responsible
Consulted
Informed

	Monitoring & Exploration			Analysis		Strategy Definition		
	Market Segmentation	Internal Monitoring	External Monitoring & Exploration	Internal Analysis	External Analysis			
Prod/Svc. House 1	C	C	A	A	A	A	A	A
Prod/Svc. House 2		A	A	A	A	A	A	A
Prod/Svc. House 3	A	C	C	A	A	A	A	A
P&S Marketing 1	I	I	R	I	R	R	I	C
P&S Marketing 2	I	I	I	A	A	C	C	C
P&S Marketing 3	I	I		I	I	I	I	C
Marketing Operations	I	I	I	I	I	C	C	C
Market Research			R		R	C	C	C
Strategic Planning	I	I	R	R	R			

No clear A and no R's (Market Segmentation, Internal Monitoring)

Inconsistency in assumed responsibilities (External Monitoring & Exploration, Internal Analysis, External Analysis)

- ***Responsible*** — identifies the function(s) that is assigned to *execute* a particular activity. The degree of responsibility is determined by the *Accountable* person. R's can be shared.
- ***Accountable*** — designates the function that is ultimately "accountable" for the completion of the activity, and who has the ability to say "yes" or "no." There can be only one "A" for a decision or activity, and accountability cannot be delegated.
- ***Consulted*** — identifies the function(s) that must be "consulted" before a decision or activity is finalized. This is a two-way communication.
- ***Informed*** — identifies the function(s) that must be notified about the completion or output of the decision or activity. This is a one-way communication.

FIGURE 14-7. RACI analysis of portfolio management process.

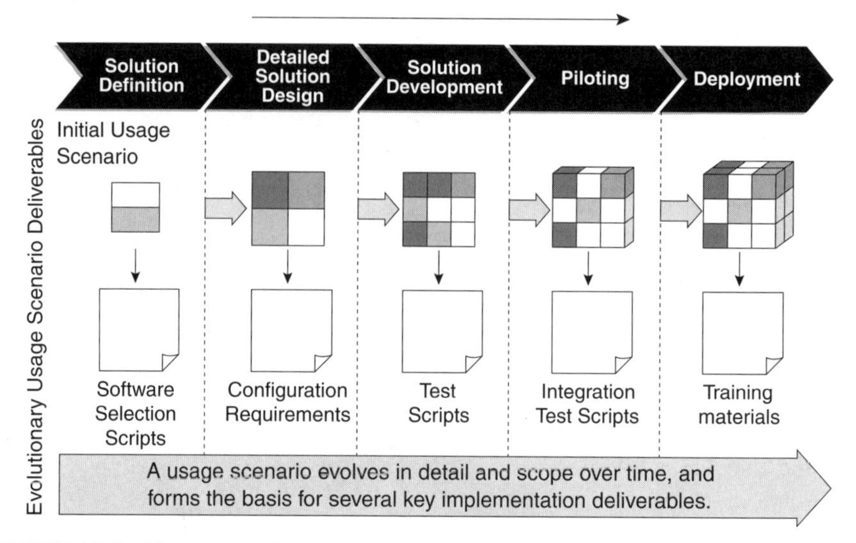

FIGURE 14-8. Usage scenarios.

take up a good deal of the design effort. Design sessions initially focus on design principles, practices, and policies before going into more details such as process flows and IT requirements.

Conference room pilots allow you to simulate a business situation using scripts—use cases—to observe and modify the process and the capabilities of the systems. Experience has shown that iterative solution design, development, and testing are the most effective way to minimize risk and time. The objectives of these multiple iterations include the following:

- Confirm usage scenarios
- Develop customizations and interfaces
- Develop and test data conversion routines
- Convert/create static data
- Design security components
- Establish user procedures and training documentation
- Plan pilots

At Avaya, these conference room pilots helped establish what functionality really needed to be implemented. For example, the system it chose was capable of tracking resources in a hierarchical fashion. Following is an example hierarchy:

Level 1. Development Engineer

Level 2. Software Developer

Level 3. Java, C++, Legacy

Level 4. Operating System, Real-time Software (SW), Device Drivers

The feedback from the conference-room pilot was that even though the system was capable of excruciating levels of detail, the only business requirement was the Level 3 detail and rough-cut capacity management capability.

Based on the feedback and experience from the pilots, the data is then collected, structured, scrubbed, and loaded into the application. The configuration of the application for the company needs is often a time-consuming task, since most enterprise applications need a great deal of configuration to enable them to perform the required business functions. Testing is conducted on this configured application to ensure that it is functioning as intended and also to ensure data integrity. This culminates in a pilot, with live data and end users who can now perform the business process using the new process and IT.

The final key activity during this phase is developing training components such as an application user's guide, a systems administrator guide, and a business process operating manual. The goal is to develop documentation that describes the functions and features of the application and also describes business processes, which are part of the solution (but not covered by the IT application) from a user's point of view. The user's guide should provide detailed information to users on how to access, navigate, and operate the application. It is usually organized functionally such that the information is presented in the same way the application actually works. This helps users understand the flow of menus and options to access desired functionality of the application. The need of the users, and also their level of sophistication, dictate the document presentation style and level of detail. The systems administrators guide is more technical in nature and is focused on providing detailed descriptions of elements like installation, security, database administration, and error handling and other application maintenance issues. The main purpose of this document is to be used as a reference guide (installation, maintenance, upgrade) by the system administrator. The business process operating manual should be a how-to guide to perform business processes that are part of the solution but are performed without using the application. Sometimes, the user's guide and the business process operating guide may be combined into one single manual.

The next set of activities is to develop the training plans and training materials that will be used to train users in the next (rollout) phase. The team can identify and document who the training audience is, their training needs, and any constraints, such as room capacity, time, logistics, and so on. Next, the team develops the training process and approach, and determines the logistical and training environment requirements. The team identifies course objectives, creates course outlines, and determines training material standards. Student and instructor training materials, promotional materials, hands-on exercises, quizzes, and case studies are developed now as well.

This task is complete when all systems operating documents such as the users guide, the systems administrators guide, and the business process-operating manual have been developed, along with all training plans and all required training materials. The tips for implementation (Phase 3) are summarized in Table 14.3.

TABLE 14-3.

Tip Number	Phase 3: Implementation
1.	Treat IT enablement like a project: use the same best-practice project governance and program management techniques: ◆ Executive governance with clear decision points and performance criteria. ◆ Prioritize implementation by highest-benefit functionality. ◆ Voice of the user/customer inspires design and implementation. ◆ Cross-functional heavy-weight teams drive to the "market window." Manage the project to maximize early benefits while minimizing complexity.
2.	Use tools such as use cases and conference room pilots to focus on the integration of process, information technology, and change management: ◆ A strong process foundation makes a big difference. ◆ Understand the difference between system and process from a user perspective. Minimize the amount of customization: ◆ New management processes are often enabled and needed.
3.	Conduct extensive data cleansing and integrated solution testing.
4.	Don't underestimate the integration complexity: Advanced integration development capabilities may be required.

Phase 4: Rollout and Training

The objective of this phase is to deploy the solution according to the rollout plan. At this stage, the solution has been piloted at a test site, it has been refined based on the results of the pilot, and it is now ready to be rolled out in the organization. This deployment/rollout usually occurs in waves: Each wave represents one complete deployment with its defined scope and objective. Just-in-time training and executive briefings should support each wave. Because it is tough to unseat homegrown approaches, executive and functional manager support is critical. The pilot implementation team should summarize lessons learned and then move to maintenance mode, where a subset of the team monitors the acceptance and use of the system and recommends any corrective action. See Table 14-4.

SUMMARY

The recent and anticipated advances in system capability mean that IT enablement will increasingly become one of the many approaches a company must master to have an effective new product development. Avoiding the mistakes

TABLE 14-4.

Tip Number	Phase 4: Rollout and Training
1.	Communicate pilot successes and the benefits ("What's it in for me") to all the stakeholders.
2.	Deliver training "just-in-time" and ensure it covers both process and the tools.
3.	Make sure the infrastructure is ready to support rollout (scaling up the servers, controlled access, etc.).
4.	Ensure effective "point/time-of-use" benefit to end users.

of past IT efforts requires a pragmatic framework for thinking about IT enablement, as well as a defined, repeatable, and structured process for implementing IT enablement at the company. This chapter provided some frameworks and tips for the product development practitioner to lead or actively participate in an IT enablement effort. The payoff for doing this right is significant: Companies will be able to achieve productivity improvements of as much as 50 percent based on improved resource utilization, increased project productivity, reduced time spent on non-value-added and administrative tasks, better management of intellectual property and knowledge, and the ability to collaborate with supplier and development partners more effectively.

Part 4

Tools for Managing the NPD Portfolio and Pipeline

This final ToolBook 2 part provides tools to be applied across the entire portfolio of NPD projects. The portfolio could include projects ranging in maturity/longevity from those that have not yet entered the FFE to those commercial products that are being renewed and improved to extend their life in the marketplace. Managing an NPD portfolio is all about making choices—which project is in, which is out; the level of resources from project to project; and how all the projects fit together in light of an organization's capability and strategy. This is a very complex managerial task, and the tools provided in this part provide insight and guidance to improve organizational performance.

Chapter 15 tackles the complex subject of technology mapping. Although mapping is a technique that may be more suited to learning by doing, the authors present this very valuable NPD technique in an understandable and implementable fashion. Mapping is a visual tool to present the time evolution of technologies and markets in a clear and simple fashion. It allows projects to be ordered and organized relative to technology and market generations. This then helps senior management grasp the important linkages between individual projects and helps guide project selection and investment decisions.

The next chapter provides the means to closely link technology development with product development. Chapter 16 describes the tools needed to assure an organization that investments in technology will result in superior new products. A number of decision support tools will help an NPD organization make sure that technology development is time-phased with the eventual needs of product development.

Finally, Chapter 17 addresses an organization's portfolio process with a maturity model. Most firms have implemented an NPD portfolio prioritization process. The primary tool in this chapter enables an organization to assess the maturity of its portfolio process and then offers effective practice techniques to improve the process, maturity level by maturity level. Many hints and tips for improving portfolio prioritization performance are offered throughout the chapter.

15 Product and Technology Mapping Tools for Planning and Portfolio Decision Making

Richard E. Albright and Beebe Nelson

PART 1: THE ROLE OF MAPPING IN NEW PRODUCT PLANNING AND STRATEGY

Fuel technology is changing. So is the cultural and regulatory environment for transportation in the United States. Our hypothetical example, Acme Motors, is a leading producer of autos. Acme's product developers are faced with questions of how they should position Acme's products. Should they join the technology leaders such as Toyota? Should they enter as fast-fuel-cell followers? Should they bet on an alternative strategy, for instance that the internal combustion engine might become so efficient that the electric/internal combustion hybrid engine, or even the internal combustion engine alone, might prove competitive with the fuel cell? As they better understand these issues, they will have to make more specific plans. Will they focus on the hybrid, or will they skip that and go directly to the fuel cell? Will they focus on being a leader in developing the efficiency of the internal combustion engine, or will they be content to play follow the leader? At the project level, what product and technology projects will be commissioned, and what will be the actual targets for the development teams?

Throughout the chapter we show the use of maps in the hypothetical case of Acme Motors and its plans for the next several generations of its passenger car product line. To succeed, Acme must take into account the evolving market, regulatory, and competitive environment, along with technology innovations and changing customer needs.

This chapter provides examples of many of the tools and maps that a product developer needs to address and unscramble questions similar to those facing Acme's developers. Some of these maps are very high level and strategic. Although they can be backed up by lots of detail, we call them "back-of-the-envelope" maps, because the logic is transparent and the amount of information is small.[1] A good place for a company to start is with these high-level maps, and we describe several of them, including an event map, a technology map, a product line map, and a value chain map. The more tactical, team-based maps can have a more complex logic and generally include much more detail. We describe how the Acme product team used a number of tactical maps, including experience curves, product and market architectures, and product-technology roadmaps. These maps expand on the mapping tools described in Wheelwright and Clark (1992) and in the first PDMA Toolbook (Meadows, 2002).

When mapping is fully integrated in a company's planning and delivery processes, the high-level and tactical maps are tied together to provide a through-line from strategy to execution (Albright, 2002). An example of a set of linked maps is shown in Figure 15-1. By the end of the chapter, you will be able to see how these maps can be constructed individually and tied together to provide a comprchensive picture of the market, customer, product, and technology space in which the teams execute and the company strives to be competitive.

The thumbnails in Figure 15-1 sketch how a company might produce individual maps of the market, of products, and of technology, and then link them together to produce an action plan for a function or a team. On the other hand, back-of-the-envelope maps by themselves are particularly good at combining roadmap information. For example, market, product, and technology information can be shown together in a map that connects specific market needs with existing or planned products and technology. One of the most useful roadmaps, the product-technology roadmap, is discussed at some length in Part 5.

This chapter also reflects on the place of mapping in strategic and project decision making. Mapping can introduce a sea change in a corporation's ability to understand its resources and link those to market needs over time. Mapping will help Acme address such questions as: How are your markets and technologies evolving? If you could take a snapshot in time, what would you see right now? Will you be in a competitive position five or ten years out? What are your fundamental technologies and product lines? Do your products and technologies support each other? Or do they conflict, overlap, or simply exist in isolation as if divided by corporate firewalls? Maps provide companies with a way of identifying their assets—including what they *have*, what they *know*, what they're *good at*, and what might be called their *social capital* (channels, brand, etc.).

What Are Maps and How Do They Work?

Product and technology maps provide a rich catalogue of a corporation's product and technology assets that can be sequenced on a timeline and linked with other assets. At their best, maps tie together market, technology, capabil-

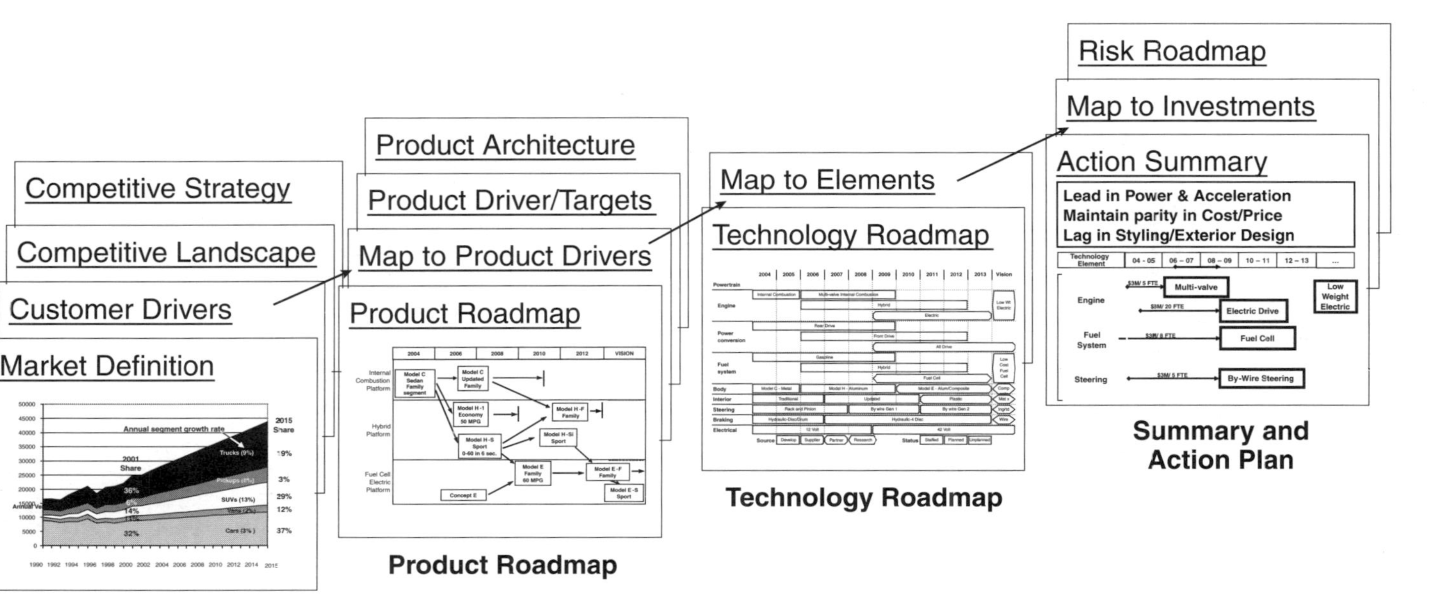

FIGURE 15-1. Linked maps.

THE BENEFITS OF ROADMAPPING

1. Roadmapping is *good planning.*
2. Roadmaps incorporate an explicit element of *time.*
3. Roadmaps *link* business strategy and market data with product and technology decisions.
4. Roadmaps reveal *gaps* in product and technology plans.
5. Roadmaps *prioritize* investments.
6. Roadmapping helps set more competitive and realistic *targets.*
7. Roadmaps provide a *guide* that allows the team to recognize and act on events that require a change in direction.
8. Sharing roadmaps allows *strategic* use of technology across product lines.
9. Roadmapping *communicates* business, technology, and product plans to team members, management, customers, and suppliers.
10. Roadmapping *builds* the team.

ity, and product/service information so that real-time decisions can be made to allocate the firm's resources and direct its efforts to meet short- and long-term goals. The essence of mapping lies in the creation of graphic presentations of information that have been built from the frequently tacit information residing in different functional areas. Maps create a syntax and a vocabulary that allow functions to communicate, and they provide evolving pictures of market-product-technology links. Good maps connect their elements to help both map makers and map readers understand the whys, whats, and hows of product and technology plans.

Mapping activities are inherently cross-functional exercises. They can be carried out by a product team, by a product management team, or by the functional heads of a strategic business unit (SBU) or a corporation. They are fractal in nature. When the data are sufficiently robust, and when the maps are well linked, maps can allow teams and managers to zoom in or zoom out to the appropriate level of complexity and detail. The information at the team level carries up to senior decision-making levels, and the strategy maps at higher business levels guide mapping structures and decisions at the team level. Mapping plays a crucial role in portfolio decision making; the maps relate information about projects, values, and assets to create a context for portfolio decisions that cannot be achieved by other tools typical of portfolio management, such as bubble charts, strategic buckets, or financial measures.

Maps can be developed by a few people who come together to create the core distinctions, or by teams that have maintenance of the map as an ongoing task. Maps can function as a guide to business decision makers, as an essential input to strategy, and as a planning tool for technology and product development.

In this chapter we review methods for informal mapping processes as well as methods for more formal and fine-grained processes. It's important to remember to use the right tool for the job, matching the sophistication and complexity of a map to the context of its use. A product team may require

THE LANGUAGE OF ROADMAPPING

The *vocabulary* of a map refers to how the elements are represented, the signs and symbols of the map. If you represent platforms as solid rectangles on map A, they should be represented the same way on map B. If "O" signifies "advanced development project" and "□" signifies "focused development project" on map A, you should use that same coding on map B. *No map branding by business units, functions, or teams!*

Map syntax refers to the rules and norms that relate the elements. It includes the map architecture. For instance, we normally read time from left (early) to right (late). Although we would think that someone who represented the time scale from right to left was simply wrong, in fact this is just a well-embedded convention. If your business can embed other conventions, so that map readers can easily compare data from one map to another, you will find maps more and more useful at strategic levels of decision making. For example, some firms show product lines as left-to-right arrows; many use the BCG grid or the S curve for life cycle. Adopt or invent useful structures, and reuse them.

The map maker uses other mapping conventions to codify and convey meaning, such as using *color* and *whitespace*. Be sure to use them judiciously, however; never use them merely decoratively; and use them consistently. Keep *labeling* to a minimum. If your syntax and vocabulary are clear, labels will be less necessary. Establish company norms around labeling, including type size/font and label placement (e.g., left of the y-axis and below the x-axis).

much more concrete detail to map the elements of a product platform, while an executive strategy team may find the more informal maps not only adequate but more helpful than a detailed, project-focused map. If the firm employs a consistent syntax and language, and if the maps are shared among functions and levels, then the firm will build a valuable knowledge resource over time in which the less formal maps are grounded in the maps with finer detail and the fine-grained maps can be read in the context of overall strategic importance.

There are three useful rules for successful integrated product and technology mapping:

1. *Use cross-functional teams or groups to build and maintain the maps.* The relationships among the information in a product or technology map provide market and technology *contexts* for planning and decision making. By building up the data from different areas in the corporation, the map makers can display market information in the context of technology assets and gaps, product line information in the context of market need, and so on.
2. *Iterate across groups and over time.* The first maps will be first drafts. As people in the business become more familiar with the process, they can find and integrate more detail and edit the categories of the first-draft map. Even when the mapping process has become quite mature, maps need to be maintained because the information will change over time—because of changing internal and external conditions, new information, and increased sophistication in the use of the maps.

3. *Develop a consistent syntax and vocabulary*. For maps to be a useful tool, the syntax—which displays the relationship of the parts—and the vocabulary—which identifies the elements of the map—must remain constant, or evolve to meet changing circumstances. If syntax and vocabulary change arbitrarily or are out of step, the maps will lose their utility as communication tools and map makers in one part of the organization will not be able to use information from others.

Good map making and map using enables firms to identify and preserve their core assets and recognize key gaps in relation to product plans and market needs. Maps create a rational, discussable plan for the timing of product and technology efforts, and provide a compelling format for difficult decisions about what to keep, what to develop, what to discard, and what to outsource. Good mapping practices build consensual decision making at the team, the management, and the executive levels of the business.

PART 2: IMPLEMENTATION ALERT

Pictures drive knowledge management, decision making, planning, and follow-through. In implementing market, technology, and product mapping, it is important to recognize that the *form* in which the data are presented is an important part of their utility. This *implementation alert* is intended to alert practitioners of mapping to some subtleties that will help them get the most benefit from the practice. We recommend Edward Tufte's books on the visual display of quantitative information for much more on the subject (Tufte, 1983).

1. Data that are presented linearly, for instance, in a spreadsheet or list, invite us to think of them separately and lead to conflict over individual merit rather than comparison and investigation of the potential of combining options. Data that are presented relationally invite, even provoke, relational thinking. Furthermore, the thinking and pattern seeking is suggested by the way the data are presented: the "playing field" for interpretation (the categories of interpretation) is part of the picture. This has two important implications, one about building maps and one about using maps. The map builders set the categories within which the maps will be used or interpreted. So, for example, the map builders decide on the time frame for planning and strategic thinking. An electronic device roadmap, for instance, will typically have a short horizon, a pharmaceutical roadmap a much longer one. Target markets will be represented in the map: Who decides on the segmentation? Who selects the key customer and market needs among the many possibilities? Not every technology or product will be displayed; how will the decisions be made about selecting and grouping them for display? Map builders are pattern seekers and category builders; they provide

the fundamental distinctions against which critical decisions will be made. Companies need to take seriously the building of the mapping framework, ask searching questions, and challenge the categories so that categorization becomes a part of the iterative process of developing the best and most useable maps.

In practice, the fact that maps define the interpretative field and invite relational thinking blocks anyone's attempt to argue from one salient data point to a conclusion. For example, if a product development project has the highest expected ROI, that is part of the picture to be looked at in relationship with all the other parts and not by itself a reason to choose to do it. In this way mapping discourages decision making by political influence or position of power—and mapping can be difficult to implement in a culture where top-down decision making or executive intuition plays an overly large role.

2. Data mapped on a visual field are ideal for group inspection, discussion, and decision making. For the maps to provide support for the group's decisions, decision makers must agree that the data that are available are all the relevant data. In using roadmaps, when data gaps or distortions are found, this is cause for discussion—and for a decision to proceed or to improve the data before a decision can be made. Data in someone's hip pocket may emerge, but it must take its place on the map with the other data, not as a reason to override or to circumvent the group decisions.

Making maps is not something to be assigned to a person with nothing else to do, or to be pulled together just before a meeting or to impress a group of stakeholders. Businesses would do well to develop the art and technique of map building, and embed rules for best practice such as those found in Tufte's work. In his book *The Visual Display of Quantitative Information,* he reviews methods of relational graphics. The best, he tells us, invite causal and relational thinking, and the best foreground the data, not the method of its presentation.

PART 3: PRODUCT AND TECHNOLOGY MAPPING ON THE BACK OF THE ENVELOPE

A high-level strategy team at Acme Motors has chartered a cross-functional team—the New Technologies Team—with exploring the evolution of automobile fuel sources so that the company can address the strategic question of how it might position itself vis à vis the emerging technology. It quickly becomes apparent that the team needs to understand at a macro level the evolution of technology, of regulations, and of market drivers, and how this evolution may intersect with or be influenced by other economic, social, and technology factors. It needs to know when the technology is likely to become competitive and cost-effective and how much pressure there is likely to be to develop alternative fuels. It should also be aware of the likely changes to transportation systems, as well as other political, market, and competitive dynamics.

BACK OF THE ENVELOPE MAP 1: THE EVENT MAP—MAPPING THE STRATEGIC GEOGRAPHY. It is frequently appropriate to begin mapping with a very high level map that draws out and sorts what the team knows and also provides a map of the information gaps. We will call this the "event map," a simple grid that allows mapping of the "strategic geography." (See Gill, Nelson, Spring, 1996.) The Acme team will enlist the help of other functions, including marketing, and plot such things as their best guess about how much the cost of fossil fuel is likely to increase and how efficient internal combustion engines are likely to become. The team is also interested in how consistent the switch to SUV's will prove to be. If there is another trend, similar to the shift in the 1970s away from the gas-guzzlers to the small imports, how will that affect the overall construction of the fuel cell opportunity? What if the congestion on our highways spurs a trend toward public transportation in cities and suburban areas? Anticipating possible market trends and making informed guesses about technology evolution, the team will begin to pinpoint the potential pitfalls and opportunities, and will begin to identify where it needs more precise or sophisticated information. The team's event map of the environment surrounding Acme's passenger car development is shown in Figure 15-2.

The event map shows several trends that will impact one another over time. On the face of it, replacement of fossil fuel (internal combustion) engines with hydrogen fuel cells seems like an obvious move. However, the efficiency of the internal combustion engine, which has stayed pretty level for decades, has begun to increase markedly in the past several years. How economical will the fuel cell technology have to be in order to compete with an optimized internal combustion engine in 20 years? This back-of-the-envelope mapping raises questions for the next level of mapping, in which the team will have to work hard to make projections and assumptions about the relative costs/mile as precise as possible. For instance, it might be a good guess to assume that customers will continue to prefer the heavier, larger SUV-type vehicle, but before Acme targets technology development, it should also explore the impact of a possible trend to "vehicle downsizing." The team decides to launch a market research project to explore customer preferences and market trends, perhaps uncovering latent needs that might enable Acme to shape the future market.

BACK OF THE ENVELOPE MAP 2: A PRODUCT LINE MAP—IDENTIFYING PRODUCT FAMILIES. The Acme New Technologies Team realized that it needed to understand the history and current conditions surrounding Acme's product development to provide a context for the next step, which would be to map the current and future technology. The team called a meeting to discuss just what products should be included in the project scope and decided to map the company's small, midsize, and luxury cars, but not vans, SUVs, and light trucks. After about a half an hour, the team realized that most of the product differ-

Market	↑ Use of SUVs ↑ Safety concerns	Auto companies selling used cars	↑ Smaller cars? “green” becomes significant market driver	“I’d take the train if it were a bit more convenient”	
Economics Politics Social trends	↑Traffic/congestion ↑ Energy prices ↑ Mideast unrest	Infrastructure for electric cars	↑ Government support for fuel cell development	↑Public transportation? ↓ energy prices	Highways make great parks in inner city
Regulations	Stricter emission control	Rebates encourage hybrids	California requirements spur hybrid	Zero emissions policy in all urban areas	
Technology	IC engine at x gal/mile ↑ IC efficiency Hybrid engine	All-electric car for short trips Ford releases hybrid SUV	Every car maker has hybrid option		Will IC efficiency make fuel cell less interesting?
	2000	**2005**	**2010**	**2015**	**2020**

Event map: Best guesses: indicate confidence, list assumptions

Best constructed by a cross-functional group over a period of time. Each mapping session should resemble “brainstorming,” with group members adding thoughts and information using sticky notes or pictures. Frequent discussion breaks keep members up to speed; the knowledge gaps are recorded, and members leave with research assignments to be completed before the next session.

FIGURE 15-2. Acme’s passenger car environment event map.

CASE NOTE

The Photographic Imaging business unit at a leading photographic company realized that every time a new camera or media project was started, the tendency was to reinvent or redesign not as little as possible but as much as possible. They brought together a cross-functional group and addressed the question: What are our product platforms, and what defines a product line? Their product roadmap is shown in Figure15-3. They began by mapping the product that the customer first decides to buy—cameras—but for the company the revenue-producing product was media, and the cameras were essentially boxes that enabled customers to use the media. Mapping the media streams as product lines brought new clarity into questions of what products to develop and when. It also highlighted the consistency of some of the product lines over time, including their "600 line," which has been a consistent seller in Asia, and encouraged the business unit to support these lines even though they were not new and interesting.

What they learned from what they mapped may have been obvious in the sense that there was no new information. But before the mapping project, it was not unusual to find that a new camera project would require unexpected changes to media, raising project complexity and cost and introducing all kinds of supply chain and distribution/logistics problems. With the product line map, these questions could be identified ahead of time. Product line mapping supported portfolio decisions—which lines to support, eliminate, or bolster—as well as decisions at the level of the individual project.

The first decision the team had to make was what would form the basic categorization for product line and platform mapping. The fact that media was the revenue driver won out, but there was in fact much more attention at the time on hardware, in particular a very new and different small-format camera. The team also chose to map products back in time, and the historical view brought older successful lines more clearly into view. If mapping had begun closer to the present, they might have missed the important point that the 600 line had been and was still a successful product, particularly in the Far East.

entiation among the end-user vehicles relied on size, styling/design, interiors, and added features, and that relatively little depended on the parameters in which the team was interested. The team shifted its focus to mapping the engines and the fuel technologies, and came up with a map that identified *engine families* rather than *product families.*

Getting started with product and technology mapping can seem overwhelming. There is too much information, and too many information gaps as well. Good mapping requires time, and it takes resources from other planning and development efforts. The best way to start mapping in your company is to begin with a question you need to answer or a decision that you need to make. This will give you guidance on how to structure the data, what data needs to be included, and who should be involved.

The team now wants to get a better picture of the technology that is currently available. At some point it will want to look at competitors' technology, and technology trends, but for now, it simply wants to begin an inventory of the relevant technology assets of the company, whether historical, current, or

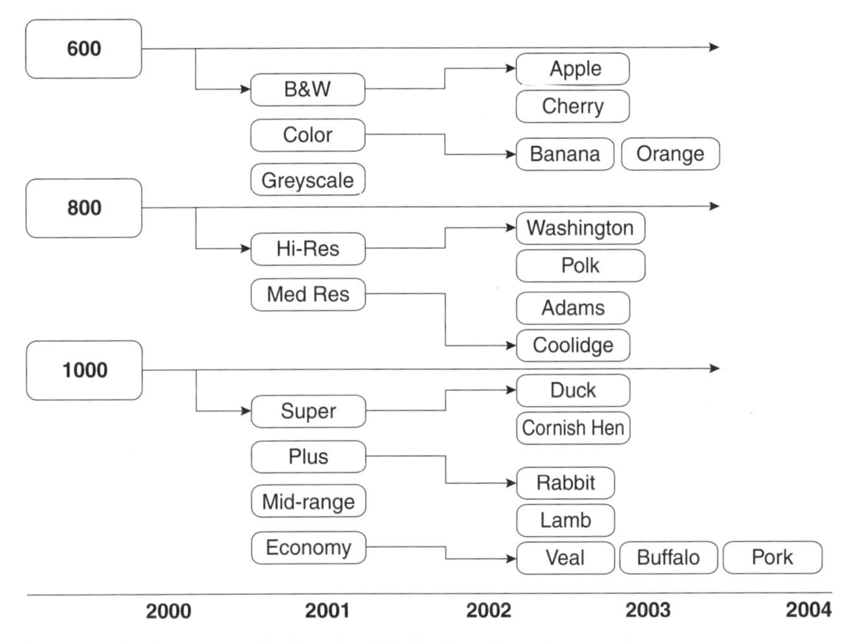

FIGURE 15-3. Photographic Imaging Media-Based product roadmap.

CASE NOTE

A manufacturer of heating and air-conditioning products simply wanted to create an inventory of its technology assets and to begin to discover how those capabilities were focused on market and product needs. It was driven by the common concern that the innovation needed by one project might be already "on the shelf" of the project in the next building. Technology projects in this company were assigned based on the needs of product development projects, including those that might have long-range implications. The technology group seldom sat together to explore how the different projects might overlap or support each other. Since technology projects had been launched and subsequently identified with business or product issues, the function did not have its own technology categories, so the manufacturer's first question was this: What are the basic categories for our technology projects?

The facilitator suggested using a bottom-up, or inductive, way of defining its technology categories. First planners wrote all the technology projects on sticky notes and assembled them in rough logical order. These projects were then broken down into technology elements—called "technology building blocks"—which they then assembled using an affinity map. (See Burchill and Brodie, 1997.) What they ended up with was a map of projects organized so that they could see overlaps, relationships, and synergies. This map is shown in Figure 15-4.

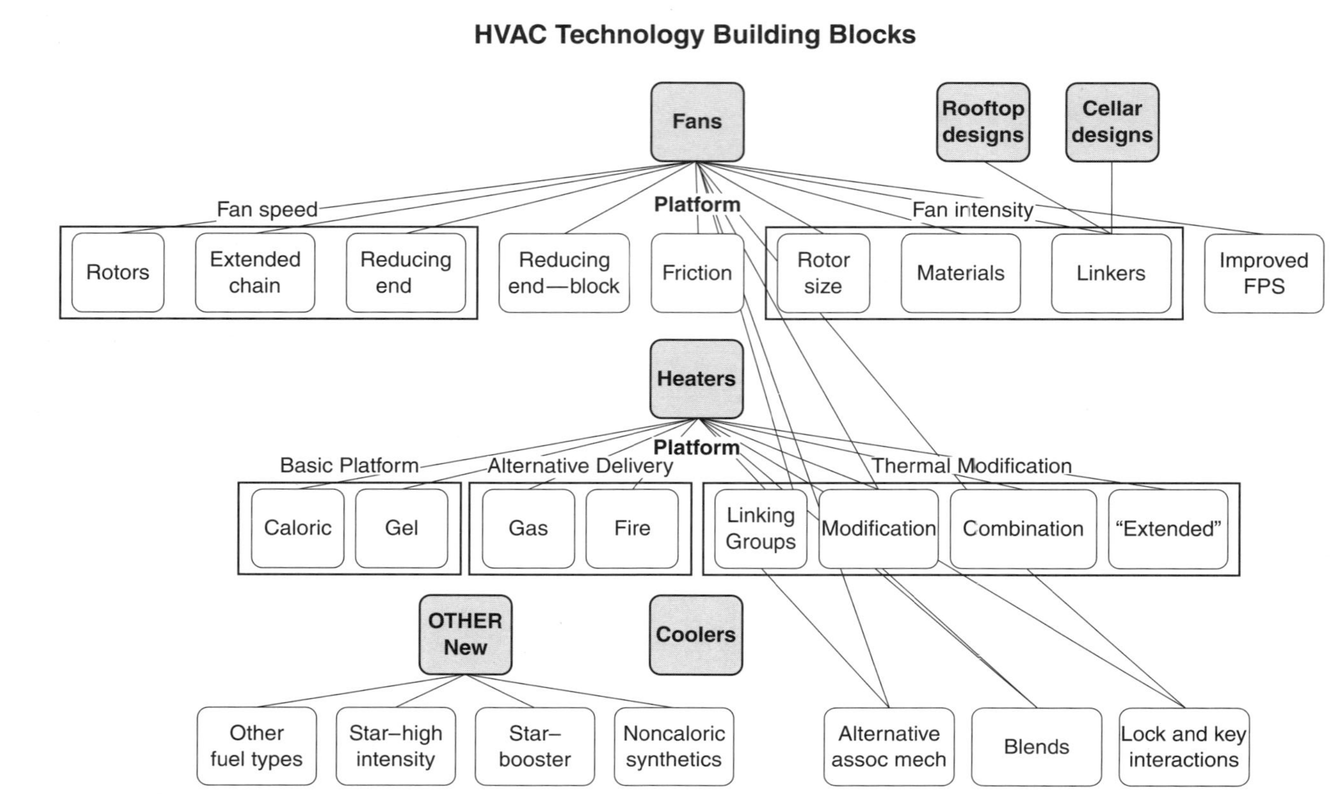

FIGURE 15-4. Technology project map.

CASE NOTE

A chemical formulator wanted to explore how its product fit into a larger value chain. It divided the space, from raw materials to retailers and distributors, and used the graphic analysis to uncover where it could add the most value. The resulting map is shown in Figure 15-5 (Taytelbaum, 2000). This fairly quick cross-functional mapping exercise helped the company recognize that where it traditionally added value gave it limited opportunity to leverage its core competencies. It began to explore how its competencies could address more attractive and profitable issues further down the value chain.

planned. This will give the team a good baseline for deciding on likely areas for investigating in order to fill out the strategic picture.

The team's next question had to do with how the engine, and the fuel technology, fit into the overall supplier-to-customer value chain. For this mapping exercise, the team brought together a cross-functional group including sales, marketing, procurement, manufacturing, contract manufacturing, distribution, and service. Focusing on the passenger car engine, the team used informal process mapping to locate the contributors along the value chain, including suppliers of raw materials and engine parts, engine assembly and installation, and engine use, maintenance, and repair.

In all of these examples, a small group came together to map issues of key strategic importance. They worked with limited information and created the categories that would help them think through the issues. In some cases, they met over time to make their initial hunches more precise and to develop more information.

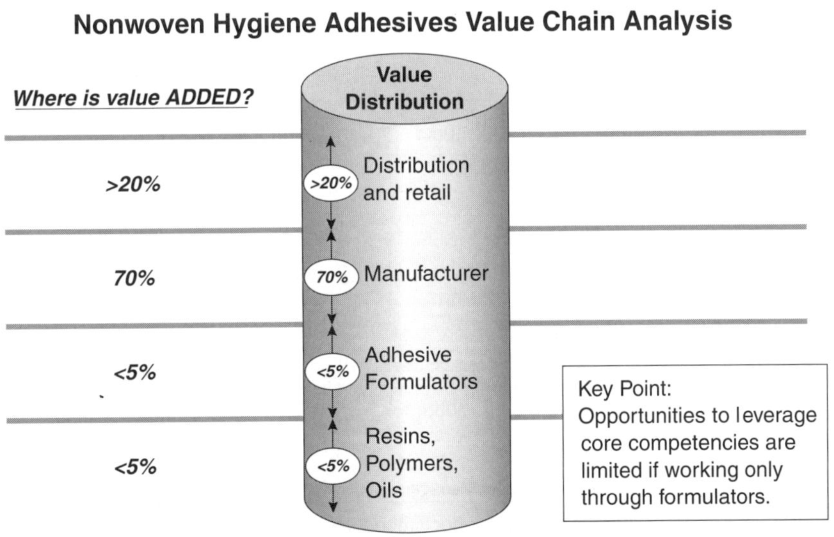

FIGURE 15-5. Chemical formulator value chain analysis.

These back-of-the-envelope maps can be used in companies with a relatively low level of process maturity, but companies that make the best use of mapping then go on to implement mapping across businesses, product lines, and functions, making the mapping exercise a cross-company endeavor that is used at both the project and the portfolio level. Before detailing the tools and methods for the more mature and sophisticated uses of mapping, we will discuss the relationship of a company's process maturity to the choice of mapping tools.

PART 4: IMPLEMENTATION ALERT

Planning tends to follow the structure of an organization. This section describes the four key dimensions of mapping and how maps must fit an organization and match its planning maturity. The barriers between functional units or business units are an important factor in determining what type of planning will be done and what types of maps may be created and maintained.

Mapping is most easily done within the bounds of a structure where relationships are solid and well defined, and it can be impeded by the barriers created by organization structure. Most organizations are structured in a matrix of functions (marketing, research and development, manufacturing, etc.) and businesses (such as product lines or profit centers). Figure 15-6 shows a simple view of planning in the matrix organization of a corporation, with functional organizations laid out horizontally and product lines or businesses vertically.

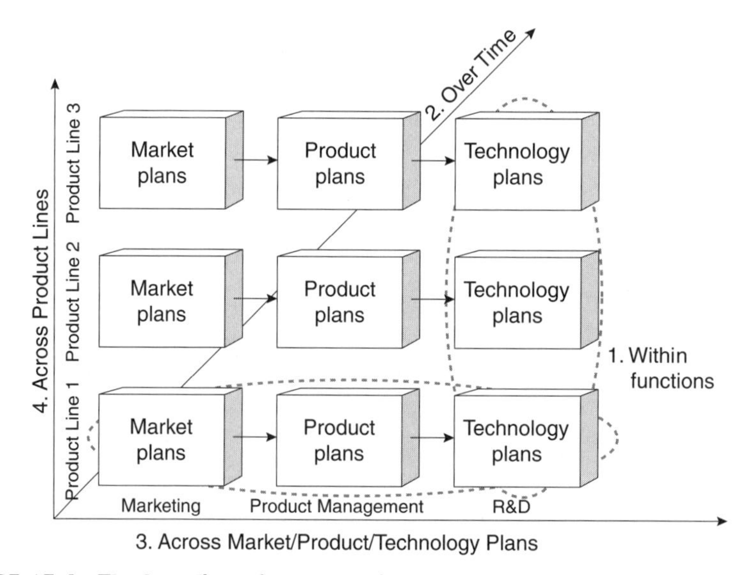

FIGURE 15-6. The four dimension of mapping.

Four basic dimensions of planning and mapping are shown: (1) planning within functions, (2) planning over time, (3) planning across functions, and (4) planning across product lines.

Within each business unit or product line, marketing, product management, and research and development organizations typically have functional responsibility for planning and execution in their respective areas. Each functional group creates its own plan over time; product line plans may then be constructed from functional plans for the market, for the product, and for technology to implement the product. Within each function, maps may be created and used to show the evolution over time of some aspect of the plan—for example, how the market will grow over time, or the cost targets that the product must meet over time. Also within functions, architectural maps may be made—for example, how the market is segmented, how the product is constructed of components or subsystems, or how the product is expected to evolve against the life cycle S curve. These maps of architecture across time, created for functions and product lines, represent the most basic level of mapping sophistication.

At the next level of sophistication, consistent with an organization that has implemented cross-functional product teams and phase/gate processes, maps may show the linkages across the functions—for example, how customer needs determine key product characteristics and how key product characteristics determine the most important technologies for success in the marketplace. For example, a product-technology roadmap integrates architecture, time, and functional and/or business linkages to provide a powerful product line planning story. These roadmaps help a team agree and align around a plan, provide a framework to tell the team's story to others, set priorities for development and market actions, and help guide the team during the development process. These types of maps flow across the horizontal dimension (as in the dashed horizontal oval), linking the functional elements and incorporating time. For example, the product-technology roadmap of Figure 15-1 mirrors the functional elements (market/strategy, product, and technology) that contribute to the plan.

Finally, in organizations with a well-developed portfolio management process, maps may be made to show the connections across product lines. For example, maps may show where two or more product lines are addressing common market segments, where there are key technologies whose development may be shared among product lines, or where technologies needed by one development project may be found already developed in another product line. These maps link plans in the vertical dimension (as in the vertical dashed oval), linking functions and groups of functions across product lines or business units. Maps at this level promote rich discussion in portfolio decision making.

To summarize, the four dimensions of mapping capture information within functions, over time, across functions, and across product lines. The complexity of the mapping process increases as more of these dimensions are incorporated. It is important to match the sophistication of your maps to the

Maturity / Mapping	Within Functions	Over Time	Across Functions	Across Product Lines
1. Limited planning within functions. Initial, ad hoc integration.	◒	○	○	○
2. Architectures within functions, but not integrated. Managed integration depends on individuals.	●	◒	○	○
3. Planned evolution over time, some integration. Defined.	●	●	◒	○
4. Product plans with integrated market, product, technology plans over time. Quantitatively managed.	●	●	●	◒
5. Integrated plans across product lines, driving portfolio decisions. Optimizing.	●	●	●	●

FIGURE 15-7. Mapping and product development maturity.

maturity and sophistication of the surrounding processes, such as the strategic planning, marketing, portfolio management, phase/gate, and life cycle management processes. The objective of developing and using maps is to aid decision making in the organization's processes, and a map should use the information available but should not imply greater precision than is found in the data or provide more information than is appropriate for the decisions being made.

The maturity of product development and management planning and mapping may be measured by the degree of integration of functions, product management and development, and management across product lines (including portfolio management processes). The levels of maturity are shown in Figure 15.7. At the first level of maturity, limited planning exists within functions, with virtually no cross-functional planning. Integration takes place on an ad hoc, informal basis. At this level, mapping is useful to a limited extent within functions, and the time basis for plans and maps is often ill-defined or not considered. At maturity level 2, architectures are defined and planned within functions, along with some thinking about evolution over time. Integration across functions is managed, but it depends on individuals rather than systematic planning. Architecture maps within functions begin to take on characteristics that enable planning for significant product evolution. At maturity level 3, planning and mapping consistently consider evolution over time, and some systematic integration across functions is performed. Maps at level 3 firmly include a sense of evolution and begin to bridge the functions in an informal way. At maturity level 4, plans are well integrated over functions, producing integrated market, product, and technology plans within product lines or business units. Maps take on the full product-technology roadmap of Figure 15-1. Finally, at the most mature level, 5, plans are integrated across product lines and business units. At this level, mapping in standard formats allows systematic comparison of all aspects of plans and provides a foundation for portfolio management.

A team should manage the level of detail of its maps to be consistent with the maturity of the planning process or the level of detail required in decision making. Early in planning stages, maps should be high level, sketching the level of knowledge available. As development and planning proceed, maps can become more detailed. For decision making involving many portfolio elements (projects, products, etc.), maps should fit the decisions that are needed. For example, in a corporate product portfolio process, maps should focus on decision criteria such as value creation, fit with strategy, and balance of the portfolio, rather than on detailed feature evolution or target setting. As we have already stressed, a good corporate approach is to layer maps within a common structure, so that decision makers can start at a high, broad level of information, drilling down for greater detail as needed.

PART 5: OPENING THE ENVELOPE

As the portfolio and management processes in an organization become more mature, they can use and support more sophisticated mapping tools that may be used together to create an integrated product and technology roadmap. Several tools and methods are described in this section, using Acme Motors to illustrate.

Acme's back-of-the-envelope maps (Part 3) laid out critical events in a possible evolution of the industry and the next several generations of vehicles. Senior management has studied the strategic and tactical implications of this work and has assigned to a new product team the challenge of developing and incorporating new technologies in Acme's new passenger vehicles. The team will need to incorporate new architectures and designs for power trains, fuel sources, steering, braking, body systems, and more. At the same time the team must keep an eye on competitors and competing technologies to be sure that Acme's cars are competitive and differentiated. The team must also develop its passenger car line in conjunction with its other product lines, gaining efficiency from collaboration and sharing across the lines. (*Note:* Acme's team can begin the planning process developing maps for each functional area; but Acme's planning processes are quite mature, so the cross-functional team is able to develop an integrated roadmap.)

Based on information and decisions from the back-of-the-envelope maps, Acme's strategy board has defined the following vision for their passenger car product line: "Acme Motors will develop and market an efficient and competitive fuel-cell-powered vehicle, with attractive styling and well-integrated and efficient subsystems for passenger comfort and safety." This vision is a number of years and multiple product generations away, and Acme's passenger car product line will likely have to evolve through multiple powertrain configurations and related subsystem changes to arrive at a competitive, efficient, and cost-effective vehicle. All of these product and technology evolutions must be coordinated to produce a total package that will achieve the vision while remaining competitive over time. As the team progresses in developing its maps and plans, it will keep the strategic maps updated so that senior management and the entire organization can continue to be included in this strategically important and risky venture.

Because of the very large scope of its charter, the team has decided to create the following maps to guide its work: (1) maps of functional architectures, (2) maps that set targets or objectives over time, (3) maps that link market drivers to product drivers and technology drivers, (4) maps that show product or technology evolution, (5) maps that link architecture, targets, and plans over time to make an integrated product line and technology strategy, and (6) maps across product lines to identify key technologies that can be shared or potential points for collaboration. Maps 1, 2, and 4 may be drawn up first by functional subteams and then linked together as maps 3 and 5 are created by the cross-functional team. Finally, this team can integrate its work with other product teams to create map 6.

(1) Maps of Functional Architectures

These maps define how all the parts of the product or service fit together and interact. They can be as simple as the component layout of a product or as complex as the interactions among product subsystems. The old adage "a picture is worth a thousand words" applies here in that the picture can quickly and effectively define the relationships of the parts and give a reader a rapid understanding of the parts of the problem.

An architecture can take a pictorial form, showing the assembly of the physical product, perhaps in a cutaway view, or it can take a logical form, showing a hierarchical assembly structure or a layered block diagram of the product. Physical products are often best architected by a picture or a subsystem/component layout. Many software and service products can be best shown in a layered view, indicating how modules or functions are connected or exchange information. The architecture for process-produced products, such as many chemicals, includes the manufacturing process steps. The key determining factor in constructing an architecture is that it should include all the major components that determine the value of the product.

Product planning needs to understand market architecture as well. Luke Hohmann distinguishes between "tarchitecture," or the technical architecture, and "marchitecture," which "embodies the complete business model, includ-

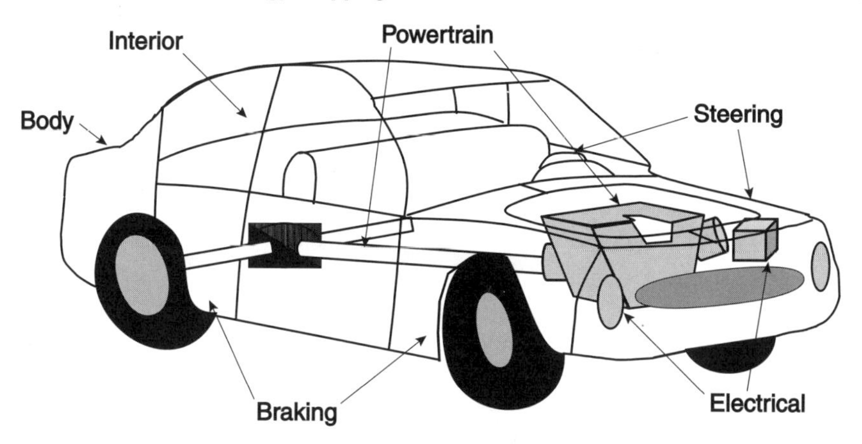

FIGURE 15-8A. Cutaway view of Acme's product architecture.

ing the licensing and selling models, value propositions, technical details relevant to the customer, data sheets, competitive differentiation, brand elements, [and] the mental model marketing is attempting to create for the customer"(Hohmann, 2003; p. 51).

The first map created by Acme's cross-functional planning team is an architecture that defines the subsystems, or technology elements, of the passenger car product line (see Figure 15-8A). A cutaway view shows the engine compartment and powertrain components in greater detail. In this first-level view of the car, it is difficult to show many components; a more detailed view might include several subdiagrams showing details for each subsystem. For example, the powertrain diagram would show fuel source, engine, power conversion, and emissions subsystems.

Another, hierarchical, architecture view for Acme's passenger car is shown in Figure 15-8B. The subsystems are grouped and defined in increasing detail. While the figure shows two layers, the team will go on to develop several more layers for detailed discussion and analysis. This architectural structure allows each subsystem to be evaluated, performance criteria to be defined, performance targets set, and the design process structured. The architecture keeps the team focused on the key marketing and design problems, and helps the team organize its thinking about how the subsystems of the product fit together.

An issue with architectures is how broad they should be—what they should include. For all the issues related to getting a car to market successfully to be understood, Acme's passenger car architecture will likely need to go beyond the physical product to include the fueling infrastructure—as it changes from gasoline to alternative fuels such as hydrogen or methanol. It might also be extended to include marketing programs and servicing infrastructure. Acme's passenger car product line team has decided to develop a

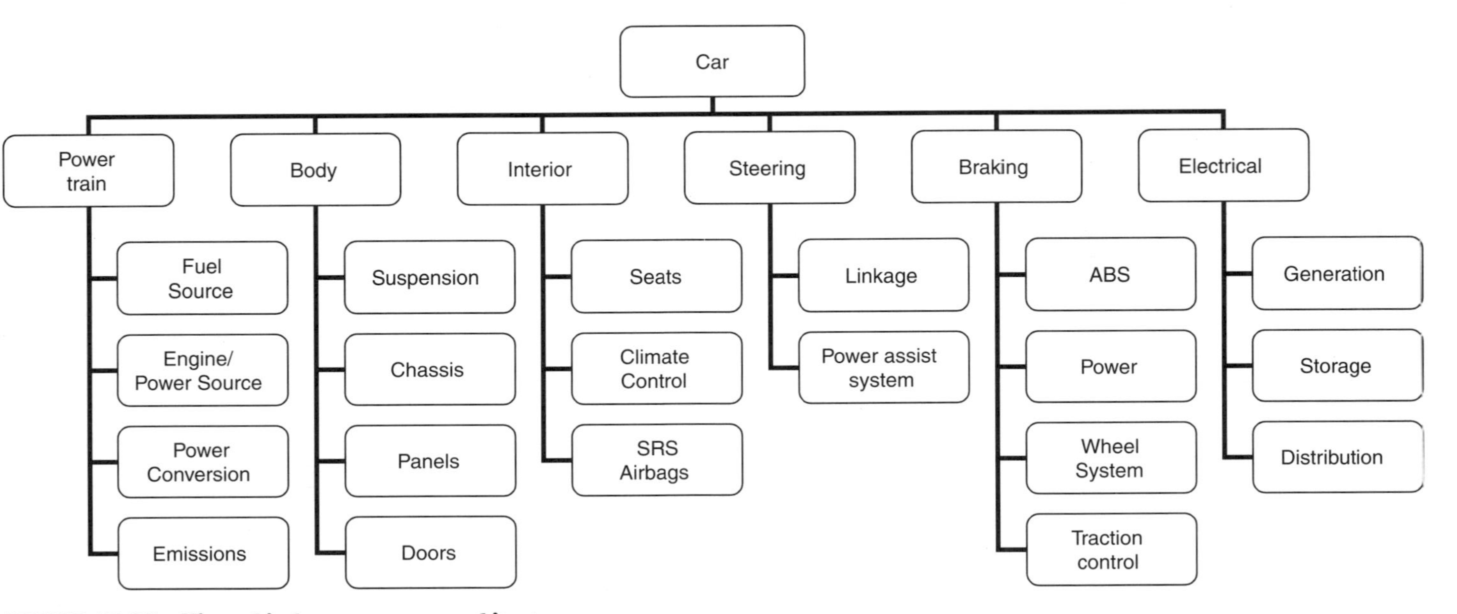

FIGURE 15-8B. Hierarchical passenger car architecture.

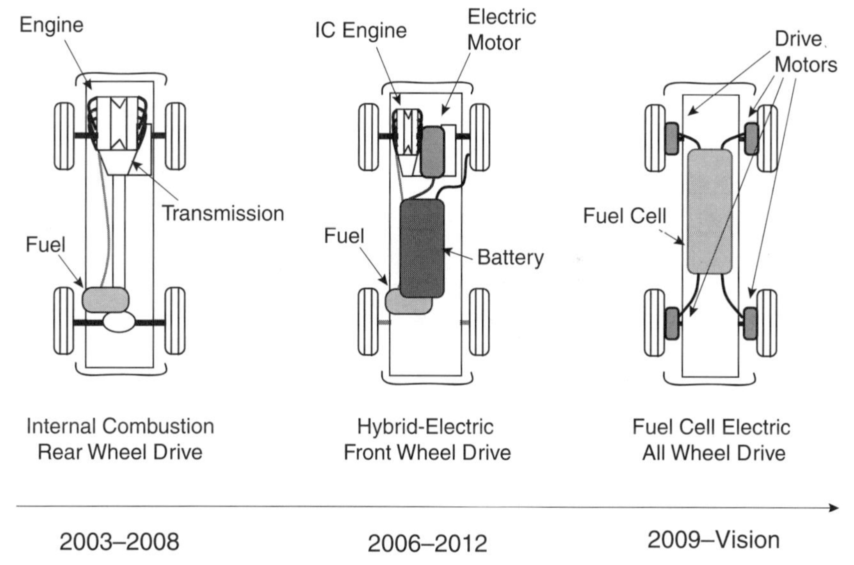

FIGURE 15-8C. Architectural evolution.

"tarchitecture" for its initial plan for development of the car. In subsequent planning steps the team will build a "marchitecture" with a broader scope, including marketing and distribution issues, fueling and maintenance infrastructure, supply chain issues, and regulatory issues.

The architecture diagrams in Figures 15-8A and Figure 15-8B are static views. For many products where there is a single dominant architecture—for example, vehicles with traditional, internal combustion engine powertrains—this may be sufficient. However, many system architectures evolve, introducing the time dimension. The Acme team's ultimate target is a radically different fuel-cell-powered, electric drive architecture. The team has also determined that an evolutionary approach is needed to reach the target, and they have defined an architectural evolution of the powertrain subsystem, shown in Figure 15-8C. In the case of Acme's passenger car, the company plans to evolve the powertrain from a gasoline-powered engine rear wheel drive configuration to a hybrid gasoline-electric front wheel drive model to a fuel-cell-powered all wheel electric drive system. The hybrid phase, to be introduced in 2006, will enable learning and development of electric motor technology to progress while fuel cell costs are reduced to the point where they will be competitive with internal combustion technology. Finally, the fuel cell electric powertrain will be introduced in 2009, after which the other architectures will be phased out.

Maps may also be used to describe the evolution of plans in other functional areas. For example, the team may map market segmentation, market size, and customer needs prioritization. Figure 15-9 shows the evolution of the market for light vehicles in the United States (Hellman and Heavenrich, 2001).

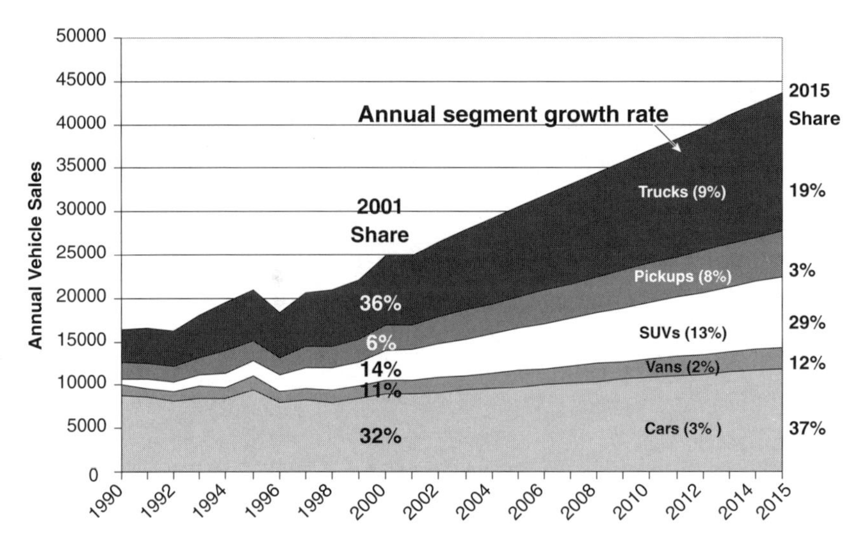

FIGURE 15-9. Market segmentation/structure.

Acme's marketing department has projected the historical market forward to help the team size its expectations and future production forecasts. Acme's marketing department has placed the targeted market segment for passenger cars in the context of the overall market for light vehicles and has forecast market segment sizes by projecting from the market growth of the late 1990s using "best-fit curves." The marketing department projects the car segment to grow at about 3 percent per year, while several segments grow at faster rates: up to 29 percent per year for SUVs. The forecast shows that the car segment declines from about 36 percent of the market in 2001 to under 20 percent by 2015.

Historically, the market has been highly variable from year to year, driven by economic conditions and also by changing tastes of the public, so substantial debate may ensue among members of the planning team about the future of the overall market and of the growth or decline of the passenger car segment. That debate might include discussion of whether forecast growth is too conservative or too optimistic, whether the growth in the various segments will remain at current rates, or whether economic or social conditions might bring about a shift in consumer preferences for cars versus SUVs or pickup trucks. The team might even call into question whether its focus on the car segment is the right one. It may also decide to create several scenarios, each with its own map, using different assumptions about growth and consumer preferences.

Market maps may also identify the important customer needs for a product line. Acme's marketing department has identified two significant customer subsegments in the passenger car market: the family sedan segment and the performance segment. The segments have similar needs, but with differing pri-

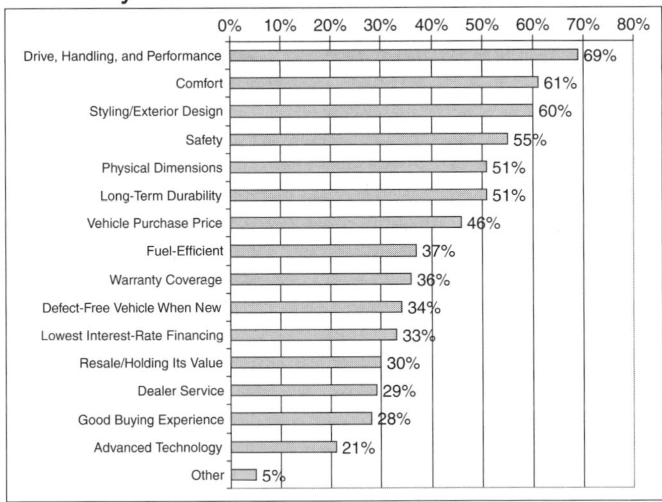

FIGURE 15-10. Customer drivers.

orities. The marketing department has obtained independent research on customer's buying decision priorities, and this information is shown in Figure 15-10. The figure lists the key customer drivers, ordered by the frequency with which they are cited by survey respondents (McManus, 2003). Drive, handling, and performance are cited as important by nearly 70 percent of potential buyers, followed by comfort, design, and safety. The list of customer drivers will be used by the team to focus their product line development. Note that the highest-priority drivers focus on the car itself, while dealer service and good buying experience are further down the list, confirming the team's decision to concentrate first on creating a roadmap for the car itself and later on the larger "marchitecture."

(2) Maps That Set Targets or Objectives Over Time

Mapping over time is especially important for product line development. A key element of nearly every product line strategy is pricing and costing, and industry learning and experience curves can help set bounds and competitive targets. The Acme team must set price and cost targets for its passenger car line over the planning horizon to 2015. To make these projections, the Acme team uses an experience curve of the historical average car selling price/horsepower versus the cumulative U.S. industry production of passenger cars since 1975, shown in Figure 15-11. This experience curve, created by Acme's product managers using government data and market surveys, helps the team determine price and cost targets in the context of industry competition (Davis and

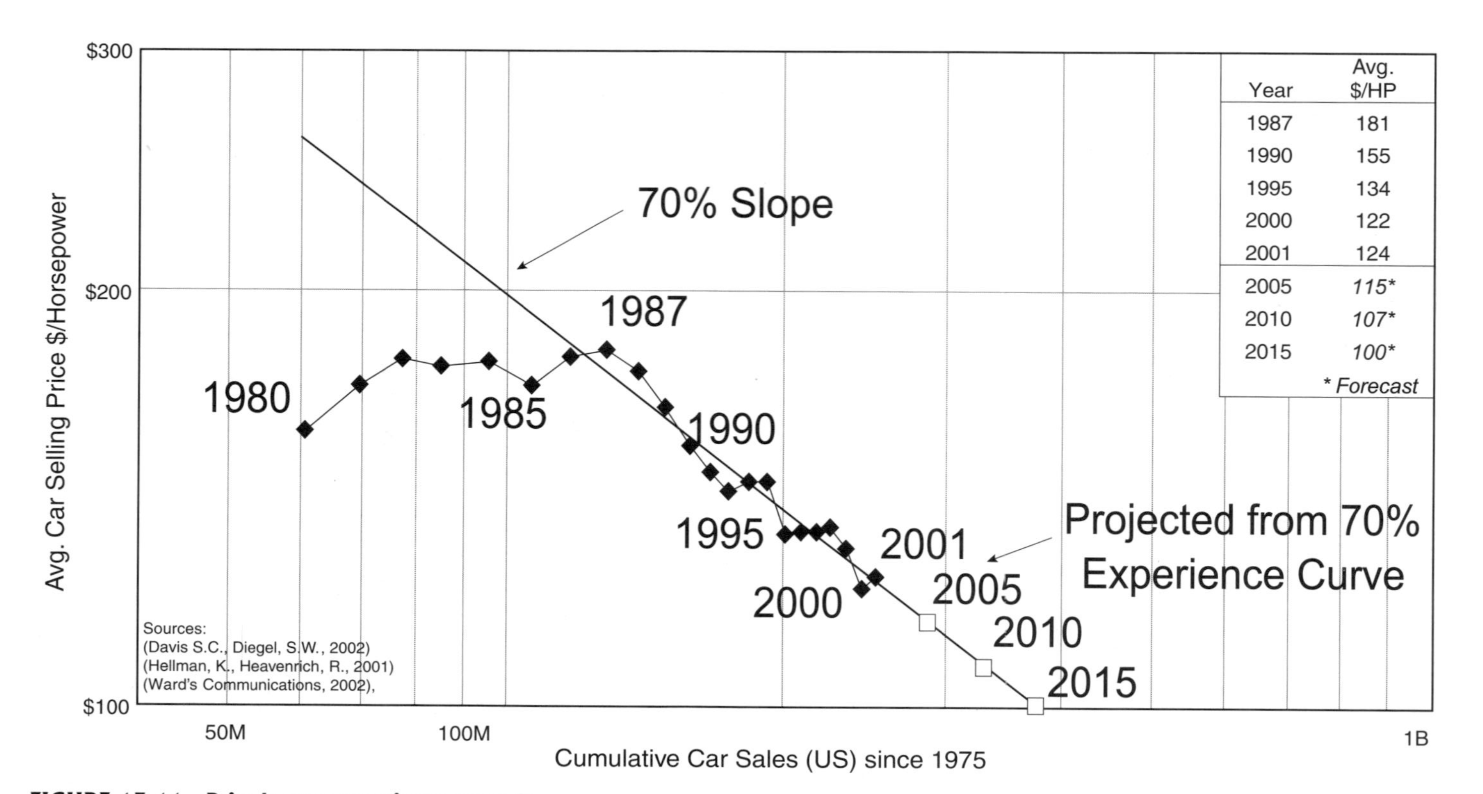

FIGURE 15-11. Price/power experience curve for passenger cars.

Diegel, 2002, Hellman and Heavenrich, 2001, Ward's Communications, 2002). To help set targets for vehicle pricing, product managers have extrapolated the historical data using the industry learning rate and the marketing department's assumptions about market production.

As it introduces alternative powertrain architectures, Acme's team must make sure that the new configurations are competitive with the moving target of internal combustion engine technology. The experience curve shows that the 1980s were a period of little learning or reduction in price per horsepower. Then, during the 1990s, industry learning moved ahead at a rapid pace, reducing the price per horsepower, most likely due to the introduction of computer-controlled multivalve, fuel-injected engine technology and increased global competition. This learning changed the industry experience curve to a 70 percent slope, where a cumulative doubling of volume results in prices (and costs) declining to 70 percent of what they were. If this experience is extrapolated assuming industry production at levels similar to the past few years, the industry average price per horsepower will decline from about $124/horsepower in 2001 to about $100/horsepower in 2015.

An experience curve captures many forces that will impact prices and costs. It reflects the advance of technology innovation, conditions of industry competitiveness and culture (of sharing and exchanging learning), and the drivers of market demand. Forecasts using the slope of the experience curve are based on a number of assumptions that can be varied to study their impacts on targeting. For example, expected future market demand affects targets by determining how far along the forecast path the market will be.

The experience curve may also be used to evaluate the viability of competing technologies—to understand which technologies are most likely to win out over competition from a cost perspective. If the cost of a new competing technology declines at a steeper slope than the existing dominant technology, experience curves help forecast at what point the new technology will overtake the existing technology in the market. For example, the experience curve for price per horsepower tells Acme where it must set its cost targets for new powertrain technologies in order to be competitive in the future. In 2003, for example, the cost of a fuel cell alone was estimated to be about $1,300/horsepower.[2] If the price of the entire car is expected to be about $100/horsepower in 2015, auto manufacturers must reduce the fuel cell cost to the range of $20 to $30/horsepower by 2015 in order to be competitive. This sets a high bar for management's vision. The team will have to set aggressive performance targets for all subsystems, thereby setting the stage for competitive targets for the highest-priority product drivers, the key performance characteristics of the company's product line. It also signals a priority for Acme's research organization to focus on fuel cell cost reduction.

Maps described so far have largely been focused on one function in the product development chain. Maps across functions create a more complete product line story by connecting and integrating the functions.

(3) Maps That Link Market Drivers to Product Drivers and Technology Drivers

Linking functional plans together is a powerful way to focus development on the most important features serving the highest-priority needs. This linkage connects the most important product features to the architectural elements that will implement those features. Linking maps makes sure that product development priorities are focused on the things that are important to the customer and that will provide differentiation.

Figure 15-12 is a map developed by Acme's cross-functional product team to show a part of the linkage of customer drivers to product drivers to technology elements of the architecture for Acme's passenger car product line. Customer drivers might be called the "know-whys" for the product line—the customers' priorities tell the Acme team why they should take a particular development course or introduce certain features. The product drivers are the "know-whats," identifying key product features. And the technology elements are the "know-hows" for the product line, showing how technologies can realize the product vision. The connections among drivers help the development team trace their designs and decisions back to fundamental customer needs. The linkages can also help establish priorities and timelines. For example, the technology elements with the greatest number of links to the high-priority product drivers will often become the focus of implementation. To develop the linking map, the team starts with the customer drivers, the key product performance dimensions, and the architectural elements; makes connections; and then prioritizes the drivers in each category. The Acme team finds that product drivers of power, acceleration, frame stiffness, suspension characteristics, and steering are the keys to the most important customer driver for handling and performance. The team's analysis proceeds until it has determined a small number of key drivers for which performance targets will be set. The team also determines the characteristics in which Acme will seek to lead competitors, perform at parity with the industry average, or lag the industry. Finally, the team links the product drivers to the elements of the architecture, indicating where in the architecture of the product new technologies must be introduced in order to meet performance targets. The Acme team determines that the powertrain is the key area of the architecture, and labels the powertrain as its attack technology, the area on which it will concentrate.

(4) Maps That Show Product or Technology Evolution

The next set of maps that the Acme team develops involves the display of product or technology evolution over time. Maps of this type are often called roadmaps, because they indicate a direction for product and technology devel-

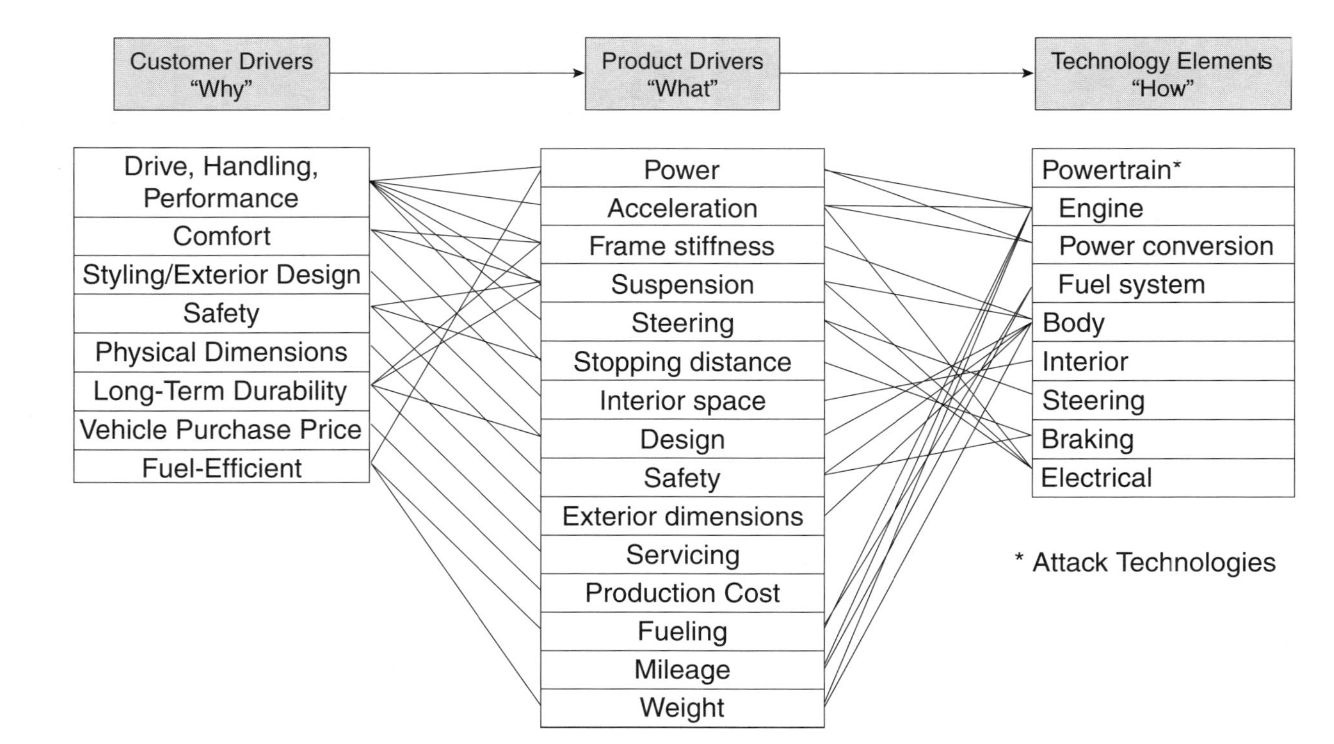

FIGURE 15-12. Driver mapping for cars.

opment and document the decisions the team has made to pursue one of many possible routes.

Maps may be used to show the evolution of the product line over time. The "product roadmap" shows how the product line evolves and branches to introduce new models or releases and when certain models or platforms are to be discontinued. This is a key output of the cross-functional team, and Figure 15-13 shows a product roadmap developed for Acme's passenger car product line. The roadmap shows the introduction of new platforms serving specific segments and branches to include new models targeted at specific segments. As the market evolves, Acme's product line will develop to include models targeted at the family and performance segments of the market. The current internal combustion engine, rear-wheel-drive platform will be renewed for the 2004 model year with a new body, the Model C, aimed at the family sedan segment. This platform will be updated once for the 2006 model year and then discontinued. Development will focus initially on a hybrid-electric platform, using the body components of the Model C to produce a no-frills model with high gas mileage that appeals to buyers looking for an economy sedan—the Model H-1. Late in 2006, the sporty Model H-S will be introduced, with fast acceleration (0 to 60 miles per hour in 6 seconds). The H-S will use the same body and other components as the H-1. Meanwhile, research will continue on a fuel cell-electric platform, with a concept car due in 2006. If the concept car meets market appeal and technical targets, the Model E will be introduced in 2009. The Model E will be the basis of the continuing product line with family and sport variations, and is indicated as the "vision." To consolidate the product line, the Model C will be phased out in 2010, and the Model H platform will be discontinued in 2012.

The product roadmap shows the current plan for the product line. Events will undoubtedly cause these plans to change, so the roadmap must be a living document, revisited periodically (yearly or twice yearly, for example) or when there are major changes in the marketplace. For example, if the Chinese market were to suddenly open up, a model to meet the needs of this market might be added to the roadmap.

Acme's product roadmap includes a vision at the end of the time frame. This allows the team to describe the ultimate goal for the product line. In Acme's case, the fuel cell-electric platform will be the ultimate destination of the product line.

Acme's product roadmap is organized by powertrain platform. In alternative formulations, a critical dimension of product performance may be added. In the computer industry for example, roadmaps for successive generations of computers often use clock speed or some other measure of processing throughput versus time to show the expected evolution. In a market-driven company, the product roadmap might be organized by market segment to highlight the key differentiating features provided by marketing campaigns.

With the product roadmap in hand and key product drivers and targets set, the Acme team is ready to develop its "technology roadmap," building a

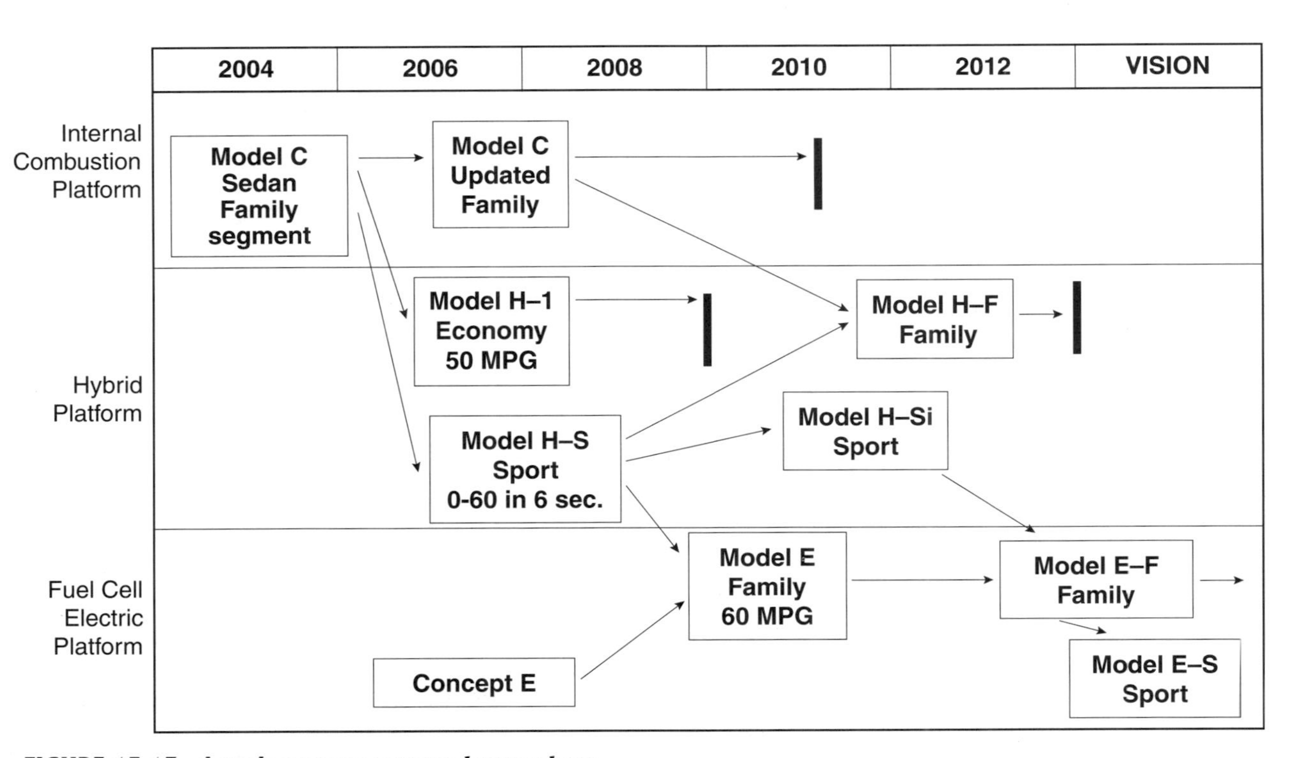

FIGURE 15-13. Acme's passenger car product roadmap.

picture of the technologies that will be used over time in the product line. The technology roadmap is organized by the product architecture and helps make sure the right technologies, resources, and competencies will be ready when they are needed. A technology roadmap for Acme's passenger vehicle product line is shown in Figure 15-14. The team constructs the technology roadmap from alternatives presented by the R&D organization, and the roadmap is developed to be consistent with the product roadmap.

Acme's passenger car roadmap shows the evolution of technologies to match the platform introduction shown in the product roadmap. Each row in the roadmap shows an element of the architecture, and the bars in each row identify the technologies that will be used during specified years of the plan. The current internal combustion engine design is used from 2004 through 2005, and an upgraded multivalve engine is used from 2006 until the Model C's production is discontinued in 2009. A hybrid engine is used in the Model H from 2006 through 2012, and electric drive is used in the Model E from 2009 through 2013. Each line of the technology roadmap includes space for a vision. For the engine, Acme's vision is to develop a very low weight electric drive system. The source of each technology is indicated on the roadmap by the shape of its bar (color may also be used). The internal combustion and hybrid engines will be developed by Acme's R&D organization, while the electric drive motors will be developed with a partner. The low-weight electric motor of the vision will be a project of Acme's research organization. The status of each technology is indicated by the thickness of the outline of its bar. The development of the internal combustion engine is staffed, the development of the hybrid engine is not yet staffed but it is planned, and the development of the electric drive motors is not yet planned.

Additional rows of the technology roadmap show the evolution of key technology elements of Acme's passenger car product line. The technology roadmap in Figure 15-14 shows the first level of technology detail. Each row may be expanded into many items or components as the team develops greater detail. In the end, the technology roadmap may extend to several pages of information. The technology roadmap will be a living document, revisited by the planning team periodically or as priorities and plans change.

(5) Maps That Integrate Architecture, Time, and Linkages to Make a Product Line Story

The power of mapping becomes evident when functional maps are linked and integrated with each other to form a complete view of the product line evolution: the product-technology roadmap. Figure 15-15 shows an overview of the Acme team's complete product-technology roadmap. The figure shows thumbnail images of the maps described previously, along with several additional maps that complete the product line roadmap.

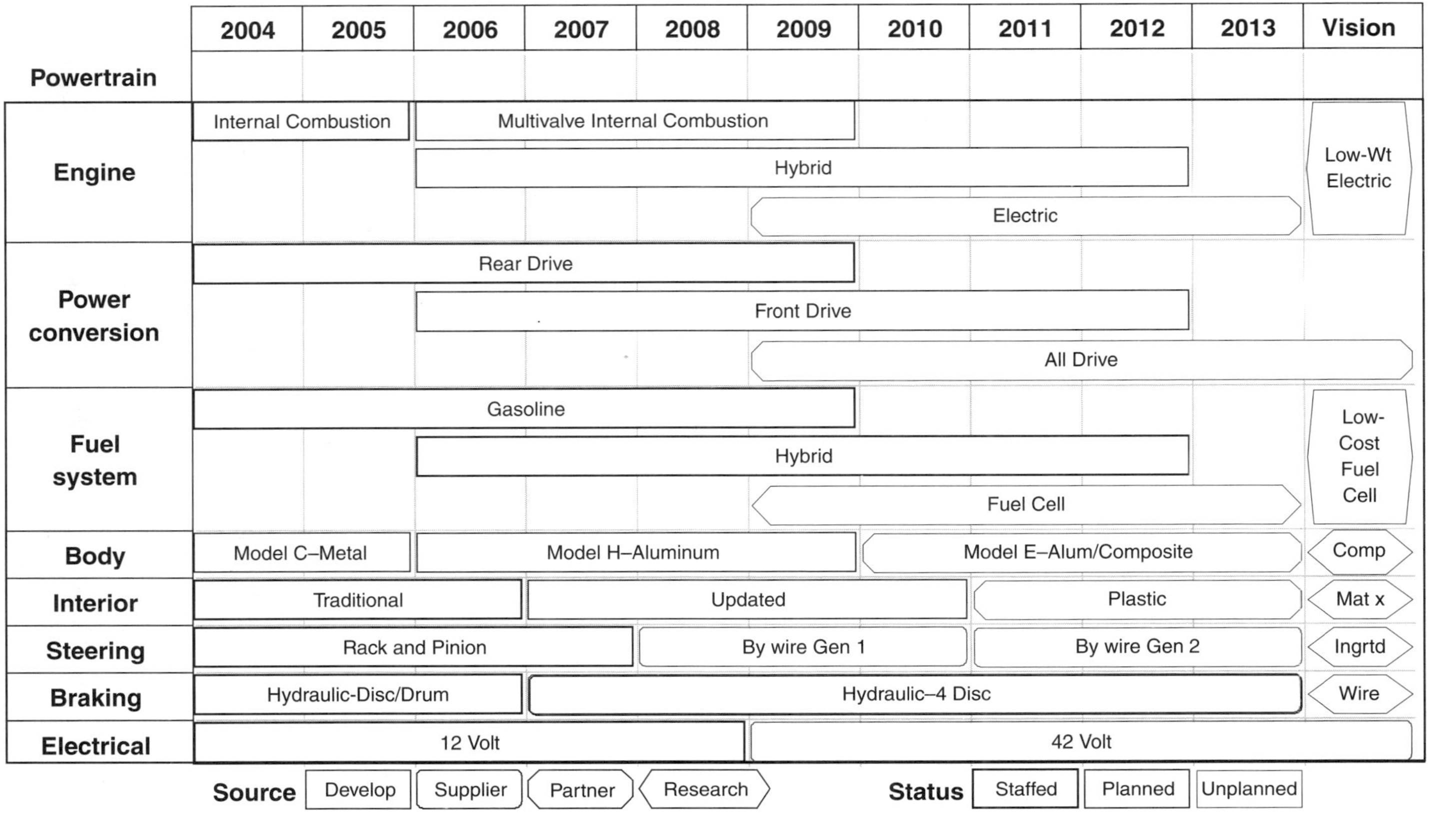

2004 2005 2006 2007 2008 2009 2010 2011 2012 2013 Vision

Powertrain

Engine
Internal Combustion
Multivalve Internal Combustion
Hybrid
Electric
Low-Wt Electric

Power conversion
Rear Drive
Front Drive
All Drive

Fuel system
Gasoline
Hybrid
Fuel Cell
Low-Cost Fuel Cell

Body
Model C–Metal
Model H–Aluminum
Model E–Alum/Composite
Comp

Interior
Traditional
Updated
Plastic
Mat x

Steering
Rack and Pinion
By wire Gen 1
By wire Gen 2
Ingrtd

Braking
Hydraulic-Disc/Drum
Hydraulic–4 Disc
Wire

Electrical
12 Volt
42 Volt

Source: Develop, Supplier, Partner, Research

Status: Staffed, Planned, Unplanned

FIGURE 15-14. Acme's passenger car technology roadmap.

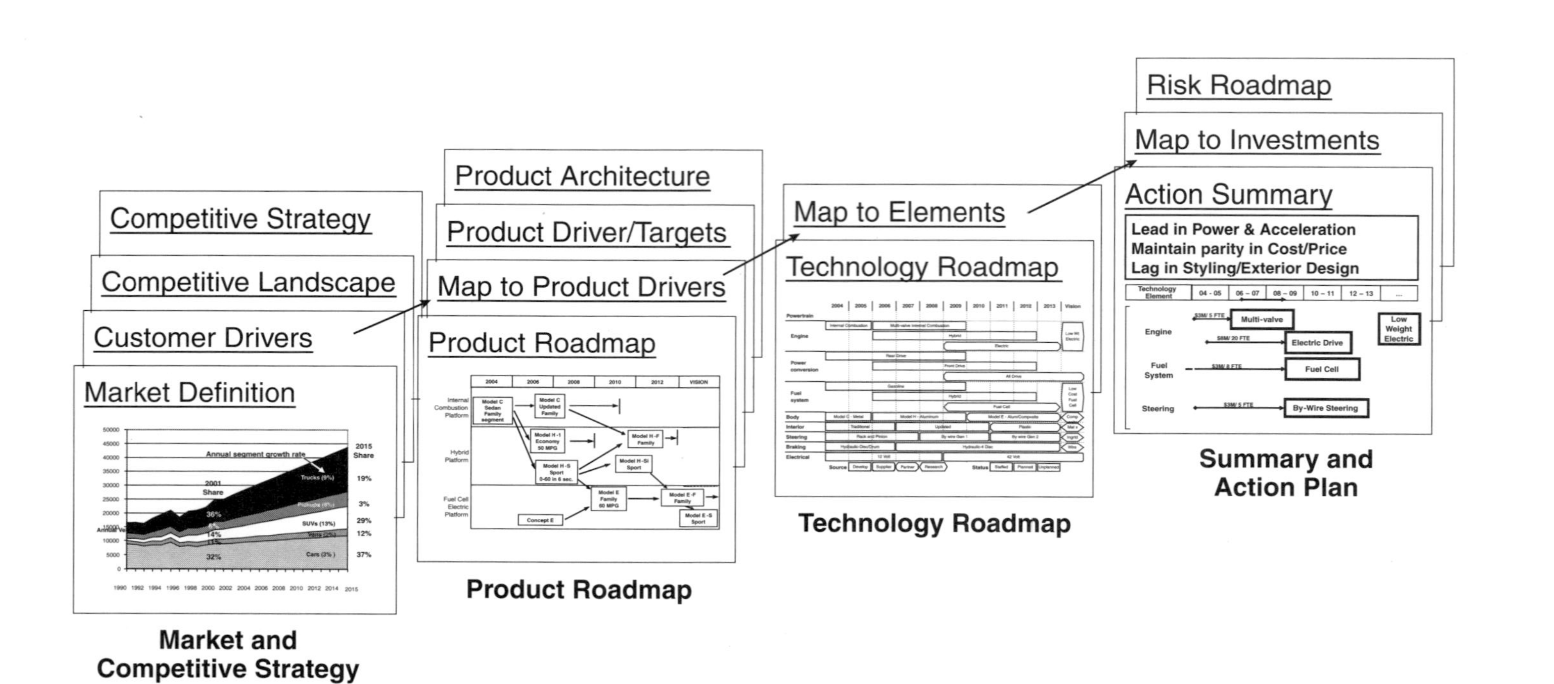

FIGURE 15-15. Putting it all together—Acme's passenger car product-technology roadmap.

Acme's integrated product-technology roadmap is composed of four sections: market and competitive strategy, product definition and evolution, technology plans, and action plans. The arrows linking the parts together indicate the links formed by the driver maps. The market and competitive strategy section of the product technology roadmap includes a description of the market (including segmentation, growth, and key issues); a definition and prioritization of customer and market drivers; a competitive landscape that outlines the strengths, weaknesses, and strategies of competitors; and a statement of Acme's competitive market strategy defining how Acme plans to win in the marketplace. The second section of the integrated roadmap (product definition) includes the product roadmap, the product architecture, product drivers and targets, and a mapping of customer drivers to product drivers. The third section includes a mapping of product drivers to technology elements and the technology roadmap.

The first three parts of the product-technology roadmap lay out the team's plan for the product line. This information is brought to action in the fourth section, the summary and action plan. This section of the roadmap identifies key technology developments or acquisitions that must take place to realize the plan. The action plan also summarizes the key decisions of the team. Acme's team has decided to lead competitors in power and acceleration, maintain cost and price parity, and lag in styling/exterior design. To achieve the lead in power and acceleration, Acme will concentrate on engine design, incorporating multivalve technology in its internal combustion engine and then focusing on fuel cell and electric motor technologies, which can yield high torque for fast acceleration. Acme will seek to maintain cost/price parity, using the industry experience curve to target its price points. Acme's styling has never been leading-edge, and Acme's team has decided to cede leadership to other manufacturers in this area, planning to make up for Acme's pedestrian styling with driving excitement. In turn, based on these decisions, the Acme team has identified several key development projects that must be undertaken. Based on when these key technologies must be ready for use, the team has indicated when development must begin and provided a rough idea of resources (budget and staffing) that will be required. In this way, the action summary presents a plan for key actions.

The summary may also include a technology investment map that identifies all the planned technologies by their competitive impact. This map helps manage the technology portfolio by identifying each technology by its competitive potential (base, differentiating, or disruptive) and source (development, partnering, or acquisition). The objective of this mapping is to make sure that Acme's technology investment portfolio is achieving the right mix of base and differentiating technologies and that development resources are appropriately allocated. Often, portfolio managers find that resources are heavily devoted to base technologies (widespread and well-developed), rather than to differentiating and disruptive technologies. Maps help identify the right balance.

Finally, the summary may include a risk roadmap. A risk roadmap charts signals of game-changing events to watch for, those things that will require a change of plans. It is often derived from the event map developed in the early stages of planning, and it identifies key events or conditions for the team to monitor during the development period. Categories of risk may include market, technical, schedule, economic, political, resource, and other areas. The team will review and update the risk roadmap in their periodic review sessions as product developments progress.

(6) Maps Across Product Lines

The final application of maps described here crosses product lines to promote a rich product and technology portfolio process. Using its product-technology roadmap, Acme's passenger car product line team can compare its strategies, product driver targets, and technology needs with the roadmaps of Acme's other product teams. With a set of roadmaps in common markets or using common technologies, a cross-roadmap analysis can identify areas where development can be shared, technologies can be reused, or collaboration can benefit the corporation. A particularly effective framework for a cross-roadmap analysis uses a matrix of architectural elements versus market segments.

Figure 15-16 shows Acme's cross-roadmapping analysis, identifying common technologies across its several product lines, including passenger cars, vans, SUVs, pickups, and trucks. The product lines are listed across the page, while the architectural elements are listed vertically. In a cross-roadmap review, each product line team describes its plan and its key technology needs. These are then compared for commonality and summarized on the matrix, organized by the architectural elements. Each of the product lines, except for the truck line, need multivalve engine technology, so this development may be shared across several product lines. On the other hand, only the passenger car and van product lines have a need for front-wheel-drive technology. These two teams may collaborate to develop this technology. Several of the product lines have similar, but slightly different, needs for fuel cell technology, and they have each been asking Acme's small fuel cell research team for help. With the recognition of needs across several product lines, Acme's chief technology officer may give the research team a charter to develop fuel cells that can meet several units' needs.

Roadmaps are used for many purposes in corporations. The origin of roadmap is usually attributed to Motorola Corporation in the 1980s (Williard and McClees, 1989). Roadmaps have come to see many applications within corporations (Albright et al., 2003), as well as for industry planning and technology foresight (Kostoff and Schaller, 2001).

Roadmaps may be used as nested sources of information. A view at the top level of a roadmap gives a concise, graphical summary of plans for upper

<table>
<tr><th></th><th>Car</th><th>Van</th><th>SUVs</th><th>Pickups</th><th>Trucks</th></tr>
<tr><td>Powertrain</td><td></td><td></td><td></td><td></td><td></td></tr>
<tr><td rowspan="3">Engine</td><td colspan="4">Multivalve internal combustion</td><td></td></tr>
<tr><td colspan="5">Hybrid</td></tr>
<tr><td>Electric</td><td></td><td colspan="3">Electric</td></tr>
<tr><td rowspan="3">Power conversion</td><td></td><td></td><td></td><td colspan="2">Rear-Wheel Drive</td></tr>
<tr><td colspan="2">Front-Wheel Drive</td><td></td><td></td><td></td></tr>
<tr><td colspan="4">All-Wheel Drive</td><td></td></tr>
<tr><td>Fuel System</td><td>Fuel Cell</td><td></td><td colspan="3">Fuel Cell</td></tr>
<tr><td rowspan="2">Body</td><td colspan="2">Unitized Body</td><td></td><td></td><td></td></tr>
<tr><td></td><td></td><td colspan="3">Body on frame</td></tr>
<tr><td>Interior</td><td></td><td></td><td></td><td></td><td></td></tr>
<tr><td rowspan="2">Steering</td><td colspan="5">Rack-and-Pinion</td></tr>
<tr><td colspan="3">By-Wire</td><td></td><td></td></tr>
<tr><td>Braking</td><td colspan="3">Disc</td><td></td><td></td></tr>
<tr><td>Electrical</td><td colspan="3">42 Volt</td><td></td><td></td></tr>
</table>

FIGURE 15-16. Acme's technology cross-roadmap framework.

management review, or for communicating product strategy and directions to customers. Probed to greater depth, the roadmap reveals detailed information that guides the product development team, the life cycle management team, suppliers, and other stakeholders. Roadmaps can be key sources for product portfolio management, providing a common format for information that the portfolio management team can use to assess trade-offs. Key success factors for roadmapping include processes for getting started, dealing with sharing and secrecy, identifying situations for which roadmapping is a good fit (and those that are not), finding the right time horizon, making roadmaps compelling (Kappel, 2001), and measuring the value created by roadmapping (Albright, 2003).

EPILOGUE: OVERCOMING THE CORPORATE LEARNING DISABILITY

Maps in planning and portfolio management are graphic presentations of information that quickly and effectively communicate plans, objectives, and expected results. But just as important as the story the finished maps tell are the benefits of learning to the map makers as they construct, debate, and redefine their maps in the planning process. Maps give teams a way to identify knowledge gaps, set targets, and document results of research. They also easily communicate the work at the team level to the executive level, so strategic decisions can be made that reach across projects, markets, businesses, and time frames.

Peter Senge speculated over a decade ago that most—perhaps all—companies have "learning disabilities" (Senge, 1990: pp. 17–18). The practices of product and technology mapping provide a way for corporations to document intellectual property, describe the gaps between what they have and what they need to meet market needs, and turn this knowledge into a portfolio of product and technology development projects. Maps facilitate communication across the system boundaries that so often block learning and knowledge.

The important thing to remember about mapping is that you can begin implementation at a very modest level and continue to build on existing maps year after year. And another equally important thing to remember about mapping is that it does require attention and resources—it can be done part-time, but not in your spare time.

NOTES

1. Hohmann (2003) calls these "lo-fi" maps. Although the form of what we are calling "back-of-the-envelope" maps may remain high level, the depth of data and information that supports them can become very sophisticated over time.
2. The estimated cost of an automotive fuel cell in 2003 was about $1,000/kilowatt. A kilowatt is equivalent to 1.341 horsepower.

REFERENCES

Albright, Richard E. 2003. "A Unifying Architecture for Roadmaps Frames A Value Scorecard." IEEE International Engineering Management Conference, Albany, NY. November 2–4, 2003

Albright, Richard E. 2002. "Roadmapping for Global Platform Products." Product Development and Management Association. *Visions Magazine* 26, 4: 19–22.

Albright, Richard E., et al. 2003. "Technology Roadmapping," *Research Technology Management* 46, 2: 26–59.

Burchill, Gary, and Christina Hepner Brodie. 1997. *Voices into Choices: Acting on the Voice of the Customer*. Madison, WI: Joiner Publications (Oriel Inc.).

Davis, S. C., and S. W. Diegel. 2002. *Transportation Energy Data Book, Ed. 22*. US Department of Energy, Oak Ridge National Laboratory. http://www-cta.ornl.gov/cta/data.

Gill, R., B. Nelson, and S. Spring. 1996. "Seven Steps to Strategic New Product Development." In *PDMA Handbook of New Product Development*. Ed. Milton Rosenau et al. New York: John Wiley & Sons.

Hellman, K., and R. Heavenrich. 2001, *Light-Duty Automotive Technology and Fuel Economy Trends, 1975–2001*. US Environmental Protection Agency, EPA420-R-01-008, Appendix F. September, 2001, http://www.epa.gov.otag/fetrends.htm.

Hohmann, Luke. 2003. *Beyond Software Architecture: Creating and Sustaining Winning Solutions*. Reading, MA: Addison-Wesley.

Kappel, Thomas A. 2001. "Perspectives on Roadmaps: How Organizations Talk about the Future." *Journal of Product Innovation Management* 18, 1: 39–50.

Kostoff, R. N., and R. R. Schaller. 2001. "Science and Technology Roadmaps." *IEEE Transactions on Engineering Management* 48, 2: 132–143.

McManus, W.. 2003. "Consumer Demand for Alternative Powertrain Vehicles." Tenth Annual Automotive Outlook Symposium. Federal Reserve Bank of Chicago, May 29, 2003. http://www.chicagofed.org/news_and_conferences/conferences_and_events/files/tenth_automotive_outlook_consumer_demand_for_alternative.pdf.

Meadows, L. 2002. "Lead User Research and Trend Mapping." In *PDMA ToolBook for New Product Development*. Ed. Milton Rosenau et al. New York: John Wiley & Sons.

Senge, Peter. 1990. *The Fifth Discipline: The Art and Practice of the Learning Organization*. New York: Currency/Doubleday.

Taytelbaum, M. 2000. "Market Segmentation and Technology Mapping: Focusing Product Strategy on the Real Opportunities." PDMA International Conference, New Orleans, October, 2000.

Tufte, Edward R. 1983. *The Visual Display of Quantitative Information*. Cheshire, CT: Graphics Press.

Ward's Communications, 2002. *Ward's Motor Vehicle Facts & Figures*. Southfield, Michigan, 2002.

Wheelwright, S., and K. Clark. 1992. *Revolutionizing Product Development*. New York: The Free Press.

Williard, C. H., and McClees, C. W. 1987. "Motorola's Technology Roadmap Process." *Research Management* 30, 5, September–October: 13–19.

16 Decision Support Tools for Effective Technology Commercialization

Kevin J. Schwartz, Ed K. Yu, and
Douglas N. Modlin

In *The PDMA ToolBook 1*, the authors presented various tools for managing the development of new products and technologies. The first volume included a discussion of how to adapt standard Stage-Gate™ processes for product development management to the differing requirements of advanced technology development. In this chapter, we expand on that concept by providing an additional set of tools not only for effective management of a single technology project (i.e., technology staging) but also of multiple technology projects (i.e., technology portfolio management). The emphasis in this chapter will be on tools that ensure that a technology development effort will not merely meet some predefined set of performance criteria but ultimately will translate into a commercialized product. This focus on *technology commercialization* has become more vital to various organizations as research budgets have come under closer scrutiny for their ultimate value and return on investment (ROI).

First, you explore the framework for synchronizing technology development with product development. Then, you look at tools to effectively manage a technology-focused project for efficient commercialization. The discussion of decision tools will begin with effective technology portfolio management. Next, decision tools for effective technology staging are discussed.

Technology portfolio management and *technology staging* are two categories of practices with significant impact on the likelihood of a new technology's commercial success. The reason is simply that these practices are designed to address fundamental questions faced by every technology-based business. They include the following:

- How do we decide which technologies to invest in or focus on?
- How much should we invest in "blue-sky" versus sustaining technologies?

- Why does so much technology effort so seldom result in a real product?
- Why can't we ever predict what technologies will be ready and when?

Although these issues seem daunting, the potential of effective technology portfolio management and technology staging offers incredible returns. Benchmarking studies (Yu, Schwartz, Carrascosa, 2002) indicate that top-performing technology leaders (those that invest heavily in technology and that successfully leverage these efforts in their commercially released products) realize as much as 80 percent higher-than-average profitability and 40 percent higher-than-average growth.

This promise is particularly relevant for new companies or business units working to commercialize new technologies as the underpinning of their entire product line, but it is also relevant for established entities that want to stay ahead of the competition through constant innovation. Our work with several companies dependent on new technology commercialization and our study of top-performing companies show that two key practices can make the difference between success and failure in realizing the benefits of heavy technology investment.

Difference between Technology Development and Product Development

Before delving deeper into the tools for managing technology commercialization, let's first make a clear distinction between what we mean by products and technologies. In *The PDMA ToolBook 1*, the differences between technology and product development were discussed. The attributes of each are summarized in Figure 16-1.

Though the distinctions are intuitive, it remains difficult to know with certainty whether a technology-based project should be managed with a serial product development process or a less structured technology development process. Realistically, the choice is often based on management judgment rather than on a quantitative evaluation. For example, consider the case of an aerospace contractor that is funded by the government to develop a new material for the surface of radar-avoiding or "stealth" aircraft. The project involves an unexplored technology and therefore presents a high level of technology risk. It also may be the case that too much structure would inhibit the creativity required for the team's success in developing a new material. Yet the project's end result could be considered a product that will be sold to a specific customer with predefined expectations. Similarly, you could look at a new product line within a semiconductor company—perhaps the first Bluetooth chip developed by an established telecommunication chip maker. Development of that first product is a highly risky, technology-driven project with a much lower level of predictability than the organization usually faces. However,

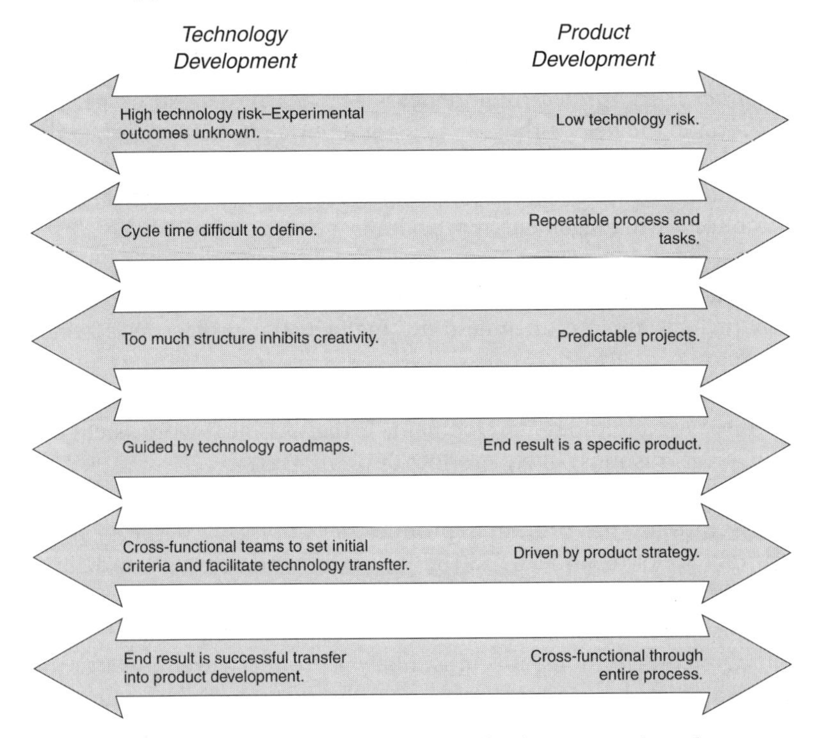

FIGURE 16-1. Differing attributes of technology development and product development.

once again, the project's desired outcome is a specific product, driven by a product roadmap that likely involves defined product and schedule requirements. In either case, whether to manage these projects under a formal product development process or a less structured technology development process becomes a deliberate management decision.

To translate the differences described in the preceding text into a practical set of criteria for making this management trade-off, first you need to think about the management objectives for a given technology-based development project. Seven dimensions, shown in Figure 16-2, should be considered when you are determining whether a project should be managed as a technology development project or a product development project. They are as follows:

- *Set goals.* If the project objective is delivering a discrete service or hardware to a clearly defined set of customers, the project is best managed as a product development effort. However, if the project objective is driven by the desire to shape or create a market, the project is better managed as a technology development effort.
- *Strategize.* If the project strategy is shaped by other products or competition, the project would be more successfully managed as a product

development effort. If the strategy is shaped by technology trends and suspected disruptions, the project is a technology development effort.

- *Define.* If the project requirements are clearly defined as performance criteria in the field, either for a customer or a marketplace, then the project should be addressed as a product. If the requirements are determined by internal capability needs, the project is generally better suited to a technology management approach.
- *Manage.* If a project has clear end goals that can be scheduled, use a product development process. If the end goal is difficult to define or is shaped by an interim milestone, manage the project as a technology development effort.
- *Design.* If the nature of the technical work is scientific discovery, the project is technology development. If the technical work is engineering, the project is most likely product development.
- *Aim.* If the project seeks to demonstrate repeatability to meet market expectations, the project is product development. If the project more simply seeks to understand the technical fundamentals to demonstrate feasibility, the project is technology development.
- *Value.* If the project is expected or committed to make revenue or margin contributions, it is most likely a product development project. If the value of the project is more broadly focused on the product options that the project will enable, the project is a technology development effort.

In summary, the choice of which process to apply to a given technology-based project is driven by the management objectives. Ultimately, this choice still will remain a somewhat subjective one made by informed managers, but the preceding criteria can provide some objective guidance in making this selection. The remainder of this chapter presents tools and techniques to help you manage a portfolio of projects on the "technology-focused" end of the spectrum for effective overall commercialization.

SYNCHRONIZING TECHNOLOGY AND PRODUCT DEVELOPMENT

When researchers discuss technology management, it is not uncommon to hear, "You can't schedule innovation." Even the most process-oriented managers of technology organizations will, on occasion, echo this sentiment when confronted with inquiries about schedules or status from a product development team. And while there is some truth in this statement, it is ultimately deceiving. For although it is correct that you never can truly predict when an innovative breakthrough or technological solution will be found, you can manage the environment within which technology development projects are conducted to foster creativity, recognize results, and capitalize on innovations

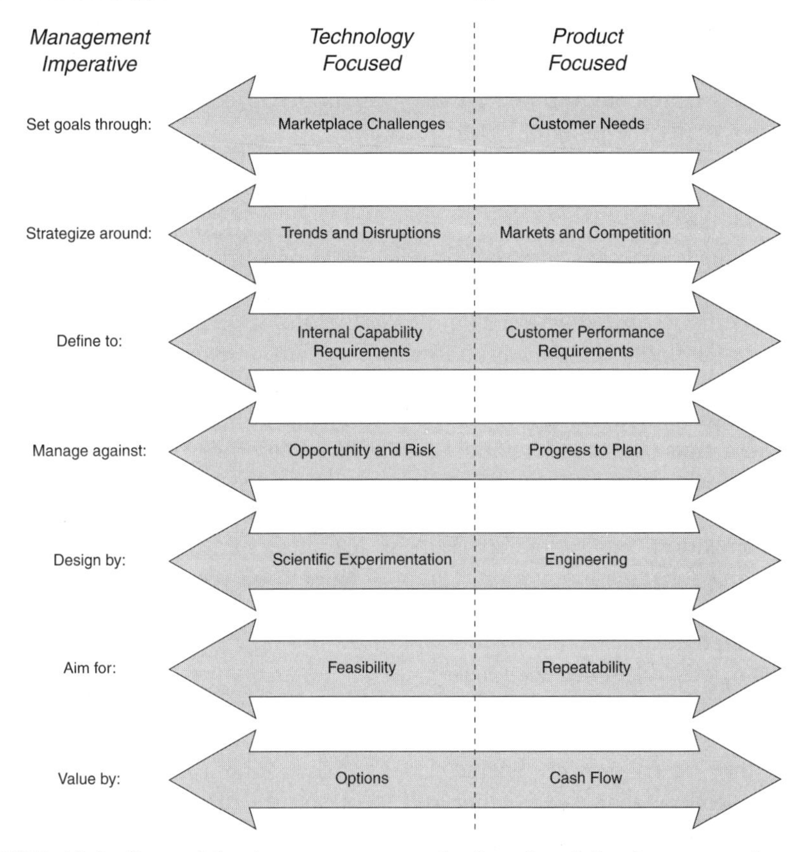

FIGURE 16-2. Determining how to manage technology-based development projects.

rapidly when they do occur. Consequently, "You cannot schedule innovation breakthroughs, but you can schedule innovation projects."

The heart of the conflict between management of technology and product development lies in this fundamental issue of predictability. As shown in Figure 16-3, the final intention of most technology development efforts is to

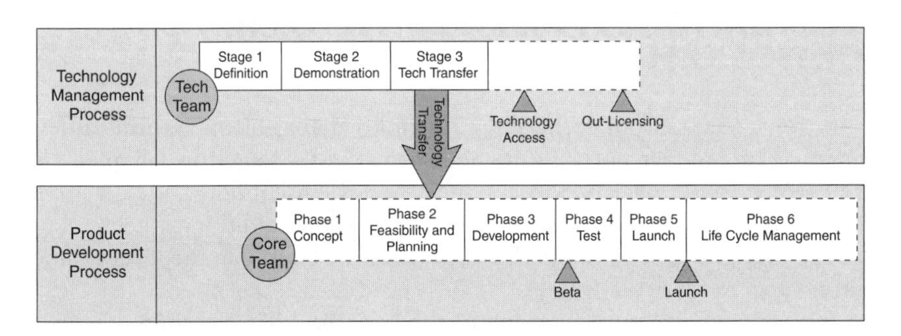

FIGURE 16-3. Linkage of technology and product development processes.

transfer a new capability or component into a product development work stream. However, product development tends to function in an extremely schedule-oriented environment, and the natural unpredictability of new technologies creates the potential for disruption.

Similarly, product development and technology development projects usually have fundamentally different definitions of project *requirements* that make it exceedingly difficult to achieve smooth transitions. Product developers tend to think in terms of distinct performance requirements within the boundaries of specific market or customer operating conditions. On the other hand, technologists, by nature, must focus more on high-level performance criteria that allow for flexibility of application in various environments. Often, the requirements are not even defined using the same terminology. Recognizing this difference and addressing it as an issue of product-technology transfer, rather than as a sign of misunderstanding or technology immaturity, is one of the keys to synchronizing product and technology efforts.

The tools presented in this chapter help you deal with the various issues of synchronization and ultimately address the false element in the statement, "You can't schedule innovation." Each of these tools, combined with general management awareness and cultural sensitivity to commercialization, help create an environment that *fosters creativity and capitalizes on innovations as rapidly as possible when they occur.*

We would be remiss if we did not include the role of intellectual property (IP) management as an important consideration in how one manages a given technology-based project. Businesses that place high value on patents will want to ensure that their technology projects are driving patent filings to achieve domain leadership. Once product development projects reach the product launch phase, certain patent rights may be lost after they are offered for sale. Please consult your patent attorney for advice on your specific situation. Therefore, in managing both a core research function and the transition period between technology and product development, companies must be actively aware of their IP strategy and approach. Although the concept is too expansive to address adequately in this chapter, good examples are provided in publications such as "Patent Law Essentials" Durham, A. L. 1999.

DECISION TOOLS FOR EFFECTIVE TECHNOLOGY PORTFOLIO MANAGEMENT

At its root, the issue of technology portfolio management is little different from that of product portfolio management. As presented in Chapter 13 of *The PDMA ToolBook 1* (Cooper, Edgett, and Kleinschmidt, 2002), it is still a question of learning how to *do the right projects* in addition to being able to *do projects right*. The fundamental difference lies simply in the level of uncertainty.

In product development, many of the variables that determine the potential value of new projects can be reasonably well understood and quantified. Though there are always unknowns, standard cost-benefit analyses (such as net present value, or NPV; expected commercial value, or ECV; productivity indices; and scoring matrices) can be applied with reasonable confidence to evaluate the relative importance of individual projects. These evaluations, combined with bubble charts and other techniques (Cooper, Edgett, and Kleinschmidt, 2002), support logical portfolio decisions around new product development efforts.

However, with technology management, many of these tools lose their validity. Whereas the likely revenue from a new product may have a range of potential values, the likely return on a technology development project is often a binary variable. The project either will be successful, returning huge, breakthrough benefits across numerous products, or it will fail to produce adequate results and be dropped before product development. Even if we assume success, the potential returns from a new technology tend to be so distant and uncertain that it's difficult to apply traditional cost-benefit tools.

Despite this difficulty, the need for portfolio management tools remains. Most companies are presented with a plethora of opportunities for new technology exploration and investment. Some are simple, near-term technologies for cost reduction or manufacturing process improvements on current products. Others are revolutionary, theoretically possible technologies that will open new markets or fundamentally change the nature of the company or its products. Of course, most companies have only a limited amount of resources to devote to exploring these various opportunities; therefore, they must carefully select the most promising ones for investment. Yet, if the valuation of technology projects is so uncertain, how do you decide where to place your bets? Luckily, tools exist for addressing this challenge. The three tools presented in this section provide frameworks and analytical methods to allow managers to derive the value-based trade-offs of portfolio management in the uncertain world of new technology.

Tool 1—Component Technologies and Technology Platforms

One of the first pitfalls that organizations encounter when applying typical portfolio management approaches to technology efforts involves project prioritization. In a product development environment, companies tend to develop an algorithm or analytical approach to prioritizing individual projects, placing them into rank order as a guide for day-to-day management trade-offs. While this may work for product development, it does not work for technology development. Why? First, technology projects often are so disaggregated that it is difficult to assign a financial return or end benefit to any individual project. These returns are also so vague and uncertain that it is also difficult, if not impossible, to gain consensus on the values assigned to any individual project. Instead, technology innovators often rely on the "gut feel" of their lead researchers—their true visionaries. This reliance is not something that easily can be formalized into a quantitative model for project prioritization.

One solution is to ignore the prioritization effort at the component technology level. Instead, technology efforts can be grouped into logical families or technology *platforms* that enable new product lines or streams of products. Once grouped this way, it becomes much easier to make prioritization decisions without a complex prioritization algorithm. Instead, logical priority decisions can be made based on a few quantitative criteria (e.g., economic return, risk level) and on alignment of the platforms to specific goals of the corporate strategic vision.

Each of the three technology platforms in Figure 16-4 forms the basis for a unique lead product. A *lead product* is the first of a potential series of products that serves the need of a specific set of customers. (These groupings of products are sometimes referred to as *product platforms*, which are out of the scope of this chapter. Product platforms are addressed in another *ToolBook* chapter.) The technology platforms become the fulcrum of the business from a strategic business-making perspective and can be used to ensure that technology work supports the corporate strategy. These platforms facilitate investment decisions that are influenced both by technical know-how, based on technology maturity, and by market intelligence, through the lead product requirements.

In general, most of the technology modules underlying the product platforms are not usable across platforms; however, sometimes they are, and this can save time and money in the development of new products. It is good to stay on the lookout for such opportunities. Inevitably, a limited number of technology efforts will not fall into any one platform. They may be supporting technologies necessary to enable or improve the company's product offerings

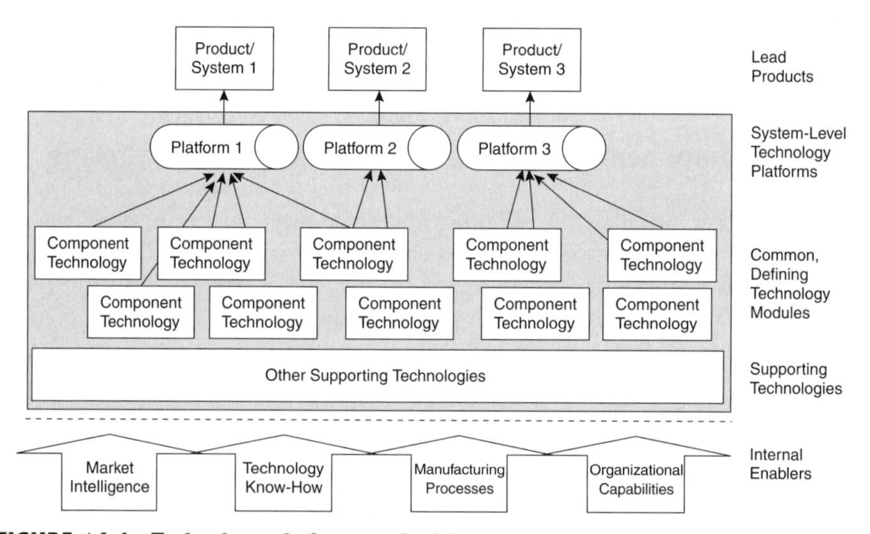

FIGURE 16-4. Technology platform methodology.

across the board, or they may be true blue-sky/bet-the-farm research efforts that cannot be tied to a near-term product opportunity. A defined percentage of resources or budget can be set aside for this type of work. For example, a company may choose to invest 30 percent of its total technology development budget toward research efforts. This is an investment toward evaluating long-term commercialization opportunities. The remainder of the 70 percent investment is focused on near-term, product-specific technology development. The actual ratio will vary from company to company, but the optimal ratio is one that balances a company's long- and short-term business needs.

Additional technology efforts may be focused on targeted technologies that do not fit in one of the platforms but are needed for a specific product. These projects can be reviewed individually and prioritized as appropriate, but generally they should be kept to a minimum to maintain a platform view of the technology portfolio. The portfolio of platforms becomes the plan-of-record set of investment choices reflecting senior management's vision.

In Figure 16-5a, the application of the technology platform methodology is shown for a hypothetical toaster company with three technology platforms: low-end, midrange, and high-end. In this example, the Thermostat Control technology module can be used with two of the three platforms. This could reduce cost and time to market for the Classic and Deluxe lead products.

Through the utilization of both internal and external technical know-how, new technology modules can be developed that can enable new products based on each technology platform. Underlying each technology module are supporting elements or functions such as specific design concepts and integrated functional blocks. At the next layer down, in Figure 16-5b, you find components and discrete devices, moving to materials, and finally processes. In the end, what is important for a successful business opportunity is that all the

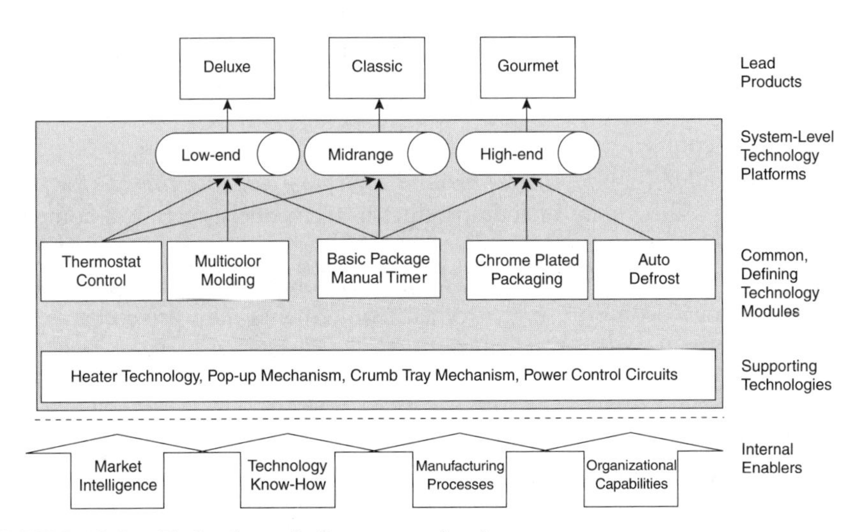

FIGURE 16-5a. Technology platform example: the toaster.

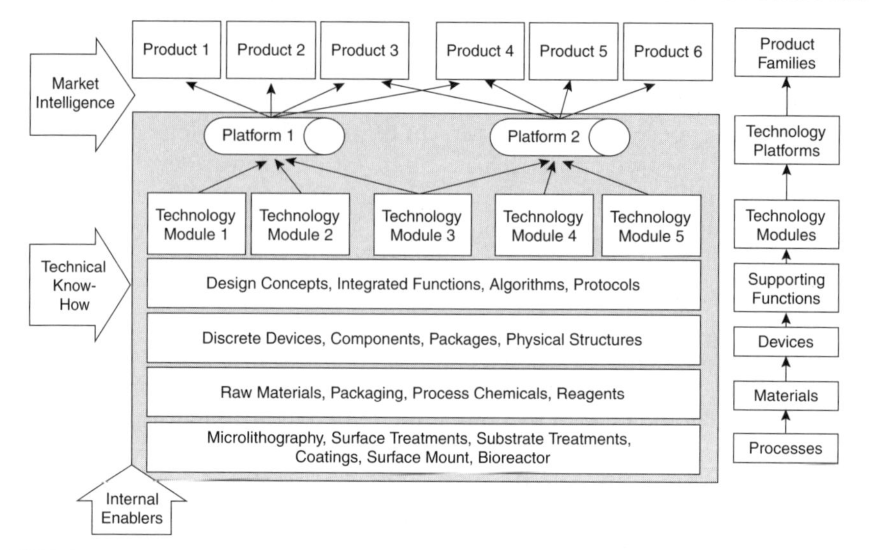

FIGURE 16-5b. Technology platform example.

technologies required by the customer to realize the value of the product are accessible. The customer ultimately desires a solution that meets his or her needs rather than a repackaged technology module.

Following are a few tips on this tool:

- *Keep the number of platforms to a minimum.* A proliferation of platforms dilutes their value for driving prioritization decisions.
- *Select your technology platforms intelligently.* Definition of technology platforms reflects strategic decisions as well as technical ones and should be reviewed by a cross-functional management team.
- *Leverage technologies both internally and externally.* It is not essential that your company supplies all elements required to enable a platform or product; however, it is essential that all elements are available from a reliable source.
- *Focus technology platforms around logical organizing themes for your business.* These may include product usage models, technical compatibility/linkages, or cost points.
- *Review your platform strategy frequently.* Changes in business strategy, technology evolution, or customer requirements may drive changes in your optimal platform structure.

Tool 2—Applying Real Options Theory to Aid Technology Path Selection

Platform technologies are inherently desirable because, in principle, investment in a technology platform can be leveraged in creating multiple products.

Microelectronic fabrication processes provide good examples, since once the process is developed, new products can be created with relatively small changes in the baseline processes or platform. Other examples of platform technologies include certain computer programs and operating systems. Each new generation leverages the previous one, and many products can be leveraged from each new version. Following are key questions that must be answered by organizations seeking to develop multiple products using platform technologies:

- How should development resources be deployed to yield the highest return on investment?
- How many technologies and products should be developed in parallel?
- How can the risks associated with parallel technology and product development programs be effectively managed?

We assume that while an organization's development resources are limited, generally they can simultaneously support more than one product in the development pipeline. In the model, each phase of the product development process is equivalent to taking out an option to find the result of the next phase. This idea is very similar to drilling an oil well. An investment is made to secure the right to drill for oil on a certain piece of land. If no oil is found, the value of the project is zero. If oil is found, the project can be quite valuable, depending on how much oil is discovered. In product development, an investment is made to explore the feasibility of a new product idea. Depending on the results of the feasibility experiments, the product could have no value or possibly a very high value if the results are good. A negative result adds value to the organization; if resources are quickly redeployed to other projects with better prospects, the overall return on investment will be higher than if the product development team continued to work on the project in the hope that something would come of it. The key to realizing the value of this structured decision process is in setting up specific metrics for project performance in advance, making timely decisions, and quickly redeploying resources from killed projects.

The value of the options approach also can be appreciated when viewed in contrast to the limitations of conventional NPV and ROI analysis. The following list illustrates portfolio and project management aspects that are difficult to realize with conventional approaches:

1. NPV is probably not an accurate model for product development, but it is acceptable because systematic error is the same for all projects.
2. It is hard to model risk with NPV. This is sometimes done by increasing cost of money (i.e., the discount rate used in the NPV calculation) or by padding the schedule.
3. Net present value and internal rate of return (IRR) methods do not model the effect of redeployment of resources at predetermined decision points.

4. Using NPV or IRR to compare multiple projects doesn't seek to maximize the value of the portfolio, whereas use of the options approach tends to do so.
5. Using the options approach to optimize value creates opportunities to capture value generated in evolutionary processes.

It is useful to consider the idea of a lattice model as a simple way to represent options-based structured decision processes (Luenberger, 1998). A lattice model can be constructed using a simple spreadsheet with the states of the project represented in individual cells as either inputs or outputs (see Figure 16-6). As mentioned, the feasibility stage of a technology project can be viewed as "taking out an option" to determine the feasibility of proceeding with the development with a product or technology if the feasibility investigation is successful. If the feasibility investigation fails, the value of the project is low or zero. If feasibility is established, the value is increased and the risk of failure in product development is reduced. The probability of success or failure of a given feasibility study is dependent on the specifics of the project, and assigning a number to the probability depends on the intuition and judgment of management, both from the technical and business perspectives.

Part A in Figure 16-7 illustrates a case where the cost of a feasibility study is $200,000 and the probability of success is estimated at 0.5 (50:50 chance of success). Here, the company believes that $20 million will be required to develop the product once the feasibility study is completed. In the best-case

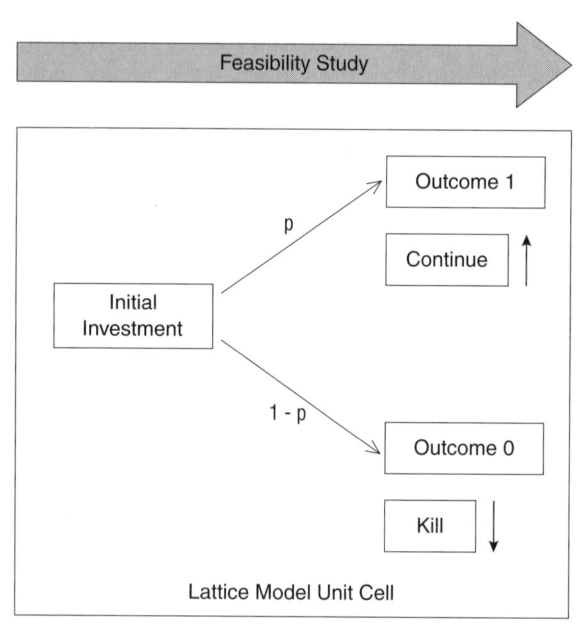

FIGURE 16-6. Lattice model.

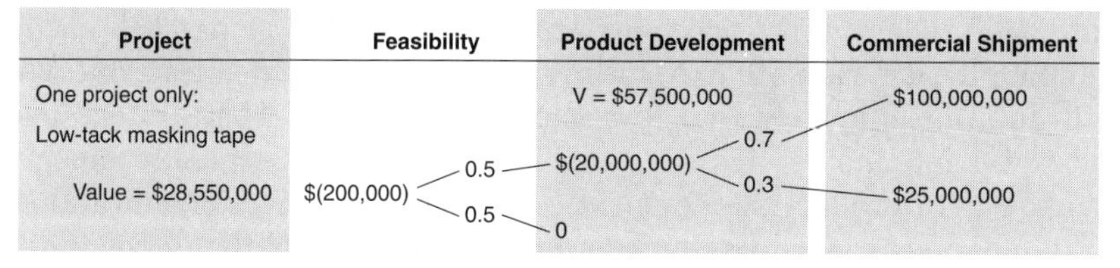

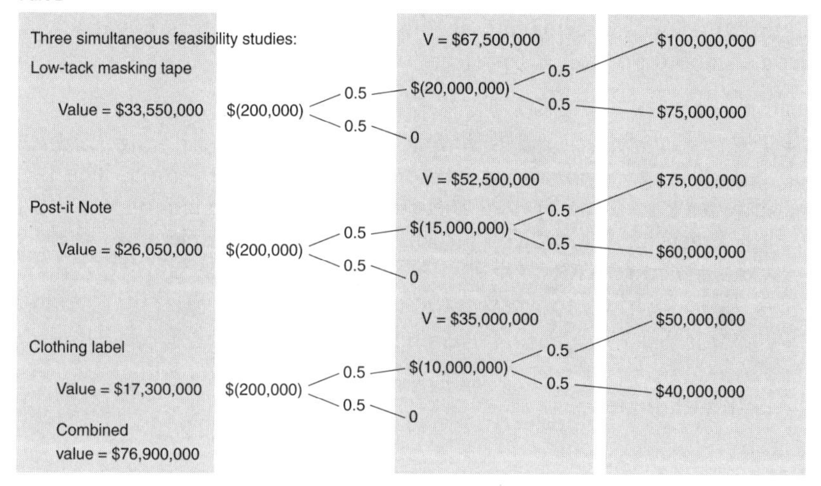

FIGURE 16-7. Options analysis examples.

scenario, management believes that the project could result in a $100 million revenue opportunity for the company. This scenario assumes that the product hits the market on schedule and on budget, that it meets all marketing specs, and that competition is minimal. Management believes that the best-case scenario has a 70 percent chance of occurring.

On the other hand, if things go according to a worst-case scenario, the result will be only a $25 million opportunity. The lattice model simultaneously calculates the effects of both scenarios and predicts a net value for the project of $28 million at the start of the project. Once the feasibility study is successfully completed, the model predicts the value will be $57.5 million. This value is calculated by subtracting the investment required in development ($20,000,000) from the probability weighted expected outcomes: V = (–$20,000,000) + $100,000,000*0.7 + $25,000,000*0.3 = $57,500,000. Similarly, the original option value of the project prior to the feasibility study is calculated as: V = (–$200,000) + $57,500,000*0.5 + $0*0.5 = $28,550,000.

One of the attractive features of this model is that, by staging the decisions at the ends of each phase, the value of a portfolio can be optimized by factoring in all that has been learned during each phase and by redeploying funds toward projects with higher value rather than continuing projects by default—

a practice that has relatively lower value. This idea is illustrated in Part B of Figure 16-7, where three feasibility studies are funded in parallel, each for $200,000. If each has a 50:50 chance, it is highly likely that at least one will be successful and ultimately lead to a successful product. This strategy makes good sense for another reason. Feasibility studies are generally much less expensive than product development programs. Even if an organization can afford only one product development program, it is likely that it could afford more than one feasibility study. The three cases modeled here include high, medium, and low scenarios and are illustrative of how it is possible to construct a balanced portfolio of projects.

Following are some tips on this tool:

- *Try to keep your models simple; simple models are more transparent and easier to communicate throughout the organization.* The biggest drawback to using options theory is that excessive complexity can rapidly creep in. Try to keep models conceptually simple and use them to see big-picture issues. Given all the real-world uncertainties, it may not make sense to try to work in too much detail.
- *Do not try to replace NPV and ROI models with options models.* Rather, use them in a complementary fashion so you will get the best of both techniques.

Tool 3—Technology and Product Roadmap Integration

Tools 1 and 2 provide some management structure around the organic and fluid nature of innovation work. By establishing a portfolio of platforms, you have essentially created an organizing framework that articulates the motivation for all technology projects. However, the missing element in this framework is timing. The graphic representation of planned technologies (both at a component and a platform level) by objective and relative time frames is called a *technology roadmap*. The concept of a technology roadmap serves a basic purpose for a research organization in identifying the various projects that are being worked (or are planned for work) over a given period of time. As with a product roadmap, this information can help align individuals on work plans and keep organizational confusion to a minimum. However, the real power of the technology roadmap comes when it is integrated with the corresponding product roadmaps.

Fundamentally, the integration of technology and product roadmaps enables the coordination of technology development with product development. This integration provides a powerful tool in two ways: as a communication tool and as a portfolio management tool. As a communication tool, the integrated roadmaps provide a clear and common view for the organization of what technologies are being researched and, more importantly, how they are intended to support planned or projected product development efforts. As a portfolio management tool, these roadmaps help management make intelligent decisions about deployment of resources and authorization of new projects. For example, if management sees that a project on the company's product development roadmap requires a new technology component that will not possibly be ready until near the end of the product's design phase, it can push the product team to take corrective action in advance. This might mean delaying the product development project, dropping the new technology from its scope, or realigning performance expectations to allow for better synchronization.

When actively reviewed by a management team, an integrated product and technology roadmap helps the organization frame (and hopefully answer) key questions to drive more value from its technology investments:

- What technologies are in development for future product use, and where will they be leveraged?
- Will the required technologies be ready for a given product development project on an appropriate timeline?
- Where do gaps exist in the technology needs for a future product?
- How are technologies that are nearing feasibility being leveraged into current products?
- Are there opportunities for additional leverage of mature (or almost mature) technologies that can increase your return on the technology investment?

Figure 16-8 shows a simplified representation of an integrated product and technology roadmap. The key characteristic is that each planned or cur-

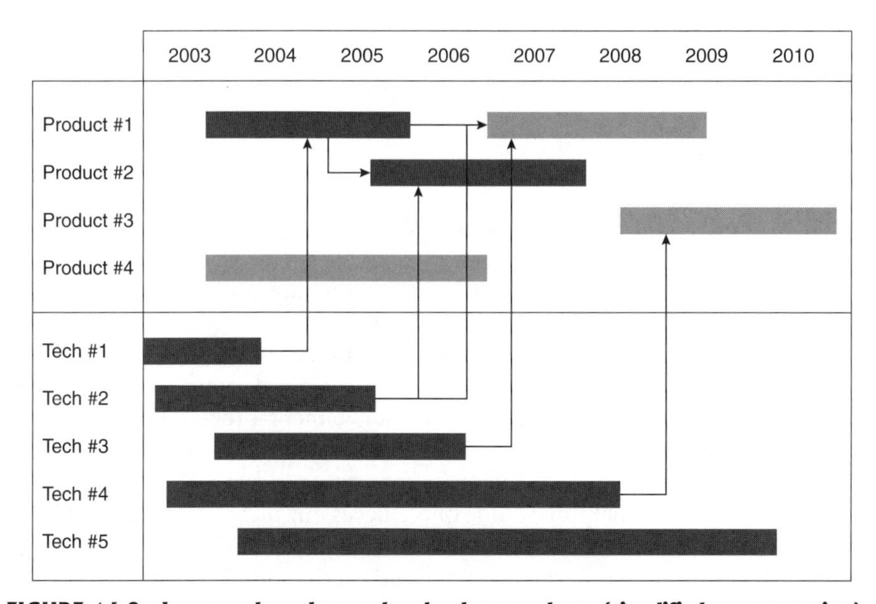

FIGURE 16-8. Integrated product and technology roadmap (simplified representation).

rent technology and product development project is shown on a timeline indicating linkages between the technology projects and those product development efforts that will be the leading projects to take those new technologies to market. Key milestones—including rough project start and completion targets—also are included.

In a realistic application, it is often impossible to show all the technology and product development efforts (and the appropriate linkages) for a company on a single chart. Therefore, it usually makes sense to create distinct charts by business unit, technology application area, or some other appropriate segmentation.

Following are some tips on this tool:

- *Keep the roadmaps simple and intuitive.* One pitfall of technology roadmaps is that they may lead to complex constructions that are difficult to communicate. Avoid the temptation to generate very granular roadmaps. One way to maintain simplicity is to create roadmaps of platforms only (Tool 1), rather than of technology modules.
- *Segment the roadmap around logical organizing groups for your business.* Do not try to put so much information on one graphic that it becomes cluttered and impedes effective communication.
- *Make the connections clear in both directions.* (a) All products that plan to leverage a given technology, and (b) all technologies needed to support a planned product development project.
- *Update and maintain synchronization of the roadmaps to preclude nonproductive debate during reviews.* Update roadmaps on a regular basis and review them on a periodic basis, often quarterly or twice a year. The periodic reviews must be tied to a company's annual planning cycle so that one of the periodic reviews is linked to the start of the planning cycle. The many tools available for publishing and controlling change of roadmaps should be considered.

Decision Tools for Effective Technology Staging

In Chapter 11 of *The PDMA ToolBook 1* (Ajamian and Koen, 2002), the authors present the TechSG process, a framework for applying fundamental Stage-Gate™ techniques to technology development. The framework draws distinctions between management of new product development and advanced technology development, and presents a management process that addresses those distinctions. However, as various leading companies have gained experience with TechSG framework elements, a recurring need has emerged for additional techniques that help drive not only successful technology *development* but also ensure that those technologies will be successfully *commercialized.* The three tools presented in this section build on the initial presentation of the

TechSG process with methodologies that companies have successfully applied to drive more effective commercialization of their new technologies.

Tool 4—Simple and Practical Application of Stage Gating to Technology Development

The TechSG process refers to an indeterminate number of technology stages (TR_0 to TR_N). Though this approach still may apply at the researcher's level, we have found that applying multiple stage-gate reviews at the management level can be overly confusing and eventually ineffective.

As an alternative, leading companies have established a streamlined high-level set of standardized stage-gates, thus providing a consistent management framework across all technology projects. Given the uncertain—often unstructured—nature of research and innovation, these standardized gates must allow for the flexibility required by technologists and the creative process. And given the often antiprocess orientation of researchers and technologists, the framework also must be kept extremely simple and straightforward.

The three-stage model in Figure 16-9 has been applied successfully to address these issues. Though this model does not contradict the basic TechSG presentation—in which TR_0 is intended as a planning stage and TR_N represents a technology transfer period—this three-stage treatment both simplifies the Stage-Gate™ approach for management-level discussion and places additional emphasis on the first and last stages—those areas that are particularly critical for successful commercialization.

Stage 1—Definition

The Definition stage represents the beginning of a technology development effort. Unlike product development, it is often difficult to identify up front exactly what you are looking for in a new technology. For example, the performance criteria—or even the definition of "performance"—may be unclear until some initial experimentation is complete. Therefore, it is best to think of the initial period of a new technology effort as an unstructured experimenta-

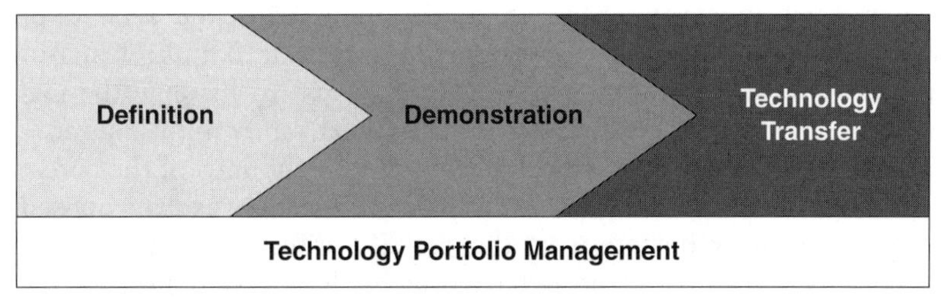

FIGURE 16-9. Standardized technology Stage-Gate™ model.

tion stage aimed at identifying potential investigation paths and hypothetically possible technology objectives.

When presented to a technologist, this initial stage of technology development may seem intuitive. It is essentially what researchers have long considered their core mode of operation—unstructured experimentation aimed at identifying potential breakthroughs or capabilities. The problem with this interpretation is that it leaves open the risk of long, unmonitored periods of spending without any guarantees of a result for the management team. To address the issue, this stage should include some structured limits on the maximum duration or budget expended before the research team returns to management for a review.

We have heard of companies that do not pull the plug on a project fast enough; they end up wasting money and losing the opportunity to redeploy resources to more valuable endeavors. Structured limits create a mechanism to define formal communication and business review between a technology project team and senior management. The reviews ensure that the progress made on a technology project remains in synch with management expectations and company priorities. The limits should be based on business strategy and budget availability, as well as on the technical team's estimate of its budget requirements. As always in a Stage-Gate™ model, if the project team feels that additional experimentation and planning effort are warranted when the initial stage limits are reached, the stage may be extended at the management team's discretion.

Prior to exit from this stage, a basic technology planning exercise must be completed. The objective is to develop the technology performance matrix (which sets the goals for the technology project) and to establish baseline boundaries for the Investigation and Maturation stage (e.g., duration, budget). The ability to distinctly describe this performance matrix (see Chapter 11, *The PDMA ToolBook 1*) is the defining criterion for readiness to exit this stage. Essentially, this means that the project team has enough understanding of where it is going with a technology to reach management consensus about the criteria for successful completion of the technology development effort.

Stage 2—Demonstration

The Demonstration stage focuses on research, analysis, prototyping, and staged maturation of the technology or capability to achieve the performance criteria set forth in the preceding stage. As discussed in the TechSG process (Ajamian and Koen, 2002), this stage may be divided into multiple maturation stages with interim technical reviews (TR_0 to TR_N) to manage the high-risk nature of technology development. This model, shown in Figure 16-10, remains relevant for use at a project level within the bounds of the core period of technology investigation and development. The approach proposed here does not significantly deviate from the TechSG process, and all the elements described in that process (project charter, technology review process, technology development teams, structured planning, and technology review committee) remain applicable.

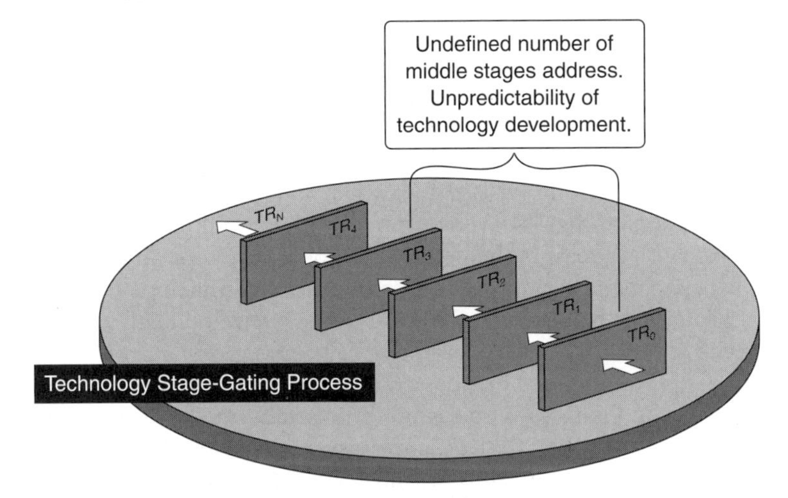

FIGURE 16-10. TechSG model for technical reviews.

Stage 3—Technology Transfer

This final stage represents a focused period of knowledge transfer that achieves a handoff of the technology into the product development organization for incorporation into a new product. This period involves a highly cross-functional team to verify that performance objectives have been achieved and to ensure effective knowledge transfer. An effective Technology Transfer stage ensures that technology know-how is transferred into the product development cycle, which provides the opportunity for eventual payback. Many research investment losses result from the ineffective transfer of reasonable technologies into products.

Following are some tips on this tool:

- *Keep the process structured, yet simple.* Walk the fine line of acceptability for both management and researchers.
- *Define clear stage-exit criteria and clearly communicate the definitions.*
- *Enforce appropriate limitations on the duration of each stage.* Try to avoid the classic pitfall of open-loop experimentation or investigation that never seems to produce results, yet always receives continuing budget approval.

Tool 5—Use of Technology Summit Meetings to Guide Technology Maturation

A technology summit meeting serves a complementary purpose to the standard stage-gate reviews discussed in the TechSG process (Ajamian and Koen, 2002). The latter is conducted on an event-driven basis and is focused on reviewing an individual technology development effort against its target performance criteria and development plan. It is weighted toward management topics that involve expected business return and schedule as much as (or more than) technical status. The stage-gate review is an essential element in ensuring that the research teams' efforts and spending stay in line with the strategic direction from management. Conversely, technology summit meetings can be held on an event-driven opportunistic basis, either when the project leader and management feel that something discovered in the project could be utilized elsewhere or when the project may be helped by collaborating with another unrelated project or department within the company. This could include collaborations between groups such as marketing or business development and R&D to match needs in the marketplace with new developments in a project.

Technology summits are highly technical in nature and involve senior individuals across the business with visibility not only into the current research projects but also into the strategic imperatives across the business (and perhaps even into multiple divisions within the company). The objective of the technology summit is to discuss the current status of various research efforts and quite literally look for the "revolutionary accidents" that inevitably occur on occasion in early-stage technology work. It is often the recognition of a revolutionary accident that provides the impetus to exploit an opportunity created when a development in one project is found to meet either a need in another project or in the marketplace.

Alternatively, synergies between projects identified at technology summits could result in an exchange of enabling information that could ultimately reduce the time to market for a new product. Thus, the goal of this review is to help drive technologies to the fastest-possible point of commercialization, whether or not that will be achieved by following the original target goal or development path laid out when the technology project passed through the planning stage of the TechSG process. In essence, the focus is trying to build an organization that embodies the adage, "Every problem is really an opportunity in disguise."

Figure 16-11 illustrates the point. In a traditional research center, the technology portfolio or set of technologies under development would be seen as $Tech_1$ through $Tech_3$. In this alternate approach, however, the set of technologies in development includes those three technology paths, but also includes any "accidental" developments that emerge from these efforts at any point along the way (A_1 to A_9).

This way of thinking represents a dramatic shift for many researchers and directors of research organizations. It is the kind of thinking, however, that allowed 3M to create one of the most successful products—the Post-it® Note. The technology that enabled the product's introduction was supposed to be a new type of glue, but the researchers found that it wasn't sticky enough for its original purpose and would release its grip under a bit of lateral force. Rather than discard this creation as a dead end on the desired development path, the 3M researchers realized that they had stumbled onto something with an alternate application. So the "sticky tab" was born.

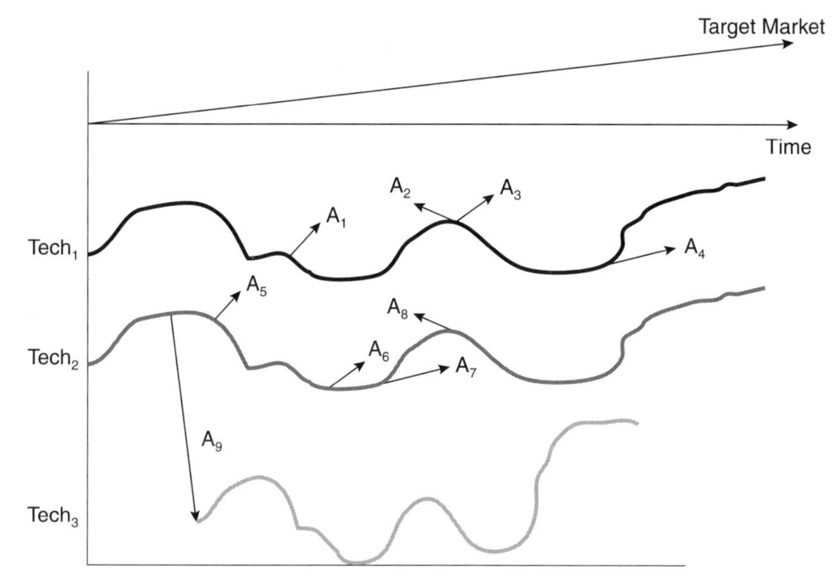

FIGURE 16-11. Technology evolution diagram.

In contrast, the Xerox Corporation arguably has been the original source of some of the last century's greatest electronics breakthroughs without realizing the commercial benefits of its innovation. On technologies as diverse as Ethernet protocols, the graphical user interface (GUI), and personal computers, Xerox had technology breakthroughs—and in some cases patents—that put it well ahead of the competition. Yet, possibly because these technologies did not appear to support Xerox's core product lines, these breakthroughs were never commercialized. So, Xerox ultimately watched competitors develop these markets.

Now, the question for other technology-oriented companies large and small is how to build an organization primed to identify these "revolutionary accidents" when they occur and take advantage of them. One can always rely on luck or the insight of an individual technologist, but it is also possible to build a structure (and drive a corresponding culture) that makes luck more likely to strike. The technology summit serves this role. The benefits of structure ultimately manifest either in a reduction of development cost or time to market or an increase in product quality. The technology summit allows synergistic interactions between multiple projects in large companies or simply increases efficiencies in smaller companies with fewer projects and more focused development programs. Alternatively, out-of-the-box benefits can occur—such as technology out-licensing opportunities.

Let's momentarily consider a company that sets out to invent an erasable ink for use on a newly developed surface, the whiteboard. The project starts with a series of experiments that identify a set of chemicals that shows exciting potential for producing an erasable ink. So, a dedicated technology development team is formed, performance criteria are established, and the team begins its core period of investigation. After the first few weeks of effort, the team shows exciting results and continues down its primary investigation path. Eventually, it demonstrates an ink that meets all the specified performance criteria for erasability. Unfortunately, it turns out to be so toxic that anyone spending more than a few minutes in a conference room with an open pen would risk passing out from the fumes. Obviously, this is a core design flaw, given the team's charter. So, if left to themselves, the team members would likely view the toxic ink as an undesirable outcome and continue working to eliminate the toxic characteristic while maintaining the other performance elements.

This is where the technology summit plays a valuable role. In this example, now assume that the development team is required to report, on a periodic basis, the results of its investigation and development work to an independent technical review committee. This committee is aware of the team's mandate and its target performance criteria, but it is also aware of other parallel technology efforts the company may be pursuing or may have a strategic interest in pursuing in the future. So, when the erasable ink team reports on its development of a toxic ink at the technology summit meeting, the committee might raise the idea of transferring this chemical formulation to the company's pesticide division. Within the pesticide division, this "undesirable outcome" may quickly demonstrate its feasibility as a profitable product or product component, yielding a return on the initial technology development investment on a much faster schedule than the erasable ink program would achieve.

The original development path toward an erasable ink could then continue in parallel or be canceled by the technology review committee, as appropriate, based on its own merits. But regardless of the committee's decision, the technology will be *commercialized* in a shorter time span than originally expected. In a small company, out-licensing the toxic ink might be the preferred approach, with the development team refocusing as directed by the management committee. Either outcome illustrates the power of creating a shift to viewing problems or challenges as opportunities.

Following are some tips on this tool:

- *Make technology summits supplemental to stage-gate reviews.* They need to be distinct—but linked—events that serve different purposes.
- *Moderate the frequency of summits.* They must occur often enough to capture interim results from research efforts but not so frequently as to create a "meetings burden." Typically, about one technology summit is conducted prior to the completion of each stage-gate. Other events may also justify holding a summit, such as significantly new information from a technology development project.

- *Keep the focus of the meeting on technology leverage.* Look for opportunities to quickly commercialize whatever capabilities your team's research has revealed, not on how to achieve the original project goals (that's the purpose of technology stage reviews).
- *Involve a cross-functional audience.* Marketers who can identify potential new product opportunities are as vital as technologists who can interpret research results or chart future development paths.
- *Consider business strategy, but don't let it handcuff you.* Ignoring strategy can send research efforts in too many directions, but over-adherence to established paths can lead to missed opportunities (e.g., Post-it® Notes and Xerox).

Tool 6—Use of Transfer Teams to Drive Successful Handoff

The TechSG process, particularly if used with the two additions mentioned previously, provides a useful structure for managing the development of a new technology. The last step in that process is generally referred to as *technology transfer*. At this stage of development the results of the technology development effort are documented or otherwise handed off to a new product development team for use in a commercial product. Unfortunately, this last step—often an area of difficulty for many companies—makes the difference between a truly promising technology and commercial success.

As leading technology organizations have tried to address this issue over the past few years, the transfer team concept has become much more refined. Traditionally, technology or research organizations often maintained a silo-like separation from the development groups that led product design and launch efforts. The tech transfer concept was generally considered the act of documenting basic design elements or performance characteristics and literally passing them "over the wall" to a product team. The risk inherent in this approach, and one that has been realized repeatedly by product launch teams, is that the product development team may not have the expertise (or be prepared for the required effort) to refine a given new technology for use in a product in the field. In the worst case, the technology element is simply dropped from the product or the project is canceled.

In response to this problem, the technology transfer team concept was developed. In practice, it means that key product development team members (representing engineering and marketing/sales, at a minimum) are added to the technology development team during the final phase of the technology project. These additional members become essential in ensuring that the technology is ready for handoff into a product development effort. They also provide a measure of continuity for the product development effort once the technology team members have moved on to a new area of work. Additionally, key members from the core technology development team may be selected to temporarily transfer onto the lead product development effort leveraging the new discovery.

The other element of a successful transfer process revolves around duration. Traditionally, most organizations have viewed the technology feasibility point as the point in time when technology transfer occurs. In this traditional view, once the technology team has demonstrated that a technology can work under a defined set of conditions and has adequately documented the technology for use by product development teams, the technology development effort is over. What has been discovered over time, however, is that such an approach often leads to downstream problems that can delay the release of the first product to incorporate the new technology. Engineering usually will blame these problems on "immature technology" and point the finger back at the research organization. The technologists, in turn, will point to the engineers and say—rightly so—no one identified the given set of operating conditions or interfaces as a requirement for the technology effort.

The difficult truth is that both groups are right. The fact is that you can never hope to fully prove out a technology under all conditions before transfer to product development. If you did, the technology effort would take so long, it would be pointless. Similarly, engineers never can fully identify in advance the complete set of operating conditions that will be needed once a product specification is developed. The outcome of this catch-22 is that a new technology will only reach a reasonable level of performance stability after the product development team has conducted some amount of integration work, and sometimes not even until initial customer usage data have been collected from the field. This point in time, shown in Figure 16-12, represents what we will call the *true transfer point*.

This true time to transfer (TTT) represents the period during which the transfer team needs to focus its activities. It also represents the time when a

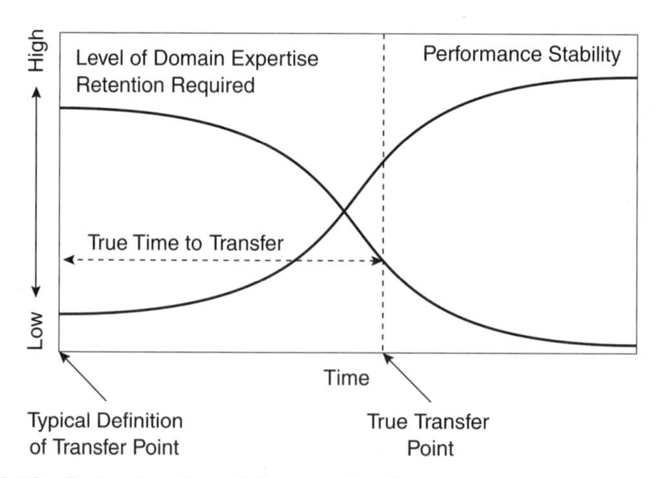

FIGURE 16-12. Technology knowledge transfer diagram.

significant level of technical domain expertise is made accessible to the product development team. In practice, this may correspond to one or more members of the technology development team being transferred to the product development team as full-time members. Or it might simply mean that key members of the technology development team will be released—on an as-needed basis—from any other work to rapidly resolve any technology performance stability issues. When done right, this represents a significant commitment of resources to the transfer process, not just "spare time" support.

If you consider that a new product only can be released or perhaps successfully marketed once its component technology elements have reached a point of relative performance stability, then time to market can be seen as directly dependent on true time to transfer. So, driving TTT reductions provides real benefits to the product development organization through faster product launches. Consequently, best-in-class organizations have recognized the value of a real emphasis on transfer teams.

Following are some tips on this tool:

- *Design transfer team membership to be more cross-functional than a pure technology development team.* The team should represent a hybrid between the technology team and the product development core team.
- *Transfer team members should have a core understanding of the technology and/or the product integration requirements.* Therefore, pick experienced individuals, not your most junior individuals.
- *Don't cut the cord too soon.* Maintain a product development team's access to technology expertise until the technology has reached a relatively high level of performance stability.
- *Time to transfer is a function of the performance stability of the product when transfer begins.* Make a realistic assessment of the state of the project, and try to use metrics such as "bug rate" or number of complaints per unit time to monitor stability over time. Targets can be set as appropriate for each project.
- *Time to transfer will usually correlate to the complexity and magnitude of the product development effort: the bigger the project, the longer the transfer time.* A reasonable expectation is that transfer should take less than half the time required for the development phase of the project.
- *Lastly, since time to transfer will be a strong function of manufacturing process complexity, we recommend that manufacturing processes be kept as simple as possible.* In addition, adequate time should be allowed to establish and stabilize critical processes.

IMPLEMENTATION THOUGHTS AND CONCLUSION

The tools offered in this chapter are intended to provide technology-based businesses with more concrete tools to leverage the organization's innovation assets. These tools will help the management team establish an organized dialogue around the direction of a single technology project as well as the commercial implications of the entire technology portfolio.

- *Tool 1.* Component technologies and technology platforms
- *Tool 2.* Real options theory
- *Tool 3.* Technology and product roadmap integration
- *Tool 4.* Simplified technology stage-gating
- *Tool 5.* Technology summit meetings
- *Tool 6.* Technology transfer teams

We have provided examples where each tool has demonstrated management value. However, it is unlikely that a management team will use all these tools. So the key to success is judgment of which tools to use when, and how to apply them in the context of a specific organizational culture. The answer usually starts with a conversation about technology management practices with your technology management group.

BIBLIOGRAPHY

Ajamian, G. M., and P. A. Koen. 2002. "Technology Stage-Gate™: A Structured Process for Managing High-Risk New Technology Projects." In *The PDMA ToolBook for New Product Development,* ed. P. Belliveau, A. Griffin, and S. Somermeyer. New York: John Wiley & Sons.

Cooper, R. G., Edgett, S. J., and Kleinschmidt, E. J. 2002. "Portfolio Management: Fundamental to New Product Success." In *The PDMA ToolBook for New Product Development,* ed. P. Belliveau, A. Griffin, and S. Somermeyer. New York: John Wiley & Sons.

Luenberger, D. G. 1998. *Investment Science.* New York: Oxford University Press.

Durham, A. L. 1999. *Patent Law Essentials.* Westport, CT: Quorum Books.

Luenberger, D. G. 1998. *Investment Science.* New York: Oxford University Press.

Yu, E., Schwartz, K., and Carrascosa, M. 2002/2003. "Balancing Creativity with Discipline in Tech Development." *PRTM's Insight* Winter: 39–42.

17 Spiral-Up Implementation of NPD Portfolio and Pipeline Management

Paul O'Connor

Almost all organizations have benefited from doing more product development projects faster. Undoubtedly, there is much to gain by speeding product development (Smith and Reinertsen, 1998 Yet, for those organizations that have made significant strides in accelerating NPD, notable gains are diminishing. It seems that most companies, along with their competitors, keep trying to accelerate their development efforts. But new product development can only go so fast without increasing the probability of commercial failure and decreasing the expected return from a project. In response to market conditions, though, it is still common to see senior managers push harder on the same "more projects faster" approach.

Effective portfolio and pipeline management (PPM) will help overcome this difficult development challenge. There are many discussions and descriptions of PPM in the literature (see Cooper, Edgett, and Kleinschmidt, 2001, as well as other ToolBook chapters). As a process, it pulls together project selection, project mix management, and resource allocation management. When executed well, PPM enables the organization to reach beyond just trying to do more projects faster. It enables the organization to emphasize a balance between speed to market, strategic impact, and resource use efficiency (see Figure 17-1). The objective of an effective PPM process is to optimize these three goals simultaneously for the greatest economic gain. It is the economic gain that management wants, not the process and not just speed to market.

Portfolio and pipeline management is run as a process because markets, technologies, competitors, and the development projects themselves never stand still. A dynamic exists that requires manipulate of the set of projects, so that overall development efforts are optimal. The PPM process, through various tools and practices, is a proactive approach to do this. When organizations establish links between PPM and predevelopment or front-end processes, it can also drive, and not just react to, the overall mix of projects. For some organizations, PPM initiates desirable projects in the front end. This link makes PPM a powerful complement to new product development. Unfor-

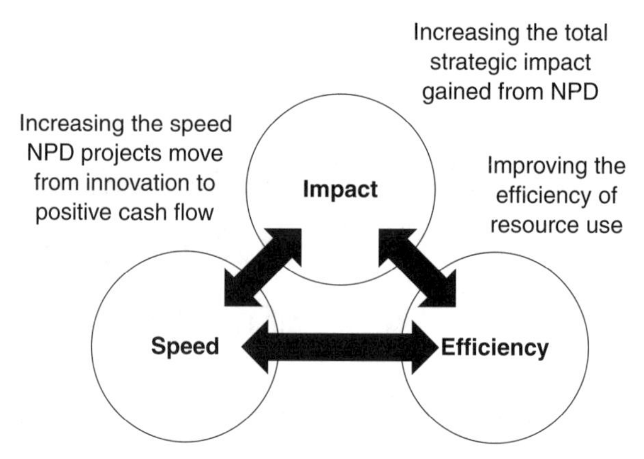

FIGURE 17-1. Three complementary orientations of portfolio and pipeline management.

tunately, understanding what PPM is and getting an organization to do it well are two entirely different issues. This chapter of the *ToolBook* deals solely with implementing PPM. It is intended to help organizations establish a sound PPM process and accrue benefits from it quickly.

IMPLEMENTATION CHALLENGE

Implementing good PPM practices is a notable challenge for even the most progressive company. Managers naturally want to make PPM decisions based on reliable, timely, and insightful data or metrics. But gathering and then updating this information can be a very difficult challenge. At the start of PPM, data are seldom reliable or timely. Often, the exact metrics needed to communicate with management may not be known. Even with seemingly good data in-hand, some managers may perceive it as undermining current decision-making practices and may criticize the quality of the data. Other managers may simply not perceive the need for PPM and therefore choose to ignore requests for data or, for that matter, the entire PPM implementation.

In addition, many other factors impede PPM implementation and slow down benefit gains. A summary of stated or implied factors from literature and conference proceedings is shown in Table 17-1. Some issues are deeply rooted in the organization. Perhaps the most revealing of these is when the new process forces an explicit and unified understanding of business and product line strategies. While managers embrace their own perceptions of their organization's product development strategy, a unified cross-organizational understanding of the strategy is more difficult to come by. During PPM implementation, in fact, it is common to discover that an explicit new product development strategy may not exist. Yet this is fundamental to good PPM. Without an agreed-upon strategy, it is impossible to delineate what the opti-

TABLE 17-1.
Factors Cited in Literature and Implied to Retard PPM Implementation

Factors Hindering PPM Implementation
◆ No explicitly stated and unified understanding of strategy.
◆ Poor data/metrics on project characteristics: They are either too old, not reliable, or do not match decisions.
◆ Lack of criteria for guiding decisions toward an optimal mix of projects.
◆ Organization structures changing, key people leaving.
◆ Inadequate or outdated resource assignment and usage data.
◆ No ability to forecast resource bottlenecks.
◆ No historic data on which to establish norms.
◆ Poor foundation in project management.
◆ Project and commercial risks are unknown.
◆ Project tasks uncertainties (outcome, work load and duration) are unknown.
◆ Poor financial forecasting.
◆ Embedded single-project decisions mechanisms.
◆ No tools for easing the gathering, analysis, or communication of data and metrics.
◆ No central repository for data.
◆ Lack of management involvement.
◆ No perceived need to establish PPM (no value proposition for PPM).
◆ Inadequate development processes (e.g., phase-gate, front-end).
◆ No implementation team, no person responsible.
◆ Organizational structure causes conflicting interests.

mal mix of projects should be, or which projects are of higher priority for receiving resources.

When several hindering factors combine, implementations may even reverse momentum and erode benefits. This is particularly true should top management's focus or interest wane. In such cases, the organization as a whole will decrease its support for PPM and lose interest in implementation. An inability to understand and address hindering factors makes PPM implementations very challenging.

HOW YOU IMPLEMENT PPM MATTERS

Changing how an organization carries out work and makes decisions is never easy (Hammer and Champy, 1994; Kotter, 1996; Tichy, 1999). For new product development, establishing new practices can present some very unique challenges (O'Connor, 1994; Repenning, 2001). Consider some of the real concerns stated by managers trying to implement PPM (see Table 17-2). Notice the variety of the statements. It would be impossible to address all of these challenges with a single action. Some form of a broad, coordinated approach must be taken.

TABLE 17-2.
Managers' Statements of Top Challenges in PPM Implementation

- Managing reduced resources and still maintaining an aggressive launch schedule.
- Trying to think and act long term (strategically), but managing quarter-to-quarter to satisfy Wall Street.
- Gaining buy-in from the entire organization.
- Poor analysis because of emphasis on speed and lack of good information is forcing a reevaluation of current practices.
- Systematically looking at strategic objectives and the resource pool to understand how to deploy available resources and set priorities.
- Not having consistent metrics to measure key project characteristics.
- Lacking commitment to spend sufficient time to manage the portfolio.
- We have pockets of excellence in various aspects of portfolio management. Our challenge is to transfer these best practices broadly across the organization.
- The time it takes to get up the curve.
- Getting and maintaining top management support.
- Making adjustments because of our ever-changing environment; adjusting for changing resource needs and availability, along with a changing competitive landscape.
- Having better resource-planning tools.
- Consistent use of the established process and consistent data assumptions across new product development opportunities.
- Pulling product management, marketing management, and top management out of the daily fire fighting and moving them fully into a strategic view.
- Too many new project requests (many driven by customers) coming into an already full pipeline.
- We have a culture of wanting to do everything.
- Defining and getting accurate input for metrics.
- Integrating project management and portfolio management into a seamless system with a single point of data entry.
- Articulating the strategic success measures that are expected from implementing portfolio management.
- Having the right NPD portfolio analysis tools and resource management tools.

What, then, is the best way for an organization to deploy PPM? What implementation practices will deliver the most benefits fastest? Where should an organization start? These questions led to research specific to PPM implementation (Adept Group, 2002). Approximately 175 companies participated, representing a wide distribution of company types, duration of PPM practices, and implementation experiences (see Figure 17-2).

Many academic papers, newsletter articles, and conference presentations, along with a few books, have been published that shed light on the "whats" and, to some degree, the "hows" of PPM (PDMA and IIR 2002; MRT 2001 The first step in this implementation research was to compile and categorize these "whats" and "hows." The result was a list of 26 different components of PPM that, for organizational purposes, were divided into seven logical group-

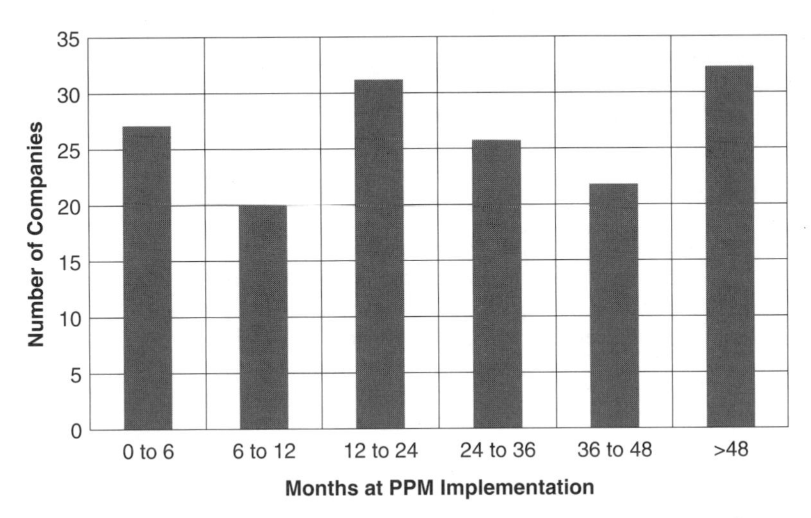

FIGURE 17-2. Survey respondents stated their organizations had been working at PPM implementation for varying durations.

ings (see Table 17-3). A list of descriptions of these components helps explain what they are in more detail (see Table 17-4).

COMPONENTS OF PPM

The seven component groups offer a means to categorize all of the pieces that make up PPM. Two of the groups are primary or fundamental to PPM:

- Mix management
- Throughput management

Combined, the components within these two groups form the essence of PPM: identifying which projects to work on and determining the resources assigned to each. The other five groups are secondary or supportive to the primary groups. Nonetheless, all component groups have notable influence on benefit accrual. Another way of looking at this relationship is that organizations want the gains from mix management and throughput management. To realize these gains, however, progress must also be made with the components in the supportive component groups.

In the implementation research, many organizations were well along the way to proficient use of some components and realizing notable benefit gains. Yet, exactly what they were doing with regard to each component differed significantly. This makes discerning a single, specific "best practice," applicable to all organizations, improbable. Rather, the best practice is situationa—a function of business and product strategy, corporate culture, and readiness of the organization.

TABLE 17-3.
Portfolio and Pipeline Management Components

Component Grouping	Components
Primary	
Mix Management	Project Selection Criteria Mix Criteria Strategic Buckets Project Impact Dependencies Mix Optimization Analysis
Throughout Management	Project Management Foundation Project Prioritization Resources to Project Assignments Resource Use Forecasts Critical Chain Buffer Management
Supportive	
Measures/Methods	Metrics Financial Priority Listing Risk Assessment Project Complexity
Software/Data	Data Gathering and Handling View Creation Software Enterprise Software
NPD Processes	Portfolio Objects Being Managed Stage-Gate Redesign Front-End Concept Generation Product Line Planning
Top Management	Top Management Involvement Top Management Proficiency Top Management Focus
Implementation Focus	Organizational Challenges Implementation Team Focus

CAPABILITY MATURITY

A particularly interesting finding in the implementation research is the wide distribution of both experience and benefits accrued from PPM as revealed by the respondents. In complex organizational systems, such distribution of performance should be expected. IBM experienced this issue with software development (Paulk, 1993). What it learned was that the quality of resulting software (the output of the system) is dependent upon the maturity of the capabilities a group has in developing software. More importantly, it identified maturity levels through which a software development group had to progress in order to improve its output. IBM's work led to what is com-

TABLE 17-4.
Description of PPM Components

Component	Description
Project selection criteria	Conditions for selecting or screening projects. Usually established as part of the Phase-Gate or phase review process. As PPM is deployed and front-end processes become established, such conditions may need to be altered.
Mix criteria	Guidelines for how the portfolio should look. Some measures are made in terms of percentages, others in terms of absolute values (e.g., count, risk, dollars, durations). Time periods may also be captured, e.g., current versus 6 months out, versus 24 months out.
Strategic buckets	Grouping of projects based on strategic rationale.
Project impact dependencies	One project may influence another project either positively or negatively. For example, a positive influence might be "Project A must succeed for project B to succeed." An example of a negative influence is "If project A succeeds, then project B cannot succeed."
Mix optimization analyses	An in-depth review of how the blend of projects matches the objectives of the portfolio, as well as a recommendation of how the blend should be changed to best fit the objective.
Project management foundation	The basic people and systems skills of project management required to carry out an NPD project.
Project prioritization	A specific rank order of projects that enables management to assign resources in that order and, by doing so, achieve the desired objective of the portfolio.
Resources to project assignments	The manner in which resources are assigned and conduct work to projects
Resource use forecasts	The ability to predict when each appropriate resource will be used on a specific project and when there will be shortfall or excess of the appropriate resource for each project.
Critical chain buffer management	A method of project management that accounts for the uncertainty of tasks and the dependency of specific tasks on certain resources.
Metrics	Measurements of PPM.
Financial priority listing	Specific rank order of projects based on financial measures such as IIR, NPV, or EVA.
Risk assessment	The likelihood of a desired outcome happening. This may be a task, a metric, or even the commercial success of a project. See the "Special Consideration to Risk" sidebar.
Data gathering and handling	The manner or method in which data is collected, stored, and submitted for analysis.

TABLE 17-4. *(continued)*

Component	Description
View creation software	Software that enables managers to create graphs and charts for displaying trade-offs that are central to PPM decision making. The trade-offs are always among multiple PPM metrics.
Enterprise system	A software structure that pulls together all data gathering, storage, analysis, and display from across an entire organization.
Portfolio objects being managed	The projects included in PPM: "Products-in-Development," "Products-in-the-Market," "Product Innovation Charters," "Product-Mapping-Projects," "Cost Reduction Projects," and "Special Customer Request Projects."
Phase-gate redesign	The development process that, as PPM and front-end processes become established, must be revamped to new aspects of NPD.
Front-end concept generation	The process immediately preceding the development (Phase-Gate) process, and immediately following Product Line Planning. The input to this process is a target for innovation (Product Innovation Charter). The output from this process is a concept that may be developed into a product (Product in Development).
Product line planning and mapping	A process that scrutinizes and visually maps the market and technology influences on a product line. This process follows business strategy and precedes the front-end process. The inputs to this process are target market segments and a product line. The outputs from this process are specific targets for innovation called product innovation charters.
Top management Involvement	The actual time spent by top management from a business entity on the development and use of PPM.
Top management proficiency	The ability and adeptness of top management to carry out PPM decision making and follow though on these decisions.
Organizational challenges	The hindrances that are caused by organizational factors.
Implementation team focus	The specific focus of the implementation teams at any given time in spiral-up implementation.

monly known as a "Capability Maturity Model" (CMM) for software development. The Software Engineering Institute at Carnegie Mellon Institute later put CMM into broad practice across the software industry (Paulk, 1993).

Rather than seeking overall best practices, CMM suggests that, in complex systems, organizations should focus on attaining each maturity level, one at a time. All requisite practices must be in place, in the right order, for the system to be effective. CMM recognizes that certain practices rely on certain other practices. For example, it is difficult to do (1) software platform planning without first understanding (2) customer requirements. It is not sufficient to establish only a couple of the practices in the complex system of software development.

The same is true for PPM implementation. Consistent practices on certain PPM components are necessary for other components to be effective. In PPM, for instance, establishing the project mix component without establishing a data gathering component would not yield much benefit to the organization. Such out-of-sync deployment of practices may even cause harm to the initiative. The implementation of PPM components must be coordinated. The analysis sought to do just this. By using implementation experience, logic, and trial and error to determine required relationships of components, the analysis identified five critical maturity levels for PPM implementation. The matrix in Table 17-5 details each component within each maturity level.

Applying CMM to portfolio and pipeline management implementation yields much different insights than suggested by other, more traditional research. A typical research effort in PPM would review the NPD performance of many organizations and then divide the companies into three tiers: top performers, mean performers, and bottom performers. The analysis would then compare the practices of top performers against the practices of the bottom performers and the mean performer. These findings can be insightful, but managers have a problem applying the learnings. The research implies that to improve, an organization must change its practices from those of bottom performance companies to the practices of top-performing companies. But the path between the two is not straightforward. Breaking through PPM implementation impediments, it would seem, is not about a concerted push to the end point of a single "best practice." Doing it all at once is a surefire way to encumber a PPM implementation initiative and deter the accrual of benefits.

SPIRAL-UP APPROACH

PPM implementation teams can accelerate the accrual of benefits and leverage the knowledge of the Capability Maturity Model simply by addressing one maturity level at a time. The idea is to (1) execute the mechanics of all components, (2) develop organizational learning about the components, and (3) estab lish "consistency-in-use" of the components during each maturity level. The implementation can proceed to the next maturity level, once a team establishes the components for a current maturity level. This iterative approach to

TABLE 17-5.
Matrix of PPM Components by Spiral-Up Maturity Level

	Maturity Level 1	Maturity Level 2	Maturity Level 3	Maturity Level 4	Maturity Level 5
Software/Data					
Data Gathering and Handling	Using MS Office (Excel/Access)	Using MS Office (Excel/Access)	Online forms/XML to MS Office (Excel/Access)	Using online forms/XML to SQL Database	Integrating to enterprise system
View Creation Software	Using MS Office: Excel, PowerPoint; data gathering/handling is challenge	Point solution software used, political agreement on views explored	Views agreed upon, point solutions software linked to data/metrics	Point solution integrated with data gathering/handling methods	Integrating to enterprise system
Enterprise System	Observing who offers what, who uses what; understanding why enterprise system	Learning about alternatives available, attributes, and features, building value case	Selecting system, crafting customization/deployment/training plan	Piloting system/customizing a full enterprise system	Deploying and training all aspects of enterprise system
Processes					
Portfolio Objects being Managed	PIDs ('Products in Development', those in Phase-Gate), SCRPs (special customer request projects)	Adding PICs (Product Innovation Charters: Targets for innovation in the Front-End process)	Adding PIMs (Products in the market—life cycle management)	Adding platforms, market segments, technology building blocks.	Continuing, integrating with enterprise system
Phase-Gate redesign	Using as-is phase-gate process, integrating gates with PPM decision-flow	Revamping gates, stages: reflect PPM mix criteria/resource decision making	Reworking framework to reflect front-end process linkage	Building new framework into all systems supports/data handling	Integrating to enterprise system

Front-End concept generation	Ad hoc approach, establishing value case for proactive FE	Designing and deploying FE process, linking to phase-gate process	Trying out point solution support software for FE	Training users of selected point solution support software	Integrating to enterprise system
Product line mapping	Ad hoc approach, establishing value case for product line mapping	Exploring market segmentation schemes, technology roadmap	Laying out ground work; exploring segmentation strategy and PIC creation	Designing and deploying product line mapping process	Point solution support to PLM, portfolio includes PICs, PIDs, and PIMs
Management					
Top Management Involvement	4–8 hours per month: likely focus on portfolio triage, criteria definitions; desired views	2–6 hours per month, all resource owners participating, concern with details	2–4 hours per month; political agreement on criteria, decision making established	2–4 hours per month; focus shifts to use PPM to influence strategic outcomes	2–4 hours per month; managing at strategic level, not in details; top management owns PPM
Top Management Proficiency	Learning and questioning. Low consensus, yet concerned with all details of PPM.	Learning and influencing changes, establishing consensus	Gaining full understanding of whole PPM process; peer-to-peer influence	Using and driving PPM as key tool in actualizing strategy	High trust in/ dependency on PPM
Top Management Focus	Leading organization to make improvement in overall NPD productivity	Developing management proficiency/skills in NPD portfolio management	Streamlining decision making and follow-through; instilling org. discipline	Learning, using automated supports; customizing enterprise system	Driving greater strategic impact from all efforts, demanding org. proficiency

TABLE 17-5. *(continued)*

	Maturity Level 1	Maturity Level 2	Maturity Level 3	Maturity Level 4	Maturity Level 5
Focus					
Organizational Challenges	Org. structure versus authority to assign resource; terminating seemingly good projects that no longer fit	Political issues on resource assignments and project decisions being worked out	Testing strategy; developing cross-org. understanding of PPM; training the org.: honing behavior with metrics	Training systems and methods, training automated supports; embracing discipline	Improving training systems, finding and evaluating new methods that improve productivity
Implementation Team Focus	Establishing value case, gaining top management involvement	Determining key challenges; driving/ sustaining implementation	Assuring consistency in approach/ sustaining implementation	Automating work flow and decision making/sustaining implementation	Integrating full enterprise/front-to-back system

addressing the component groupings implies that PPM implementation is much like moving up a spiral (see Figure 17-3).

A spiral-up implementation enables organizations to gain benefits faster and to build steadily on investments in each PPM component. Consider the progression of data storage (see Table 17-6). For most organizations, data storage will advance from having none at all to having a central repository, supporting a Web-based system. But the steps in between are very important. The use of Microsoft Excel and Access in maturity levels 2 and 3 enables great flexibility. The tools are easy to use and do not require the IT department's involvement. The interim practice, in effect, helps teams establish other components such as specific metrics, strategic buckets, and criteria and guidelines. "Hard-wiring" these components into a central data repository before they are both known and accepted would be significantly more difficult and time-consuming.

"Consistency-in-use" of a component is the most important factor driving the accrual of benefits. The implementation research shows that the total number of components carried out consistently by an organization correlates strongly with the total benefits accrued by the organization. A key factor in spiral-up implementation, therefore, is getting people across the organization to use each component consistently—that is, at an appropriate frequency and in a quality manner.

For each component within each maturity level, the spiral-up implementation goal should be to

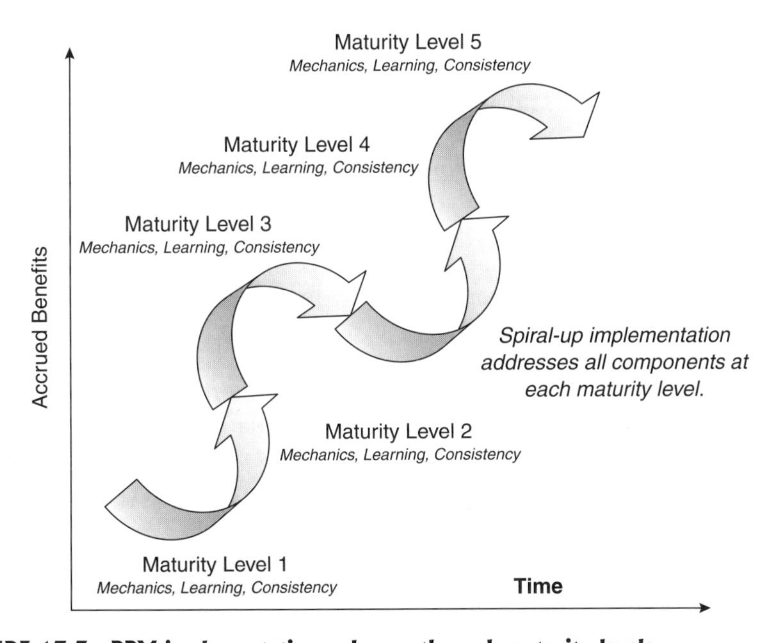

FIGURE 17-3. PPM implementations advance through maturity levels.

TABLE 17-6.
Progression of Data Storage across Maturity Levels

Maturity Level 1:	No data storage
Maturity Level 2–3:	Data storage in multiple MS Excel/Access sources
Maturity Level 4:	Data storage in a single database
Maturity Level 5:	Central storage for a Web-based system

1. iron-out the mechanics of the component's use;
2. learn and understand the influences on the component, and alter or adapt it as needed; and
3. increase the consistency with which the organization uses the component.

In fact, this makes up much of the work of the implementation team throughout all maturity levels. A key to realizing notable benefits is to get to consistency-in-use as quickly as possible.

MATURITY LEVELS

Awareness of the orientation of each maturity level is very helpful in understanding the spiral-up implementation. It enables a focus for the organization and conceptual guidelines for all involved.

Maturity Level 1: Establishing the Groundwork

During this first stage, the PPM implementation team is trying out and experimenting with different components. Very little is firmly anchored. The work effort of the team focuses on gaining full buy-in to the implementation by top and middle management. The main challenge in this initial level is to avoid spending too many person-hours over too long a duration without making progress. Because little if any benefit will accrue and a lot of time will be spent in this maturity level, many people across the organization will perceive PPM as more of a liability than an asset. PPM implementation teams should try to get their management to expect this before the initiative begins.

At this point, the organization's NPD process most likely includes only a staged- or phased-type development process. A defined front end and/or product line planning process may not exist. PPM efforts, therefore, are oriented to only those projects within the development process (products in development) and do not extend to products in the market nor to projects in front-end processes.

Top management involvement, not just their blessing, is the most critical component during this first maturity level. Their ownership of both the issues and potential benefits of PPM implementation is what provides the energy for the entire process. To get management buy-in and involvement, make sure that the value proposition for PPM (as denoted in the "change equation" described later in this chapter) is very solid. If management does not perceive significant value in a new PPM process, they will not support it whole-heartedly.

Maturity Level 2: Setting Up Decisions

The second maturity level drives the organization to an agreement on what is desired from PPM. Efforts should focus on establishing strategic groupings, or "buckets," for projects, on delineating the criteria or guidelines for exactly what the mix of projects should look like, and on how the process should flow. Attaining a political agreement from all involved managers and functions should be the key objective. There is no way to get through this maturity level without involvement by key people from across the organization.

Spiral-up implementation teams must also make sure that project management and planning skills become established across the organization in this maturity level. This includes the use of systems, the detailing of tasks, and the assignment and tracking of resources. Such skills will be requisite for later levels. These skills will be the foundation upon which pipeline throughput management will be based.

To gain benefit quickly, teams should also be exploring and trying out software tools in this level. They should begin by using "point solutions"—software that does only one or two specific tasks (e.g., graphical views, metric gathering, or resource leveling) and is not fully integrated. The team's challenge is to use these tools to help establish all of the needed components.

Maturity Level 3: Anchoring the Process

This maturity level is the most critical in PPM implementation. It is in this level that both portfolio mix and pipeline throughput components become anchored. Here the organization will manage resource assignments concurrently with the strategic mix of projects. Implementation team members should be spending much of their time on integrating project characterization data with resource use and resource availability data. This is where portfolio mix management and pipeline throughput management come together.

During this third level, the key objective should be to get the organization to recognize that PPM is delivering benefits. This is the benefit "break-even point" of PPM (shown in Figure 17-5, later in the chapter). Attention also must to shift toward other development processes. PPM should now begin to include projects within a defined front-end process as well as product line planning activities. This inclusion will set up the full, front-to-back orientation needed for overall PPM optimization.

Maturity Level 4: Turning Up the Gains

Comfort with PPM and an expectation of its benefits should set in across the organization during this maturity level. The full, front-to-back product development architecture should be established by this time, and the manageable objects within the portfolio should include projects from all of the processes (product planning, front end, phase gate, and product management). To increase benefits, teams will need to explore advanced practices within the components. Two key areas include modeling of project characterizations and resources (such as Monte Carlo simulations) and critical chain buffer management (Paulk, 1993).

The key objective in this maturity level is related to data handling. It is here that a central repository of data becomes enormously helpful. The challenge is to integrate all data and metrics on both projects and resources into a single database. Often, such integration reveals inconsistencies and inadequacies in data and data handling that must be worked out.

Maturity Level 5: Automating the Flow

Automation with real-time data and information is the key objective of the fifth level of maturity. The challenge, though, is to work out the mechanics of the system and to build consistency with its use. Because it is necessary to automate PPM for continued benefit gains, previous shortcomings will become clear in this maturity level. It is important that the system integrates data and information from across the full enterprise with all project management components and with all NPD process management components. Such systems need to pull together real-time data on project characteristics, task status, and resource use.

MOVING UP THE SPIRALS

Implementing PPM can be like navigating through a maze. But there are some important milestones. For instance, the percentage of all projects included in an organization's PPM process has significant bearing on the amount of benefit gained. According to the implementation research, the goal should be to get greater than 80 percent of all projects included in PPM efforts. This is the point at which notable benefits begin to accrue. Such inclusion requires that data on projects be meaningful, up-to-date, and of sufficient quality. Benefit gains are greatly undermined when PPM includes less than 80 percent of projects or when data on projects are poor. Another way of viewing this critical point is that if too many projects are run outside of the portfolio, PPM will serve little purpose to the organization. Indeed, consistency on 80 percent inclusion is a point of critical mass for effective PPM. Sound spiral-up implementations should recognize and target, at minimum, this inclusion percentage.

Three other critical turning points also anchor benefit accrual. Two of these turning points deal with components in maturity level three of the spiral-up implementation:

1. The establishment and consistent use of portfolio mix criteria and guidelines
2. The establishment and consistent use of resource and pipeline bottleneck forecasting

Each is within a primary component. They are fundamental to PPM.

The third critical turning point comes in maturity level 4 and is the foundation for seamless and automated system support to PPM:

3. The consistent use of a centralized data repository.

By creating a centralized data repository, teams enable their organizations to easily submit and retrieve information. Other components can then be executed more consistently, and decisions can be made more rapidly.

These three components, when taken together, offer an organization the biggest steps forward in PPM. Collectively, they anchor the PPM implementation. They are important milestones in spiral-up implementation. Yet even though they are necessary practices, they are not sufficient by themselves to sustain the initiative.

SPIRAL-UP BENEFIT ACCRUAL

Plotting the progression of benefits from PPM is insightful (see Figure 17-4). During early maturity levels, managers perceive PPM to impact speed to market negatively. This is probably due to a shift in attention required by project managers and their teams in order to get good data in place. However, positive gains appear notable in the three key benefit areas (speed to market, strategic impact, and resource use efficiency) once an organization emerges from maturity level 3. This seems to be the point of critical mass for the spiral-up implementation. If strong support for PPM had been absent, the organization's attitude toward PPM should shift during this third maturity level. Here, the benefits gains should be significant enough for the organization to not want to lose them. As a result, the actions of managers across the organization will be supportive.

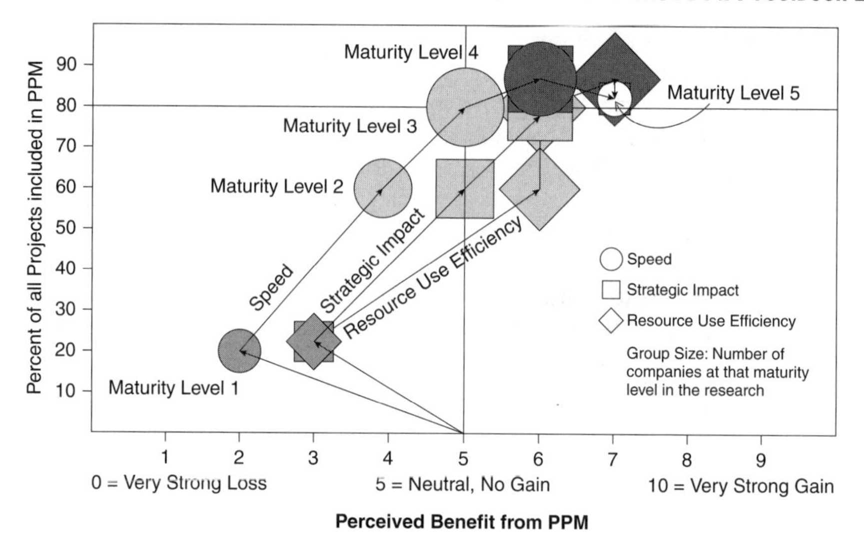

FIGURE 17-4. Benefit realized versus percent of projects included in PPM.

Another way of plotting the normal path of PPM benefits versus maturity levels is seen in Figure 17-5. This U-shaped progression implies that implementations will likely struggle through the first two maturity levels, before delivering net gains in benefits. Strong management and leadership support—in terms of participation, not just verbal commitment—is necessary to get through this negative benefit period. "Break-even" should occur in the third maturity level.

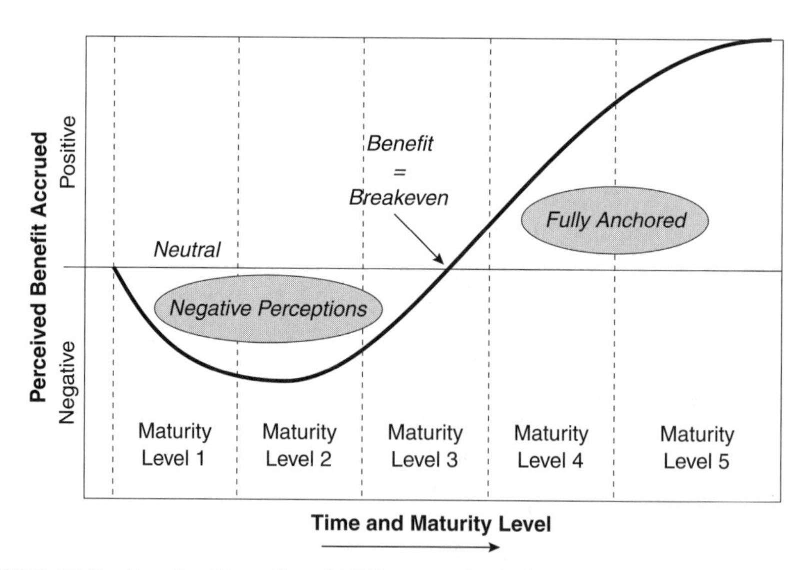

FIGURE 17-5. Perceived benefits of PPM versus both time and maturity level.

TABLE 17-7.
Range of Duration for Each Maturity Level

Maturity Level 1	Maturity Level 2	Maturity Level 3	Maturity Level 4	Maturity Level 5
2 to 4 months	2 to 6 months	2 to 9 months	4 to 9 months	6 to 9 months

DURATION OF MATURITY LEVELS

The time it takes organization to get through each maturity level varies greatly (see Table 17-7). Several factors contribute to how long it takes an organization to progress through. The PPM implementation research suggests that the larger the company (i.e., the more people contributing to new product development) or the longer the life cycle of the resulting products, the more time it takes realize benefit gains. This makes perfect sense. Gaining consistency-in-use of components, a precursor to benefit gains, will undoubtedly be more difficult and take more time in larger organizations. For companies with long product life cycles, the perception of economic benefit accrual is likely to be discounted because of the inherently long lead times. While the goal should be to gain benefits as quickly as possible, all companies will not progress through maturity levels at the same pace and the challenges for implementation teams will be different.

Some influences on the speed of benefit accrual are controllable. Perhaps the most significant are the following:

1. Understanding and counteracting the hindrances to the spiral-up implementation
2. Keeping a concerted focus on the right component practices at each maturity level

Both the implementation team and top management hold responsibility to make sure these are addressed effectively. Also, because of the significant economic value being pursued from PPM benefits, a strong case can be made for organizations to complement their internal skills and capabilities with those of an experienced outsider or consulting firm. Experts from outside the organization can help organizations drive consistency-in-use of components faster. They offer experience, insights, and an independence from organizational issues that can prove very helpful.

PPM AND A FRONT-TO-BACK ARCHITECTURE OF NPD

Most discussions and presentations on PPM address only the inclusion of those projects that are in the development or phase-gate process—that is, product in development. The problem is that the greatest influence on the mix

SPIRAL-UP IMPLEMENTATION HINTS

- Organizations are not equal, so all implementations will not proceed at the same pace. In general:
  a. The larger an organization, the longer it takes benefits to accrue. It is simply more difficult to get all projects included in the process, and it is more difficult to gain consistency-in-use of components across the organization.
  b. The longer the life cycle of an organization's products, the longer it takes to realize benefit from PPM. Conversely, the shorter the life cycle, the quicker benefits will accrue.
- Sporadic (sometimes) use of components delivers little gain of benefits. Consistent (always) use of components correlates with notable benefit gains. Consistency with one component increases the likelihood there will be consistency with other components. The greater the number of components executed consistently, the greater the benefits that an organization will realize.
- Inclusion of projects matters. The more projects that are using planning software (like MS Project), the more benefit an organization will gain.
  a. Organizations must include the majority (greater than 80 percent) of NPD projects in PPM before notable benefits accrue, but inclusion must be with up-to-date data and project plans/tasks.
  b. Once project planning consistently includes more than 80 percent of projects, organizations can realize significantly greater benefit from higher maturity level components like resource leveling and critical chain buffer management (an advanced technique for the optimization of task to resource relationships).
- Management must participate. Their proficiency in PPM practices and decision making contributes significantly. Notable gains in benefit happen once top management consistently dedicates more than two hours per month to PPM. Early levels require more time from top managers. Obviously, the greater the value of the portfolio to the core business, the greater the time management is likely to spend.
- During maturity levels 4 and 5, decisions related to front-end and product line planning processes should integrate with PPM decision making.

Metrics yield benefits once an organization knows and agrees on "most" of them. Knowing only "some" metrics does not deliver much benefit gain. Organizations gain consistency with metrics once there is cultural and political agreement on a seemingly full set of metrics, even though others may be under consideration. Organizations seldom know most metrics during early maturity levels.

of projects in development is the output of the front end of product development. Many, and perhaps most, organizations do not have a structured or proactive front end (Koen et al., 2002). Instead, development projects come from iterative practices and sources such as R&D, customers, marketing, the sales force, and even top management. Unfortunately, this "default" front end delivers development projects that can, over time, start to look similar in characteristics. This is because such default practices tend to leverage the same skills, resources, knowledge, and customer needs set to produce new concepts. Projects tend to be of similar size of opportunity, have similar degree of newness, have similar risk profiles, and even have typical or expect competitive

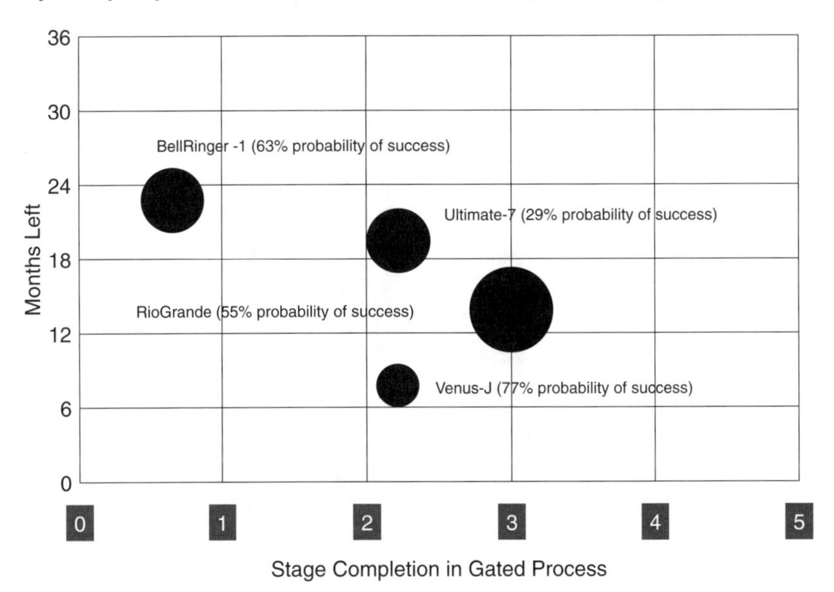

FIGURE 17-6. Portfolio view of products in development: time to launch versus stage of development.

alternatives. In this case, without an ability to guide the front end, management can only turn current phase-gate projects on or off, and increase or decrease the resources assigned to each. This is not sufficient to optimize the portfolio proactively.

The following example helps explain this. A four-project portfolio is displayed in the Figure 17-6. Each project is at a different point in the development process and will be commercialized between 8 and 24 months out. The size of the bubble represents the peak annual revenue the project should generate if it is successful. The text next to the bubble indicates the name of the project and the probability of success the organization calculated for each project. Figure 17-7 displays the risk-adjusted peak annual revenue for the projects.1 Notice, too, that only one (the number inside the bar) project from this portfolio will have launched in the year 2004. Its risk-adjusted peak annual revenue is about $5 million (peak annual revenue times risk). In 2005, three of the four products will have launched, and the summation of the risk-adjusted peak annual revenue is about $25 million. Unfortunately for this fictitious company, the financial objective of their new product development efforts requires more peak annual revenue coming from product development. The gap has to be filled somehow.

There is a dilemma here. If the company starts more projects, resources will spread thinner on the current projects and/or the probability of success will decline, causing expected peak annual revenue to decrease. Perhaps management may choose to shift resources to the two near-term projects. This

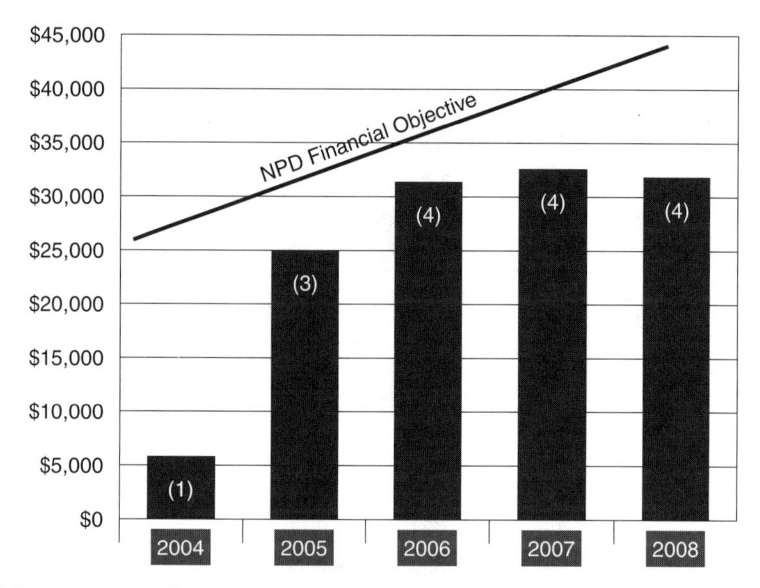

FIGURE 17-7. Portfolio view of 5 year accumulated, risk-adjusted peak revenue.

would move the bars to the left but also widen the gap in future years. The problem is that the nature of the projects in the portfolio is not adequate to reach the objective. Simply stated, this organization needs bigger, better projects to come out of the front end. The burden of the portfolio and overall product development performance, therefore, shifts to the front end. Without a proactive, structured front end, it is impossible to achieve the financial objectives. Such organizations only have the ability to optimize what is already in the phase-gate process.

The same type of dilemma exists with products that are already in the market. The launch of a new product, for example, may negatively affect the profitability and life cycle of products already in the market. Or, the product in the market may greatly hinder the diffusion of a new offering into the market. Organizations must also be able to retire products from the market. If PPM only pushes products into the market, overall profitability cannot be optimized.

During higher maturity levels, PPM should coordinate and account for all projects in the phase-gate process, plus all projects in the front end and all products already in the market. PPM in a front-to-back architecture should enable management to better shift resources across all process groups, not just among those projects in the phase-gate process. But the expansion of what is included in the portfolio does not end with just the NPD processes. The implementation research suggests that organizations that are further up the implementation spiral also capture non-new product development projects, such as manufacturing improvements and cost reductions, in their PPM processes.

This is necessary because non-NPD projects consume resources and can have significant impact on the NPD portfolio. Inclusion of both the non-phase-gate NPD projects and the non-NPD project tends not to occur until the fourth maturity level.

No matter the maturity level of front-to-back processes, organizations quickly realize the need to integrate portfolio decisions with single project decisions. Most organizations employ the majority of their resources available for product development, within their staged development process. Establishing the link between PPM decisions and phase-gate decisions is very important during early maturity levels. Previous research suggests that there are two approaches to this: management conducts gate reviews and portfolio reviews concurrently, or management conducts gate reviews and portfolio reviews at two separate meetings (Cooper and Edgett, 2002). This research concludes that both approaches work. The implementation research confirms this but also shows that at early maturity levels in a spiral-up implementation, organizations realize significant "strategic impact" benefits by conducting separate meetings. At later maturity levels, organizations tend to realize significant "speed" benefits by combining meetings. With all things considered, organizations should start a spiral-up implementation with separate meetings during maturity levels 1 through 3, and then change to combined meetings at maturity levels 4 and 5. During the first two maturity levels, this also enables management to concentrate on portfolio issues and not to get bogged down in discussions and decisions on individual projects. During the last two maturity levels, the merging of meetings creates a stability and proficiency in PPM that can be advantageous in resolving single project issues.

PPM AND ORGANIZATIONAL STRUCTURE

Organizational structure is typically designed to carry out the core business as efficiently as possible. Product development, however, may not be the core of the business. In such cases, it is common to see certain critical resources, such as R&D, shared across business units. Consider the organizational structure depicted in Figure 17-8. Here both R&D and manufacturing must participate in the NPD processes for three business units. Without adequate PPM, R&D management and manufacturing management must decide which business should be assigned what resources. By default, the R&D vice president and the manufacturing vice president control the portfolio.

Organizational structure challenges should be identified very early in spiral-up implementation. The tendency for organizations is to try to manage the portfolio mix and resource assignments at the lowest level in the organization, whether that is by product category or by business unit. When NPD relies on shared resources, organizational issues may arise during maturity levels 2 and 3 when establishing critical project mix components. The goal of the

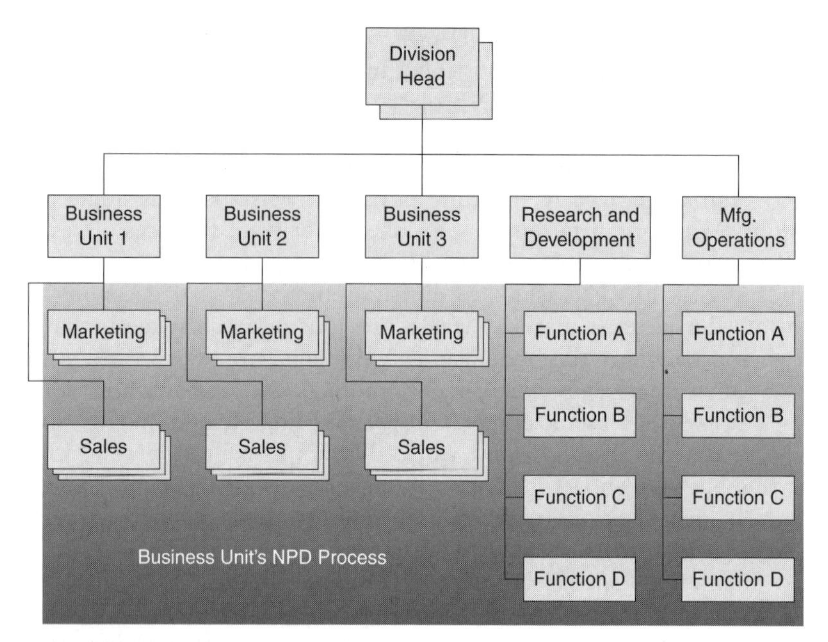

FIGURE 17-8. Each of the three business units has its own NPD process but shares resources from R&D and manufacturing.

implementation team during this work is to secure political agreements across the organizational structure on all components.

It is best to conduct PPM at the level of the organization directly above where sharing of resources occurs. For the preceding case, this is at the divisional level. While the research suggests that benefits will accrue no matter at which level of the organizational the portfolio is managed, the key is to avoid having the conflict of shared resources.

Allied Corporation (before Honeywell's acquisition) took this one step further. They created a resource pool at the enterprise level that could be shared across divisions. Following the principles of creating the next generation of technologies (Jonash and Sommerlatte, 2000), the corporate-wide portfolio included only the top 15 percent largest projects from around the organization. The corporate resources were then assigned to those projects based on expected economic value. For organizations with multiple business units, rolling up subportfolios into enterprise-wide PPM is not usually carried out until after each business unit can anchor its benefits from PPM. This would be in maturity levels 4 or 5.

CHANGE EQUILIBRIUM

Implementing any organizational process, PPM included, always requires some elements of the organization to change. Consider how PPM might

SPECIAL CONSIDERATION TO RISKS

Risk management is central to PPM. Just as in a financial portfolio, NPD portfolio management should seek to exploit the relationship between risk and reward. Organizations can do this by weeding out the projects with too little reward or too much risk, and by seeking a balance between higher risk/reward and lower risk/reward projects. Such a balance needs to be specific to the organization, its industry, and its strategy. No matter the case, some measures of risk and rewards are needed to carry out meaningful PPM. Managers should be aware of three types of risks in their portfolios:

1. *Project management and resource risks.* This is the uncertainty that tasks within a project can be completed on schedule and on budget. Think of this as an accumulation of distribution curves for likelihood of every aspect of the project plan occurring as specified. Add to this, too, the interrelationship of projects and resources, because people work on multiple projects and there is uncertainty as to whether these people will be available to do work at any given time.
2. *Project and market characterization risks.* This is an accumulation of all factors related to the market and to the project that could impact the degree of success or the degree of failure of the product once it launches. Financial projects and NewProd (Cooper, 1985)characteristics would be included in this risk category.
3. *Systemic risks.* This is when the same risk shows up across a significant number of projects in the portfolio. Usually, these risks are either strategic in nature (e.g., a competitor's action, market growth, etc.) or organizational (e.g., a shortfall in certain resources or in the quality of certain work). In either case, such risks are difficult for individual project teams to address. Rather, they should be identified and then addressed by top management through cross-project or cross organizational initiatives.

change how work and decision making occurs. It adds new work for those who gather, analyze, and present data and metrics to management. It adds other work for project managers to create and deliver the new data. It may change, for example, a vice president's current means of assigning her subordinates to projects, or picking which projects she wishes to support. The list could go on and on. The challenge during implementation is that whenever there is change, there is also some force pushing in the opposite direction. All things being equal, people in organizations would prefer not to change. The key is to manage and influence each component so that things are not equal.

Organizational behavioral professionals have known for decades that there are forces on either side of a change effort. Kurt Lewin, an MIT researcher, introduced this notion in the 1940s, but it was not employed in process implementation until the 1990s (Lewin, 1947). Lewin's force field theorem provides a means to think about what is driving and what is hindering the implementation. The idea is to first decrease the hindering forces and then to increase the driving forces. A more tactical way of recognizing and understanding the consequences of such force fields is shown in the change equilibrium, adapted to the spiral-up implementation of PPM, and displayed in Figure 17-9.

$$P_{(current)} + G_{(desired)} \geq CoC_{(psychological,\ economic)}$$

Where:

$P_{(current\ ML)}$ is the pain perceived from practices in the current maturity level.

$G_{(desired\ next\ ML)}$ is the gain perceived from practices in the next maturity level.

$CoC_{(psychological,\ economic)}$ is the perceived cost, both psychological and economic, of changing the practices from one maturity level to the next.

FIGURE 17-9. The change equilibrium for driving spiral-up implementation.

This simple equation implies that a spiral-up implementation will proceed when people (individual top managers, functional groups, and individual contributors) perceive that the "pain" from current practices plus the "gain" from desired practices of the next maturity level is, in total, greater than the cost of changing these practices. Again, it is important to emphasize that this equation applies to both individual managers and to groups, such as departments, functions, and business units. When applying the change equation, implementation teams should work to ensure that perceptions in the organization always sat-

PROJECT MANAGEMENT OFFICE AND PPM

Two worlds are converging in portfolio and pipeline management: the world of project management and the world of new product development. Just as NPD professionals have moved to embrace PPM, so too has the project management field moved to embrace the office of project management. There is much overlap between the two.

Both areas continue to evolve. In the literature, project management started with an orientation toward excellence in project execution, augmented this with an orientation toward excellence in multiple-project execution, and then added project selection and mix management. NPD started with project selection, added project mix management, and then supplemented this with multiple project execution. The difference between the two is indistinguishable from a distance.

Differences are more apparent in the details of work flow and decision flow. One would notice, for example, concern about the uncertainty of task duration in settings that evolved from project management. These concerns would manifest themselves in metrics, support tools, and even whole systems. In these settings, there is likely to be less of an emphasis on integrating PPM decisions into the phase-gate process. PPM evolving from NPD would, on the other hand, likely build off of phase-gate decision making and be concerned about commercialization risks. You would notice, also, that a project management office would emphasize the inclusion of non-NPD projects into PPM, whereas an NPD orientation may not.

Both orientations are necessary in good PPM. When observing the practices of organizations, try to understand where those practices came from and where they are going. It is likely, though, that both orientations will converge on the same track.

isfy the condition that the pain plus the gain is greater than the "cost of change." Experts in PPM from outside the organization often play a valuable role helping to assess and articulate the pain and the Gain, as well as laying out a plan that minimizes the cost of change. The initial presentation of factors contributing to the change equations is often referred to as the "value proposition" of PPM. The change equation is not intended to just start the implementation. Continually ensuring that the equation remains satisfied also ensures that PPM implementation is on track.

Notice the implication of the change equation for organizations that are doing very well. Sometimes, while experiencing an industry's growth period, managers may perceive changing or installing a PPM process is not necessary. Growth simply erases the pain. Change is also very difficult where an attitude of "we know best" exists. Here, a function or department may see little gain from any new process. In such cases, implementation teams have only one recourse: *to get top management to make the organization feel the pain and see the gain.* This a "Jack Welch" type of leadership approach. When organizations are experiencing good fortune, leaders can demand that the organization hit key performance improvement goals or practice standards. In this case, the organization's leadership induces the pain (fear of under performing) and drives the change.

The change equilibrium plays an important role in spiral-up implementations. Think for a moment about PPM implementation and the cost of change for people or groups to jump from maturity level 1 all the way to maturity level 5. This requires significant change. Such psychological and economic cost would be perceived as very high. The current pain and the perceived gain would have to be intense for the implementation to move forward. Any "push-back" would result in slowing down desired benefit accrual, which in turn would further decrease the perception of potential gains being made. The division of spiral-up implementations into maturity levels helps considerably. At the start, the pain remains high. The potential for gain after the first maturity level may decrease slightly, but this should be offset by a significant reduction in the perceived cost of change. Because of separate maturity levels, the implementation will proceed much more readily. This is particularly true through the difficult early maturity levels. The maturity levels enable teams to keep the change equilibrium tilted more definitely in favor of driving forces.

STARTING A SPIRAL-UP IMPLEMENTATION

The starting point in spiral-up implementation can be different for different organizations. Many organizations already have some PPM components in place. Others have none. To be effective, spiral-up implementations must match the organization. The key is to do only what is necessary, when it is necessary . . . but to do it as quickly and effectively as possible. The following six action steps will help you get started in your organization:

1. The first step is to figure out where your organization is. Do this by involving a cross section of people from the organization in PPM will be deployed. Share the matrix with these individuals. Ask each person to score where on the maturity levels (from 0 to 5, using whatever decimals the person wants between integers) he or she believes the organization to be for each of the 26 components. Arrange the scores in each component grouping and use a simple spider plot (see Figure 17-10) to exhibit each person's score. Next, facilitate a group dialogue on the scores and seek group agreement on where the organization is in maturity levels. In lieu of gaining absolute agreement, should that be the case, seek agreement on why there is no agreement overall. Use the dialogue to identify key issues and opportunities for spiral-up implementation.
2. Revisit the change equilibrium to determine exactly what should be done to get the organization moving. A value proposition must be made to declare the gain that the organization should realize as well as the pain that should be alleviated through PPM. Decide which components need to be addressed and how best to drive the change so that costs (economic and psychological) are minimized.
3. Create the implementation team and the necessary support teams. Try to engage those people who will be using the process or contributing to the process. People are more likely to use and benefit from practices that they help create than those handed to them by others. Recognize,

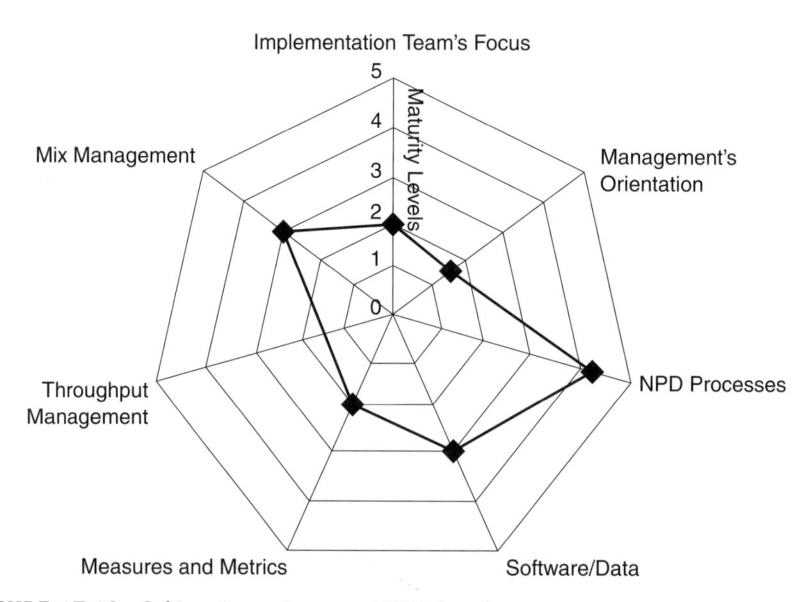

FIGURE 17-10. Spider chart of current PPM situation.

SPIRAL UP IMPLEMENTATION TEAMS

Because PPM is a cross-organizational process, setting it up requires cross-organizational involvement. How such involvement is carried out will also affect how fast benefits accrue. It is much more likely that people will use, embrace, and benefit from practices that they contribute to than practices simply handed to them. A guiding principle in spiral-up implementation, therefore, is that those who will be using or will be affected by PPM should contribute to its deployment.

A basic approach would create three different teams to implement PPM across an organization:

1. *Process implementation team.*
   - At least 15 percent of each contributor's time over duration of implementation.
   - Team members can switch out as maturity levels progress. Continuity of leadership is suggested.
   - At least 33 percent of the team leader's time over duration of implementation.
   - Four to seven members.
   - Members become champions of components.
   - Direct report to top functions managers in the business.
   - Implementation responsibilities include establishing and building consistency-in-use of all components at each maturity level, ensuring favorable change equation throughout implementation.
   - Ongoing responsibilities include managing the PPM flow, conducting portfolio and pipeline analysis work, and recommending specific decision/actions to management.
   - Team sponsor is head of business entity.
2. *Senior management team.*
   - Leader is head of the business entity.
   - All direct reports of the top person, and who have resources to be assigned through process (R&D, manufacturing, marketing, sales, etc.).
   - Includes strategic planning personnel.
   - Includes process implementation team leader.
   - Implementation responsibilities include contributing to, shaping, and specifying decision criteria and decision flow.
   - Ongoing responsibilities include reviewing and making decisions on the mix of projects underway, initiating new projects, and assigning resources to projects.
3. *Project contributors extended team.*
   - Cross-organizational representation.
   - Less than 10 percent of time.
   - Well-respected individuals.
   - Evaluating and providing feedback on PPM decision criteria.
   - Evaluating and providing feedback on PPM components and their deployment.
   - Assisting with training when necessary
   - Implementation responsibilities include providing feedback on metrics, data gathering, and PPM decision criteria.
   - Ongoing responsibilities include updating data and providing feedback to improve data gathering.

though, that time dedicated to PPM implementation is usually time away from other development activities. Consider engaging an outside expert in PPM implementations, and then use that person's experience and knowledge to avoid pitfalls and to gain benefits faster. Sometimes it is easier to rent experience than to develop it.

4. Create a six-month plan (think Gantt chart, with assigned resources) and a 24-month plan. Create a forum for all key parties, including top management, to reshape and add value to the plan. Their contribution will help decrease the cost–of change and secure their buy-in. Make sure that the plan demands that top management is involved in carrying out spiral-up implementation, not just giving it their blessing.
5. Carry out the plan. Return to the spider chart periodically to measure your progress. Make sure that your efforts do, in fact, spiral upward.

The most common mistake in starting up PPM is not giving enough attention to defining and articulating the value proposition. This is the starting point of the change equation, and if management does not believe it, the initiative will simply stop. Make sure that management will accept the case for change before it is presented. Build, test, and amend the value proposition before you present it. Management's buy-in is prerequisite to the entire initiative.

SUMMARY

There is much written on portfolio and pipeline management in new product development. The benefits of PPM can be tremendous. But putting PPM process in place is not easy. The spiral-up implementation of PPM helps organizations overcome the complexity of PPM implementations and achieve economic benefits faster. It provides the blueprint and the framework for the initiative.

Spiral-up implementation recognizes that an organization's PPM capabilities must mature over time. Some component of PPM must be in place and used consistently across the organization before other components can provide benefit. Spiral-up implementation categorizes these practices into five maturity levels, each providing a milestone for gaining benefits from PPM.

How an organization goes about putting PPM does make a difference. Spiral-up implementation provides organizations with a means to secure economic benefits from PPM quickly.

NOTE

1. Peak annual revenue is an estimate of the revenue during the product's peak year in the market. It is dependent upon the expected life cycle of the product.

BIBLIOGRAPHY

Adept Group Limited, Inc. 2002. Year 2002 research.

Cooper, Robert G. 1985. "Selecting Winning New Product Project: Using the NewProd System." *Journal of Product Innovation Management iss:2 pgs:34–44*

Edgett, S, Kleinschmidt, and E. Cooper, R, Portfolio Management for New Products, 2001 2nd edition, Perseus Publishing) Cooper, R, and S. Edgett. 2002. *Portfolio Management: Fundamentals to New Product Success.* Chapter 13 in *The PDMA ToolBook for New Product Development,* ed. Paul Belliveau, Abbie Griffin, and Stephen Somermeyer. New York: John Wiley & Sons.

Hammer, Michael, and James Champy. 1994. *Reengineering the Corporation: A Manifesto for Business Revolution.* New York: HarperBusiness.

Jonash, R., and T. Sommerlatte, 2000. *The Innovation Premium: How Next Generation Companies Are Achieving Peak Performance and Profitability.* Cambridge, MA: Perseus Publishing.

Koen, P. 2002. Fuzzy Front End: Effective Methods, Tools, Techniques Chapter 1 in *The PDMA ToolBook for New Product Development,* ed. Paul Belliveau, Abbie Griffin, and Stephen Somermeyer. New York: John Wiley & Sons.

Kotter, John P. 1996. *Leading Change.* Boston: Harvard Business School Press.

Lewin, K. Research Center on Group Dynamics at Massachusetts's Institute of Technology (M.I.T.) 1947. O'Connor, P. 1994. "Implementing a Stage-Gate Process: A Multi-Company Perspective." *Journal of Product Innovation Management.* 1994: vol 11 iss 3 June, pp. 183-Paulk, M. 1993. Capability Maturity Model for Software, Version 1.1. Pittsburgh: Carnegie Mellon University, Software Engineering Institute.

Product Development and Management Association and The Institute of International Research, Conference proceedings "Pragmatic Portfolio Management for Product Development" Scottsdale, AZ Feb. 2002; Management Roundtable Conference proceedings "Pipeline and Portfolio Management: Balancing Multiple Projects with Limited Resources" Chicago, IL Sept 2001. Repenning, N. 2001. *Past the Tipping Point: The Perspective of Firefighting in Product Development.* Cambridge, MA: Sloan School of Management.

Smith, P., and D. Reinertsen. 1998. *Developing Products in Half the Time: New Rules,* New Tools. John Wiley & Sons.

Tichy, Noel M., and Stratford Sherman. 1999. *Control Your Destiny or Someone Else Will: Lessons in Mastering Change-From the Principles Jack Welch Is Using to Revolutionize GE.* New York: HarperBusiness.

The PDMA Glossary for New Product Development

© Product Development & Management Association, 2004. Reprinted with permission.

Accidental Discovery: New designs, ideas, and developments resulting from unexpected insight, which can be obtained either internal or external to the organization.

Adoption Curve: The phases through which consumers or a market proceed in deciding to adopt a new product or technology. At the individual level, each consumer must move from a cognitive state (becoming aware of and knowledgeable about) to an emotional state (liking and then preferring the product) and into a conative, or behavioral state (deciding and then purchasing the product). At the market level, the new product is first purchased by the innovators in the marketplace, which are generally thought to constitute about 2.5 percent of the market. Early adopters (13.5 percent of the market) are the next to purchase, followed by the early majority (34 percent), late majority (34 percent) and finally, the laggards (16 percent).

Affinity Charting: A "bottom-up" technique for discovering connections between pieces of data. An individual or group starts with one piece of data (say, a customer need). They then look through the rest of the data they have (say, statements of other customer needs) to find other data (needs) similar to the first, and place it in the same group. As they come across pieces of data that differ from those in the first group, they create a new category. The end result is a set of groups where the data contained within a category is similar and the groups all differ in some way. (*See also* Qualitative Cluster Analysis.)

Alliance: Formal arrangement with a separate company for purposes of development, and involving exchange of information, hardware, intellectual property, or enabling technology. Alliances involve shared risk and reward (e.g., co-development projects). (*See also* Chapter 11 of *The PDMA HandBook,* 2nd Edition.)

Alpha Test: Pre-production product testing to find and eliminate the most obvious design defects or deficiencies, usually in a laboratory setting or in some part of the developing firm's regular operations, although in some cases it may be done in controlled settings with lead customers. (*See also* Beta Test and Gamma Test.)

Alpha Testing: A crucial "first look" at the initial design, usually done in-house. The results of the alpha test either confirm that the product performs according to its specifications or uncovers areas where the product is deficient. The testing environment should try to simulate the conditions under which the product will actually be used as closely as possible. The alpha test should not be performed by the same people who are doing the development work. Since this is the first "flight" for the new product, basic questions of fit and function should be evaluated. Any suggested modifications or revisions to the specifications should be solicited from all parties involved in the evaluation and considered for inclusion. Since the testing is done in-house, special care must be taken to remain as objective as possible.

Analytical Hierarchy Process (AHP): A decision-making tool for complex, multi-criteria problems where both qualitative and quantitative aspects of a problem need to be incorporated. AHP clusters decision elements according to their common characteristics into a hierarchical structure similar to a family tree or affinity chart. The AHP process was designed by T. L. Saaty.

Analyzer: A firm that follows an imitative innovation strategy, where the goal is to get to market with an equivalent or slightly better product very quickly once someone else opens up the market, rather than to be first to market with new products or technologies. Sometimes called an imitator or a "fast follower."

Anticipatory Failure Determination (AFD): A failure analysis method. In this process, developers start from a particular failure of interest as the intended consequence and try to devise ways to ensure that the failure always happens reliably. Then the developers use that information to develop ways to better identify steps to avoid the failure.

Applications Development: The iterative process through which software is designed and written to meet the needs and requirements of the user base, or the process of enhancing or developing new products.

Architecture: *See* Product Architecture.

Asynchronous Groupware: Software used to help people work as groups, but not requiring those people to work at the same time.

Attribute Testing: A quantitative market research technique in which respondents are asked to rate a detailed list of product or category attributes on one or more types of scales, such as relative importance, current performance, and current satisfaction with a particular product or service, for the purpose of ascertaining customer preferences for some attributes over others, to help guide the design and development process. Great care and rigor should be taken in the development of the list of attributes, and it must be neither too long for the respondent to answer comfortably or too short such that it lumps too many ideas together at too high a level.

Audit: When applied to new product development, an audit is an appraisal of the effectiveness of the processes by which the new product was developed and brought to market. (*See* Chapter 14 of *The PDMA ToolBook 1.*)

Augmented Product: The core product, plus all other sources of product benefits, such as service, warranty, and image.

Autonomous Team: A completely self-sufficient project team with very little, if any, link to the funding organization. Frequently used as an organizational model to bring a radical innovation to the marketplace. Sometimes called a "tiger" team.

Awareness: A measure of the percent of target customers who are aware that the new product exists. Awareness is variously defined, including recall of brand, recognition of brand, recall of key features, or positioning.

Back-up: A project that moves forward, either in synchrony or with a moderate time lag, and for the same marketplace as the lead project to provide an alternative asset should the lead project fail in development. A back-up has essentially the same mechanism of action performance as the lead project. Normally a company would not advance both the lead and the back-up project through to the market place, since they would compete directly with each other.

Balanced Scorecard: A comprehensive performance measurement technique that balances four performance dimensions: 1. Customer perceptions of how we are performing; 2. Internal perceptions of how we are doing at what we must excel at; 3. Innovation and learning performance; 4. Financial performance.

Benchmarking: A process of collecting process performance data, generally in a confidential, blinded fashion, from a number of organizations to allow them to assess their performance individually and as a whole.

Benefit: A product attribute expressed in terms of what the user gets from the product rather than its physical characteristics or features. Benefits are often paired with specific features, but they need not be.

Best Practice: Methods, tools, or techniques that are associated with improved performance. In new product development, no one tool or technique ensures success; however, a number of them are associated with higher probabilities of achieving success. Best practices likely are at least somewhat context-specific. Sometimes called "effective practice."

Best Practice Study: A process of studying successful organizations and selecting the best of their actions or processes for emulation. In new product development it means finding the best process practices, adapting them and adopting them for internal use. (*See* Chapters 33 and 36, in the *PDMA HandBook*, 2nd Edition; *see also* Griffin, "PDMA Research on

New Product Development Practices: Updating Trends and Benchmarking Best Practices," *JPIM* 14:6, 429–458, November 1997, and "Drivers of NPD Success: The 1997 PDMA Report," PDMA, October 1997.)

Beta Test: An external test of preproduction products. The purpose is to test the product for all functions in a breadth of field situations to find those system faults that are more likely to show in actual use than in the firm's more controlled in-house tests before sale to the general market. (*See also* Field Test.)

Beta Testing: A more extensive test than the alpha, performed by real users and customers. The purpose of beta testing is to determine how the product performs in an actual user environment. It is critical that real customers perform this evaluation, not the firm developing the product or a contracted testing company. As with the alpha test, results of the beta test should be carefully evaluated with an eye toward any needed modifications or corrections.

Bowling Alley: An early growth stage strategy that emphasizes focusing on specific niche markets, building a strong position in those markets by delivering clearly differentiated "whole products," and using that niche market strength as a leverage point for conquering conceptually neighboring niche markets. Success in the bowling alley is predicated on building product leadership via customer intimacy.

Brainstorming: A group method of creative problem solving frequently used in product concept generation. There are many modifications in format, each variation with its own name. The basis of all of these methods uses a group of people to creatively generate a list of ideas related to a particular topic. As many ideas as possible are listed before any critical evaluation is performed. (*See* Chapters 16 and 17 in *The PDMA HandBook*, 2nd Edition.)

Brand: A name, term, design, symbol, or any other feature that identifies one seller's good or service as distinct from those of other sellers. The legal term for brand is *trademark*. A brand may identify one item, a family of items, or all items of that seller.

Brand Development Index (BDI): A measure of the relative strength of a brand's sales in a geographic area. Computationally, BDI is the percent of total national brand sales that occur in an area divided by the percent of U.S. households that reside in that area.

Breadboard: A proof-of-concept modeling technique that represents how a product will work, but not how a product will look.

Break-even Point: The point in the commercial life of a product when cumulative development costs are recovered through accrued profits from sales.

Business Analysis: An analysis of the business situation surrounding a proposed project. Usually includes financial forecasts in terms of discounted cash flows, net present values, or internal rates of returns.

Business Case: The results of the market, technical, and financial analyses, or up-front homework. Ideally defined just prior to the "go to development" decision (gate), the case defines the product and project, including the project justification and the action or business plan. (*See* Chapter 21 of *The PDMA HandBook,* 2nd Edition.)

Business-to-Business: Transactions with non-consumer purchasers such as manufacturers, resellers (distributors, wholesalers, jobbers, and retailers, for example), and institutional, professional, and governmental organizations. Frequently referred to as "industrial" businesses in the past.

Buyer: The purchaser of a product, whether or not he or she will be the ultimate user. Especially in business-to-business markets, a purchasing agent may contract for the actual purchase of a good or service, yet never benefit from the function(s) purchased.

Buyer Concentration: The degree to which purchasing power is held by a relatively small percentage of the total number of buyers in the market.

Cannibalization: That portion of the demand for a new product that comes from the erosion of the demand for (sales of) a current product the firm markets (*See* Chapter 34 in *The PDMA HandBook,* 2nd Edition.)

Capacity Planning: A forward-looking activity that monitors the skill sets and effective resource capacity of the organization. For product development, the objective is to manage the flow of projects through development such that none of the functions (skill sets) creates a bottleneck to timely completion. Necessary in optimizing the project portfolio.

Centers of Excellence: A geographic or organizational group with an acknowledged technical, business, or competitive competency.

Certification: A process for formally acknowledging that someone has mastered a body of knowledge on a subject. In new product development, the PDMA has created and manages a certification process to become a New Product Development Professional (NPDP).

Champion: A person who takes a passionate interest in seeing that a particular process or product is fully developed and marketed. This informal role varies from situations calling for little more than stimulating awareness of the opportunity to extreme cases where the champion tries to force a project past the strongly entrenched internal resistance of company policy or that of objecting parties. (*See* Chapter 5 in *The PDMA ToolBook 1.*)

Charter: A project team document defining the context, specific details, and plans of a project. It includes the initial business case, problem and goal statements, constraints and assumptions, and preliminary plan and scope. Periodic reviews with the sponsor ensure alignment with business strategies. (*See* also *Product Innovation Charter.*)

Checklist: A list of items used to remind an analyst to think of all relevant aspects. It finds frequent use as a tool of creativity in concept generation, as a factor consideration list in concept screening, and to ensure that all appropriate tasks have been completed in any stage of the product development process.

Chunks: The building blocks of product architecture. They are made up of inseparable physical elements. Other terms for chunks may be modules or major subassemblies.

Clockspeed: The evolution rate of different industries. High-clockspeed industries, like electronics, see multiple generations of products within short time periods, perhaps even within 12 months. In low-clockspeed industries, like the chemical industry, a generation of products may last as long as five or even ten years. It is believed that high-clockspeed industries can be used to understand the dynamics of change that will in the long run affect all industries, much like fruit flies are used to understand the dynamics of genetic change in a speeded-up genetic environment, because of their short life spans.

Cognitive Modeling: A method for producing a computational model for how individuals solve problems and perform tasks, which is based on psychological principles. The modeling process outlines the steps a person goes through in solving a particular problem or completing a task, which allows one to predict the time it will take or the types of errors an individual may make. Cognitive models are frequently used to determine ways to improve a user interface to minimize interaction errors or time by anticipating user behavior.

Cognitive Walk-through: Once a model of the steps or tasks a person must go through to complete a task is constructed, an expert can role play the part of a user to cognitively "walk through" the user's expected experience. Results from this walk-through can help make human-product interfaces more intuitive and increase product usability.

Collaborative Product Development: When two firms work together to develop and commercialize a specialized product. The smaller firm may contribute technical or creative expertise, while the larger firm may be more likely to contribute capital, marketing, and distribution capabilities. When two firms of more equal size collaborate, they may each bring some specialized technology capability to the table in developing some highly complex product or system requiring expertise in both technologies. Collaborative product development has several variations. In customer collaboration, a supplier reaches out and partners with a key or lead customer. In supplier collaboration, a company partners with the provider(s) of technologies, components, or services to create an integrated solution. In collaborative contract manufacturing, a company contracts with a manufacturing partner to produce the intended product. Collaborative development (also known as *codevelopment*) differs from simple outsourcing in its levels of depth of partnership in that the collaborative firms are linked in the process of delivering the final solution to the intended customer.

Co-location: Physically locating project personnel in one area, enabling more rapid and frequent decision-making and communication among them.

Commercialization: The process of taking a new product from development to market. It generally includes production launch and ramp-up, marketing materials and program development, supply chain development, sales channel development, training development, training, and service and support development (*See* Chapter 30 of *The PDMA HandBook*, 2nd Edition.)

Competitive Intelligence: Methods and activities for transforming disaggregated public competitor information into relevant and strategic knowledge about competitors' position, size, efforts, and trends. The term refers to the broad practice of collecting, analyzing, and communicating the best available information on competitive trends occurring outside one's own company.

Computer-Aided Engineering (CAE): Using computers in designing, analyzing, and manufacturing a product or process. Sometimes refers more narrowly to using computers just at the engineering analysis stage.

Computer-Aided Design (CAD): A technology that allows designers and engineers to use computers for their design work. Early programs enabled two-dimensional (2-D) design. Current programs allow designers to work in 3-D (three dimensions), and in either wire or solid models.

Computer-Enhanced Creativity: Using specially designed computer software that aids in the process of recording, recalling, and reconstructing ideas to speed up the new product development process.

Concept: A clearly written and possibly visual description of the new product idea that includes its primary features and consumer benefits, combined with a broad understanding of the technology needed.

Concept Generation: The processes by which new concepts, or product ideas, are generated. Sometimes also called *idea generation* or *ideation.* (*See* Chapters 15 and 17 in *The PDMA HandBook,* 2nd Edition.)

Concept Optimization: A research approach that evaluates how specific product benefits or features contribute to a concept's overall appeal to consumers. Results are used to select from the options investigated to construct the most appealing concept from the consumer's perspective.

Concept Statement: A verbal or pictorial statement of a concept that is prepared for presentation to consumers to get their reaction prior to development.

Concept Study Activity: The set of product development tasks in which a concept is given enough examination to determine if there are substantial unknowns about the market, technology, or production process.

Concept Screening: The evaluation of potential new product concepts during the discovery phase of a product development project. Potential concepts are evaluated for their fit with business strategy, technical feasibility, manufacturability, and potential for financial success.

Concept Testing: The process by which a concept statement is presented to consumers for their reactions. These reactions can either be used to permit the developer to estimate the sales value of the concept or to make changes to the concept to enhance its potential sales value. (*See* Chapter 6 in *The PDMA HandBook,* 2nd Edition.)

Concurrency: Carrying out separate activities of the product development process at the same time rather than sequentially.

Concurrent Engineering (CE): When product design and manufacturing process development occur concurrently in an integrated fashion, using a cross-functional team, rather than sequentially by separate functions. CE is intended to cause the development team to consider all elements of the product life cycle from conception through disposal—including quality, cost, and maintenance—from the project's outset. Also called *simultaneous engineering.* (*See* Chapter 30 of *The PDMA HandBook,* 1st Edition.)

Conjoint Analysis A market research technique in which respondents are systematically presented with a rotating set of product descriptions, each of which contains a rotating set of attributes and levels of those attributes. By asking respondents to choose their preferred product and/or to indicate their degree of preference from within each set of options, conjoint analysis can determine the relative contribution to overall preference of each attribute and each level. The two key advantages of conjoint analysis over other methods of determining importance are (1) the variables and levels can be either continuous (e.g., weight) or discrete (e.g. color), and (2) it is just about the only valid market research method for evaluating the role of price—that is, how much someone would pay for a given feature (*See* Chapter 18 of *The PDMA HandBook,* 2nd Edition.)

Consumer: The most generic and all-encompassing term for a firm's targets. The term is used in either the business-to-business or household context and may refer to the firm's current customers, competitors' customers, or current non-purchasers with similar needs or demographic characteristics. The term does not differentiate between whether the person is a buyer or a user target. Only a fraction of consumers will become customers.

Consumer Market: The purchasing of goods and services by individuals and for household use (rather than for use in business settings). Consumer purchases are generally made by individual decision makers, either for themselves or for others in the family.

Consumer Need: A problem the consumer would like to have solved; what a consumer would like a product to do for them.

Consumer Panels: Specially recruited groups of consumers whose longitudinal category purchases are recorded via the scanner systems at stores.

Contextual Inquiry: A structured qualitative market research method that uses a combination of techniques from anthropology and journalism. Contextual inquiry is a customer needs discovery process that observes and interviews users of products in their actual environment.

Contingency Plan: A plan to cope with events whose occurrence, timing, and severity cannot be predicted.

Continuous Improvement: The review, analysis, and rework directed at incrementally improving practices and processes. Also called *Kaizen*.

Continuous Innovation: A product alteration that allows improved performance and benefits without changing either consumption patterns or behavior. The product's general appearance and basic performance do not functionally change. Examples include fluoride toothpaste and higher computer speeds.

Continuous Learning Activity: The set of activities involving an objective examination of how a product development project is progressing or how it was carried out to permit process changes to simplify its remaining steps or improve the product being developed or its schedule. (*See also* Learning Organization.)

Contract Developer: An external provider of product development services.

Controlled Store Testing: A method of test marketing where specialized companies are employed to handle product distribution and auditing rather than using the company's normal sales force.

Convergent Thinking: A technique generally performed late in the initial phase of idea generation to help funnel the high volume of ideas created through divergent thinking into a small group or single idea on which more effort and analysis will be focused.

Cooperation (Team Cooperation): The extent to which team members actively work together in reaching team-level objectives.

Coordination Matrix: A summary chart that identifies the key stages of a development project, the goals and key activities within each stage, and who (what function) is responsible for each.

Core Benefit Proposition (CBP): The central benefit or purpose for which a consumer buys a product. The CBP may come either from the physical good or service or it may come from augmented dimensions of the product. (*See also* Value Proposition.) (*See* Chapter 3 of *The PDMA ToolBook 1*.)

Core Competence: That capability a company does better than other firms, which provides them with a distinctive competitive advantage and contributes to acquiring and retaining customers. Something that a firm does better than other firms. Can include technical, organizational, supply chain, operational, financial, marketing, partnership, or other capabilities. The purest definition adds "and is also the lowest cost provider."

Corporate Culture: The "feel" of an organization. Culture arises from the belief system through which an organization operates. Corporate cultures are variously described as being authoritative, bureaucratic, and entrepreneurial. The firm's culture frequently impacts the organizational appropriateness for getting things done.

Cost of Goods Sold (COGS or CGS): The direct costs (labor and materials) associated with producing a product and delivering it to the marketplace.

Creativity: "An arbitrary harmony, an expected astonishment, a habitual revelation, a familiar surprise, a generous selfishness, an unexpected certainty, a formable stubbornness, a vital triviality, a disciplined freedom, an intoxicating steadiness, a repeated initiation, a difficult delight, a predictable gamble, an ephemeral solidity, a unifying difference, a demanding satisfier, a miraculous expectation, and accustomed amazement" (George M. Prince, *The Practice of Creativity*, 1970). Creativity is the ability to produce work that is both novel and appropriate.

Criteria: Statements of standards used by decision makers at decision gates. The dimensions of performance necessary to achieve or surpass for product development projects to continue in development. In the aggregate, these criteria reflect a business unit's new product strategy. (*See* Chapters 21 and 29 of *The PDMA Handbook*, 2nd Edition.)

Critical Assumption: An explicit or implicit assumption in the new product business case that, if wrong, could undermine the viability of the opportunity.

Critical Path: The set of interrelated activities that must be completed for the project to be finished successfully can be mapped into a chart showing how long each task takes, and which tasks cannot be started before which other tasks are completed. The critical path is the set of linkages through the chart that is the longest. It determines how long a project will take.

Critical Path Scheduling: A project management technique, frequently incorporated in various software programs, which puts all important steps of a given new product project into a sequential network based on task interdependencies.

Critical Success Factors: Those critical few factors that are necessary for, but don't guarantee, commercial success. (*See* Chapter 1 of *The PDMA Hand-Book,* 2nd Edition.)

Cross-Functional Team: A team consisting of representatives from the various functions involved in product development, usually including members from all key functions required to deliver a successful product, typically including marketing, engineering, manufacturing/operations, finance, purchasing, customer support, and quality. The team is empowered by the departments to represent each function's perspective in the development process. (*See* Chapters 9 and 10 in *The PDMA HandBook,* 2nd Edition and Chapter 6 in *The PDMA ToolBook 1.*)

Crossing the Chasm: Making the transition to a mainstream market from an early market dominated by a few visionary customers (sometimes also called *innovators* or *lead adopters*). This concept typically applies to the adoption of new market creating technology-based products and services.

Customer: One who purchases or uses a firm's products or services.

Customer-based Success: The extent to which a new product is accepted by customers and the trade.

Customer Needs: Problems to be solved. These needs, either expressed or yet to be articulated, provide new product development opportunities for the firm. (*See* Chapter 14 in *The PDMA HandBook,* 2nd Edition.)

Customer Perceived Value (CPV): The result of the customer's evaluation of all the benefits and all the costs of an offering as compared to that customer's perceived alternative. It is the basis on which customers decide to buy things. (*See* Chapter 4 of *The PDMA ToolBook 1.*)

Customer Site Visits: A qualitative market research technique for uncovering customer needs. The method involves going to a customer's work site, watching as a person performs functions associated with the customer needs your firm wants to solve, and then debriefing that person about what he or she did and why, the problems encountered as the customer was trying to perform the function, and what worked well. (*See* Chapters 15 and 16 of *The PDMA HandBook,* 2nd Edition.)

Customer Value-Added Ratio: The ratio of WWPF (worth what paid for) for your products to WWPF for your competitors' products. A ratio above 1 indicates superior value compared to your competitors'.

Cycle Time: The length of time for any operation, from start to completion. In the new product development sense, it is the length of time to develop a new product from an early initial idea for a new product to initial market sales. Precise definitions of the start and end point vary from one company to another, and may vary from one project to another within the company. (*See* Chapter 12 of *The PDMA HandBook,* 2nd Edition).

Dashboard: A colored graphical presentation of a project's status or a portfolio's status by project, so called because it resembles a vehicle's dashboard. Typically, red is used to flag urgent problems, yellow to flag impending problems, and green to signal projects on track.

Database: An electronic gathering of information organized in some way to make it easy to search, discover, analyze, and manipulate.

Decision Screens: Sets of criteria that are applied as checklists or screens at new product decision points. The criteria may vary by stage in the process. (*See* Chapter 7 in *The PDMA ToolBook 1* and Chapter 21 of *The PDMA HandBook,* 2nd Edition).

Decision Tree: A diagram used for making decisions in business or computer programming. The "branches" of the tree diagram represent choices, with associated risks, costs, results, and outcome probabilities. By calculating outcomes (profits) for each of the branches, the firm can determine the best decision.

Decline Stage: The fourth and last stage of the product life cycle. Entry into this stage is generally caused by technology advancements, consumer or user preference changes, global competition, or environmental or regulatory changes (*See* Chapter 34 of *The PDMA HandBook,* 2nd Edition).

Defenders: Firms that stake out a product's turf and protect it by whatever means, not necessarily through developing new products.

Deliverable: The output (such as test reports, regulatory approvals, working prototypes, or marketing research reports) that shows a project has achieved a result. Deliverables may be specified for the commercial launch of the product or at the end of a development stage.

Delphi Processes: A technique that uses iterative rounds of consensus development across a group of experts to arrive at a forecast of the most probable outcome for some future state.

Demographic: The statistical description of a human population. Characteristics included in the description may include gender, age, education level, and marital status, as well as various behavioral and psychological characteristics.

Derivative Product: A new product based on changes to an existing product that modifies, refines, or improves some product features without affecting the basic product architecture or platform.

Design for the Environment (DFE): The systematic consideration of environmental safety and health issues over the product's projected life cycle in the design and development process.

Design for Excellence (DFX): The systematic consideration of *all* relevant life cycle factors, such as manufacturability, reliability, maintainability, affordability, testability, and so on, in the design and development process.

Design for Maintainability (DFMt): The systematic consideration of maintainability issues over the product's projected life cycle in the design and development process.

Design for Manufacturability (DFM): The systematic consideration of manufacturing issues in the design and development process, facilitating the fabrication of the product's components and their assembly into the overall product.

Design of Experiments (DOE): A statistical method for evaluating multiple product and process design parameters simultaneously rather than one parameter at a time.

Design to Cost: A development methodology that treats costs as an independent design parameter, rather than an outcome. Cost objectives are established based on customer affordability and competitive constraints.

Design Validation: Product tests to ensure that the product or service conforms to defined user needs and requirements. These may be performed on working prototypes or using computer simulations of the finished product.

Development: The functional part of the organization responsible for converting product requirements into a working product. Also, a phrase in the overall concept-to-market cycle where the new product or service is developed for the first time.

Development Change Order (DCO): A document used to implement changes during product development. It spells out the desired change, the reason for the change and the consequences to time-to-market, development cost, and the cost of producing the final product. It gets attached to the project's charter as an addendum.

Development Teams: Teams formed to take one or more new products from concept through development, testing, and launch.

Digital Mock-Up: An electronic model of the product created with a solids modeling program. Mock-ups can be used to check for interface interferences and component incompatibilities. Using a digital mock-up can be less expensive than building physical prototypes.

Discontinuous Innovation: Previously unknown products that establish new consumption patterns and behavior changes. Examples include microwave ovens and cellular phones.

Discounted Cash Flow (DCF) Analysis: One method for providing an estimate of the current value of future incomes and expenses projected for a project. Future cash flows for a number of years are estimated for the project and then discounted back to the present using forecast interest rates.

Discrete Choice Experiment: A quantitative market research tool used to model and predict customer buying decisions.

Dispersed Teams: Product development teams that have members working at different locations, across time zones, and perhaps even in different countries.

Distribution: The method and partners used to get the product (or service) from where it is produced to where the end user can buy it.

Divergent Thinking: Technique performed early in the initial phase of idea generation that expands thinking processes to generate, record, and recall a high volume of new or interesting ideas.

Dynamically Continuous Innovation: A new product that changes behavior but not necessarily consumption patterns. Examples include personal digital assistants (PDAs), electric toothbrushes, and electric haircurlers.

Early Adopters: For new products, these are customers who, relying on their own intuition and vision, buy into new product concepts very early in the life cycle. For new processes, these are organizational entities that were willing to try out new processes rather than just maintaining the old.

Economic Value Added (EVA): The value added to or subtracted from shareholder value during the life of a project.

Empathic Design: A five-step method for uncovering customer needs and sparking ideas for new concepts. The method involves going to a customer's work site, watching as he or she performs functions associated with the customer needs your firm wants to solve, and then debriefing the customer about what he or she did and why, the problems the customer encountered when trying to perform the function, and what worked well. By spending time with customers, the team develops empathy for the problems customers encounter trying to perform their daily tasks. (*See also* Customer Site Visits.)

Engineering Design: A function in the product creation process where a good or service is configured and specific form is decided.

Engineering Model: The combination of hardware and software intended to demonstrate the simulated functioning of the intended product as currently designed.

Enhanced New Product: A form of derivative product. Enhanced products include additional features not previously found on the base platform, which provide increased value to consumers.

Entrance Requirement: The document(s) and reviews required before any phase of a stages and gates development process can be started. (*See* Chapter 7 of *The PDMA ToolBook 1*.)

Entrepreneur: A person who initiates, organizes, operates, assumes the risk, and reaps the potential reward for a new business venture.

Ethnography: A descriptive, qualitative market research methodology for studying the customer in relation to his or her environment. Researchers spend time in the field observing customers and their environment to acquire a deep understanding of the lifestyles or cultures as a basis for better understanding their needs and problems. (*See* Customer Site Visits and Chapter 15 in *The PDMA HandBook*, 2nd Edition.)

Event: Marks the point in time when a task is completed.

Event Map: A chart showing important events in the future that is used to map out potential responses to probable or certain future events.

Excursion: An idea generation technique to force discontinuities into the idea set. Excursions consist of three generic steps: (1) step away from the task, (2) generate disconnected or irrelevant material, and (3) force a connection back to the task.

Exit Requirement: The document(s) and reviews required to complete a stage of a stages and gates development process. (*See* Chapter 7 of *The PDMA ToolBook 1* and Chapter 21 of *The PDMA HandBook*, 2nd Edition.)

Exit Strategy: A preplanned process for deleting a product or product line from the firm's portfolio. At its most basic level, it includes plans for clearing inventory out of the supply chain pipeline with a minimum of losses, continuing to provide for after-sales parts supply and maintenance support, and converting customers of the deleted product line to a different one. (*See* Chapter 34 of *The PDMA HandBook,* 2nd Edition.)

Explicit Customer Requirement: What the customer asks for in a product.

Factory Cost: The cost of producing the product in the production location, including materials, labor, and overhead.

Failure Mode Effects Analysis (FMEA): A technique used at the development stage to determine the different ways in which a product may fail, and evaluating the consequences of each type of failure.

Failure Rate: The percentage of a firm's new products that make it to full market commercialization but that fail to achieve the objectives set for them.

Feasibility Determination: The set of product development tasks in which major unknowns (technical or market) are examined to produce knowledge about how to resolve or overcome them or to clarify the nature of any limitations. Sometimes called *exploratory investigation.*

Feature: The solution to a consumer need or problem. Features provide benefits to consumers—for instance, the handle (feature) allows a laptop computer to be carried easily (benefit). Usually any one of several different features will be chosen to meet a customer need. For example, a carrying case with shoulder straps is another feature that allows a laptop computer to be carried easily.

Feature Creep: The tendency for designers or engineers to add more capability, functions, and features to a product as development proceeds than were originally intended. These additions frequently cause schedule slip, development cost increases, and product cost increases.

Feature Roadmap: The evolution over time of the performance attributes associated with a product. Defines the specific features associated with each iteration/generation of a product over its lifetime, grouped into releases (sets of features that are commercialized). *See also* Product Life Cycle Management.

Field Testing: Product use testing with users from the target market in the actual context in which the product will be used.

Financial Success: The extent to which a new product meets its profit, margin, and return-on-investment goals.

Firefighting: An unplanned diversion of scarce resources, and the reassignment of some of them to fix problems discovered late in a product's development cycle. (*See* N. Repenning, *JPIM*, September 2001.)

Firm-Level Success: The aggregate impact of the firm's proficiency at developing and commercializing new products. Several different specific measures may be used to estimate performance. (*See* Chapter 36 in *The PDMA HandBook,* 2nd Edition.)

First-to-Market: The first product to create a new product category or a substantial subdivision of a category.

Flexible Gate: A permissive or permeable gate in a Stage-Gate™ process that is less rigid than the traditional "go-stop-recycle" gate. Flexible gates are useful in shortening time to market. A permissive gate is one where the next stage is authorized although some work in the almost-completed stage has not yet been finished. In a permeable gate, some work in a subsequent stage is authorized before a substantial amount of work in the prior stage is completed. (See Robert G. Cooper, *JPIM*, 1994.)

Focus Groups: A qualitative market research technique where 8 to 12 market participants are gathered in one room for a discussion under the leadership of a trained moderator. Discussion focuses on a consumer problem, product, or potential solution to a problem. The results of these discussions are not projectable to the general market.

Forecast: A prediction, over some defined time, of the success or failure of implementing a business plan's decisions derived from an existing strategy. (*See* Chapter 23 of *The PDMA HandBook,* 2nd Edition.)

Function: (1) An abstracted description of work that a product must perform to meet customer needs. A function is something the product or service must do. (2) Term describing an internal group within which resides a basic business capability, such as engineering.

Functional Elements: The individual operations that a product performs. These elements are often used to describe a product schematically.

Functional Pipeline Management: Optimizing the flow of projects through all functional areas in the context of the company's priorities.

Functional Reviews: A technical evaluation of the product and the development process from a functional perspective (such as mechanical engineering or manufacturing), in which a group of experts and peers review the product design in detail to identify weaknesses, incorporate lessons learned from past products, and make decisions about the direction of the design going forward. The technical community may perform a single review that evaluates the design from all perspectives, or individual functional departments may conduct independent reviews.

Functional Schematic: A schematic drawing that is made up of all of the functional elements in a product. It shows the product's functions, as well as how material, energy, and signal flow through the product.

Functional Testing: Testing either an element of or the complete product to determine whether it will function as planned and as actually used when sold.

Fuzzy Front End (FFE): The messy "getting started" period of product development, when the product concept is still very fuzzy. Preceding the more formal product development process, it generally consists of three tasks: strategic planning, concept generation, and, especially, pre-technical evaluation. These activities are often chaotic, unpredictable, and unstructured. In comparison, the subsequent new product development process is typically structured, predictable, and formal, with prescribed sets of activities, questions to be answered, and decisions to be made. (*See* Chapter 6 of *The PDMA HandBook,* 2nd Edition.)

Fuzzy Gates: Fuzzy gates are conditional or situational, rather than full "go" decisions. Their purpose is to try to balance timely decisions and risk management. Conditional go decisions are "go," subject to a task being successfully completed by a future, but specified, date. Situational gates have some criteria that must be met for all projects, and others that are only required for some projects. For example, a new-to-the world product may have distribution feasibility criteria that a line extension will not have. (See R.G. Cooper, *JPIM*, 1994.) (*See also* Flexible Gates.)

Gamma Test: A product use test in which the developers measure the extent to which the item meets the needs of the target customers, solves the problems(s) targeted during development, and leaves the customer satisfied.

Gamma/In-Market Testing: Not to be confused with *test marketing* (which is an overall determination of marketability and financial viability), the in-market test is an evaluation of the product itself and its marketing plan through placement of the product in a field setting. Another way of thinking about this is to view it as an in-market test using a real distribution channel in a constrained geographic area or two, for a specific period of time, with advertising, promotion, and all associated elements of the marketing plan working. In addition to an evaluation of the features and benefits of the product, the components of the marketing plan are tested in a real-world environment to make sure they deliver the desired results. The key element being evaluated is the synergy of the product and the marketing plan, not the individual components. The market test should deliver a more accurate forecast of dollar and unit sales volume, as opposed to the approximate range estimates produced earlier in the discovery phase. It should also produce diagnostic information on any facet of the proposed launch that may need adjustment, be it product, communications, packaging, positioning, or any other element of the launch plan. (*See also* Test Marketing.)

Gantt Chart: A horizontal bar chart used in project scheduling and management that shows the start date, end date, and duration of tasks within the project.

Gap Analysis: The difference between projected outcomes and desired outcomes. In product development, the gap is frequently measured as the difference between expected and desired revenues or profits from currently planned new products if the corporation is to meet its objectives.

Gate: The point at which a management decision is made to allow the product development project to proceed to the next stage, to recycle back into the current stage to better complete some of the tasks, or to terminate. The number of gates varies by company. (*See* Chapter 21 in *The PDMA HandBook,* 2nd Edition.)

Gatekeepers: The group of managers who serve as advisors, decision-makers, and investors in a Stage-Gate™ process. Using established business criteria, this multifunctional group reviews new product opportunities and project progress, and allocates resources accordingly at each gate. This group is also commonly called a product approval committee or portfolio management team.

Graceful Degradation: When a product, system, or design slides into defective operation a little at a time, while providing ample opportunity to take corrective preventative action or protect against the worst consequences of failure before it happens. The opposite is catastrophic failure.

Gross Rating Points (GRPs): A measure of the overall media exposure of consumer households (reach times frequency).

Groupware: Software designed to facilitate group efforts such as communication, work flow coordination, and collaborative problem solving. The term generally refers to technologies relying on modern computer networks (external or internal).

Growth Stage: The second stage of the product life cycle. This stage is marked by a rapid surge in sales and market acceptance for the good or service. Products that reach the growth stage have successfully "crossed the chasm."

Heavyweight Team: An empowered project team with adequate resourcing to complete the project. Personnel report to the team leader and are as co-located as is practical.

Hunting for Hunting Grounds: A structured methodology for completing the fuzzy front end of new product development (*see* Chapter 2 of *The PDMA ToolBook 1.*)

Hunting Ground: A discontinuity in technology or the market that opens up a new product development opportunity.

Hurdle Rate: The minimum return on investment or internal rate of return percentage a new product must meet or exceed as it goes through development.

Idea: The most embryonic form of a new product or service. It often consists of a high-level view of the envisioned solution needed to solve the problem identified by a person, team, or firm.

Idea Generation (Ideation): All of those activities and processes that lead to creating broad sets of solutions to consumer problems. These techniques may be used in the early stages of product development to generate initial product concepts, in the intermediate stages for overcoming implementation issues, in the later stages for planning launch, and in the postmortem stage to better understand success and failure in the marketplace. (*See* Chapter 17 in *The PDMA HandBook,* 2nd Edition.)

Idea Exchange: A divergent thinking technique that provides a structure for building on different ideas in a quiet, non-judgmental setting that encourages reflection.

Idea Merit Index: An internal metric used to impartially rank new product ideas.

Implementation Team: A team that converts the concepts and good intentions of the "should-be" process into practical reality.

Implicit Product Requirement: What the customer expects in a product, but does not ask for, and may not even be able to articulate.

Importance Surveys: A particular type of attribute testing in which respondents are asked to evaluate how important each of the product attributes are in their choice of products or services.

Incremental Improvement: A small change made to an existing product that serves to keep the product fresh in the eyes of customers.

Incremental Innovation: An innovation that improves the conveyance of a currently delivered benefit but produces neither a behavior change nor a change in consumption.

Individual Depth Interviews (IDIs): A qualitative market research technique in which a skilled moderator conducts an open-ended, in-depth, guided conversation with an individual respondent (as opposed to in a focus group format). Such an interview can be used to better understand the respondent's thought processes, motivations, current behaviors, preferences, opinions, and desires.

Industrial Design (ID): The professional service of creating and developing concepts and specifications that optimize the function, value, and appearance of products and systems for the mutual benefit of both user and manufacturer [Industrial Designers Society of America]. *See* Chapters 24 and 25 of *The PDMA HandBook,* 2nd Edition.)

Information: Knowledge and insight, often gained by examining data.

Information Acceleration: A concept testing method employing virtual reality. In it, a virtual buying environment is created that simulates the information available (product, societal, political, and technological) in a real purchase situation at some time several years or more into the future.

Informed Intuition: Using the gathered experiences and knowledge of the team in a structured manner.

Initial Screening: The first decision to spend resources (time or money) on a project. The project is born at this point. Sometimes called *idea screening.*

In-Licensed: The acquisition from external sources of novel product concepts or technologies for inclusion in the aggregate NPD portfolio.

Innovation: A new idea, method, or device. The act of creating a new product or process. The act includes invention as well as the work required to bring an idea or concept into final form.

Innovation-Based Culture: A corporate culture where senior management teams and employees work habitually to reinforce best practices that *systematically* and *continuously* churn out valued new products to customers.

Innovation Engine: The creative activities and people that actually think of new ideas. It represents the synthesis phase when someone first recognizes that customer and market opportunities can be translated into new product ideas.

Innovation Steering Committee: The senior management team or a subset of it responsible for gaining alignment on the strategic and financial goals for new product development, as well as setting expectations for portfolio and development teams.

Innovation Strategy: The firm's positioning for developing new technologies and products. One categorization divides firms into Prospectors (those who lead in technology, product and market development, and commercialization, even though an individual product may not lead to profits), Analyzers (fast followers, or imitators, who let the prospectors lead, but have a product development process organized to imitate and commercialize quickly any new product a Prospector has put on the market), Defenders (those who stake out a product turf and protect it by whatever means, not necessarily through developing new products), and Reactors (those who have no coherent innovation strategy). (*See* Chapter 2 of *The PDMA HandBook,* 2nd Edition.)

Innovative Problem Solving: Methods that combine rigorous problem definition, pattern-breaking generation of ideas, and action planning that results in new, unique, and unexpected solutions.

Integrated Architecture: A product architecture in which most or all of the functional elements map into a single or very small number of chunks. It is difficult to subdivide an integrally designed product into partially functioning components.

Integrated Product Development (IPD): A philosophy that systematically employs an integrated team effort from multiple functional disciplines to develop effectively and efficiently new products that satisfy customer needs.

Intellectual Property (IP): Information, including proprietary knowledge, technical competencies, and design information, that provides commercially exploitable competitive benefit to an organization.

Internal Rate of Return (IRR): The discount rate at which the present value of the future cash flows of an investment equals the cost of the investment. The discount rate with a net present value of 0.

Intrapreneur: The large-firm equivalent of an entrepreneur. Someone who develops new enterprises within the confines of a large corporation.

Introduction Stage: The first stage of a product's commercial launch and the product life cycle. This stage is generally seen as the point of market entry, user trial, and product adoption.

ISO-9000: A set of five auditable standards of the International Standards Organization that establishes the role of a quality system in a company and that is used to assess whether the company can be certified as compliant to the standards. ISO-9001 deals specifically with new products.

Issue: A certainty that will affect the outcome of a project, either negatively or positively. Issues require investigation as to their potential impacts, and decisions about how to deal with them. Open issues are those for which the appropriate actions have not been resolved, while closed issues are ones that the team has dealt with successfully.

Journal of Product Innovation Management: The premier academic journal in the field of innovation, new product development, and management of technology. The journal, which is owned by the PDMA, is dedicated to the advancement of management practice in all of the functions involved in the total process of product innovation. Its purpose is to bring to managers and students of product innovation the theoretical structures and the practical techniques that will enable them to operate at the cutting edge of effective management practice. Web site: www.jpim.org.

Kaizen: A Japanese term describing a process or philosophy of continuous, incremental improvement.

Launch: The process by which a new product is introduced into the market for initial sale. (*See* Chapter 30 of *The PDMA HandBook,* 2nd Edition.)

Lead Users: Users for whom finding a solution to one of their consumer needs is so important that they have modified a current product or invented a new product to solve the need themselves because they have not found a supplier who can solve it for them. When these consumers' needs are portents of needs that the center of the market will have in the future, their solutions are new product opportunities.

Learning Organization: An organization that continuously tests and updates the experience of those in the organization, and transforms that experience into improved work processes and knowledge that is accessible to the whole organization and relevant to its core purpose. (*See* Continuous Learning Activity.)

Life Cycle Cost: The total cost of acquiring, owning, and operating a product over its useful life. Associated costs may include purchase price, training expenses, maintenance expenses, warranty costs, support, disposal, and profit loss due to repair downtime.

Lightweight Team: New product team charged with successfully developing a product concept and delivering to the marketplace. Resources are, for the most part, not dedicated, and the team depends on the technical functions for resources necessary to get the work accomplished.

Line Extension: A form of derivative product that adds or modifies features without significantly changing the product functionality.

Long-Term Success: The new product's performance in the long run or at some large fraction of the product's life cycle.

"M" Curve: An illustration of the volume of ideas generated over a given amount of time. The illustration often looks like two arches from the letter M.

Maintenance Activity: That set of product development tasks aimed at solving initial market and user problems with the new product or service. (*See* Chapter 33 of *The PDMA HandBook,* 2nd Edition.)

Manufacturability: The extent to which a new product can be easily and effectively manufactured at minimum cost and with maximum reliability.

Manufacturing Assembly Procedure: Procedural documents normally prepared by manufacturing personnel that describe how a component, subassembly, or system will be put together to create a final product.

Manufacturing Design: The process of determining the manufacturing process that will be used to make a new product. (*See* Chapter 23 of *The PDMA HandBook,* 1st Edition.)

Manufacturing Test Specification and Procedure: Documents prepared by development and manufacturing personnel that describe the performance specifications of a component, subassembly, or system that will be met during the manufacturing process, and that describe the procedure by which the specifications will be assessed.

Market Conditions: The characteristics of the market into which a new product will be placed, including the number of competing products, level of competitiveness, and growth rate.

Market Development: Taking current products to new consumers or users. This effort may involve making some product modifications.

Market-Driven: Allowing the marketplace to direct a firm's product innovation efforts.

Market Research: Information about the firm's customers, competitors, or markets. Information may be from secondary sources (already published and publicly available) or primary sources (from customers themselves). Market research may be qualitative in nature or quantitative (*see* entries for these two types of market research).

Market Segmentation: A framework by which to subdivide a larger heterogeneous market into smaller, more homogeneous parts. These segments can be defined in many different ways: *demographic* (men vs. women, young vs. old, or richer vs. poorer), *behavioral* (those who buy on the phone vs. the Internet vs. retail, or those who pay with cash vs. credit cards), or *attitudinal* (those who believe that store brands are just as good as national brands vs. those who don't). There are many analytical techniques used to identify segments such as cluster analysis, factor analysis, or discriminate analysis. But the most common method is simply to hypothesize a potential segmentation definition and then to test whether any differences that are observed are statistically significant. (*See* Chapter 13 of *The PDMA HandBook,* 2nd Edition.)

Market Share: A company's sales in a product area as a percent of the total market sales in that area.

Market Testing: The product development stage when the new product and its marketing plan are tested together. A market test simulates the eventual marketing mix and takes many different forms, only one of which bears the name *test market.*

Maturity Stage: The third stage of the product life cycle. This is the stage where sales begin to level off because of market saturation. It is a time when heavy competition, alternative product options, and (possibly) changing buyer or user preferences start to make it difficult to achieve profitability.

Metrics: A set of measurements to track product development and allow a firm to measure the impact of process improvements over time. These measures generally vary by firm but may include measures characterizing both aspects of the process, such as time-to-market and duration of particular process stages, as well as outcomes from product development, such as the number of products commercialized per year and percentage of sales due to new products.

Modular Architecture: A product architecture in which each functional element maps into its own physical chunk. Different chunks perform different functions, the interactions between the chunks are minimal, and they are generally well defined.

Monitoring Frequency: The frequency with which performance indicators are measured.

Morphological Analysis: A matrix tool that breaks a product down by needs met and technology components, allowing for targeted analysis and idea creation.

Multifunctional Team: A group of individuals brought together from the different functional areas of a business to work on a problem or process that requires the knowledge, training, and capabilities across the areas to successfully complete the work. (*See* Chapters 9 and 10 in *The PDMA HandBook,* 2nd Edition and Chapter 6 in *The PDMA ToolBook 1.*) (*See also* Cross-functional Team.)

Needs Statement: Summary of consumer needs and wants, described in customer terms, to be addressed by a new product. (*See* Chapter 14 of *The PDMA HandBook,* 2nd Edition.)

Net Present Value (NPV): Method to evaluate comparable investments in very dissimilar projects by discounting the current and projected future cash inflows and outflows back to the present value based on the discount rate, or cost of capital, of the firm.

Network Diagram: A graphical diagram with boxes connected by lines that shows the sequence of development activities and the interrelationship of each task with another. Often used in conjunction with a Gantt chart.

New Concept Development Model: A theoretical construct that provides for a common terminology and vocabulary for the fuzzy front end. The model consists of three parts: the uncontrollable influencing factors, the controllable engine that drives the activities in the fuzzy front end, and five activity elements: Opportunity Identification, Opportunity Analysis, Idea Generation and Enrichment, Idea Selection, and Concept Definition. (*See* Chapter 1 of *The PDMA ToolBook 1.*)

New Product: A term of many opinions and practices, but most generally defined as a product (either a good or service) new to the firm marketing it. Excludes products that are only changed in promotion.

New Product Development (NPD): The overall process of strategy, organization, concept generation, product and marketing plan creation and evaluation, and commercialization of a new product. Also frequently referred to just as *product development*.

New Product Introduction (NPI): The launch or commercialization of a new product into the marketplace. Takes place at the end of a successful product development project. (*See* Chapter 30 of *The PDMA HandBook,* 2nd Edition.)

New Product Development Process (NPD Process): A disciplined and defined set of tasks and steps that describe the normal means by which a company repetitively converts embryonic ideas into salable products or services. (*See* Chapters 4 and 5 of *The PDMA HandBook,* 2nd Edition.)

New Product Development Professional (NPDP): A New Product Development Professional is certified by the PDMA as having mastered the body of knowledge in new product development, as proven by performance on the certification test. To qualify for the NPDP certification examination, a candidate must hold a bachelor's or higher university degree (or an equivalent degree) from an accredited institution and have spent a minimum of two years working in the new product development field.

New Product Idea: A preliminary plan or purpose of action for formulating new products or services.

New-to-the-World Product: A good or service that has never before been available to either consumers or producers. The automobile was new-to-the-world when it was introduced, as were microwave ovens and pet rocks.

Nominal Group Process: A brainstorming process in which members of a group first write their ideas out individually, and then participate in a group discussion about each idea.

Non-Destructive Test: A test of the product that retains the product's physical and operational integrity.

Non-Product Advantage: Elements of the marketing mix that create competitive advantage other than the product itself. These elements can include marketing communications, distribution, company reputation, technical support, and associated services.

Operational Strategy: An activity that determines the best way to develop a new product while minimizing costs, ensuring adherence to schedule, and delivering a quality product. For product development, the objective is to maximize the return on investment and deliver a high-quality product in the optimal market window of opportunity.

Operations: A term that includes manufacturing but is much broader, usually including procurement, physical distribution, and, for services, management of the offices or other areas where the services are provided.

Operator's Manual: The written instructions to the users of a product or process. These may be intended for the ultimate customer or for the use of the manufacturing operation.

Opportunity: A business or technology gap that a company or individual realizes, by design or accident, that exists between the current situation and an envisioned future in order to capture competitive advantage, respond to a threat, solve a problem, or ameliorate a difficulty.

Outsourcing: The process of procuring a good or service from someone else, rather than the firm producing it themselves.

Outstanding Corporate Innovator Award: An annual PDMA award given to firms acknowledged through a formal vetting process as being outstanding innovators. The basic requirements for receiving this award are (1) sustained success in launching new products over a five-year time frame, (2) significant company growth from new product success, (3) a defined new product development process, that can be described to others, and (4) distinctive innovative characteristics and intangibles.

Pareto Chart: A bar graph with the bars sorted in descending order used to identify the largest opportunity for improvement. Pareto charts distinguish the "vital few" from the "useful many."

Participatory Design: A democratic approach to design that does not simply make potential users the subjects of user testing but empowers them to be a part of the design and decision-making process.

Payback: The time, usually in years, from some point in the development process until the commercialized product or service has recovered its costs of development and marketing. While some firms take the point of full-scale market introduction of a new product as the starting point, others begin the clock at the start of development expense.

Payout: The amount of profits and their timing expected from commercializing a new product.

Perceptual Mapping: A quantitative market research tool used to understand how customers think of current and future products. Perceptual maps are visual representations of the positions that sets of products hold in consumers' minds.

Performance Indicators: Criteria on which the performance of a new product in the market are evaluated. (*See* Chapter 29 of *The PDMA HandBook,* 2nd Edition.)

Performance Measurement System: The system that enables the firm to monitor the relevant performance indicators of new products in the appropriate time frame.

Performance/Satisfaction Surveys: A particular type of market research tool in which respondents are asked to evaluate how well a particular product or service is performing and/or how satisfied they are with that product or service on a specific list of attributes. It is often useful to ask respondents to evaluate more than one product or service on these attributes in order to compare them and to better understand what they like and dislike about one versus the other. In this way, this information can become a key input to the development process for next-generation product modifications.

PERT (Program Evaluation and Review Technique): An event-oriented network analysis technique used to estimate project duration when there is a high degree of uncertainty in estimates of duration times for individual activities.

Phase Review Process: A staged product development process in which one function completes a set of tasks, then passes the information generated sequentially to another function, which in turn completes the next set of tasks and then passes everything along to the next function. Multifunctional teamwork is largely absent in these types of product development processes, which may also be called *baton-passing processes.* Most firms have moved from these processes to Stage-Gate™ processes using multifunctional teams.

Physical Elements: The components that make up a product. These can be both components (or individual parts) in addition to minor subassemblies of components.

Pilot Gate Meeting: A trial, informal gate meeting usually held at the launch of a Stage-Gate™ process to test the design of the process and familiarize participants with the Stage-Gate™ process.

Pipeline (Product Pipeline): The scheduled stream of products in development for release to the market.

Pipeline Alignment: The balancing of project demand with resource supply. (*See* Chapter 5 in *The PDMA HandBook,* 1st Edition and Chapter 3 in *The PDMA HandBook,* 2nd Edition.)

Pipeline Inventory: Production of a new product that has not yet been sold to end consumers but that exists within the distribution chain.

Pipeline Loading: The volume and time phasing of new products in various stages of development within an organization.

Pipeline Management: A process that integrates product strategy, project management, and functional management to continually optimize the cross-project management of all development-related activities. (*See* Chapter 5 in *The PDMA HandBook,* 1st Edition and Chapter 3 in *The PDMA HandBook,* 2nd Edition.)

Pipeline Management Enabling Tools: The decision-assistance and data-handling tools that aid managing the pipeline. The decision-assistance tools allow the pipeline team to systematically perform trade-offs without losing sight of priorities. The data-handling tools deal with the vast amount of information needed to analyze project priorities, understand resource and skill-set loads, and perform pipeline analysis.

Pipeline Management Process: Consists of three elements: pipeline management teams, a structured methodology, and enabling tools.

Pipeline Management Teams: The teams of people at the strategic, project, and functional levels responsible for resolving pipeline issues.

Platform Product: The design and components that are shared by a set of products in a product family. From this platform, numerous derivative products can be designed. (*See* also Product Platform.)

Platform Roadmap: A graphical representation of the current and planned evolution of products developed by the organization, showing the relationship between the architecture and features of different generations of products.

Porter's Five Forces: Analysis framework developed by Michael Porter in which a company is evaluated based on its capabilities versus competitors, suppliers, customers, barriers to entry, and the threat of substitutes. (*See* Michael Porter 1998, *Competitive Strategy,* New York: The Free Press.)

Portfolio: Commonly referred to as a set of projects or products that a company is investing in and making strategic trade-offs against. (*See also* Project Portfolio and Product Portfolio.)

Portfolio Criteria: The set of criteria against which the business judges both proposed and currently active product development projects to create a balanced and diverse mix of ongoing efforts.

Portfolio Management: A business process by which a business unit decides on the mix of active projects, staffing, and dollar budget allocated to each project currently being undertaken. *See also* Pipeline Management. (*See* Chapter 13 of *The PDMA ToolBook 1* and Chapter 3 of *The PDMA HandBook,* 2nd Edition.)

Portfolio Map: A chart or graph that graphically displays the relative scalar strength and weakness of a portfolio of products, or competitors in two orthogonal dimensions of customer value or other parameters. Typical portfolio maps include Price vs. Performance, Newness to Company vs. Newness to Market, and Risk vs. Return.

Portfolio Rollout Scenarios: Hypothetical illustrations of the number and magnitude of new products that would need to be launched over a certain time frame to reach the desired financial goals; accounts for success/failure rates and considers company and competitive benchmarks.

Portfolio Team: A short-term, cross-functional, high-powered team focused on shaping the concepts and business cases for a portfolio of new product concepts within a market, category, brand, or business to be launched over a two- to five-year time period, depending on the pace of the industry.

Preproduction Unit: A product that looks like and acts like the intended final product but is made either by hand or in pilot facilities rather than by the final production process.

Process Champion: The person responsible for the daily promotion of and encouragement to use a formal business process throughout the organization. The champion is also responsible for the ongoing training, innovation input, and continuous improvement of the process.

Process Managers: The operational managers responsible for ensuring the orderly and timely flow of ideas and projects through the process.

Process Map: A work flow diagram that uses an x-axis for process time and a y-axis that shows participants and tasks.

Process Mapping: The act of identifying and defining all of the steps, participants, inputs, outputs, and decisions associated with completing any particular process.

Process Maturity Level: The amount of movement of a reengineered process from the "as-is" map, which describes how the process operated initially, to the "should-be" map of the desired future state of the operation.

Process Owner: The executive manager responsible for the strategic results of the NPD process. This includes process throughput, quality of output, and participation within the organization. (*See* Section 3 of *The PDMA ToolBook 1* for four tools that process owners might find useful, and *see* Chapter 5 of *The PDMA HandBook,* 1st Edition.)

Process Re-engineering: A discipline to measure and modify organizational effectiveness by documenting, analyzing, and comparing an existing process to "best-in-class" practice, and then implementing significant process improvements or installing a whole new process.

Product: Term used to describe all goods, services, and knowledge sold. Products are bundles of attributes (features, functions, benefits, and uses) and can be tangible, as in the case of physical goods; intangible, as in the case of those associated with service benefits; or a combination of the two.

Product and Process Performance Success: The extent to which a new product meets its technical performance and product development process performance criteria.

Product Approval Committee (PAC): The group of managers who serve as advisors, decision-makers, and investors in a Stage-Gate™ process: a company's NPD executive committee. Using established business criteria, this multifunctional group reviews new product opportunities and project progress, and allocates resources accordingly at each gate. (*See* Chapter 7 of *The PDMA ToolBook 1* and Chapters 21 and 22 of *The PDMA HandBook,* 2nd Edition.)

Product Architecture: The way in which the functional elements are assigned to the physical chunks of a product and the way in which those physical chunks interact to perform the overall function of the product. (*See* Chapter 16 of *The PDMA HandBook,* 1st Edition.)

Project Decision Making and Reviews: A series of go/no-go decisions about the viability of a project that ensure the completion of the project provides a product that meets the marketing and financial objectives of the company. This includes a systematic review of the viability of a project as it moves through the various stages in the development process. These periodic checks validate that the project is still close enough to the original plan to deliver against the business case. (*See* Chapters 21 and 22 of *The PDMA HandBook,* 2nd Edition.)

Product Definition: Defines the product, including the target market, product concept, benefits to be delivered, positioning strategy, price point, and even product requirements and design specifications.

Product Development: The overall process of strategy, organization, concept generation, product and marketing plan creation and evaluation, and commercialization of a new product. (*See* Chapters 19 to 22 of *The PDMA HandBook,* 1st Edition.)

Product Development & Management Association (PDMA): A not-for-profit professional organization whose purpose is to seek out, develop, organize, and disseminate leading-edge information on the theory and practice of product development and product development processes. The PDMA uses local, national, and international meetings and conferences; educational workshops; a quarterly newsletter (*Visions*); a bimonthly scholarly journal (*Journal of Product Innovation Management*), research proposal and dissertation proposal competitions, *The PDMA HandBook of New Product Development,* 1st and 2nd Editions, and *The PDMA ToolBook 1* and *Toolbook 2 for New Product Development* to achieve its purposes. The association also manages the certification process for New Product Development Professionals (www.pdma.org).

Product Development Check List: A pre-determined list of activities and disciplines responsible for completing those activities used as a guideline to ensure that all the tasks of product development are considered prior to commercialization. (*See* Ray Riek, *JPIM*, 2001.)

Product Development Engine: The systematic set of corporate competencies, principles, processes, practices, tools, methods, and skills that combine to define the "how" of an organization's ability to drive high-value products to the market in a competitive timely manner.

Product Development Portfolio: The collection of new product concepts and projects that are within the firm's ability to develop, are most attractive to the firm's customers, and deliver short- and long-term corporate objectives, spreading risk and diversifying investments. (*See* Chapter 13 in *The PDMA ToolBook 1* and Chapter 3 of Chapters 21 and 22 of *The PDMA HandBook,* 2nd Edition.)

Product Development Process: A disciplined and defined set of tasks, steps, and phases that describe the normal means by which a company repetitively converts embryonic ideas into salable products or services. (*See* Chapters 4 and 5 of *The PDMA HandBook,* 2nd Edition.)

Product Development Strategy: The strategy that guides the product innovation program.

Product Development Team: A multifunctional group of individuals chartered to plan and execute a new product development project.

Product Discontinuation: A product or service that is withdrawn or removed from the market because it no longer provides an economic, strategic, or competitive advantage in the firm's portfolio of offerings. (*See* Chapter 28 of *The PDMA HandBook,* 1st Edition.)

Product Discontinuation Timeline: The process and time frame in which a product is carefully withdrawn from the marketplace. The product may be discontinued immediately after the decision is made, or it may take a year or more to implement the discontinuation timeline, depending on the nature and conditions of the market and product.

Product Failure: A product development project that does not meet the objective of its charter or marketplace.

Product Family: The set of products that have been derived from a common product platform. Members of a product family normally have many common parts and assemblies.

Product Innovation Charter: A critical strategic document, the product innovation charter (PIC) is the heart of any organized effort to commercialize a new product. It contains the reasons the project has been started, as well as the goals, objectives, guidelines, and boundaries of the project. It is the "who, what, where, when, and why" of the product development project. In the discovery phase, the charter may contain assumptions about market preferences, customer needs, and sales and profit potential. As the project enters the development phase, these assumptions are challenged through prototype development and in-market testing. While business needs and market conditions can and will change as the project progresses, one must resist the strong tendency for projects to wander off as the development work takes place. The PIC must be constantly referenced during the development phase to make sure it is still valid, that the project is still within the defined arena, and that the opportunity envisioned in the discovery phase still exists.

Product Interfaces: Internal and external interfaces impacting the product development effort, including the nature of the interface, action required, and timing.

Product Life Cycle: The four stages that a new product is thought to go through from birth to death: introduction, growth, maturity, and decline. Controversy surrounds whether products go through this cycle in any predictable way.

Product Life Cycle Management: Changing the features and benefits of the product, elements of the marketing mix, and manufacturing operations over time to maximize the profits obtainable from the product over its life cycle. (*See* Chapter 33 of *The PDMA HandBook,* 2nd Edition.)

Product Line: A group of products marketed by an organization to one general market. The products have some characteristics, customers, and uses in common and may also share technologies, distribution channels, prices, services, and other elements of the marketing mix.

Product Management: Ensuring over time that a product or service profitably meets the needs of customers by continually monitoring and modifying the elements of the marketing mix, including the product and its features, the communications strategy, distribution channels and price.

Product Manager: The person assigned responsibility for overseeing all of the various activities that concern a particular product. Sometimes called a *brand manager* in consumer packaged goods firms.

Product Plan: Detailed summary of all the key elements involved in a new product development effort, such as product description, schedule, resources, financial estimations, and interface management plan.

Product Platforms: Underlying structures or basic architectures that are common across a group of products or that will be the basis of a series of products commercialized over a number of years.

Product Portfolio: The set of products and product lines the firm has placed in the market. (*See* Chapter 13 of *The PDMA ToolBook 1.*)

Product Positioning: How a product will be marketed to customers. The product positioning refers to the set of features and benefits that is valued by (and therefore defined by) the target customer audience, relative to competing products.

Product Rejuvenation: The process by which a mature or declining product is altered, updated, repackaged, or redesigned to lengthen the product life cycle and in turn extend sales demand.

Product Requirements Document: The contract between, at a minimum, marketing and development, describing completely and unambiguously the necessary attributes (functional performance requirements) of the product to be developed, as well as information about how achievement of the attributes will be verified (i.e., through testing).

Product Superiority: Differentiation of a firm's products from those of competitors, achieved by providing consumers with greater benefits and value. This is one of the critical success factors in commercializing new products.

Program Manager: The organizational leader charged with responsibility of executing a portfolio of NPD projects. (*See* Part 4 of *The PDMA ToolBook 1* for four product development tools a program manager may find helpful.)

Project Leader: The person responsible for managing an individual new product development project through to completion. He or she is responsible for ensuring that milestones and deliverables are achieved and that resources are utilized effectively. *See also* Team Leader. (See Parts 1 and 2 of *The PDMA ToolBook 1* for eight product development tools for project leaders.)

Project Management: The set of people, tools, techniques, and processes used to define the project's goal, plan all the work necessary to reach that goal, lead the project and support teams, monitor progress, and ensure that the project is completed in a satisfactory way.

Project Pipeline Management: Fine-tuning resource deployment smoothly for projects during ramp-up, ramp-down, and mid-course adjustments.

Project Plan: A formal, approved document used to guide both project execution and control. Documents planning assumptions and decisions, facilitates communication among stakeholders, and documents approved scope, cost, and schedule deadlines.

Project Portfolio: The set of projects in development at any point in time. These will vary in the extent of newness or innovativeness. (*See* Chapter 13 in *The PDMA ToolBook 1* and Chapter 3 of *The PDMA HandBook*, 2nd Edition.)

Project Resource Estimation: This activity provides one of the major contributions to the project cost calculation. Turning functional requirements into a realistic cost estimate is a key factor in the success of a product delivering against the business plan.

Project Sponsor: The authorization and funding source of the project. The person who defines the project goals and to whom the final results are presented. Typically a senior manager.

Project Strategy: The goals and objectives for an individual product development project. It includes how that project fits into the firm's product portfolio, who the target market is, and what problems the product will solve for those customers. (*See* Chapter 2 in *The PDMA HandBook*, 2nd Edition.)

Project Team: A multifunctional group of individuals chartered to plan and execute a new product development project.

Prospectors: Firms that lead in technology, product and market development, and commercialization, even though an individual product may not lead to profits. Their general goal is to be first to market with any particular innovation.

Protocol: A statement of the attributes (mainly benefits; features only when required) that a new product is expected to have. A protocol is prepared prior to assigning the project to the technical development team. The benefits statement is agreed to by all parties involved in the project.

Prototype: A physical model of the new product concept. Depending upon the purpose, prototypes may be nonworking, functionally working, or both functionally and aesthetically complete.

Psychographics: Characteristics of consumers that, rather than being purely demographic, measure their attitudes, interests, opinions, and lifestyles.

Pull-Through: The revenue created when a new product or service positively impacts the sales of other, existing products or services (the obverse of cannibalization). *See* Cannibalization.

Qualitative Cluster Analysis: An individual- or group-based process using informed intuition for clustering and connecting data points. *See* also Informed Intuition.

Qualitative Marketing Research: Research conducted with a very small number of respondents, either in groups or individually, to gain an impression of their beliefs, motivations, perceptions, and opinions. Frequently used to gather initial consumer needs and obtain initial reactions to ideas and concepts. Results are not representative of the market in general or projectable. Qualitative marketing research is used to show why people buy a particular product, whereas quantitative marketing research reveals how many people buy it. (*See* Chapters 14 to 16 of *The PDMA HandBook,* 2nd Edition.)

Quality: The collection of attributes that when present in a product means a product has conformed to or exceeded customer expectations.

Quality Assurance/Compliance: Function responsible for monitoring and evaluating development policies and practices, to ensure they meet company and applicable regulatory standards.

Quality-by-Design: The process used to design quality into the product, service, or process from the inception of product development.

Quality Control Specification and Procedure: Documents that describe the specifications and the procedures by which they will be measured that a finished subassembly or system must meet before judged ready for shipment.

Quality Function Deployment (QFD): A structured method employing matrix analysis for linking what the market requires to how it will be accomplished in the development effort. This method is most frequently used during the stage of development when a multifunctional team agrees on how customer needs relate to product specifications and the features that deliver those needs. By explicitly linking these aspects of product design, QFD minimizes the possibility of omitting important design characteristics or interactions across design characteristics. QFD is also an important mechanism in promoting multifunctional teamwork. Developed and introduced by Japanese auto manufacturers, QFD is widely used in the automotive industry.

Quantitative Market Research: Consumer research, often surveys, conducted with a large enough sample of consumers to produce statistically reliable results that can be used to project outcomes to the general consumer population. Used to determine importance levels of different customer needs, performance ratings of and satisfaction with current products, probability of trial, repurchase rate, and product preferences. These techniques are used to reduce the uncertainty associated with many other aspects of product development. (*See* Chapter 18 of *The PDMA HandBook,* 2nd Edition.)

Radical Innovation: A new product, generally containing new technologies, that significantly changes behaviors and consumption patterns in the marketplace.

Rapid Prototyping: Any of a variety of processes that avoid tooling time in producing prototypes or prototype parts and therefore allow (generally nonfunctioning) prototypes to be produced within hours or days rather than weeks. These prototypes are frequently used to test quickly the product's technical feasibility or consumer interest.

Reactors: Firms that have no coherent innovation strategy. They only develop new products when absolutely forced to by the competitive situation.

Realization Gap: The time between first perception of a need and the launch of a product that fills that need.

Render: Process that industrial designers use to visualize their ideas by putting their thoughts on paper with any number of combinations of color markers, pencils, and highlighters, or computer visualization software.

Reposition: To change the position of the product in the minds of customers, either on failure of the original positioning or to react to changes in the marketplace. Most frequently accomplished through changing the marketing mix rather than redeveloping the product.

Resource Matrix: An array that shows the percentage of each nonmanagerial person's time that is to be devoted to each of the current projects in the firm's portfolio.

Resource Plan: Detailed summary of all forms of resources required to complete a product development project, including personnel, equipment, time, and finances.

Responsibility Matrix: This matrix indicates the specific involvement of each functional department or individual in each task or activity in each stage.

Return on Investment (ROI): A standard measure of project profitability, this is the discounted profits over the life of the project expressed as a percentage of initial investment.

Rigid Gate: A review point in a Stage-Gate™ process at which all the prior stage's work and deliverables must be complete before work in the next stage can commence.

Risk: An event or condition that may or may not occur, but if it does, will impact the ability to achieve a project's objectives. In new product development, risks may take the form of market, technical, or organizational issues. For more on managing product development risks, *see* Chapters 8 and 15 in the *PDMA ToolBook 1* and Chapter 28 in *The PDMA HandBook,* 2nd Edition.

Risk Acceptance: An uncertain event or condition for which the project team has decided not to change the project plan. A team may be forced to accept an identified risk when it is unable to identify any other suitable response to the risk.

Risk Avoidance: Changing the project plan to eliminate a risk or to protect the project objectives from any potential impact due to the risk.

Risk Management: The process of identifying, measuring, and mitigating the business risk in a product development project.

Risk Mitigation: Actions taken to reduce the probability and/or impact of a risk to below some threshold of acceptability.

Risk Tolerance: The level of risk that a project stakeholder is willing to accept. Tolerance levels are context-specific. That is, stakeholders may be willing to accept different levels of risk for different types of risk, such as risks of project delay, price realization, and technical potential.

Risk Transference: Actions taken to shift the impact of a risk and the ownership of the risk response actions to a third party.

Roadmapping: A graphical multi-step process to forecast future market and/or technology changes, and then plan the products to address these changes.

Robust Design: The design of products to be less sensitive to variations, including manufacturing variation and misuse, increasing the probability that they will perform as intended.

"Rugby" Process: A product development process in which stages are partially or heavily overlapped rather than sequential with crisp demarcations between one stage and its successor.

S-Curve (Technology S-Curve): Technology performance improvements tend to progress over time in the form of an "S" curve. When a technology is first invented, its performance improves slowly and incrementally. Then, as experience with a new technology accrues, the rate of performance increase grows and technology performance increases by leaps and bounds. Finally, some of the performance limits of a new technology start to be reached and performance growth slows. At some point, the limits of the technology may be reached and further improvements are not made. Frequently, the technology then becomes vulnerable to a substitute technology that is capable of making additional performance improvements. The substitute technology is usually on the lower, slower portion of its own S-curve and quickly overtakes the original technology when performance accelerates during the middle (vertical) portion of the "S."

Scanner Test Markets: Special test markets that provide retail point-of-sale scanner data from panels of consumers to help assess the product's performance. First widely applied in the supermarket industry.

Scenario Analysis: A tool for envisioning alternate futures so that a strategy can be formulated to respond to future opportunities and challenges. (*See* Chapter 16 of *The PDMA ToolBook 1*.)

Screening: The process of evaluating and selecting new ideas or concepts to put into the project portfolio. Most firms now use a formal screening process with evaluation criteria that span customer, strategy, market, profitability, and feasibility dimensions.

Segmentation: The process of dividing a large and heterogeneous market into more homogeneous subgroups. Each subgroup, or segment, holds similar views about the product, and values, purchases, and uses the product in similar ways. (*See* Chapter 13 of *The PDMA HandBook*, 2nd Edition.)

Senior Management: That level of executive or operational management above the product development team that has approval authority or controls resources important to the development effort.

Sensitivity Analysis: A calculation of the impact that an uncertainty might have on the new product business case. It is conducted by setting upper and lower ranges on the assumptions involved and calculating the expected outcomes. (*See* Chapter 16 of *The PDMA ToolBook 1*.)

Services: Products, such as an airline flight or insurance policy, that are intangible or at least substantially so. If totally intangible, they are exchanged directly from producer to user, cannot be transported or stored, and are instantly perishable. Service delivery usually involves customer participation in some important way. Services cannot be sold in the sense of ownership transfer, and they have no title of ownership.

Short-Term Success: The new product's performance shortly after launch, well within the first year of commercial sales.

Should-Be Map: A version of a process map depicting how a process will work in the future. A revised "as-is" process map. The result of the team's re-engineering work.

Simulated Test Market: A form of quantitative market research and pre-test marketing in which consumers are exposed to new products and to their claims in a staged advertising and purchase situation. Output of the test is an early forecast of expected sales or market share, based on mathematical forecasting models, management assumptions, and input of specific measurements from the simulation.

Six Sigma: A level of process performance that produces only 3.4 defects for every 1 million operations.

Slip Rate: Measures the accuracy of the planned project schedule according to the formula: Slip Rate = ([actual schedule/planned schedule] –1) * 100%.

Specification: A detailed description of the features and performance characteristics of a product. For example, a laptop computer's specification may read as a 1.7 gigahertz Pentium M processor, with 256 megabytes of SDRAM and 80 gigabytes of hard disk space, 4.5 hours of battery life, weight of 6.9 pounds, with an active matrix 1680×1050 color screen.

Speed to Market: The length of time it takes to develop a new product from an early initial idea for a new product to initial market sales. Precise definitions of the start and end point vary from one company to another and may vary from one project to another within a company. (*See* Chapter 12 of *The PDMA HandBook*, 2nd Edition.)

Sponsor: An informal role in a product development project, usually performed by a higher-ranking person in the firm who is not directly involved in the project, but who is ready to extend a helping hand if needed or provide a barrier to interference by others.

Stage: One group of concurrently accomplished tasks, with specified outcomes and deliverables, of the overall product development process.

Stage-Gate™ Process: A widely employed product development process that divides the effort into distinct time-sequenced stages separated by management decision gates. Multifunctional teams must successfully complete a prescribed set of related cross-functional tasks in each stage prior to obtaining management approval to proceed to the next stage of product development. The framework of the Stage-Gate™ process includes work flow and decision flow paths and defines the supporting systems and practices necessary to ensure the process's ongoing smooth operation.

Staged Product Development Activity: The set of product development tasks commencing when it is believed there are no major unknowns and that result in initial production of salable product, carried out in stages.

Standard Cost: *See* Factory Cost.

Stoplight Voting: A convergent thinking technique by which participants vote their idea preferences using colored adhesive dots. Also called *preference voting*.

Strategic Balance: Balancing the portfolio of development projects along one or more of many dimensions, such as focus versus diversification, short versus long term, high versus low risk, and extending platforms versus development of new platforms.

Strategic New Product Development (SNPD): The process that ties new product strategy to new product portfolio planning. (*See* Chapter 2 of both editions of *The PDMA HandBook*.)

Strategic Partnering: An alliance or partnership between two firms (frequently one large corporation and one smaller, entrepreneurial firm) to create a specialized new product. Typically, the large firm supplies capital and the necessary product development, marketing, manufacturing, and distribution capabilities, while the small firm supplies specialized technical or creative expertise.

Strategic Pipeline Management: Strategic balancing, which entails setting priorities among the numerous opportunities and adjusting the organization's skill sets to deliver products.

Strategic Plan: Establishes the vision, mission, values, objectives, goals, and strategies of the organization's future state.

Strategy: The organization's vision, mission, and values. One subset of the firm's overall strategy is its innovation strategy.

Subassembly: A collection of components that can be put together as a single assembly to be inserted into a larger assembly or final product. Often the subassembly is tested for its ability to meet some set of explicit specifications before inclusion in the larger product.

Success: A product that meets its goals and performance expectations. Product development success has four dimensions. At the project level, there are three dimensions: financial, customer-based, and product technical performance. The fourth dimension is new product contribution to overall firm success. (*See* Chapters 1, 29, 32, 35, and 36 of *The PDMA HandBook,* 2nd Edition.)

Support Service: Any organizational function whose primary purpose is not product development but whose input is necessary to the successful completion of product development projects.

SWOT Analysis: For "Strengths, Weaknesses, Opportunities, and Threats." A SWOT analysis evaluates a company in terms of its advantages and disadvantages versus those of competitors, customer requirements, and market/economic environmental conditions.

System Hierarchy Diagram: The diagram used to represent product architectures. This diagram illustrates how the product is broken into its chunks.

Systems and Practices: Established methods, procedures, and activities that either drive or hinder product development. These may relate to the firm's day-to-day business or may be specific to product development.

Systems and Practices Team: Senior managers representing all functions who work together to identify and change those systems and practices hindering product development and who establish new tools, systems, and practices for improving product development.

Task: The smallest describable unit of accomplishment in completing a deliverable.

Target Cost: A cost objective established for a new product based on consideration of customer affordability. Target cost is treated as an independent variable that must be satisfied along with other customer requirements.

Target Market: The group of consumers or potential customers selected for marketing. This market segment is most likely to buy the products within a given category. These are sometimes called *prime prospects.*

Team: That group of persons who participate in the product development project. Frequently each team member represents a function, department, or specialty. Together they represent the full set of capabilities needed to complete the project. (*See* Chapter 9 in *The PDMA HandBook,* 2nd Edition and Chapter 6 in *The PDMA ToolBook 1.*)

Team Leader: The person leading the new product team. Responsible for ensuring that milestones and deliverables are achieved, but may not have any authority over project participants. (*See* Parts 1 and 2 of *The PDMA ToolBook 1* for eight product development tools for team leaders.)

Team Spotter's Guide: A questionnaire used by a team leader (or team members) to diagnose the quality of the team's functioning. (*See* Chapter 6 in the *PDMA ToolBook 1.*)

Technology-Driven: A new product or new product strategy based on the strength of a technical capability. Sometimes called *solutions in search of problems.*

Technology Road Map: A graphic representation of technology evolution or technology plans mapped against time. It is used to guide new technology development for or technology selection in developing new products.

Technology Stage-Gate (TSG): A process for managing the technology development efforts when there is high uncertainty and risk. The process brings a structured methodology for managing new technology development without thwarting the creativity needed in this early stage of product development. It is specifically intended to manage high-risk technology development projects when there is uncertainty and risk that the technology discovery may never occur and therefore the ultimate desired product characteristics might never be achieved. (*See* Chapter 11 in *The PDMA ToolBook 1.*)

Technology Transfer: The process of converting scientific findings from research laboratories into useful products by the commercial sector. May also be referred to as the process of transferring technology between alliance partners.

Test Marketing: The launching of a new product into one or more limited geographic regions in a very controlled manner, and measuring consumer response to the product and its launch. When multiple geographies are used in the test, different advertising or pricing policies may be tested and the results compared.

Think Links: Stimuli used in divergent thinking to help participants make new connections using seemingly unrelated concepts from a list of people, places, or things.

Think Tank: Environments, frequently isolated from normal organizational activities, created by management to generate new ideas or approaches to solving organizational problems.

Thought Organizers: Tools that help categorize information associated with ideas such that the ideas can be placed into groups that can be more easily compared or evaluated.

Three Rs: The fundamental steps of Record, Recall, and Reconstruct, which most creative minds go through when generating new product ideas.

Threshold Criteria: The minimum acceptable performance targets for any proposed product development project.

Thumbnail: The most minimal form of sketching, usually using pencils, to represent a product idea.

Time to Market: The length of time it takes to develop a new product from an early initial idea for a new product to initial market sales. Precise definitions of the start and end point vary from one company to another, and may vary from one project to another within the company.

Tone: The feeling, emotion, or attitude most associated with using a product. The appropriate tone is important to include in consumer new product concepts and advertising.

Tornado: A mid to late growth stage strategy that follows the "bowling alley" and that describes an often frenzied period of rapid growth and acceptance for a product category. Activities of the tornado phase include commoditization of a product to become an industry standard, competitive pricing to maximize share, and low-cost volume distribution channels. Success in the tornado is related to maintaining previously established product leadership and complementing it with operational excellence in a variety of strategic areas.

Total Quality Management (TQM): A business improvement philosophy that comprehensively and continuously involves all of an organization's functions in improvement activities.

Tracking Studies: Surveys of consumers (usually conducted by telephone) following the product's launch to measure consumer awareness, attitudes, trial, adoption, and repurchase rates.

TRIZ: An acronym for the Theory of Inventive Problem Solving, a Russian systematic method of solving problems and creating multiple-alternative solutions. It is based on an analysis and codification of technology solutions from millions of patents. The method enhances creativity by getting individuals to think beyond their own experience and to reach across disciplines to solve problems using solutions from other areas of science.

Uncertainty Range: The spread between the high (best-case) and low (worst-case) values in a business assumption.

User: Any person who uses a product or service to solve a problem or obtain a benefit, whether or not he or she purchases it. Users may consume a product, as in the case of people using shampoo to clean their hair or eating a potato chip to assuage hunger between meals. Users may not directly consume a product but may interact with it over a longer period of time, like a family owning a car, with multiple family members using it for many purposes over a number of years. Products also are employed in the production of other products or services, where the users may be the manufacturing personnel who operate the equipment.

Utilities: The weights derived from conjoint analysis that measure how much a product feature contributes to purchase interest or preference.

Value: Any principle to which a person or company adheres with some degree of emotion. It is one of the elements that enter into formulating a strategy.

Value-Added: The act or process by which tangible product features or intangible service attributes are bundled, combined, or packaged with other features and attributes to create a competitive advantage, reposition a product, or increase sales.

Value Analysis: A technique for analyzing systems and designs. Its purpose is to help develop a design that satisfies users by providing the needed user requirements in sufficient quality at an optimum (minimum) cost.

Value Chain: As a product moves from raw material to finished good delivered to the customer, value is added at each step in the manufacturing and delivery process. The value chain indicates the relative amount of value added at each of these steps.

Value Proposition: A short, clear, and simple statement of how and on what dimensions a product concept will deliver value to prospective customers. The essence of "value" is embedded in the trade-off between the benefits a customer receives from a new product and the price a customer pays for it. (*See* Chapter 3 of *The PDMA ToolBook 1*.)

Vertical Integration: A firm's operation across multiple levels of the value chain. In the early 1900s, Ford Motor Company was extremely vertically integrated, as it owned forests and operated logging and wood-finishing and glass-making businesses. They made all of the components that went into automobiles, as well as most of the raw materials used in those components.

Virtual Customer: A set of Web-based market research methods for gathering voice-of-the-customer data in all phases of product development. (*See* Dahan and Hauser, *JPIM*, July 2002.)

Virtual Product Development: Paperless product development. All design and analysis is computer-based.

Virtual Reality: Technology that enables a designer or user to "enter" and navigate a computer-generated 3-D environment. Users can change their viewpoint and interact with the objects in the scene in a way that simulates real-world experiences.

Virtual Team: Dispersed teams that communicate and work primarily electronically may be called virtual teams.

Vision: An act of imagining, guided by both foresight and informed discernment, that reveals the possibilities as well as the practical limits in new product development. It depicts the most desirable, future state of a product or organization.

Visionary Companies: Leading innovators in their industries, they rank first or second in market share, profitability, growth, and shareholder performance. A substantial portion (e.g., 30 percent or more) of their sales are from products introduced in the last three years. Many firms want to benchmark these firms.

Visions: The new product development practitioner-oriented magazine of the PDMA.

Voice of the Customer (VOC): A process for eliciting needs from consumers that uses structured in-depth interviews to lead interviewees through a series of situations in which they have experienced and found solutions to the set of problems being investigated. Needs are obtained through indirect questioning by coming to understand how the consumers found ways to meet their needs and, more important, why they chose the particular solutions they found. (*See* Chapter 14 of *The PDMA Handbook,* 2nd edition.)

Whole Product: A product definition concept that emphasizes delivering all aspects of a product that are required for it to deliver its full value. This would include training materials, support systems, cables, how-to recipes, additional hardware/software, standards and procedures, implementation, applications consulting—any constitutive elements necessary to ensure the customer will have a successful experience and achieve at least minimum required value from the product. Often elements of the whole product are provided via alliances with others. This term is most often used in the context of planning high-technology products.

Work Flow Design Team: Functional contributors who work together to create and execute the work flow component of a Stage-Gate™ system. They decide how the firm's Stage-Gate™ process will be structured, what tasks it will include, what decision points will be included, and who is involved at all points.

Work Plan: Detailed plan for executing the project, identifying each phase of the project, the major steps associated with them, and the specific tasks to be performed along the way. Best-practice work plans identify the specific functional resources assigned to each task, the planned task duration, and the dependencies between tasks. *See also* Gantt Chart.

Worth What Paid For (WWPF): The quantitative evaluation by a person in your customer segment of the question: "Considering the products and services that your vendor offers, are they worth what you paid for them?"

ACKNOWLEDGMENT

Some of the definitions for terms in this glossary have been adapted from the glossary in *New Products Management*, by C. Merle Crawford and C. Anthony Di Benedetto. Terms, phrases, and definitions generously have been contributed to this list by the PDMA Board of Directors, the design teams for the PDMA Body of Knowledge, the editors and authors of *The PDMA ToolBooks 1 and 2 for New Product Development* (John Wiley & Sons, 2002, 2004), the editors and authors of *The PDMA Handbook of New Product Development,* both 1st and 2nd editions (John Wiley & Sons, 1996, 2004), and several other individuals knowledgeable in the science, skills, and art of new product development. We thank all of these volunteer contributors for their continuing support.

About PDMA

Founded in 1976, PDMA is a volunteer-driven, not-for-profit organization. About 80 percent of its members are corporate practitioners of new product development, with the remaining 20 percent split evenly between academics and service providers.

PDMA's mission is to be the thought leaders in new product development, to improve the effectiveness of people engaged in developing and managing new products—both new manufactured goods and new services. This mission includes facilitating the generation of new information, helping convert this information into knowledge that is in a useable format, and making this new knowledge broadly available to those who might benefit from it. A basic tenet of the Association is that enhanced product innovation represents a desirable and necessary economic goal for firms that wish to achieve and maintain a profitable competitive advantage in the long term.

PDMA actively supports several knowledge-generating activities. It sponsors annual research competition and rewards up to three proposals with financial support and research access to PDMA members. PDMA has sponsored a yearly Ph.D. Proposal Competition since 1991 to encourage young academics to engage in new product development research, to reward the best in the field with financial support for completing their projects, and to ensure that the knowledge they develop gets distributed to the PDMA membership. Finally, PDMA has directly supported three streams of research over the last several years, resulting in a number of papers and many presentations of the findings: profiles and compensation of new product development professionals, measuring new product development success, and trends in effective practices in the management of new product development.

Knowledge-dissemination activities include an annual international conference on the general subject of new product development, incorporating a research section devoted to the presentation of the latest academic research in new product development. PDMA also conducts many shorter-stay, regionally initiated conferences on special topics of current interest to practitioners. In addition, the Association disseminates research findings in its award-winning *Journal of Product Innovation Management*, published six times per year, and provides additional knowledge in its quarterly *Visions Magazine*.

This book, *The PDMA ToolBook 2 for New Product Development* (John Wiley & Sons, 2004), follows the Association's *The PDMA ToolBook 1 for New Product Development* (John Wiley & Sons, 2002). These have brought together practical, authoritative approaches to every aspect of the new product development process, from idea generation to delivery of the final product and commercialization. The *ToolBooks* are companions to *The PDMA Handbook of New Product Development,* Second Edition (John Wiley & Sons, 2004) and its predecessor, *The PDMA Handbook of New Product Development* (John Wiley & Sons, 1996). The *Handbooks* are new product development primers from a managerial point of view.

Index

Abstract participation, 213
Access to help, 98
Ackoff, Russell L., 238
Advertising, 208
Affinity diagrams, 186–187
Agar, Mike, 203, 204
Aided questions, 179
Aim, project, 438
Akgun, A. E., 97
Alberto-Culver Co., 7
Allied Corporation, 485
AMR Research, 385
Analog Devices, 60–61
Analysis:
  cluster, 188, 226
  cross-roadmapping, 430–431
  of customer problems, 26
  of ethnographic research, 225–230
  gap, 381, 382
  ROI, 445
  tips for, 231
  value chain, 409
Anchoring (maturity level 3), 476
Anonymity, 250
Application migration, 60–61
AskMe, 110
Assessment, requirements, 335–336
Assess Project Transition Readiness, 45, 47
Audiotaping, 181, 212
Automating the flow (maturity level 5), 477
Avaya, 383, 385, 390

B2B (business-to-business), 217
Back-of-the-envelope maps, 397, 403–410, 432
Benchmarks, 11
Bernick, Carol, 7
Bernick, Howard, 7
Berra, Yogi, 235
Berschied, John, 23
Brown, John Seely, 227, 228
Browse capability, 112
Business contracts, 157, 158
Business model, 59–60
Business pains:
  developing ideas for mitigation of, 307–308, 314–316, 324–325
  identification of, 305, 307, 314–316, 322–324
Business process modeling, 303–306, 313–314, 319–322
Business strategy, 457
Business-to-business (B2B), 217
Buy-in, 332

Capabilities:
  defining product development, 381, 383–387
  features vs., 79–80
  planning of product development, 381, 382
  product features from, 83–84
  quantify value of new product, 325–327
  technical specifications transformed into, 78–80
Capability development, 78–80
Capability mapping, 104–108
Capability maturity, 466, 467, 469–473
Capability Maturity Model (CMM), 383, 466, 467
Capital One, 4, 30
Caring organizations, 98
Carstedt, Goran, 239–240
Caterpillar, 299–301
Cemex, 29–30
Center for Quality of Management, 339
Central location interviewing, 175, 176
Change equilibrium, 486–488
Charters, innovation, 7–9
Children's play cards case study, 275, 277–282, 284–285
CID, *see* Customer idealized design
Cisco Systems, 150, 154–155
Clarity, vision, 98
Cluster analysis, 188, 226
CMII, 383
CMM, *see* Capability Maturity Model

Codevelopment, 149–163
key elements for success in, 161, 163
model for, 151–152
partner assessment/selection for, 154–156
relationship initiation for, 156–159
relationship management for, 157, 160–162
structured approach to, 152–154
Collaborative environment, 98–100
Collage boards, 232
Commitment, 44, 96, 97
Communication, 97
Communication modalities, 123–124
and creativity, 127–128
and virtual teams, 137, 140–142
Communities of practice (COPs), 100–104
mechanics of, 102–103
at P&G, 110
structure of, 101–102
success factors for, 103–104
Compensatory behavior, 218
Competence/capability mapping, 104–108
Competitive advantage, 370–371
Conferences, interviewing at, 175
Conjoint approaches, 298
Connection, 224, 225
Consensus, 174
Consistency-in-use, 473, 474
Consumer needs, 110
Consumer voices, 336–337
Continuous improvements, 335
Contracts, 157, 158
Controls, 304, 321, 327
Cook, H. E., 298
Cooper, Robert, 27–28, 236, 351
COPs, *see* Communities of practice
Corning Glass Works, 42
Courage, 98
Crawford, Merle, 7
Creativity, 122, 124–130
and communication modality, 127–128
definition of, 125
and flexibility, 130
individual vs. team, 126
and proximity of team members, 123–127
and team task structure, 129–130
Creativity vignettes, 131
Credibility, 102
Cross functionality, 25–27, 401, 457
Cross-roadmapping analysis, 430–431
Culture:
and change, 350
and global research, 196
innovation-based, *see* Innovation-based culture
Customer idealized design (CID), 238–262
application of, 239–243
design phase of, 255–256
facilitation of, 247–248, 250–256
facilities for, 246, 247
follow-up to, 257–262
GlaxoSmithKline example of, 243–244
IKEA example of, 239–243
in NPD process, 244–250
number of participants, 249–250
number of sessions, 246
observation of, 250
planning/setting up sessions for, 245–250
scope of, 246, 247
screening/recruiting participants for, 248–249
specifications phase of, 253–255
Customer ideas, 30, 235–268
and customer idealized design, *see* Customer idealized design
from daily usage, 262–268
wish mode approach to, 236–238
Customer interviews, 172–180
guide for, 177–178
location of, 175–177
one-on-one vs. focus-group, 173–175
quantity of, 177
questions for, 178–180
selection step of, 172
Customer needs, 181–193
capturing, 181–183
categorizing, 183–185
definition of, 170
editing, 185
organizing, 171, 186–188
prioritizing, 171–172, 188–193
Customer requirements, 338–340
Customers:
business processes of, in quantitative economic valuation, 303–306, 313–314, 319–322
identifying types of, 207
and innovation-based culture, 4, 5, 27–31
prioritization/selection of, in quantitative economic valuation, 303, 313, 317–319
research of, 30–31
and stretching boundaries, 28–30
Customer Visits (Edward McQuarrie), 175
Customer voices, identification of, in requirements process, 336–337

Daily usage ideas, 237–238, 262–268
in NPD process, 244–250

observation/usability testing of, 26–265, 268
Daimler Chrysler, 101, 102
Data architecture, 383
Data collection methods, 211–213
Data forms, 402–403
Data handling, 476–477
Data storage maturity levels, 474
Debriefing, 221–225
Decision setup (maturity level 2), 475–476
Decision tools, 440–441
Deductive market research, 205
Definition:
project, 438
technology, 451, 452
Delegating innovation, 6–10
Deming, W. Edwards, 167
Demonstration stage, 452–453
Dendrograms, 188
Description, ethnographic data, 226
Design:
project, 438
research, 207–213
of toolkits for user innovation, 357–368
user, 236–237
"Design document," 257–258
Development teams, 20–22
Digital Equipment, 97
Diller, Barry, 232, 233
Dionne, Keith, 150
Direct questions, 180
Discontinuities in system patterns, 217
Disequilibrium in system, 217
Disintermediation opportunities, 218
Documentation, 346–348
Dow Corning, 42
Driver maps, 422, 423
DuPont, 42

Eastman Kodak, 65–67
Eli Lilly, 161, 162
Empathy, 98
Employee ideas, 18–19
"Enabling" competencies and capabilities, 106, 107
Engineering characteristics, 183, 184
Engine families, 406
Environment, collaborative, 98–100
ESA (European Space Agency), 138
Ethnography, 201–233
analysis step of, 225–230
applications of, 204–206
definition of, 202
elements of, 202–204
fieldwork step of, 219, 221–225
implementation step of, 214–220
planning step of, 205–214
utilization step of, 229–232
European Space Agency (ESA), 138
Event maps, 404, 405
Evolution maps, 422, 424–427
Evolving participation, 213
Expectations:
market development, 62
organizational, 43
Expert-guided view, 112
Explicit customer requirements, 339
Explicit knowledge, 94–95

Face-to-face communication, 127, 128
Features, 324
capabilities vs., 79–80
technical capabilities transformed into, 83–84
FFE, *see* Fuzzy front end
Field journals, 211–212
Field research teams, 214–225
debriefing, 221–225
developing, 214–216
observation/discussion guide for, 217–220
rules for, 219, 221
tips for, 225
training/coaching, 216–217
Fingerprint, virtuality, *see* Virtuality fingerprint
Fixes identification (SWIFT), 280–282
Flexibility, 128, 130–135
Flexibility scores, 131
Flextronics, 383, 384
Flow automation (maturity level 5), 477
Flow chains, 304–305, 321, 322
Focus:
on areas of interest, 102
on context, 202–203
of product team, 20
of summit meetings, 457
Focus groups, 173–175, 245
Foust, Greg, 67
Frame of reference, 288
Franklin, Benjamin, 118
Frendo, Michael, 150, 154, 155
Front-to-back architecture, 480–484
Fry, Art, 262
Full participation, 213
Functional architecture maps, 414–419
Functional requirements, 343–344
Functional specifications, 344
Funding providers, 43–44

Fuzzy front end (FFE), 165, 269–271. *See also* SWIFT

Gains acceleration (maturity level 4), 476–477
Gap analysis, 381, 382
Gartner, 385
Gate arrays, 359
Gearin, Mike, 9–10
Geertz, Clifford, 202
General Electric (GE), 354
General Mills, 31, 224
Generativity, 96
GE Plastics, 357
GlaxoSmithKline (GSK), 243–244, 261, 263
Goal setting, 437
Go/no-go criteria, 14–18
Griffin, Abbie, 28, 32, 168, 173, 177, 187
Groundwork (maturity level 1), 475
Groupthink, 126–127
Growth, 34–35
GSK, *see* GlaxoSmithKline
Guided conversation, 217
Gum, Ernie, 366, 367
Gundry, L. K., 99

Hamel, Gary, 34
Hard-wiring (of requirements process), 349–350
Hauser, John, 168, 173, 177, 187, 303
Help, access to, 98
Hewlett-Packard (HP), 351, 379
Highlight videos, 231
House of Quality, 167, 298
HP, *see* Hewlett-Packard

IBM, 42, 466
Icihjo, K., 96
ICM (Institute of Configuration Management), 383
Identification-of-Function (IDEF) process, 304–306, 313, 314
IFF, *see* International Flavors and Fragrances
IKEA, 239–243
Implementation:
  of IT-enablement, 379–393
  of PPM, *see* Portfolio and pipeline management
  of quantitative economic valuation, 316–327
Importance Survey, 341
Indirect questions, 180
Individuality identification (SWIFT), 280
Inductive market research, 202, 205
Industrial Research Institute (IRI), 34, 63–65
Information:
  capturing, 111
  management of, 111–112
Innovation:
  delegating, 6–10
  focused/specific criteria for, 10–18
  toolkits for, *see* Toolkits
Innovation-based culture, 3–32
  and customers, 4, 5, 27–31
  and new product teams, 4, 18–27
  and senior management, 3–18
Innovation charter, 7–9
Innovation cycles, 231
Innovation intranets, 108–113
  P&G example of, 109–110
  principles of, 110–112
InnovationNet, 111
Innovation steering committee, 8, 12
Innovation vision, 96–98
Inputs, 304, 320, 321, 327
Insight, 224, 225
Institute of Configuration Management (ICM), 383
Intellectual property (IP), 440
Internal rate of return (IRR), 445, 446
International Flavors and Fragrances (IFF), 355–356, 363, 371
Interpretive analysis, 227–230
Interview guides, 177–178
Interviewing, 209–211, 244–245
Intranets, innovation, 108–113
Involvement, sponsor, 213–214
IP (intellectual property), 440
IRI, *see* Industrial Research Institute
IRR, *see* Internal rate of return
IT-enablement (of NPD), 375–394
  and defining of NPD process, 376–378
  implementation of, 379–393
  pilot implementation phase of, 387–392
  product development capability definition phase of, 381, 383–387
  product development capability planning phase of, 381, 382
  rollout and training phase of, 392, 393
  stages of, 378–379
Iterative decision tools, 72

"Jack Welch" leadership approach, 487
Janis, Irving, 126
Japan, 167–168
Jargon, 171, 181
JDAs, *see* Joint development agreements
Johnson & Johnson, 379

Joint development agreements (JDAs), 157, 159
Journals, 211–212

Kano model, 193
Kano survey, 341
Kelleher, Herb, 4
Kimberly-Clark, 211
KJ approach, 298
Kleinschmidt, Elko, 236
Knowledge creation, 94–95
Knowledge sharing, 111
Kodak, 42
Kotler, P., 300, 303
Kratzer, J., 125

Lack of balance in system, 217
Language processing, 226, 227
Latent customer requirements, 339
Lattice model, 446–447
Lead products, 442
Learning by doing, 361–362
Leenders, R., 125
Leniency, 98
Leonard, D., 99
Lewin, Kurt, 486
Linkage maps, 398, 399, 422, 423
Linux, 121
Lipstick case study, 206, 207, 209, 210, 212, 214, 220, 223, 225, 227, 228, 230, 232
Listening, 336
Location. *See also* Proximity
  of customer interviews, 175–177
  of R&D project, 42–43
  of virtual team members, 120
Lo-fi maps, 432
Longitudinal studies, short vs., 210
Lynn, G. S., 97, 98

Management. *See also* Senior management
  information, 111–112
  partner relationship, 157, 160–162
  PPM, *see* Portfolio and pipeline management
  project, 438
  transition, *see* Transition management
  virtual team, 135–145
The Management Roundtable (MRT), 150
Managerial access, 309–311
Manufactured goods, 195
Manufacturing, 61–63
Mapping, 397–432
  across product lines, 430–431
  benefits of, 400
  of categories of uncertainty, 45, 48–49
  competence/capability, 104–108
  dimensions of, 410–413
  event map, 404, 405
  and form of presentation, 402–403
  of functional architectures, 414–419
  and functioning of maps, 398, 400–402
  implementation alerts, 402–403, 410–413
  language of, 401
  and linkage of market drivers to product/technology drivers, 422, 423
  product line map, 404, 406–410
  product line story created via, 426, 438–430
  role of, in NPD, 397–402
  setting targets/objectives over time by, 419–421
  showing product/technology evolution via, 422, 424–427
  technology, 448–450
Map syntax, 401
Market adoption, 11
Market development expectations, 62
Market drivers, 422, 423
Market entry and development strategy, 60–63
Market uncertainty, 59–62
Mass customization, 363, 373
Matsushita Electric, 95
Maturity:
  capability, 466, 467, 469–473
  of product development, 412–413
  spiral-up levels of, 475–477
  of technology development, 453–457
McDermott, R., 103
McGrath, M. E., 97
McQuarrie, Edward, 175
Mechanisms, 304, 321, 327
Meetings, technology summit, 453–457
Microsoft, 265
Miles, R. E., 81
Milestones, 22–25
Millennium Pharmaceuticals, 149–150
Mix management, 464
Module libraries, 365–366
"Mother role void," 203
Motorola, 61, 430
MRT (The Management Roundtable), 150

NASA, 19, 138
Nestlé FoodServices, 366–368, 371
Net present value (NPV), 445, 446
"New-new" products, 194–195

New product teams, 18–27
cross functionality of, 25–27
and innovation-based culture, 4
recognition of, 22–25
research by, 30–31
short-term/high-powered, 19–22
virtual, *see* Virtual NPD teams
Nonaka, I., 95, 96
NPD portfolio teams, 30
NPV, *see* Net present value

Objectives, research, 206
Objectives-over-time maps, 419–421
Observation, 224, 225
of CID sessions, 250
participant, 209
research, 217–220
and usability testing, 264–268
Olsen, Ken, 97
One-on-one interviews, 173–175
On-site interviewing, 175–177
Open-ended questions, 180, 244
Openness, 99
Operational metric, 324–326
Operational requirements, 342
Opinions, customer, 184
Organizational structure, 484–486
Organization uncertainty, 42–43, 59–60
Outcomes, research, 229–232
Outputs, 304, 319–321, 327
Oversight board, *see* Transition oversight board
Owens Corning Fiberglas, 42
Owens Glass, 42

Pain tables, 323
PAMM worksheet, *see* Product Attribute and Market Matrix Worksheet
Paperware, 258–259
Participant observation, 209
Partners, codevelopment:
assessment/selection, 154–156
relationship initiation, 156–159
relationship management, 157, 160–162
Partnerships:
codevelopment, *see* Codevelopment
with stakeholders, 110
Passion, 102, 104
PDMA, *see* Product Development & Management Association
PDMA ToolBook, 440, 442, 450
Personal contact, 103–104
Personnel, transition-management, 87–90
P&G, *see* Procter & Gamble

Photography, 212
Phrasing, interview, 185
Planning:
of CID sessions, 245–250
of ethnographic research, 205–214
of product development capabilities, 381, 382
project, 205–214
for VOC, 172–177
Platforms, technology, 335, 336, 442–444
Polanyi, M., 95
Polaroid, 105
Portfolio and pipeline management (PPM), 461–491
and capability maturity, 466, 467, 469–473
challenge in implementation of, 462–463
and change equilibrium, 486–488
components of, 465, 466
and maturity levels, 475–477
method of implementation, 463–469
and organizational structure, 484–486
spiral-up approach to implementation of, 473–484, 488–491
Portfolio rollout scenarios, 11–14
Portfolio teams, 20–22
Post-it Notes, 94, 185, 186, 262, 287, 290, 340, 454
Powerpointware, 259, 260
PPM, *see* Portfolio and pipeline management
Prather, C. W., 99
Presentation, research, 224, 225
Price, value vs., 299–301
Prioritization:
of customer needs, 171–172, 188–193
of customer problems, 27
of customers, in quantitative economic valuation, 303, 313, 317–319
of plan requirements, 50, 58
in technology development, 441–442
Privacy, 209
"Probe and learn" process, 62
Problem identification, 28–30
Process architecture, 383, 384
Process highlights, 317, 318, 320
Procter & Gamble (P&G), 109, 110, 232
Product Attribute and Market Matrix (PAMM) Worksheet, 85–86
Product configurator Web sites, 361–362
Product development:
synchronization of technology development with, 438–440, 448–450
technology development vs., 436–439
Product Development & Management Association (PDMA), 100, 106, 150

Product evolution maps, 422, 424–427
Product families, 406
Product features, 83–84
Product Features Worksheet, 83–84
Product line map, 404, 406–410
Product lines, 430–431
Product line story, 426, 438–430
Product platforms, 442
Product requirements, 343–344
Products, lead, 442
Product solution guidance tables, 324, 325
Product specifications, 344
Product-technology maps, 426, 428–429
Product to market, 84–86
Progressive Insurance, 4
Project Management Institute, 106
Project planning, 205–214
Proust, Marcel, 201
Proximity:
  and creativity, 123–127
  and virtual teams, 136–140
PRTM, 150, 155
"Pushback," 488

QFD, *see* Quality Function Deployment
Qualitative cluster analysis, 226
Qualitative market research, 171, 203
Quality, 167
Quality Function Deployment (QFD), 167–168
Quantitative economic valuation, 297–328
  business pains, identification of, 305, 307, 314–316, 322–324
  business process, modeling of, 303–306, 313–314, 319–322
  implementation procedure, 316–327
  method design and benefits, 297–299
  numerical worth of features, validation of, 311–312
  pain mitigation ideas, development of, 307–308, 314–316, 324–325
  and price vs. value, 299–301
  prioritization/selection of customers, 303, 313, 317–319
  results, validation of, 327
  services company example, 312–316
  telecom example, 301–312
  value propositions, development of, 308–311, 325–327
Quantitative market research, 171, 203
Question and answer tools, 110
Questionnaires, 189–192
Quote boards, 232
Quotes, verbatim, 336

RACI, *see* Responsible, Accountable, Consulted, and Informed
Radical innovations, 70
  importance of, 34–35
  management framework for, 35–37
Readiness assessment, *see* Transition Readiness Assessment Tool
Real options theory, 444–448
"Receiving unit," 43, 45
Recognition, team, 22–25
Recruiting, 207–208, 248–249
Reilly, R., 98
Reliability, 309, 310
Reporting, 110
Requests for proposal (RFPs), 385–387
Requirements:
  project- vs. technology-development, 440
  uniquely identifying, 346, 347
Requirements documentation, 346–347
Requirements management policy, 346
Requirements process (for NPD), 331–351
  assessment, 335–336
  and buy-in from executive leadership, 332
  challenges in establishment of, 331–332
  concept screening, 342–343
  creation of custom, 332
  customer voices, identification of, 336–337
  defining terminology for, 332
  and defining the process, 333–334
  hard-wiring of, 349–350
  and importance of requirements, 350–351
  listening to users, 336
  operational requirements, establishment of, 342
  piloting of, 348
  product/functional requirements, generation of, 343–344
  refinement of, 349
  roles and responsibilities, establishment of, 347–348
  selection, 339–341
  solution concepts, development of, 341–342
  specifications, generation of, 344–345
  testing, preparation for, 341
  and tracking evolution of requirements, 345–347
  translation, 337–339
  validation of requirements, 341
Research:
  on customers, 30–31
  and customer type, 207
  design of, 207–213
  longitudinal vs. short, 210

Research (*continued*)
methods of, 208–211
number of visits/interviews, 210, 211
objectives of, 206
recruitment methods for, 207–208
tips for successful, 214
Resource uncertainty, 43–44, 62
Respondent homework, 213
Responsible, Accountable, Consulted, and Informed (RACI), 388, 389
Review committee, 347
RFPs, *see* Requests for proposal
Roadmapping, *see* Mapping
ROI analysis, 445
Royal Philips Electronics, 120
Rubbermaid, 216

S. C. Johnson & Son, 22
Safety, 209
Sales teams, 215
Scheduled ideation, 237–239
Schlumberger, 104
Scope of project, 172
Screening:
for CID, 248–249
concept, 342–343
for interviews, 208
Seagate, 350
Search capability, 111–112
Seasonality, 11
"Seeding," 22
Self-Stated Importance Survey, 341
Senge, Peter, 96, 432
Senior management:
buy-in from, in requirements process, 332
and communities of practice, 103
and delegating innovation, 6–10
and focused/specific innovation criteria, 10–18
funding/commitment of, 43–44
and innovation-based culture, 3–18
and transition management, 38–39, 41
Services:
company example of, 312–316
and VOC, 195
Shell, 102
Silver, Spencer, 94, 262
Site visit profile sheets, 223
Snow, C. C., 81
Software (SW), 375, 377
Software Engineering Institute, 383, 466, 467
Solution concepts, 341–342
"Solution space" (toolkits for user innovation), 362–364
Southwest Airlines, 4
Specifications:
capabilities transformed from technical, 78–80
as CID phase, 253–255
generation of, in requirements process, 344–345
technical, 61, 78–80
and transition plan, 61
Spiral-up approach (to implementation of PPM), 473–484
benefit accrual, 478–479
duration of maturity levels, 479–480
and front-to-back architecture, 480–484
matrix of PPM components by maturity level, 470–472
maturity levels in, 473–477
moving up the spirals, 477, 478
steps of, 488–491
Sponsor involvement, 213–214
Springboard stories, 231
Stability, vision, 98
StageGate, 269, 450–453
Stakeholders, 110
Storytelling, 224
Strategy, project, 437–438
Strengths, weaknesses, individuality, fixes, and transformation (SWIFT), 270
application of, 285–287
case example of, 274–275, 277–282, 284–285
and challenging situations, 291–292
concept-definition step of, 274–278, 285
facilities for, 273–274
fixes-identification step of, 280–282, 286–287
frame of reference clarification for, 288
group norms for, 274
individuality-identification step of, 280, 286
interpreting results of, 287
number of participants in, 273
participants in, 273
pitfalls to avoid with, 290–291
Post-it Notes used with, 290
preparing to use, 272–274
response structuring for, 289–290
roles/responsibilities, 273
sequence of, 272–274
strengths-identification step of, 278–279, 285
supplies for, 274
techniques for, 288–291
template creation for, 289
time frame/agenda for, 274
transformation step of, 282–285, 287

weaknesses-identification step of, 279–280, 285–286
worksheet for, 285–286
Strengths identification (SWIFT), 278–279
Style, 97
Subscriptions, 112
Success rates, 11
Summit meetings, 453–457
Support, vision, 98
Surveys, 189–192
SW, *see* Software
SWIFT, *see* Strengths, weaknesses, individuality, fixes, and transformation
Syntax, 401, 402

Tacit knowledge, 94, 95
Takeuchi, H., 95
Tape recordings, 181
Target maps, 419–421
Target values, 184
Team creativity, *see* Creativity
Teams:
traditional vs. virtual, 120–121
transfer, 457–459
true, 118–120
Team spaces, 109
Team task structure, 124
and creativity, 129–130
and virtual teams, 137, 142–144
Technical advantages, 74–79
Technical specifications, 61, 78–80
Technical uncertainty, 61
Technology commercialization, 435, 450, 456
Technology Description Worksheet, 76–78
Technology development, 435–460
decision tools for, 440–441
prioritization in, 441–442
product development vs., 436–439
real options theory, application of, 444–448
stage gating, application of, 451–453
staging, decision tools for, 450–451
synchronization of product development with, 438–440, 448–450
and technology platforms, 442–444
technology summit meetings, use of, 453–457
transfer teams, use of, 457–459
Technology evolution maps, 422, 424–427
Technology platforms, 442–444
Technology portfolio management, 435
Technology project map, 408
Technology roadmaps, 448–450
Technology staging, 435
Technology summit meetings, 453–457
Technology-to-product market (TPM), 71–90
capability development for, 78–80
finding technical advantages for, 74–79
keys to successful, 89–90
logic behind, 72–74
P-M elaboration of, 84–86
T-to-P elaboration of, 80–84
Technology transfer, 453
TechSG process, 450–452, 454, 457
Teece, David, 81
Telecom company example, 301–312
Template, SWIFT, 289
Testing preparation, 341
Test requirements, 344, 345
Texas Instruments (TI), 60
Themes, 228–230
3M, 94, 262, 287, 360, 454
Throughput management, 464
TI (Texas Instruments), 60
Tichy, Noel, 231
Toolkits (for user innovation), 353–373
and benefits of focusing of users, 354–357
competitive advantages of, 370–371
defined, 353
deployment of, 368–372
elements of, 359, 361
keys for success with, 371–372
and learning by doing, 361–362
module libraries for creation of, 365–366
and repartitioning of development tasks, 358–360
"solution space" of, 362–364
steps for designing, 357–358
and translation of user designs into production, 366–368
user-friendly, 364–365
Tools and techniques, 93–114
collaborative environment as, 98–100
communities of practice as, 100–104
competence/capability mapping as, 104–108
innovation intranet as, 108–113
innovation vision as, 96–98
knowledge creation as, 94–95
Schlumberger example of, 104
TPM, *see* Technology-to-product market
T-P-M linkages, 72–86
TPM Worksheet, 78, 79, 82
Tracking of requirements, 345–348
Training:
of field research teams, 216–217
and IT-enablement, 392, 393
Transcripts, interview, 181, 183, 185, 224
Transfer teams, use of, in technology development, 457–459

Transformation (SWIFT), 282–285
Transition management, 33–69
  challenges of, 35, 37–38
  Eastman Kodak example of, 65–67
  funding/commitment step of, 43–44
  and growth/renewal, 34–35
  implementation of, 89–90
  oversight-board-formation step of, 41–43
  personnel for, 87–90
  plan/business-model steps of, 58–63
  process of, 39–63
  radical innovation framework for, 35–37
  readiness-assessment step of, 45–58
  roles of, 38–39
  senior-management-involvement step of, 41
  team-formation step of, 44–45
  value of, 63–67
Transition managers, 33, 38, 41
Transition oversight board, 38, 41–43
Transition plan, 58–63
  and business model, 59–60
  evolution-of-applications/markets section of, 60–61
  manufacturing-challenges-to-market-entry section of, 62–63
  market-development-expectations section of, 62
  technical-specifications/manufacturing-issues section of, 61
Transition Plan Requirements: Priority Areas, 50, 58
Transition Readiness Assessment, 50–56
Transition Readiness Assessment Grand Score and Analysis, 50, 57
Transition Readiness Assessment Tool, 45–58
  elements of, 33–34
  IRI member experiences with, 63–65
  mapping categories of uncertainty with, 45, 48–49
  plan-requirements prioritization with, 50, 58
  preliminary assessment with, 45–47
  scoring/analysis with, 50, 57
Transition team, 38–39
  formation of, 44–45
  skills required by, 44
Translation:
  in requirements process, 337–339
  with toolkits for user innovation, 366–368
Tree diagrams, 188
Trial-and-error design, 361–362
True teams, 118–120
True time to transfer (TTT), 458–459
True transfer point, 458
Trust, 98–100, 128
TTT, *see* True time to transfer

Unaided questions, 179
Uncertainty:
  continuum of, by categories/innovation types, 35, 37
  mapping categories of, 45, 48–49
  market, 62–63
  organization, 42–43, 59–60, 62–63
  resource, 43–44
  technical, 61
  transition, 39, 40
Urban, G., 303
Urban, J. M., 122
Usability testing, 264–268
Usage scenarios, 388, 390
Useem, Jerry, 233
User-friendly toolkits, 364–365, 368
User-innovation toolkits, *see* Toolkits

Validation of requirements, 341
Valuable, rare, immutable, and nonsubstitutable (VRIN) attributes, 104, 106, 108
Value:
  price vs., 299–301
  project, 438
Value chain analysis, 409
Value propositions, 308–311, 325–327
Value pyramids, 227
Van Engelen, J. M. L., 125
Verbatim quotes, 336
Videos, highlight, 231
Videotaping, 212
Virtuality fingerprint, 117–118, 121–125
Virtuality points, 122
Virtuality Pyramid, 131–135
Virtual NPD teams, 117–145
  and creativity, 122, 124–130
  fingerprint of, 121–125
  managing, *see* Virtual team management
  traditional vs., 120–121
  and true teams, 118–120
  and Virtuality Pyramid, 131–135
Virtual team management, 135–145
  and communication modalities, 137, 140–142
  considerations for, 144–145
  and proximity, 136–140
  and task structure, 137, 142–144
Vision, 96–98
VOC, *see* Voice of the Customer

Vocabulary, 401, 402
Voice of the Customer (VOC), 167–198
  defining, 170–172
  and development team's dilemma, 168–169
  and global products/services, 195–196
  interviews for, *see* Customer interviews
  and manufactured goods vs. services, 195
  myths about, 170, 197–198
  and needs-list creation, *see* Customer needs
  and "new-new" products, 194–195
  planning for, 172–177
  time/resources needed for, 193–194
Voice over IP, 307
Voices, customer, 336–337
Von Hippel, Eric, 236
Von Krough, G., 96, 98
VRIN, *see* Valuable, rare, immutable, and nonsubstitutable attributes

Wal-Mart, 233
Warmka, Scott, 350
Weaknesses identification (SWIFT), 279–280
Weber, Mary Ellen, 19
Western Union, 232
Westwood Studios, 354
Whitespace, 298–299
Wish mode approach, 236–243
Wu, A., 298

Xerox Corporation, 455

Also available from the PDMA...

The PDMA ToolBook for New Product Development 1

Paul Belliveau, Abbie Griffin, Stephen Somermeyer

0-471-20611-3 • Cloth • 480 pages

With effective methods, tools, and techniques in every chapter, this is an essential book for new product development professionals, including Project Leaders, Process Owners, and Program or Portfolio Managers in a broad range of industries from heavy manufacturing to services.

The PDMA Handbook of New Product Development, Second Edition

By Kenneth B. Kahn, Editor
George Castellion, Associate Editor
Abbie Griffin, Associate Editor

0-471-48524-1 • Cloth • 640 pages

This completely revised and updated new edition offers practical information pertaining to every stage of the product development process-from idea generation to launch to the end of the life cycle.

For more information visit www.pdma.org

wiley.com